Mastering Radio Frequency Circuits through Projects and Experiments

Other Books by the Author

Mastering IC Electronics
Mastering Oscillator Circuits through Projects and Experiments
Mastering Solid-State Amplifiers
Old Time Radios! Restoration and Repair
Secrets of RF Circuit Design
Practical Antenna Handbook

Mastering Radio Frequency Circuits through Projects and Experiments

Joseph J. Carr

TAB Books
Division of McGraw-Hill, Inc.
New York San Francisco Washington, D.C. Auckland Bogotá
Caracas Lisbon London Madrid Mexico City Milan
Montreal New Delhi San Juan Singapore
Sydney Tokyo Toronto

FIRST EDITION
FIRST PRINTING

Library of Congress Cataloging-in-Publication Data

Carr, Joseph J.
Mastering radio frequency circuits through projects and experiments / by Joseph J. Carr
p. cm.
Includes index.
ISBN 0-07-011064-6 ISBN 0-07-011065-4 (pbk.)
1. Radio circuits—Amateurs' manual. I. Title.
TK9956.C3533 1994
621.384'12—dc20 94-7088
CIP

Acquisitions editor: Roland Phelps
Editorial team: Joanne Slike, Executive Editor
David M. McCandless, Supervising Editor
Anita Louise McCormick, Editor
Production team: Katherine G. Brown, Director
Patsy D. Harne, Desktop Operator
Nancy K. Mickley, Proofreading
Joanne Woy, Indexer
Design team: Jaclyn J. Boone, Designer
Brian Allison, Associate Designer
Cover design: Holberg Design, York, Pa.

0110654
EL1

Dedicated to the memory of
Hiram Percy Maxim

Pioneer in the technologies of automobiles, motion pictures, and radio; inventor of the Maxim silencer, which we know today as the automobile muffler. Prior to World War I, Maxim was co-founder of the American Radio Relay League. Contrary to some rumors, Hiram Percy Maxim did not invent the Maxim machine gun . . . that Maxim was his father, Hiram Hamilton Maxim. A pacifist, HPM was reportedly upset when someone fitted his "industrial engine silencer" to the barrel of his father's gun and made the infamous gun silencer. Radio enthusiasts, both amateur and professional, owe a debt of gratitude to this pioneer of the radio arts.

American Radio Relay League, 225 Main Street, Newington, CT, 06111

Hiram Percy Maxim

Contents

Preface

Radio frequency circuits are used extensively in communications, electronic instruments, and medical devices. Although superficially similar to lower frequency circuits, they are a class unto themselves for two reasons: first, the high frequencies used make design and construction difficult; second, the extensive use of inductor-capacitor tuned resonant circuits. In this book you will learn the basics of RF construction, some information about RF circuit design, and generally gain a comfortable level of understanding of radio frequency circuits. The approach is practical, rather than theoretical, so you will—as in all of the TAB/McGraw-Hill *Mastering . . .* series books—perform experiments and build useful projects.

The radio spectrum runs from a little above dc to somewhat below daylight, but the range of frequencies that we need be concerned about is somewhat narrower. Because of practical considerations, frequencies much below the AM broadcast band are difficult to work with because the inductors become unreasonably large. Except for a few cases where we can purchase inductors, as opposed to making our own, few projects are below 450 kHz. On the opposite end of the spectrum, above about 40 or 50 MHz, the stray capacitances and inductances, plus the very nature of the devices used, become a bit difficult to handle. In addition, the test equipment needed at those frequencies is a bit esoteric and costly. So only a few projects are planned for those frequencies.

The experiments in this text are designed to be performed with relatively simple RF test equipment (see chapter 3). Ordinarily, the level of equipment used in radio-TV service shops, by amateur radio operators, and electronic hobbyists is sufficient for these experiments. Where possible, simple homebrew test equipment is described. This equipment can be used for some of the experiments, or to perform tests or make measurements on common communications equipment.

The projects are designed to be used, and are "user tested" by me and other builders. Some of these designs are based on similar designs that have appeared in my articles and columns in *Popular Electronics*, *73*, *RadioFun*, *Popular Communications*, and *Communications Quarterly*. Where the original designs were modified by readers, and the "mod" was deemed reasonable, I incorporated the modification into the original project so that the final project . . . which you will see, reflects the best thinking possible.

This book is intended to be well-used, so please don't feel that it's necessary to keep it in nice shape . . . the workbench and toolbox is where I expect to see many copies (suitably splashed with coffee or other stuff). Also, please don't feel a bit bashful about modifying the circuits, or trying some different approaches. Part of the fun of dealing with RF circuits, even at the amateur radio or hobby level, is noodling around with designs to make them better. So noodle away . . . and enjoy.

A special word is in order for engineering and technology school students. Many professors at these schools disdain the "practical approach" taken in books like this one. But let them sneer away, for when you are doing a lab assignment or your senior design project, it is the practical that you need most. So, fake out your professor: hide this book in your notebook, as if it were politically incorrect *samizat*,[1] sneaking a look when the prof is out of the room. Alternatively, if you really have a devious turn of mind, make a new book cover for this book, and paste it to the outside hiding the original cover. Use a laser printer to print a new title:

Theoretical Electromagnetodynamics
by Ananias Luter
(lying lout)

Seriously, though, I've received a lot of very positive letters over the years from students in both engineering and technology who appreciate the practical, informal style of writing found in this book. Use it in good health.

Joseph J. Carr, MSEE
Falls Church, VA

[1]Underground literature.

1

Introduction to radio frequencies

Radio frequency signals are electromagnetic waves, as opposed to acoustic or mechanical waves, and the useful frequency spectrum generally inhabits the region above 10 kHz or so. However, like many rules of thumb, there are exceptions to the "10 kHz" rule: natural phenomena produce radio waves under 10 kHz ("whistlers" and others), and the U.S. Navy operates a submarine communications network near 60 Hz. Radio frequency (RF) electronics is different because the higher frequencies make some circuit operation a little hard to understand or predict. The principal reason is that stray capacitance and stray inductance afflicts these circuits. *Stray capacitance* is the capacitance that exists between conductors of the circuit, between conductors or components and ground, or from internal to components. *Stray inductance* is the normal inductance of the conductors that interconnect components, as well as internal component inductances.

These stray parameters are usually not important at low ac and dc frequencies, but as frequency increases they become a much larger proportion of the total. In some older VHF TV tuners and VHF communications receiver front ends, the stray capacitances were sufficiently large to tune the circuits, so no actual capacitors are needed.

Also, skin effect exists at RF frequencies. The term *skin effect* refers to the fact that ac flows only on the outside portion of the conductor, while dc flows through the entire conductor. As frequency increases, skin effect produces a smaller and smaller zone of conduction . . . and a correspondingly higher value of ac resistance compared with dc resistance.

Another problem with RF circuits is that the signals find it easier to radiate both from the circuit and within the circuit. Thus, coupling effects between elements of the circuit, between the circuit and its environment, and from the environment to the circuit become much more crucial at RF. Interference and other strange effects that are missing at dc and negligible in most low frequency ac circuits are found at RF.

For these reasons, RF circuits are different from dc and low frequency ac circuits. When an RF electrical signal radiates, it becomes an *electromagnetic wave*. These waves include not only radio signals but also infrared (IR) light, visible light, ultraviolet light (UV), X-rays, gamma rays and others. Before proceeding with our discussion of RF electronic circuits, therefore, we ought to take a look at the electromagnetic spectrum.

The electromagnetic spectrum (Fig. 1-1) is broken into bands for the sake of convenience and identification, although there is some fuzziness on the exact boundaries of the different bands because different authorities place the band edges at different points. The full spectrum extends from the very lowest ac frequencies and continues well past visible light frequencies into the X-ray and gamma ray region.

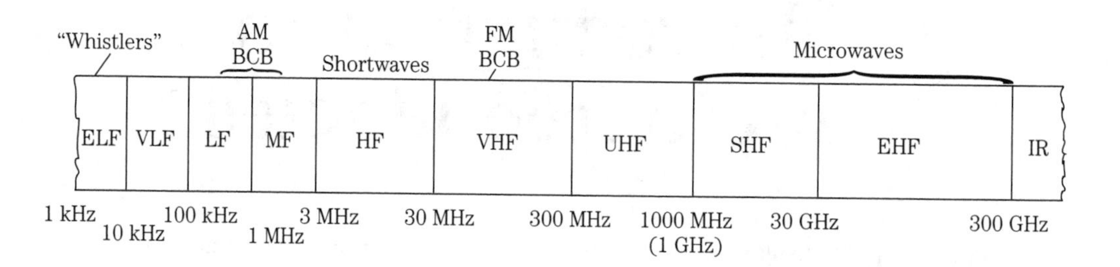

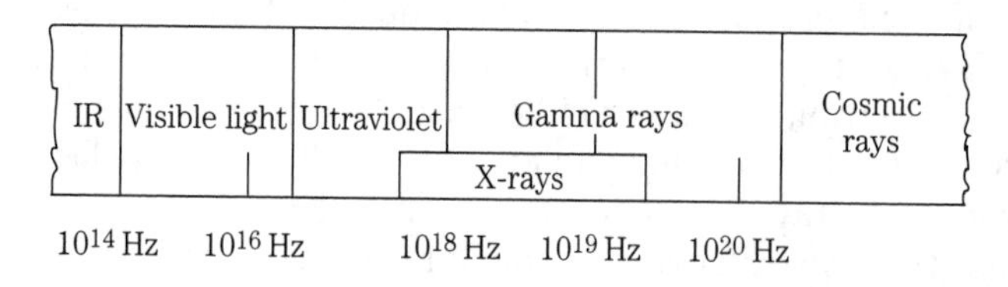

1-1 The electromagnetic spectrum extends from dc to daylight . . . and beyond. It often surprises people that visible light, infrared light, ultraviolet light, and X-rays are part of the same phenomenon as radio waves; differing only in frequency.

The *extremely low frequency* (ELF) range includes ac power line frequencies, as well as other low frequencies in the 25- to 100-hertz (Hz) region. The U.S. Navy uses these frequencies for submarine communications. The *very low frequency* (VLF) region extends from just above the ELF region, although most authorities peg it to frequencies from 10 kilohertz (kHz) to 100 kHz; at least one text lists VLF from 10 kHz to only 30 kHz. The *low frequency* (LF) region runs from 100 kHz to 1000 kHz (or 1 MHz), although again some texts put 300 kHz as the upper limit on the LF. The *medium wave* (MW) or *medium frequency* (MF) region runs from 1 MHz to 3 MHz according to some authorities and 300 kHz to 3 MHz according to others. The AM broadcast band (540 kHz to 1700 kHz)[1] spans portions of the LF and MW bands. The *high frequency* (HF) region, also called the *shortwave bands* (SW), runs from 3

[1]Originally the AM BCB in the USA was 540–1610 kHz, but that was recently changed to the 540 kHz–1700 kHz allocation. Other countries may use slightly different band edges.

MHz to 30 MHz. The *very high frequency* (VHF) region starts at 30 MHz and runs to 300 MHz. This region includes the FM broadcast band, public utilities, some television stations, aviation users, and amateur radio bands. The *ultra high frequency* (UHF) region runs from 300 to 900 MHz and includes many of the same services as VHF.

The *microwave region* officially begins above the UHF region at 900 or 1000 megahertz (MHz), depending upon source authority. One may well ask how microwaves differ from other electromagnetic waves. Microwaves almost become a separate topic of study in RF because at these frequencies the wavelength approximates the physical size of ordinary electronic components. Thus, components behave differently at microwave frequencies than they do at lower frequencies. At microwave frequencies, a half-watt metal film resistor, for example, looks like a complex RLC network with distributed L and C values . . . and a surprisingly different R value. These tiniest of distributed components have immense significance at microwave frequencies, even though they can be ignored as negligible at lower RF frequencies.

Before examining RF theory, let's first review some background and fundamentals.

Units and physical constants

In accordance with standard engineering and scientific practice, all units in this book will be in either the *CGS* (centimeter-gram-second) or *MKS* (meter-kilogram-second) systems unless otherwise specified. Because the metric (CGS and MKS) systems depend upon using multiplying prefixes in on the basic units, I've included a table of common metric prefixes (Table 1-1). Other tables are as follows: Table 1-2 gives the standard physical units; Table 1-3 shows physical constants of interest in this and other chapters; and Table 1-4 lists some common conversion factors.

Table 1-1. Metric prefixes.

Metric prefix	Multiplying factor	Symbol
tera	10^{12}	T
giga	10^{9}	G
mega	10^{6}	M
kilo	10^{3}	K
hecto	10^{2}	h
deka	10	da
deci	10^{-1}	d
centi	10^{-2}	c
milli	10^{-3}	m
micro	10^{-6}	μ
nano	10^{-9}	n
pico	10^{-12}	p
femto	10^{-15}	f
atto	10^{-18}	a

Table 1-2. Units.

Quantity	Unit	Symbol
Capacitance	farad	F
Electric charge	coulomb	Q
Conductance	mhos	
Conductivity	mhos/meter	Ω/m
Current	ampere	A
Energy	joule (watt-sec)	j
Field	volts/meter	E
Flux linkage	weber (volt-second)	
Frequency	hertz	Hz
Inductance	henry	H
Length	meter	m
Mass	gram	g
Power	watt	W
Resistance	ohm	Ω
Time	second	s
Velocity	meter/second	m/s
Electric potential	volt	V

Table 1-3. Physical constants.

Constant	Value	Symbol
Boltzmann's constant	1.38×10^{-23} J/K	K
Electric chart (e^-)	1.6×10^{-19} C	q
Electron (volt)	1.6×10^{-19} J	eV
Electron (mass)	9.12×10^{-31} kg	m
Permeability of free space	$4\pi \times 10^{-7}$ H/m	U_o
Permitivity of free space	8.85×10^{-12} F/m	ε_o
Planck's constant	6.626×10^{-34} J–s	h
Velocity of electromagnetic waves	3×10^{8} m/s	c
Pi (π)	3.1416	π

Table 1-4. Conversion factors.

1 inch	=	2.54 cm
1 inch	=	25.4 mm
1 foot	=	0.305 m
1 statute mile	=	1.61 km
1 nautical mile	=	6080 feet (6000 feet)[1]
1 statute mile	=	5280 feet
1 mil	=	0.001 in = 2.54×10^{-5} m
1 kg	=	2.2 lb
1 neper	=	8.686 dB
1 gauss	=	10,000 teslas

[1] Some navigators use 6000 feet for ease of calculation. The nautical mile is 1/360 of the earth's circumference at the equator, more or less.

Wavelength and frequency

For all forms of wave, the velocity, wavelength and frequency (see Fig. 1-2) are related such that the product of frequency and wavelength is equal to the velocity. For microwaves, this relationship can be expressed in the form:

$$\lambda F \sqrt{\varepsilon} = c \qquad \textbf{[1-1]}$$

where

λ is the wavelength in meters
F is the frequency in hertz (Hz)
ε is the dielectric constant of the propagation medium
c is the velocity of light (300,000,000 m/s)

The dielectric constant (ε) is a property of the medium in which the wave propagates. The value of ε is defined as 1.000 for a perfect vacuum and very nearly 1.0 for dry air (typically 1.006). In most practical applications, the value of ε in dry air is taken to be 1.000. For mediums other than air or vacuum, however, the velocity of propagation is slower, and the value of ε relative to a vacuum is

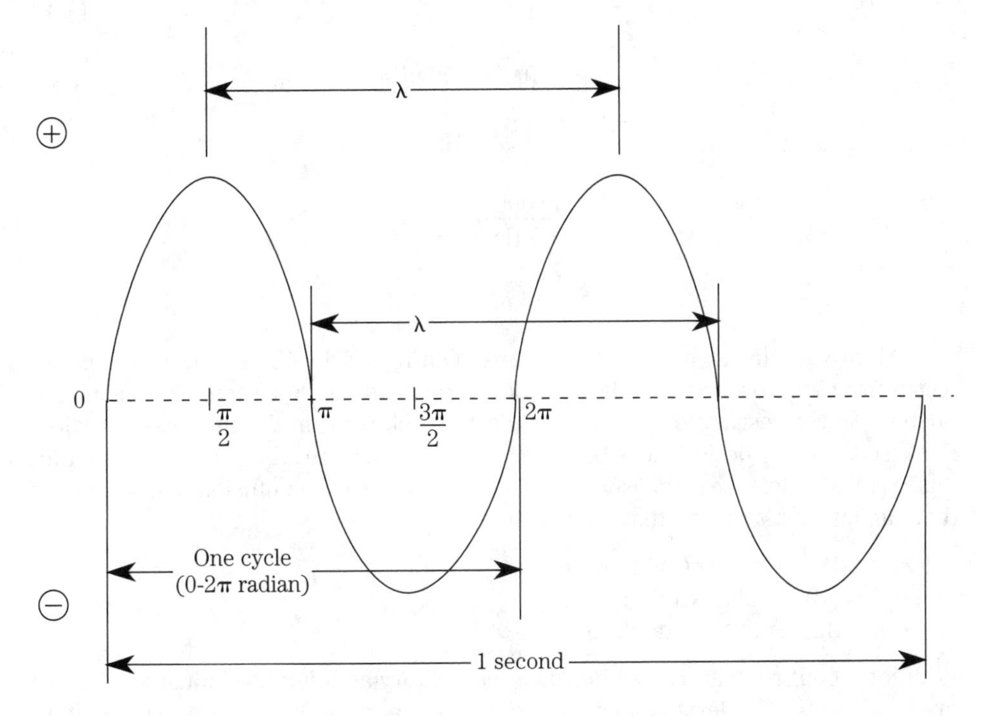

1-2 The wavelength is defined as the distance (in meters, centimeters, or millimeters) between identical points on the waveform; e.g., from one positive peak to the next positive peak, or from negative-going zero crossing to the next negative-going zero crossing. The frequency is the number of cycles per unit of time (hertz), or in this case 2 Hz.

higher. *Teflon*, for example, can be made with dielectric constant values (ε) from about 2 to 11.

Equation 1-1 is more commonly expressed in the forms of Eqs. 1-2 and 1-3:

$$\lambda = \frac{c}{F\sqrt{\varepsilon}} \qquad \textbf{[1-2]}$$

and

$$F = \frac{c}{F\sqrt{\varepsilon}} \qquad \textbf{[1-3]}$$

(All terms as defined for Eq. 1-1.)

Example 1-1

A mountaintop communications relay station receives signals on a frequency of 2.2 GHz. Calculate the wavelength of this signal.

Solution

$$\lambda = \frac{c}{F\sqrt{\varepsilon}} \qquad \textbf{[1-2]}$$

$$\lambda = \frac{300{,}000{,}000 \text{ m/s}}{2.2 \text{ GHz} \times \left(\frac{10^9 \text{ Hz}}{1 \text{ GHz}}\right) \times \sqrt{1.00}}$$

$$\lambda = \frac{300{,}000{,}000 \text{ ms}}{2.2 \times 109 \text{ Hz} \times 1}$$

$$\lambda = 1.36 \times 10^{-1} \text{ meters} = 13.6 \text{ cm}$$

Microwave letter band designations During World War II, the U.S. military began using microwaves in radar and other applications. For security reasons, alphabetic letter designations were adopted for each band in the microwave region. Because the letter designations became ingrained, they are still used throughout industry and the defense establishment. Unfortunately, some confusion exists because there are at least three systems currently in use:

- Pre-1970 military (Table 1-5)
- Post-1970 military (Table 1-6)
- An IEEE-Industry standard (Table 1-7)

Additional confusion is created because the military and defense industry use both pre- and post-1970 designations simultaneously, and industry often uses military rather than IEEE designations. The "old military" designations (Table 1-5) persist as a matter of habit.

Table 1-5.
U.S. military microwave frequency bands.

Band designation	Frequency range
P	225 - 390 MHz
L	390 - 1550 MHz
S	1550 - 3900 MHz
C	3900 - 6200 MHz
X	6.2 - 10.9 GHz
K	10.9 - 36 GHz
Q	36 - 46 GHz
V	46 - 56 GHz
W	56 - 100 GHz

Table 1-6.
New U.S. military microwave frequency bands.

Band designation	Frequency range
A	100 - 250 MHz
B	250 - 500 MHz
C	500 - 1000 MHz
D	1000 - 2000 MHz
E	2000 - 3000 MHz
F	3000 - 4000 MHz
G	4000 - 6000 MHz
H	6000 - 8000 MHz
I	8000 - 10,000 MHz
J	10 - 20 GHz
K	20 - 40 GHz
L	40 - 60 GHz
M	60 - 100 GHz

Table 1-7. IEEE/Industry standard microwave frequency bands.

Band designation	Frequency range
HF	3 - 30 MHz
VHF	30 - 300 MHz
UHF	300 - 1000 MHz
L	1000 - 2000 MHz
S	2000 - 4000 MHz
C	4000 - 8000 MHz
X	8000 - 12,000 MHz
Ku	12 - 18 GHz
K	18 - 27 GHz
Ka	27 - 40 GHz
Millimeter	40 - 300 GHz
Submillimeter	above 300 GHz

Skin effect

Ordinary lumped constant electronic components do not work well at VHF, UHF and microwave frequencies for three reasons. The first, mentioned earlier in this chapter, is that component size and lead lengths approximate those wavelengths. The second is that distributed values of inductance and capacitance become significant at these frequencies. The third is a phenomenon called *skin effect*.

Skin effect refers to the fact that alternating currents tend to flow on the surface of a conductor. While dc currents flow in the entire cross-section of the conductor, ac flows in a narrow band near the surface. Current density falls off exponentially from the surface of the conductor towards the center (Fig. 1-2).

At the *critical depth* (δ), also called *depth of penetration*, the current density is $1/e$, or 1/2.718 = 0.368, of the surface current density. The value of δ is a function of operating frequency, the permeability (μ) of the conductor, and the conductivity (σ). Equation 1-4 gives the relationship.

$$\delta = \sqrt{\frac{1}{2\pi F \sigma \mu}} \qquad \textbf{[1-4]}$$

where

δ is the critical depth
F is the frequency in hertz (Hz)
μ is the permeability in henrys per meter (H/m)
σ is the conductivity in mhos per meter

The rest of this book

In this book, you will learn the theory behind RF circuits, work some experiments based on these circuits, build some useful RF circuit projects, and generally become comfortable with their use. The approach taken is practical rather than theoretical. Those who want a much deeper approach are recommended to engineering school texts on the subject, but they should keep this book handy, as it will serve to fill some of the gaps left by those other books.

2
RF components and construction

Radio frequency components and circuits differ from those used at other frequencies—principally because unaccounted-for stray inductance and capacitance forms a significant portion of the entire inductance and capacitance in the circuit. Consider a tuning circuit consisting of a 100-picofarad (100 pF) capacitor and a 1-microhenry (1 μH) inductor. According to an equation that you will learn in a subsequent chapter, this combination should resonate at an RF frequency of about 15.92 MHz. But suppose the circuit is poorly laid out, and there is 25 pF of stray capacitance in the circuit. This capacitance could come from the interaction of the capacitor and inductor leads with the chassis, or from interaction with other components in the circuit. The input capacitance of a transistor or integrated circuit (IC) amplifier can also contribute to the total value of the strays in the circuit (one popular RF IC lists 7 pF of input capacitance). So what does this extra 25 pF do to our circuit? It is in parallel with the 100 pF discrete capacitor, so it produces a total of 125 pF. Reworking the resonance equation with 125 pF instead of 100 pF reduces the resonant frequency to 14.24 MHz.

A similar situation occurs with stray inductance. All current-carrying conductors exhibit a small inductance. In low-frequency circuits, this inductance is not sufficiently large to cause anyone concern (even in some lower HF band circuits). But as frequencies pass from upper HF to the VHF region, strays become extremely important. At those frequencies, the stray inductance becomes a significant portion of total circuit inductance.

Layout is important in RF circuits because it can reduce the effects of stray capacitance and inductance. A good strategy is to use broad-printed circuit tracks, rather than wires for RF circuit interconnection. Circuits that worked poorly when wired with #28 Kovar-covered "wire-wrap" wire sometimes become quite acceptable when redone on a printed circuit board using broad (low-inductance) tracks.

Figure 2-1 shows a sample printed circuit board layout for a simple RF amplifier circuit. The key feature in this circuit is the wide printed circuit tracks and the short distances they run. This tactic reduces stray inductance, and makes the circuit more predictable.

The top (components) side of the printed circuit board, not shown in Fig. 2-1, is all copper, except for space to allow the components to interface with the bottom side printed tracks. This layer is called the *ground plane* side of the board.

2-1 Typical RF printed circuit layout.

Impedance matching in RF circuits

In low-frequency circuits, most amplifiers are voltage amplifiers. The requirement for these types of circuit is that the source impedance is very low compared with the load impedance. For example, a sensor or signal source might have an output impedance of, say, 25 ohms. As long as the input impedance of the amplifier receiving that signal is very large relative to 25 ohms, the circuit will function. "Very large" typically means greater than 10 times, although in some cases greater than 100 times is preferred. For the 25-ohm signal source, therefore, even the most stringent case is met by an input impedance of 2500 ohms . . . which is very far below the typical input impedance of real amplifiers.

RF circuits are a little different. The amplifiers are usually specified in terms of power parameters, even when the power level is very low. In most cases, RF circuits have some fixed system impedance (50, 75, 300, and 600 ohms being common, with 50 ohms being nearly universal), and all elements of the circuit are expected to match the system impedance. While a low-frequency amplifier typically has a very high input impedance and very low output impedance, most RF amplifiers have the same impedance (usually 50 ohms) for both the input and the output.

Mismatch of the system impedance causes problems, including loss of signal, especially where power transfer is the issue (remember, for maximum power transfer, the source and load impedances must be equal). Radio frequency circuits very often use transformers or impedance-matching networks in order to effect the match between source and load impedances.

Wiring boards

Radio frequency projects are best constructed on printed circuit boards that have been specially designed for RF circuits. But that ideal is not always possible. Indeed, for many hobbyists or students, it might be impossible to obtain such boards, except for the occasional project built from a magazine article or this book. In this section, we will take a look at a couple of alternatives to printed circuit boards.

Figure 2-2 shows how perforated circuit wiring board (commonly called perfboard) are used. Electronic parts distributors, including Radio Shack and other outlets, sell various versions of this material. The products of Vector Electronics have been popular for years. The most commonly available perfboards offer 0.042-inch holes spaced on 0.100-inch centers, although other hole sizes and spacing are available. Some perfboards are completely blank, while other stock material is printed with any of several different patterns. Radio Shack perfboards are interesting because several different patterns are available. Some are designed for digital IC applications, while others are printed with a pattern of circles, one each around the 0.042-inch holes.

In Fig. 2-2, the components are mounted on the top side of an unprinted board. The wiring underneath is "point-to-point" style. Although not ideal for RF circuits, it will work throughout the HF region of the spectrum, and possibly into the low VHF frequency range (especially if lead lengths are kept short).

Note the shielded inductors on the board in Fig. 2-2. These inductors are slug-tuned through a small hole in the top of the inductor. The standard pin pattern for these components does not match the 0.100 hole pattern common to perfboard. However, if the coils are canted about half a turn from the hole matrix, the pins will fit on the diagonal. The grounding tabs for the shields can be handled in either of two ways. First, bend them 90 degrees from the shield body, and let them lay on the top side of the perfboard. Small wires can then be soldered to the tabs and passed through a nearby hole to the underside circuitry. Second, drill a pair of ⅟16-inch holes (between two of the premade holes) to accommodate the tabs. Place the coil on the board at the desired location to find the exact location for these holes.

Figure 2-3 shows another variant on the perfboard theme. In this circuit, pressure-sensitive (adhesive backed) copper foil is pressed onto the surface of the perfboard to form a ground plane. This is not optimum, but it works for "one-off" homebrew projects at HF and low VHF frequencies.

2-2 Perfboard layout.

2-3 Perfboard layout with RF ground plane.

The perfboard RF project in Fig. 2-4 is a frequency translator. It takes two frequencies (F_1 and F_2), each generated in a voltage-tuned variable-frequency oscillator (VFO) circuit, and mixes them together in a double-balanced mixer (DBM) device. A low-pass filter (the toroidal inductors seen in Fig. 2-4) select the difference frequency ($F_2 - F_1$). It is important to keep the three sections (osc 1, osc 2, and low-pass filter) isolated from each other. To accomplish this goal, a shield partition is provided. In the center of Fig. 2-4 you can see the metal package of the mixer soldered to the shield partition. This shield can be made from either 0.75-inch or 1.00-inch brass strip stock of the sort that is available from hobby and model shops.

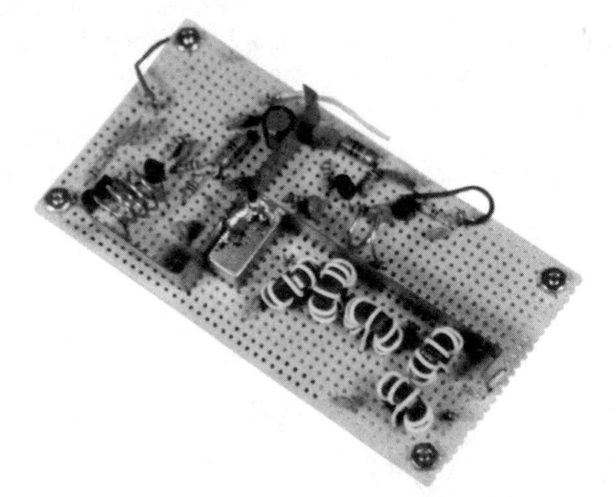

2-4 Use of shielding on perfboard.

Figure 2-5 shows a small variable-frequency oscillator (VFO) that is tuned by an air-variable capacitor. The capacitor is a 10–365 pF broadcast band variable. I built this circuit as the local oscillator for a high-performance AM broadcast band receiver project. The shielded, slug-tuned inductor, along with the capacitor, tunes the 985- to 2055-kHz range of the LO. The perfboard used for this project was a pre-printed Radio Shack board. The printed foil pattern on the underside of the perfboard is a matrix of small circles of copper, one copper pad per 0.042-inch hole. The perfboard is held off the chassis by nylon spacers and 4-40 × 0.75-inch machine screws and hex nuts.

2-5 VFO built on printed perfboard.

Chassis and cabinets

It is wise to build RF projects inside *shielded metal packages*, wherever possible. This approach to construction will prevent external interference from harming the operation of the circuit, and prevent radiation from the circuit from interfering with external devices. Figure 2-6 shows two views of an RF project built inside an aluminum chassis box; Fig. 2-6A shows the assembled box, while Fig. 2-6B shows an internal view. These boxes have flanged edges on the top portion that overlap the metal side/bottom panel. This overlap is important for interference reduction. Shun cheaper chassis boxes that use a butt fit, with only a couple of nipples and dimples to join the boxes together. Such boxes do not shield well.

The input and output terminals of the circuit in Fig. 2-6 are SO-239 UHF coaxial connectors. Such connectors are commonly used as the antenna terminal on shortwave radio receivers. Alternatives include RCA phono jacks and BNC coaxial connectors. Select the connect that is appropriate to your application.

The project in Fig. 2-7 is a small RF preselector for the AM broadcast band. It boosts weak signals and reduces interference from nearby stations. The tuning ca-

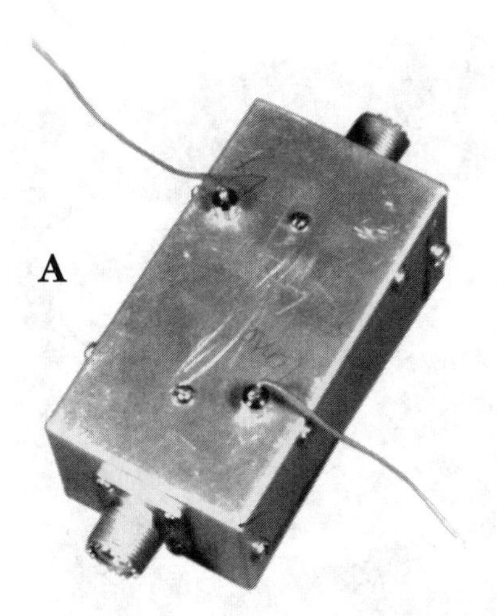

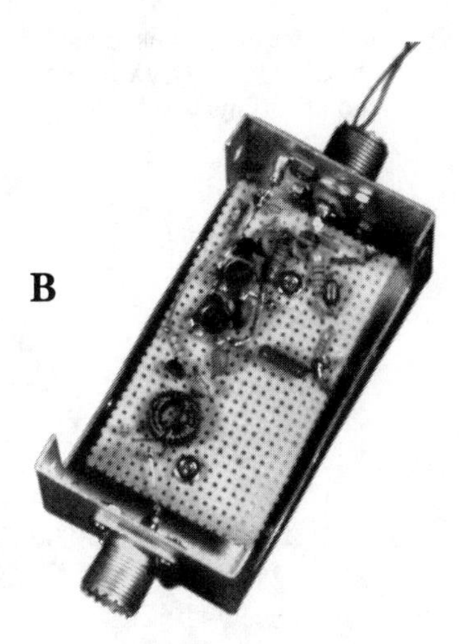

2-6 Shielded RF construction. A. Closed box showing dc connections made via coaxial capacitors. B. Box opened.

2-7 Battery powered RF project.

pacitor is mounted to the front panel, and is fitted with a knob to facilitate tuning. An on/off switch is also mounted on the front panel (a battery pack inside the aluminum chassis box provides dc power). The inductor that works with the capacitor is passed through the perfboard (where the rest of the circuit is located) to the rear panel (where its adjustment slug can be reached).

The project in Fig. 2-8 is a test bench I built for checking out direct-conversion receiver designs (see Fig. 12-17, in the chapter on direct-conversion receivers, for a view of the front panel of this project). The circuit boards are designed to be modularized so different sections of the circuit can easily be replaced with new designs. This approach allows comparison of different circuit designs on an "apples vs. apples" basis.

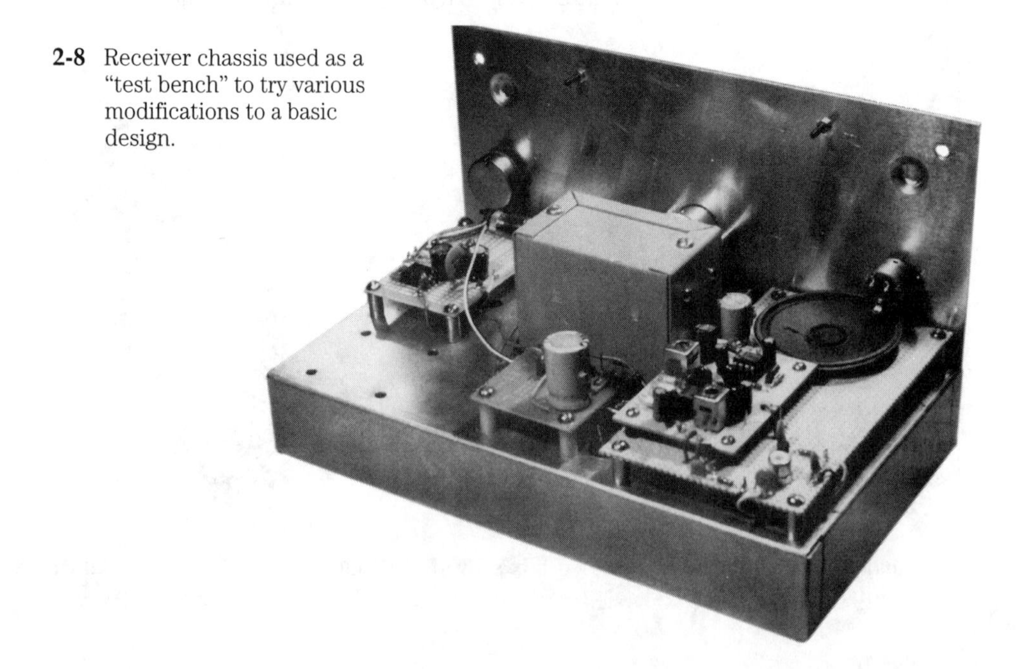

2-8 Receiver chassis used as a "test bench" to try various modifications to a basic design.

Coaxial cable transmission line (coax)

Perhaps the most common form of transmission line for shortwave and VHF/UHF receivers is *coaxial cable*. Coax consists of two conductors arranged concentric to each other, and is called "coaxial" because the two conductors share the same center axis (Fig. 2-9). The inner conductor is a solid or stranded wire, while the other conductor forms a shield. For the coax types used on receivers, the shield is a braided conductor—although some multi-stranded types are sometimes seen. Coaxial cable intended for television antenna systems has a 75-ohm characteristic impedance, and uses metal foil for the outer conductor. That type of outer conductor results in a low-loss cable over a wide frequency range, but does not work very well for most applications outside of the TV world. The problem is that the foil is made of aluminum, which doesn't take solder. The coaxial connectors used for those kinds of antennas are generally Type-F crimp-on connectors, and have too high a casualty rate for other uses.

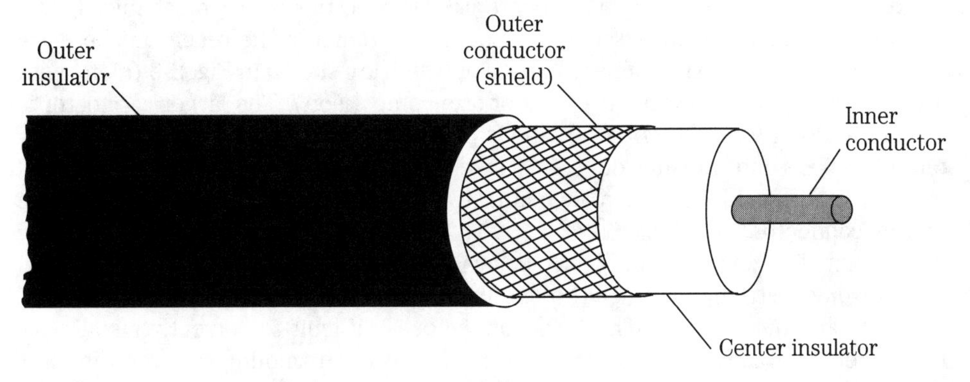

2-9 Coaxial cable (cut-away view).

The inner insulator separating the two conductors is the dielectric, of which there are several types: polyethylene, polyfoam and Teflon are common (although the latter is used primarily at high UHF and microwave frequencies). The velocity factor (V) of the coax is a function of which dielectric is used, and is

Dielectric type	Velocity factor
Polyethylene	0.66
Polyfoam	0.80
Teflon	0.70

Coaxial cable is available in a number of characteristic impedances from about 35 ohms to 125 ohms, but the vast majority of types are either 52-ohm or 75-ohm impedances. Several types that are popular with receiver antenna constructors include:

RG-8/U or RG-8/AU	52 ohm	Large diameter
RG-58/U or RG-58/AU	52 ohm	Small diameter
RG-174/U or RG-174/AU	52 ohm	Tiny diameter
RG-11/U or RG-11/AU	75 ohm	Large diameter
RG-59/U or RG-59/AU	75 ohm	Small diameter

Although the large diameter types are somewhat lower-loss cables than the small diameters, the principal advantage of the larger cable is in its power handling capability. While this is an important factor for ham radio operators, it is totally unimportant to receiver operators. Unless there is a long run (well over 100 feet), where cumulative losses become important, it is usually more practical on receiver antennas to opt for the small diameter (RG-58/U and RG-59/U) cables—they are a lot easier to handle. The tiny diameter RG-174 is sometimes used on receiver antennas, but its principal use is to make connection between devices (e.g., receiver and either preselector or ATU), in BALUN and coaxial phase shifters, and in instrumentation applications. Additional information on coaxial lines (as well as other lines) is found in chapter 16.

Installing coaxial connectors

One of the mysteries faced by newcomers to radio hobbies is the matter of installing coaxial connectors. These connectors are used to electrically and mechanically fasten the coaxial cable transmission line from the antenna to the receiver. There are two basic forms of coaxial connector, both of which are shown in Fig. 2-9 (along with an alligator clip and a banana tip plug for size comparison). The larger connector is called the *PL-259 UHF connector*, and is probably the most common form used on radio receivers and transmitters (don't take the "UHF" too seriously, it's used at all frequencies). The PL-259 is a male connector, and it mates with the SO-239 female coaxial connector.

The smaller connector in Fig. 2-10 is called a *BNC connector*. It is used mostly on electronic instrumentation, although some receiver antenna uses are seen (especially in hand-held radios). The BNC connector is difficult to correctly install, so I recommend that most readers do as I do: buy them already mounted on the wire. But the PL-259 connector is another matter. They are not readily available mounted, but are relatively easy to install.

2-10 Various types of cable ends, connectors and adapters.

Figure 2-11A shows the PL-259 coaxial connector disassembled. Also shown in Fig. 2-11A is the diameter-reducing adapter that makes the connector suitable for use with smaller cables. Without the adapter, the PL-259 connector is used for RG-8/U and RG-11/U coaxial cable, but with the correct adapter it will be used with smaller RG-58/U or RG-59/U cables (different adapters are needed for each type).

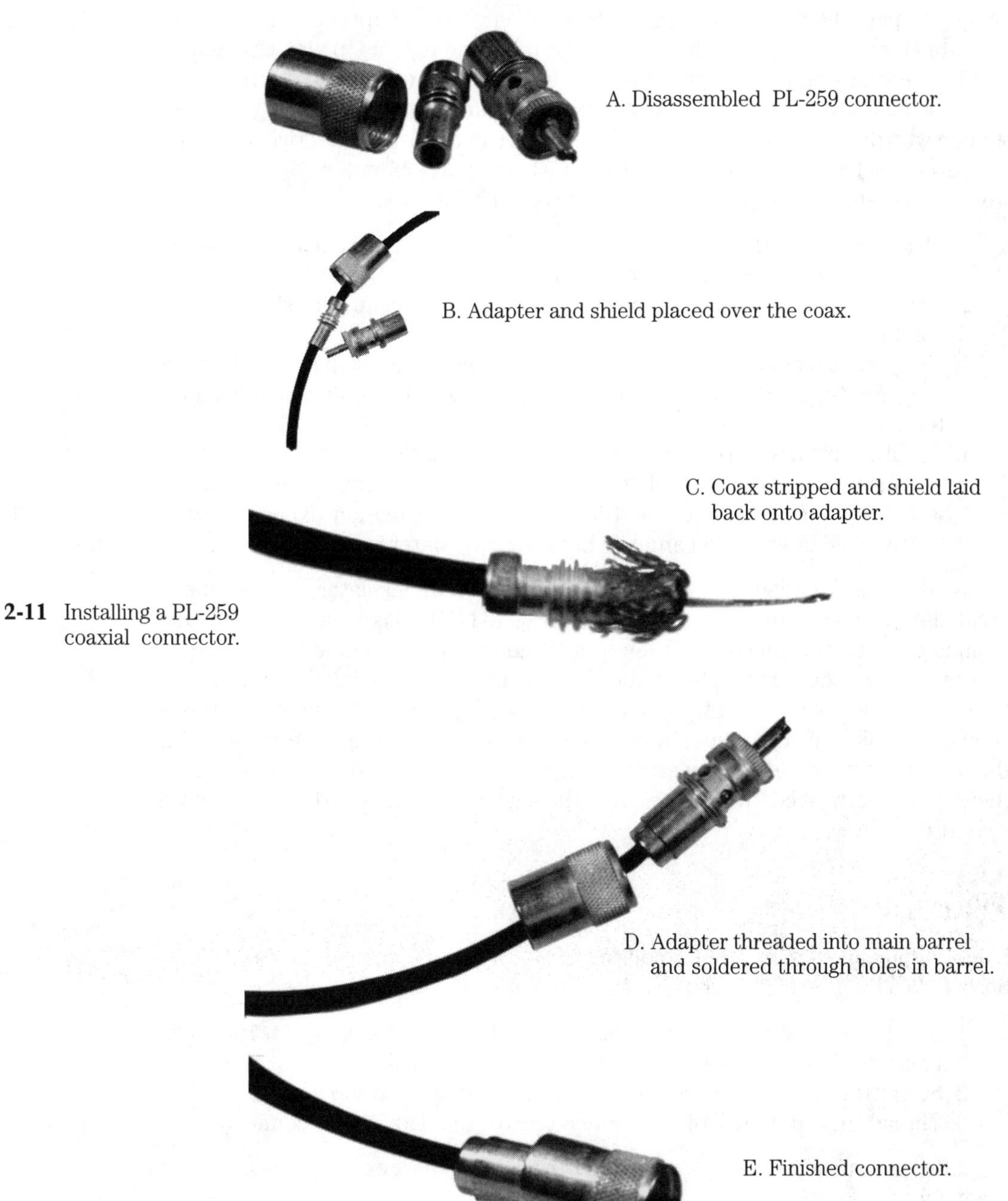

A. Disassembled PL-259 connector.

B. Adapter and shield placed over the coax.

C. Coax stripped and shield laid back onto adapter.

D. Adapter threaded into main barrel and soldered through holes in barrel.

E. Finished connector.

2-11 Installing a PL-259 coaxial connector.

The first step is to slip the adapter and threaded outer shell of the PL-259 over the end of the cable (Fig. 2-11B). You will be surprised at how many times the connector is installed, only to find that one of these components is still sitting on the workbench . . . requiring the whole job to be redone (*sigh*). If the cable is short enough that these components are likely to fall off the other end, or if the cable is dangling a particularly long distance, it is wise to trap the adapter and outer shell in a knotted loop of wire (note: the knot should not be so tight as to kink the cable).

The second step is to prepare the coaxial cable. A number of tools are available for stripping coaxial cable, but they are expensive and not terribly cost effective for anyone who doesn't do this stuff for a living. You can do just as effective a job with a scalpel or X-acto knife, either of which can be bought at hobby stores and some electronics parts stores. Follow these steps in preparing the cable:

1. Make a circumscribed cut around the body of the cable ¾ inch from the end, and then make a longitudinal cut from the first cut to the end.
2. Now strip the outer insulation from the coax, exposing the shielded outer conductor.
3. Using a small pointed tool, carefully unbraid the shield, being sure to separate the strands making up the shield. Lay it back over the outer insulation, out of the way.
4. Finally, using a wire stripper, side cutters or the scalpel, strip ⅝ inch of the inner insulation away, exposing the inner conductor. You should now have ⅝ inch of inner conductor and ⅜ inch of inner insulation exposed, and the outer shield de-stranded and laid back over the outer insulation.

Next, slide the adapter up to the edge of the outer insulator, and lay the unbraided outer conductor over the adapter (Fig. 2-11C). Make sure that the shield strands are neatly arranged, and then—using side cutters—trimmed neatly to avoid interfering with the threads. Once the shield is laid onto the adapter, slip the connector over the adapter and tighten the threads (Fig. 2-11D). Some of the threads should be visible in the solder holes that are found in the groove ahead of the threads. It is a good idea to use an ohmmeter or continuity connector to make sure there is no electrical connection between the shield and inner conductor (indicating a short circuit).

Warning

Soldering involves using a hot soldering iron. The connector will become dangerously hot to the touch. Handle the connector with a tool or cloth covering.

1. Solder the inner conductor to the center pin of the PL-259. Use a 100-watt or greater soldering gun, not a low-heat soldering pencil.
2. Solder the shield to the connector through the holes in the groove.
3. Thread the outer shell of the connector over the body of the connector.

After you run a final test to make sure there is no short circuit, the connector is ready for use (Fig. 2-11E).

3

Test equipment and workshop for RF projects

In this chapter, we will take a look at some of the test equipment and bench facilities needed for hobbyist-level work in radio frequency (RF) circuits. Most of the facilities needed for general electronic hobbyist work are suitable also for RF work, but a few special constraints and a few special types of instrumentation are necessary.

Figure 3-1 shows my own workbench (although since the photo was taken I've added several signal generators, a spectrum analyzer and a couple of other small devices). Note that the test equipment is a mix of old and new. The large black signal generator in the center is a Measurements Model 80 that was made in 1949 (and I refurbished it). It produces up to 100,000 μV of RF, amplitude-modulated or unmodulated, at frequencies from 2 MHz to 400 MHz.

I traded a local ham a "for-parts-only" single sideband (SSB) mobile rig for the signal generator. He rebuilt the SSB rig and I rebuilt the Model 80 (and we both knew the condition of the other's offering . . . lest you wonder). At hamfests I've seen these same generators in both military and civilian versions for as little as $30.

For RF circuit work, you will need several different types of instrument: multimeter, signal generators, oscilloscope, power supplies, and certain accessories which we will deal with shortly. Let's take a look at these classes, and some practical examples, each in their turn.

Kit-built instruments are a good idea, especially for newcomers. Besides saving a little bit of money, the kit-built instrument provides the owner with a project construction experience under controlled circumstances. I have built literally scores of Heathkits (The Heath Co., Benton Harbor, MI, 49022), and sorely lament their passing from the kit scene.

3-1 Workbench layout.

Multimeters

A *multimeter* is an instrument that combines into one package a multirange voltmeter, milliammeter (or ammeter), ohmmeter, and possibly some other functions. The single instrument will measure all of the basic electrical parameters the technician needs to understand in troubleshooting a receiver.

The earliest multimeters were instruments called *volt-ohm-milliammeters* (VOM). These instruments were passive devices, and the only power they used was a battery for the ohmmeter function. Because they are passive instruments, they are still preferred for troubleshooting medium-to-high-power radio transmitters. They are typically low-cost and are available from electronics stores such as Radio Shack.

The sensitivity of the VOM is a function of the full-scale deflection of the meter movement used in the instrument. *Sensitivity* is rated in terms of ohms-per-volt fullscale. Three common standard sensitivities are found: 1000 ohms/volt, 20,000 ohms/volt and 100,000 ohms/volt. These represent full-scale meter sensitivities of 1 mA, 100 μA, and 10 μA, respectively.

Voltage is read on the VOM by virtue of a multiplier resistor in series with the current meter which forms the basic instrument movement. A front panel switch selected which multiplier resistor, hence which voltage scale, was used.

The earliest form of active multimeter was the *vacuum-tube voltmeter* (VTVM). These instruments used a differential-balanced amplifier circuit based on dual triode tubes (e.g., 12AX7) to provide amplification for the instrument. As a result, the VTVM was a lot more sensitive than the VOM. It typically had an input impedance of 1 megohm, with an additional 10 megohms in the probe, for a total input impedance of 11 megohms. Compare this feature with the VOM which had a different impedance for each voltage scale.

The next improvement in meters was the solid-state meter. These instruments used a field-effect transistor in the front end, and other transistors in the rest of the circuit. As a result, these instruments were called *field-effect transistor voltmeters* (FETVM). The FETVM has a very high input impedance, typically 10 megohms or more.

Finally, the most modern form of instrument are various *digital multimeters (DMM)* in both handheld and bench (Fig. 3-2). These instruments are the meter of choice today—if you are going to purchase a new model. In fact, it might be a little difficult to find a nondigital model anyplace except Radio Shack.

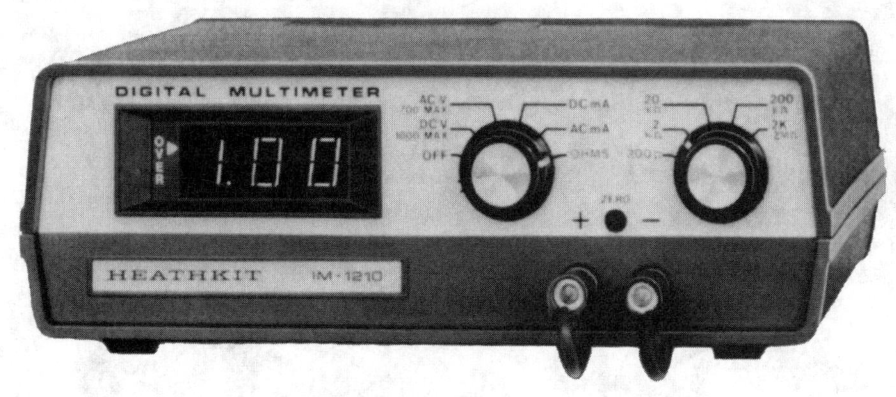

3-2 Digital multimeter.

The DMM is preferred today because it provides the same high-input impedance as the FETVM. But, because the display is numeric instead of analog, the readout is less ambiguous.

There is one bit of ambiguity, however, and that is *last digit bobble*. The digital meter only allows certain discrete states, so if a minimum value is between two of the allowed states, then the reading may switch back and forth between the two (especially if noise is present). For example, a voltage might be 1.56456 volts, but the instrument is only capable of reading to three decimal places. Thus, the actual reading might bobble back and forth between 1.564 and 1.565 volts.

Last digit bobble is not really a problem, however, unless you bought an instrument that is not good enough for your intended use. The number of digits on the meter should be one more than that required for the actual service work. Digits are specified in DMMs in terms of the number places that the meter can read. For example, the instrument in Fig. 3-2 is considered a 2½ digit model. The "½" digit refers to the fact that the most significant digit (on the left side) can only be 1 or 0 (in which case it is blanked off). Thus, the full-scale reading of this meter is always "199." The range setting controls where the decimal point is placed.

For many users, the DMM selected can be a 2½ digit model, but if you anticipate a lot of detailed solid-state work, opt instead for a 3½ digit model. These instruments will read to "1999" on each scale, and the cost difference between the two types is small.

Figure 3-3 shows a handheld digital multimeter manufactured by Hewlett-Packard for the professional market. It is extremely useful for portable servicing and other applications where a bench model instrument is not appropriate or convenient.

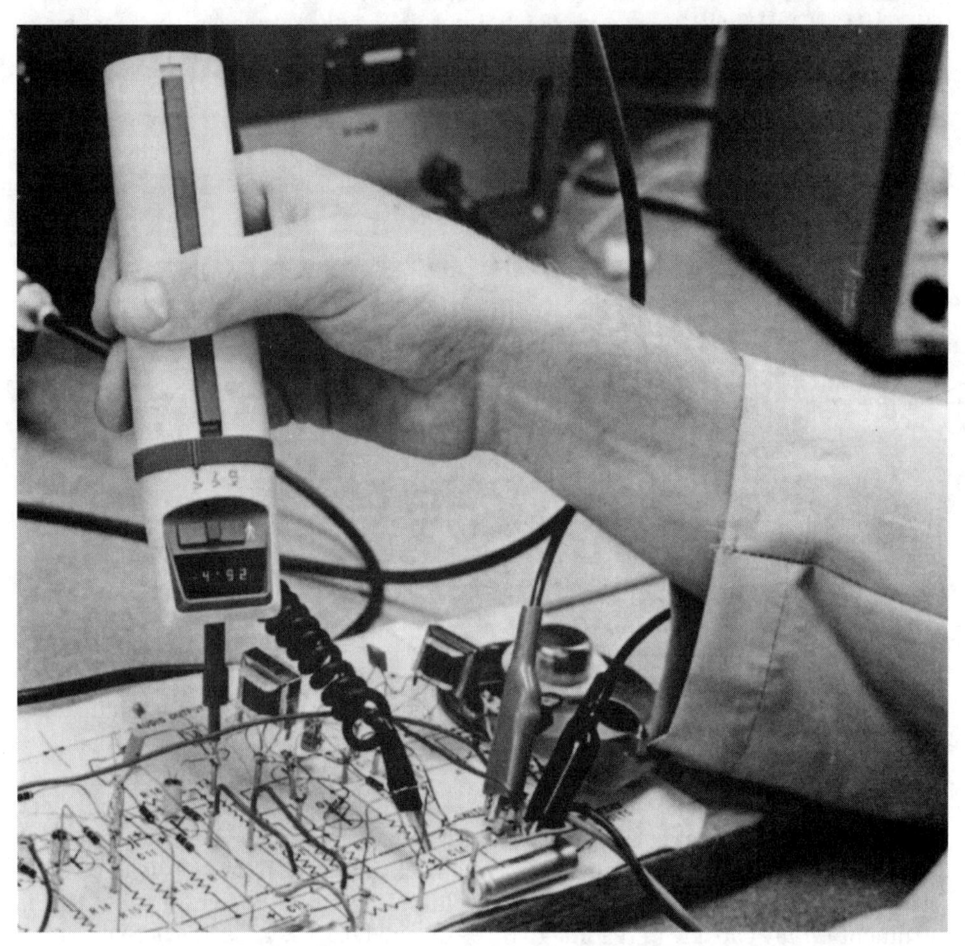

3-3 Handheld digital multimeter.

An interesting aspect of DMMs is that they use low-voltage for the ohmmeter, so the instrument can be used in the circuit on solid-state equipment. The ohmmeter or other forms of test instruments will forward-bias the pn junctions of solid-state devices, so they cannot be used for accurate measurements. However, this same feature of the DMM also means that the normal ohmmeter cannot be used to make quick tests of pn junction devices such as transistors and diodes. However, some DMMs have a special switch setting that allows this operation. In some, it is a "high power" setting—while in others, it is designated by the normal arrow diode symbol.

Another feature of the modern DMM is an aural continuity tester. It sounds a *beep* anytime the resistance between the probes is low, i.e., indicating that a short circuit exists. This feature is especially useful for testing multiconductor cables for continuity when the actual resistance measurement is not a factor.

Older instruments

As is true with other electronic equipment, the multimeter shows up on the used market (especially hamfests) quite often. It is easy to obtain workable (if older) instruments at very cheap prices. This is especially true now that everyone seems to want to dump old analog instruments in favor of the new digital varieties. However, there are some pitfalls on the pathway.

First, make sure that the instrument isn't so old that it uses a 22.5 volt (or higher) battery in the ohmmeter section. Those instruments were made prior to about 1956, and will blow most transistors or integrated circuits that you measure with them. Look out especially for the oversized RCA and Hickock instruments that were once the mainstay of Korean War era radio service shops.

Second, make sure that nothing is wrong with the instrument that cannot easily be repaired. VTVMs tend to be in good shape, or at least repairable condition. Unfortunately, a lot of VOMs are in terrible shape. Part of the problem comes from the fact that the meter will measure current easily enough, but if the operator left the meter on the current setting and then measured a voltage—*pooff!*—the instrument burns up. When examining an instrument, try to make a measurement on each scale. Also, remove the case and look for charred resistors and burnt switch contacts.

In any event, when you obtain a used meter, be sure to take it out of the case and clean the switch contacts and generally spruce it up. The usual forms of contact and tuner cleaner will work wonders on a used multimeter as well. A simple cleaning of the switches may correct intermittent or erratic operation.

Signal generators

The purpose of a *signal generator* is to produce a signal that can be used to troubleshoot, align, adjust or simply prove the performance of a piece of electronic equipment. Fortunately, only a few basic forms of signal generator are needed. Signal generators for this market come in several varieties, including: audio generators, function generators, and RF signal generators.

Audio generators produce at least sine waves on frequencies within the range of human hearing (20 Hz to 20 kHz). Some models produce frequencies over a greater range, while others produce square waves as well as sine waves. The standard audio signal generator has a variable amplitude or level control, and a fixed output impedance of 600 ohms. Audio generators may or may not have an output level meter.

The Heath IG-18 shown in Fig. 3-4 is an audio signal generator. This instrument produces sine and square wave outputs that are individually calibratible as to amplitude. The meter measures the output level of the sine wave signal. Frequency is set on this instrument by using a series of switches. The first switch is a range multiplier,

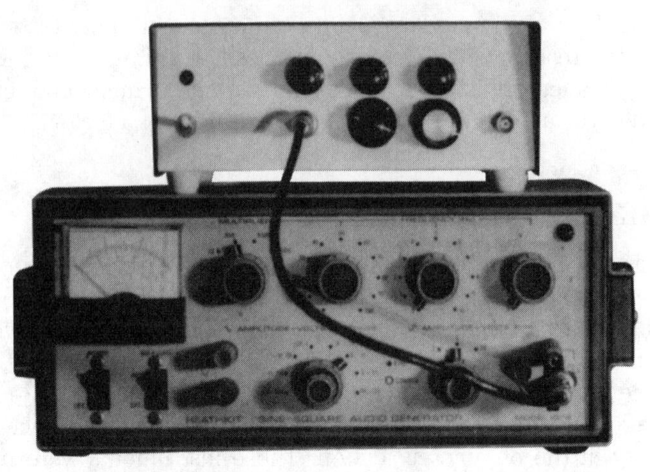

3-4 Audio signal generator with adapter.

while each successive switch is an order of magnitude less (but arranged in decade groupings). The settings shown are X10, 30, 0, and 1, so the frequency set is 10 X (30 + 0.1) or 301 Hz. The box on top of the IG-18 signal generator is a homebrew project that will produce sawtooth and triangle waves from the IG-18 outputs. Since this photo was taken, the circuitry in the small box was added to the IG-18 inside the case.

A *function generator* is much like an audio generator, but it also outputs a triangle waveform in addition to the sine and square wave signals. Some function generators also produce pulses, sawtooths, and other waveforms as well. Typical function generators operate from less than 1 Hz to more than 100 kHz. Many models have a maximum output frequency of 200 kHz, 500 kHz, 1 MHz, 2 MHz, 5 MHz, and in at least one case, 11 MHz. Like the audio generator, the standard output impedance is 600 ohms. However, some instruments also offer 50-ohm outputs (standard for RF circuits), and a TTL (digital) compatible output. The latter is a digital output that is compatible with the ubiquitous *transistor-transistor logic* (TTL or T^{2}L) family of digital logic devices (these are the ones with either 74xx- or 54xx-type numbers).

The *sweep function generator* allows the output frequency to sweep back and forth across a range around the set center frequency. As a result, you can do frequency response evaluations using the sweep function generator and an oscilloscope.

RF signal generators normally output signals in the range above 20 kHz, and typically have an output impedance of 50 ohms. This value of impedance is standard for RF circuits except in the TV industry (where 75 ohms is the standard). RF signal generators come in various types, but can easily be grouped into two general categories: service grade and laboratory grade. For service work, either type is usable.

My own bench has several different signal generator instruments: the Heath IG-18, a sweep function generator, a Model 80 (2-400 MHz) RF signal generator, and an elderly Precision Model E-200C that I refurbished after buying it at a hamfest in non-working condition. I also have a Boyd sweep generator (2-30 MHz), an analog readout audio generator, and a digital readout RF signal generator and frequency counter.

Oscilloscopes

Probably no other instrument is as useful as the *oscilloscope* in working on electronic circuits. You can look at signals and waveforms instead of just averaged dc voltages (as are measured on the DMM). Figure 3-5 shows a simple service grade instrument that can be used for most service jobs on AM, FM, and shortwave receivers, as well as most CB and amateur radio transmitters. The oscilloscope is formally called the *cathode-ray oscillograph*, or CRO. It displays the input signal on a viewing screen that is provided by a cathode-ray tube (CRT).

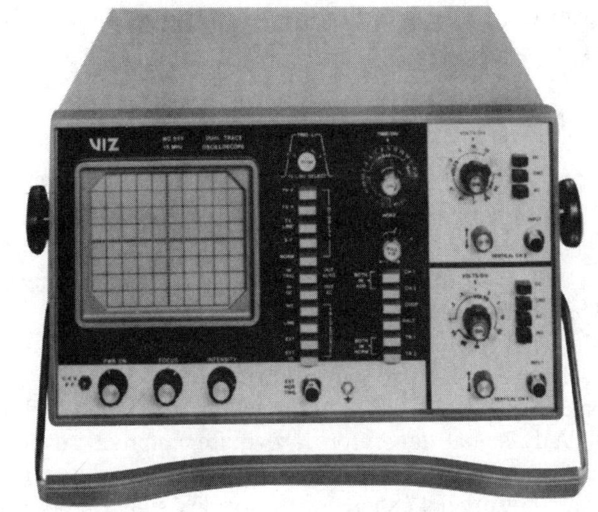

3-5 Oscilloscope.

The light on the viewing screen is produced by the CRT by deflecting an electron beam vertically (Y-direction) and horizontally (X-direction). When two signals are viewed with respect to each other on an X-Y oscilloscope, the result is a *Lissajous figure*.

The quality of oscilloscopes is often measured in terms of the *vertical bandwidth* of the instrument. This specification refers to the –3 dB frequency of the vertical amplifier (or amplifiers if it is a dual-beam 'scope). The higher the bandwidth, the higher the frequency that can be displayed and the sharper the rise time pulse that will be faithfully reproduced. For ordinary service work, a 5-MHz 'scope will suffice for most applications. However, there will be times when you will need a high frequency 'scope for RF work, so buy as much bandwidth as you can afford. Although once in the price stratosphere, even 50-MHz models can be bought relatively cheaply today . . . brand new.

dc power supplies

Although often overlooked as "test equipment," the simple dc power supply is definitely part of the test bench. In addition to powering units being repaired that either lack a dc supply or have a defective dc supply, the bench power supply can be used for a variety of biasing and other troubleshooting functions.

Signal tracers

The *signal tracer* is a high-gain audio amplifier that can be used to examine the signal at various points along the chain of stages in a radio or audio amplifier. A *demodulator probe* will permit the signal tracer to "hear" RF and IF signals. The signal tracer is a back-up replacement for the oscilloscope if it is absolutely impossible to obtain a 'scope. However, the signal tracer is such a poor back-up to the 'scope as to be deemed insufficient. However, there are advantages to these instruments and their cost is low enough to make them a worthwhile addition to the bench.

RF signal generators

Most *RF signal generators* produce only sine waves. The frequency ranges tend to be 10 kHz and up to daylight, or so. Practical signal generators tend to have frequency ranges in the 10-kHz to 50-kHz region (one common design), the 2-MHz to 400-MHz range, and so forth. When the signal generator produces signals above 1000 MHz (1 GHz), they are called *microwave generators*.

Most RF signal generators will produce a sine wave RF output, but in many cases, the operator can select either modulated or unmodulated RF. That is, the output can be either CW, or it can be modulated by an audio signal. Amplitude modulated (AM) signal generators are quite common, even in low-cost instruments. Frequency modulated (FM) signal generators are less common than AM generators, although they are not exactly rare. Most FM signal generators are also AM signal generators, the AM process being somewhat easier than the FM process. The standard impedance for RF systems is 50 ohms, so most RF signal generators have an output impedance of 50 ohms.

RF signal generators can be broken into three general classes: hobbyist/service, professional service, and engineering laboratory models. There is nothing "official" about these classes—they are only used to describe the different quality levels that one can expect. The factors that distinguish the three levels of quality are:

- The accuracy of the frequency reading
- The stability of the output frequency
- The precision and accuracy of the output level setting mechanism
- The shielding of the instrument

Shielding an RF signal generator is important because radiated leakage signal can find its way into the circuits being tested, and thus becomes an unaccounted-for stray signal. Measurements, calibrations or other settings then become suspect.

For most of the projects and experiments in this book, the hobbyist and service grade is sufficient. These instruments are made for hobbyists, ham operators, and technicians who service middle-grade consumer radio equipment. In general, the frequency accuracy, level accuracy, and other factors are not sufficient for testing or troubleshooting high-priced radio receivers or certain professional RF equipment.

Although one might expect function generators and audio generators to overlap somewhat, it is also true that RF generators and function generators also overlap. It is quite common to find function generators that offer output frequencies of 2 MHz, 5 MHz, or 10/11 MHz, which are well into the RF range. But what further characterizes these signal generators as "RF" as well as "function" generators is the existence of both 50- and 600-ohm outputs. On some instruments, there are two outputs (one 50 ohms and the other 600 ohms), while on others, a pushbutton front panel switch selects either the 50- or 600-ohm output impedance.

Other RF bench facilities

Like all electronic workbenches, the RF bench should use an ac isolation transformer between the ac power lines and the workbench power outlets. The ac power isolation transformer will prevent a possibly fatal electrical shock from occurring if you accidentally come between an ac wire and ground (through sloppy work habits or a freak accident). It isn't absolute protection (nothing ever is), and doesn't justify sloppiness, but it does provide a measure of protection for people using the bench. For most home workbenches, a 1000-volt-ampere (roughly, 1000 watts) 110:110-volt transformer will suffice.

Another useful item to have on an RF bench is one or more external antennas. These antennas are mounted outdoors, or in an attic space, and coaxial cable is brought into the workshop. The cable ends can either dangle free with PL-259 connectors, or be terminated in a panel equipped with SO-239 RF connectors. Some people like to use a coaxial switch (Fig. 3-6) so they can select from several antennas. On my bench, I have a five-band vertical ham radio antenna, a 60-foot random length end-fed wire antenna, and a VLF loop antenna terminated at the workbench; an MFJ Enterprises coax switch allows me to select the one I need. The common terminal of the coaxial switch is terminated in a single SO-239 female connector on the workbench.

A good ground is essential for RF workbenches. At lower frequencies, the ground is not terribly important. However, at RF frequencies, the ground is crucial. The same rules apply to the bench ground as apply to a ground for a ham radio installation. I use a series of three copper-clad ground rods, each eight feet long, driv-

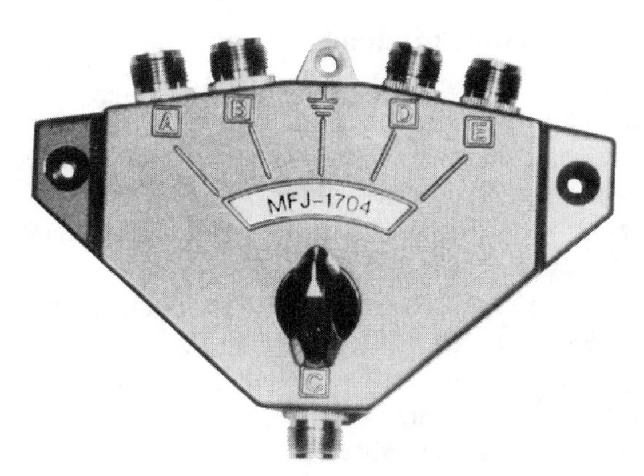

3-6 Coaxial antenna switch.

en into the soil outside my basement. A short, but very wide ground wire connects the bench to the ground.

Dummy loads

A *dummy load* is an artificial aerial that is used for making measurements and performing tests on radio transmitters without actually radiating a signal into the air. Radio operators should routinely use dummy loads to tune up on crowded channels, and only when the tuning is completed should you transfer power to the live antenna. In some countries, this procedure is not merely good manners, but required by law (even if often ignored).

Another use for dummy loads is in troubleshooting antenna systems containing multiple elements (e.g., tuners, coaxial cable, low-pass filters, and so forth). Suppose, for example, we have an antenna system in which the VSWR is high enough to adversely affect the operation of the radio transmitter. Modern transmitters, with solid-state final RF power amplifiers, have VSWR-sensitive power shutdown circuitry. We can test such a system by disconnecting each successive element in sequence, and connect the dummy load to its output. If the VSWR goes down to the normal range when a particular element is replaced with the dummy load, then the difficulty is probably distal to the point where the dummy load was inserted (i.e., towards the antenna). You will eventually find the bad element (which is usually the antenna itself).

Another use for dummy loads is in testing for television, broadcast radio or audio system interference. Once it is established that your transmitter is causing the problem, it is necessary to determine whether or not the route of transmission is through the antenna, around the cabinet flanges or through the ac power mains connection. If the offending signal is from the antenna, then the root cause might be not the signal or transmitter, but the improper filtering or shielding of the television or other appliance, or the product's inability to handle local overload conditions created (but not the fault of) your transmitter. If the interference persists in the face of using an RF dummy load substituting for the aerial, then the problem is in the power connection or flanges . . . and you must do something further to find and suppress the fault.

Forms of dummy load

A common, but irregular, form of dummy load is shown in Fig. 3-7. It consists of a 40- to 100-watt electric light bulb connected to a length of 52-ohm coaxial cable and a connector that mates with the transmitter. The center conductor of the coaxial cable is connected to the center button of the light bulb, while the shield of the coaxial cable is connected to the outer threads or base (depending on country of origin) of the bulb.

It was once common practice to paint the bulb with either aluminum or copper conductive spray paint, except for a small window for viewing the light level. The paint supposedly shields the bulb and thereby prevents RF radiation. That supposition, however, is highly optimistic. One day about 30 years ago, I used this type of dummy load to test my Heathkit DX-60B 90-watt CW transmitter. A friend of mine

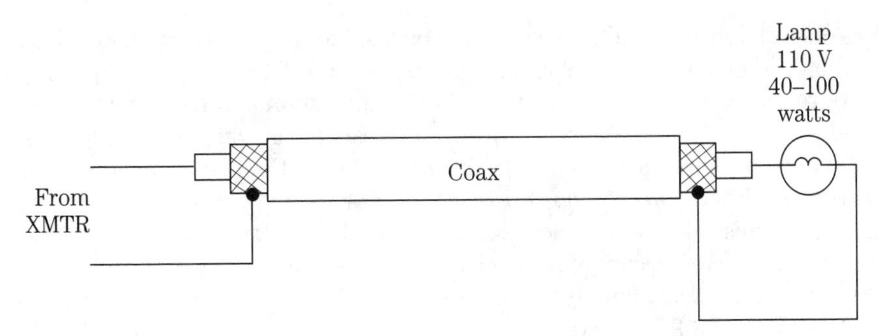

3-7 Light bulb dummy load (not recommended).

answered my "call" (supposedly made to a dummy load), and reported an S7 signal strength . . . from a distance of nearly ten kilometers! That's not exactly how a dummy load is supposed to work.

Another defect of the light bulb dummy load is that its resistance changes with light brightness, so is thus not stable enough to be seriously considered as a dummy load except in the crudest sense. The light bulb dummy load, while cheap and easy to obtain, is too much of a problem for all but impromptu emergency situations. It is *not* recommended.

Figure 3-8 shows the most elementary form of regular dummy load which consists of one or more resistors connected in series, parallel, series-parallel as needed to make the total resistance equal to the desired load impedance (usually 50 ohms).

The power dissipation of the dummy load in Fig. 3-8 is the sum of the individual power dissipations. By using ordinary 2-watt carbon composition resistors, it is possible to make reasonable dummy loads to powers of about 50 watts. Above that power level, one must consider the effects of stray capacitance and inductance from all of the resistor leads and interconnecting wires. Higher levels can be accommodated, however, if care is taken to keep capacitances and inductances low.

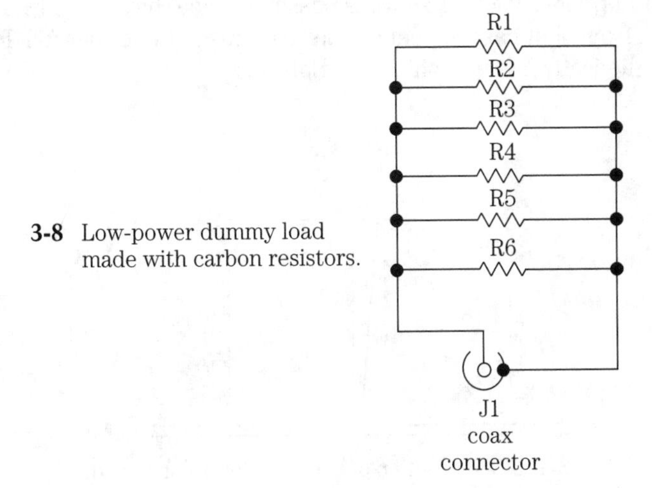

3-8 Low-power dummy load made with carbon resistors.

It is essential that noninductive resistors be used for this application. For this reason, carbon composition or metal film resistors are used. There are two forms of noninductive resistors on the market. The carbon composition and metal film types are intuitively obvious. The other form is wirewound resistors in which adjacent turns are wound in opposite directions so that their mutual magnetic fields cancel each other out. These are called *counter-wound* resistors. For very low frequency (< 20 kHz) work, it is permissible to use such counter-wound wire low-inductance resistors. These resistors, however, cannot be used at frequencies over a few hundred kilohertz.

Several commercial dummy loads are shown in Figs. 3-9 through 3-17. The dummy load in Fig. 3-9 is a 5-watt model, and is typically used in citizen's band and other low-power (QRP) HF transmitters. The resistor is mounted directly on a PL-259 coaxial connector. These loads typically work to about 300 MHz, although many are not really useful over about 150 MHz because of stray capacitance and inductance. A higher power version of the same type of dummy load is shown in Fig. 3-10. This device works from VLF to the low VHF region, and dissipates up to 50 watts of power. I have used this dummy load for servicing high VHF land mobile rigs, VHF-FM marine rigs, and low-VHF land mobile rigs, as well as ham radio rigs.

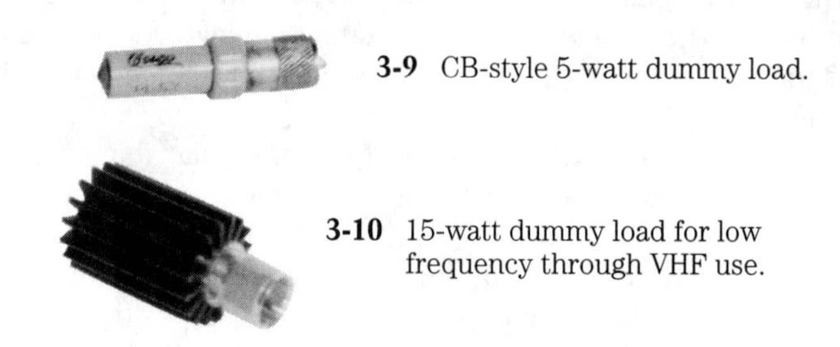

3-9 CB-style 5-watt dummy load.

3-10 15-watt dummy load for low frequency through VHF use.

Figure 3-11 shows how such a dummy load can be built from 2-watt carbon composition or metal film resistors. The end pieces are either metal disks, cut from copper or brass sheet stock, or printed circuit board material. Particularly easy to use is copper clad perforated circuit board; the holes can be used to accommodate the resistor leads. Keep those leads as short as possible. Once the assembly is constructed, make a shield from light gauge sheet copper (or brass), or copper foil. Be sure to not short-circuit the load when the shield is in place.

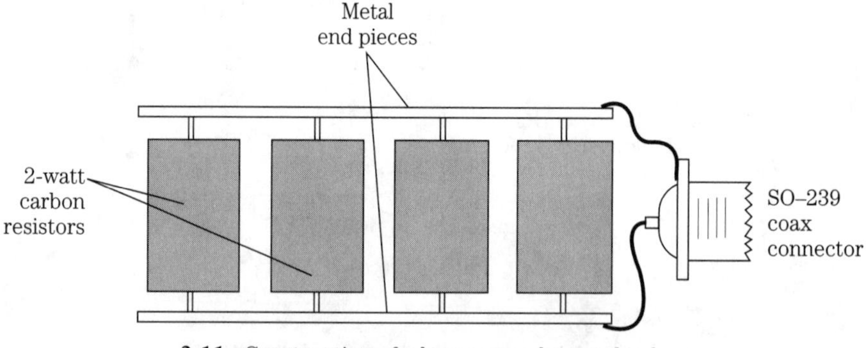

3-11 Construction of a low-power dummy load.

The load resistor in Fig. 3-12 is a Bird Electronics Termaline load that works at power levels to 50 watts, while presenting a 50-ohm load.

A 300-watt amateur radio 50-ohm dummy load is shown in Fig. 3-13; this one is made by MFJ Enterprises in the USA. Unlike the other two models, this one is built inside a sheet metal cabinet, and is inexpensive.

3-12 Bird Termaline RF wattmeter.

3-13 Moderate-power dummy load.

An oil can load is shown in Fig. 3-14. This dummy load is similar in form to the old (now off the market) Heathkit Cantenna. It consists of a high-power, 50-ohm noninductive resistor element mounted inside a paint can that is filled with ordinary motor oil. The oil increases the dissipation capability of the resistor, but tends to seep out of the load if it is not well-sealed.

Our final dummy load resistor is shown in Fig. 3-15. The actual resistor, made by R.L. Drake Co. in the USA, is shown in Fig. 3-15, while a schematic view is shown in Fig. 3-16. The long, high-power noninductive resistor element is rated at 50 ohms, and can dissipate 1000 watts for several minutes. If longer times or higher powers are antici-

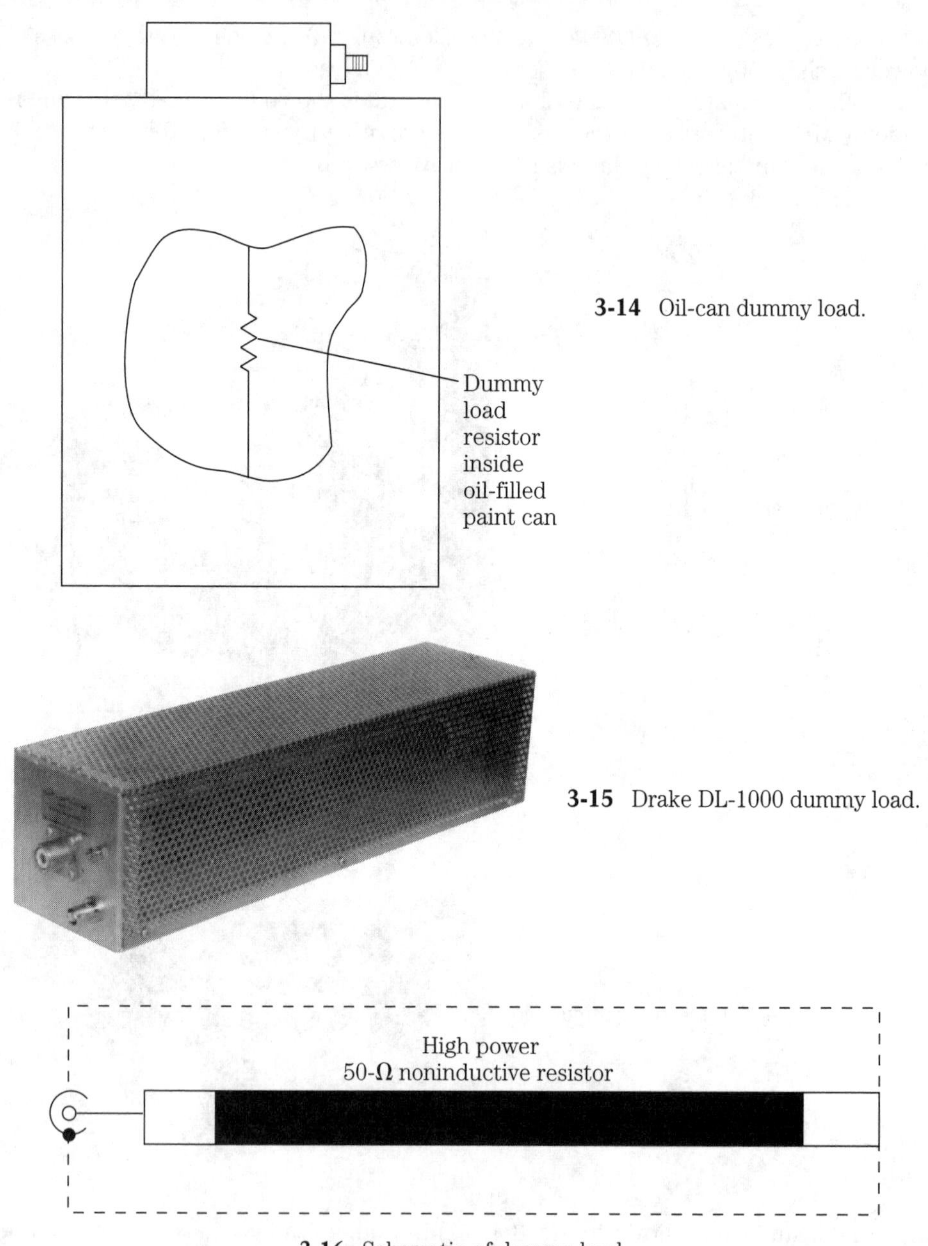

3-14 Oil-can dummy load.

3-15 Drake DL-1000 dummy load.

3-16 Schematic of dummy load.

pated, forced air cooling can be applied by adding a blower fan to one end of the cage (on the Drake product, a removable mounting plate for a 3.5-inch fan is provided).

Providing an output level indicator

The dummy load in Fig. 3-17 was modified by the author by adding the BNC jack (J2) for RF signal sampling. This jack is connected internally to either a 2-turn loop

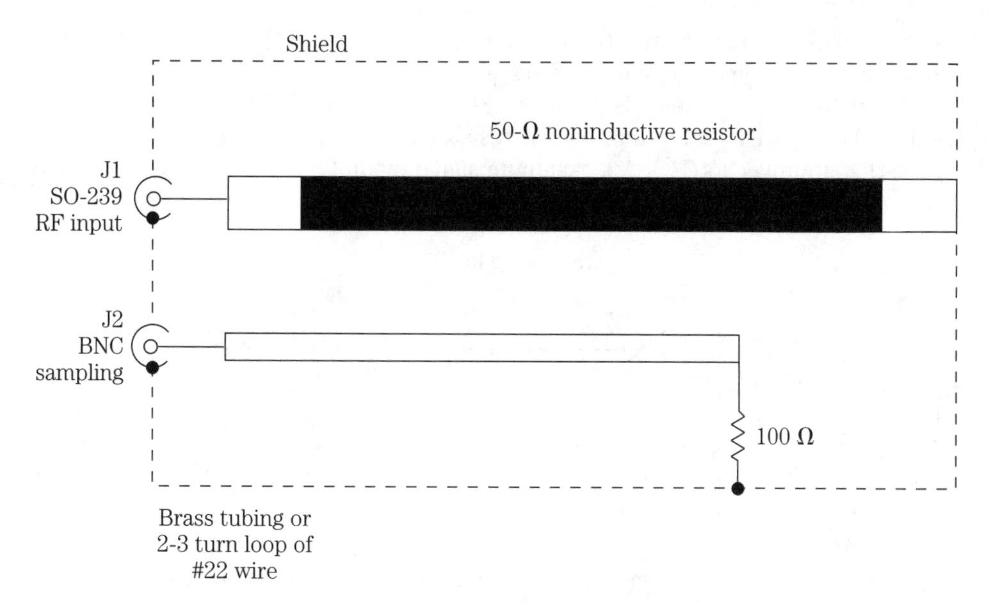

3-17 Modified dummy load circuit offers sample of output signal.

made of #22 AWG insulated hook-up wire, or a 25-cm brass rod that is positioned alongside the resistor element (as shown in Fig. 3-17). The loop or rod will pick up a sample of the signal so it can be viewed on an oscilloscope or used for other instrumentation purposes. Figure 3-18 shows an oscilloscope photograph of an amplitude modulated RF signal taken from my modified dummy load. The transmitter was a 60-watt AM rig modulated by a 400-Hz sine wave from a bench signal generator.

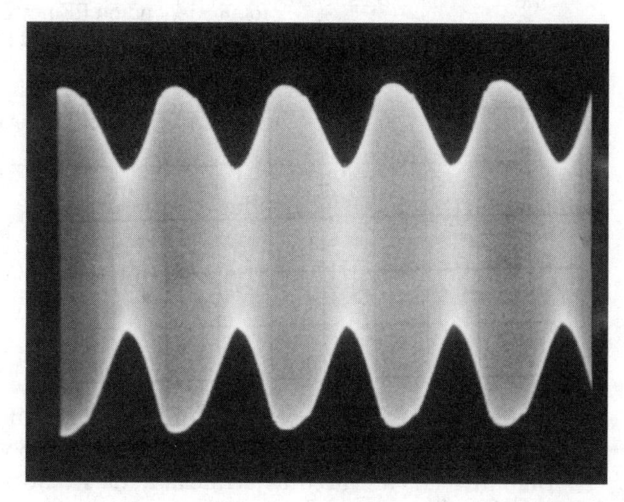

3-18 AM signal read from J2 of Fig. 3-17.

Another approach to providing an output indicator is shown in Fig. 3-19A. In this case, a germanium signal diode (1N34 or 1N60 is suitable) is connected to the end of the signal-sampling rod that connects to the output BNC jack. The diode rectifies the signal picked up by the rod (or sampling coil if one is used) and the RC network R3/C1 filters it to remove residual RF signal.

A variation on the theme is shown in Fig. 3-19B. This circuit, which is like one used in the Heathkit Cantenna, uses a resistor voltage divider (R_2/R_3) connected across the dummy load (R_1). A germanium signal diode (1N34 or 1N60) is attached to the junction of the two voltage divider resistors.

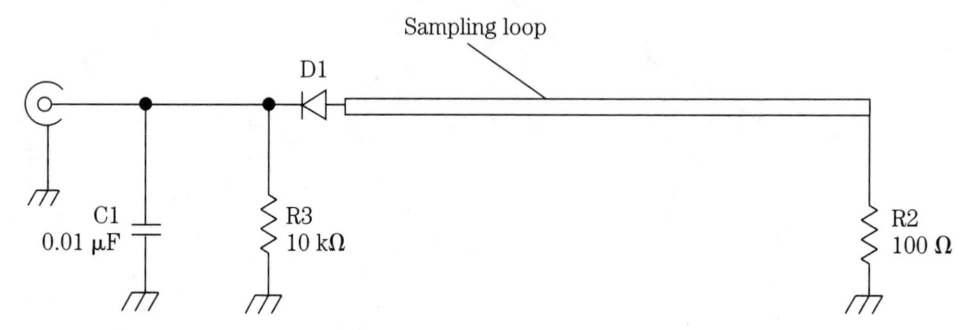

3-19 A. Pick-up loop circuit for dc power indicator output.

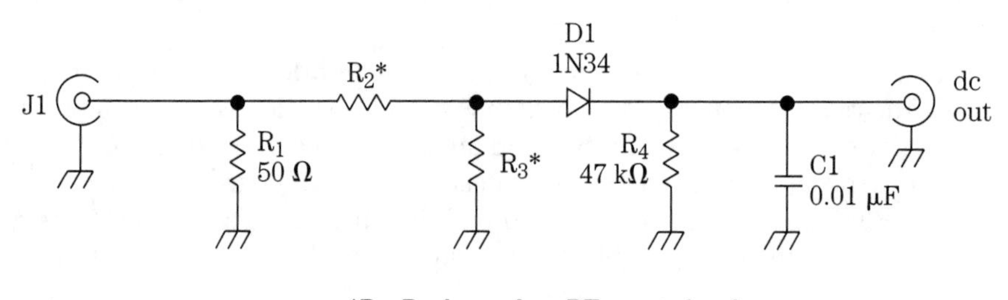

*R2, R3 depend on RF power level
R2 = 100 kΩ, R3 = 1 kΩ for full power
ham rigs

3-19 B. Voltage divider RF out-put level indicator.

The voltage at the junction (V_b) is related to the RF output power applied to the dummy load by the voltage divider equation:

$$V_b = \frac{R_3}{R_2 + R_3} \sqrt{R_1 \times \textit{power} \text{ (watts)}}$$

or

$$V_b = 0.23 \sqrt{50\text{-}\Omega \times \textit{power} \text{ (watts)}}$$

The output voltage from the circuit of Fig. 3-19B is quite reasonable. With the values shown, a 100-watt transmitter signal produces an RMS voltage on the order of 16 volts.

Dummy loads are used as artificial aerials that permit you to energize a radio transmitter for testing, troubleshooting or adjustment without actually radiating a signal. Their use is good engineering practice, is good manners for good radio neighbors, and is legally mandated in most countries. Use them, and we'll all be better off.

RF power meters

A key instrument required in checking the performance of, or troubleshooting, radio transmitters is the *RF power meter* (or wattmeter). These instruments measure the output power of the transmitter, and display the result in watts, or some related unit. Closely related to RF wattmeters is the antenna VSWR meter. These instruments also examine the output of the transmitter, and give a relative indication of output power. They can be calibrated to display the dimensionless units of voltage standing wave ratio (VSWR). Many modern instruments, a couple of which will be discussed as examples in this chapter, combine both RF power and VSWR measurement capabilities.

Measuring RF power

Measuring RF power has traditionally been notoriously difficult, except perhaps in the singular case of continuous wave (CW) sources that produce nice, well-behaved sine waves. Even in that limited case, however, some measurement methods are distinctly better than others. Table 3-1 shows RF waveforms that are output patterns of different styles of radio transmitters. This chart is provided courtesy of Bird Electronics, Inc., a supplier of professional grade RF wattmeters including the industry standard Model 43.

Table 3-1. Power readings for different RF waveforms.

Transmission Type and Scope Pattern	Frequency Spectrum (C: Carrier)	PEVrms (arbitrary)	PEP = PEV^2_{rm}/Zo	Average (Heating) Power	4380 Series CW Mode	4380 Series PEP Mode	4380 Series % MOD Mode	Model 43
Table A CW (100V)	C	$\frac{100}{\sqrt{2}}$ V	100W	100W	100W	100W	0%	100W
Table B AM 100% Mod. (200V)	C	$\frac{200}{\sqrt{2}}$ V	400W	150W	100W	400W	100%	100W
Table C AM 73% Mod. (173V)	C	$\frac{173}{\sqrt{2}}$ V	300W	127W	100W	300W	73%	100W
Table D SSB 1 tone (100V)	(C)	$\frac{100}{\sqrt{2}}$ V	100W	100W	100W	100W	0%	100W
Table E SSB 2 tone (100V)	(C)	$\frac{100}{\cdot\sqrt{2}}$ V	100W	50W	25W	100W	100%	40.5W
Table F SSB Voice (100V)	(C)	$\frac{100}{\sqrt{2}}$ V	100W	—	—	100W	—	—

Z_0 = 50 ohms

PEV: Peak Envelope Voltage. Carrier (or suppressed carrier) PEV was arbitrarily chosen at 100 volts in all examples. $PEV_{rms} = PEV/\sqrt{2}$.

The waveform in Illustration A of Table 3-1 is an ordinary unmodulated CW waveform. It has a single frequency spectrum, and is easily measured. Suppose, as shown in the example, the peak voltage of the waveform is 100 volts (i.e., peak-to-peak 200 volts). Given that the CW waveform is sinusoidal, we know that the RMS voltage is 0.707 times the peak voltage, or 70.7 volts. The output power is related to the RMS voltage across the load by:

$$P = \frac{V_{rms}^2}{Z_o}$$

where

P is the power in watts
V_{rms} is the RMS potential in volts
Z_o is the load impedance in ohms

If we assume a load impedance of 50 ohms, then we can state that the power in our hypothetical illustration A waveform (Table 3-1) is 100 watts.

We can measure power on unmodulated sinusoidal waveforms, by measuring either the RMS or peak values of either voltage or current, assuming that a constant value resistance load is present. But the problem becomes more complex on modulated signals. Several different forms of modulation are shown in illustrations B through F of Table 3-1. As you can see from comparing the various power readings on a Bird Model 4311 peak power meter, the peak (PEP) and average powers vary markedly with modulation type.

One of the earliest forms of practical RF power measurement was the *thermocouple RF ammeter* (see Fig. 3-20). This instrument works by dissipating a small amount of power in a small resistance inside the meter, and then measuring the heat generated with a thermocouple. A dc current meter monitors the output of the thermocouple device, and indicates the level of current flowing in the heating element.

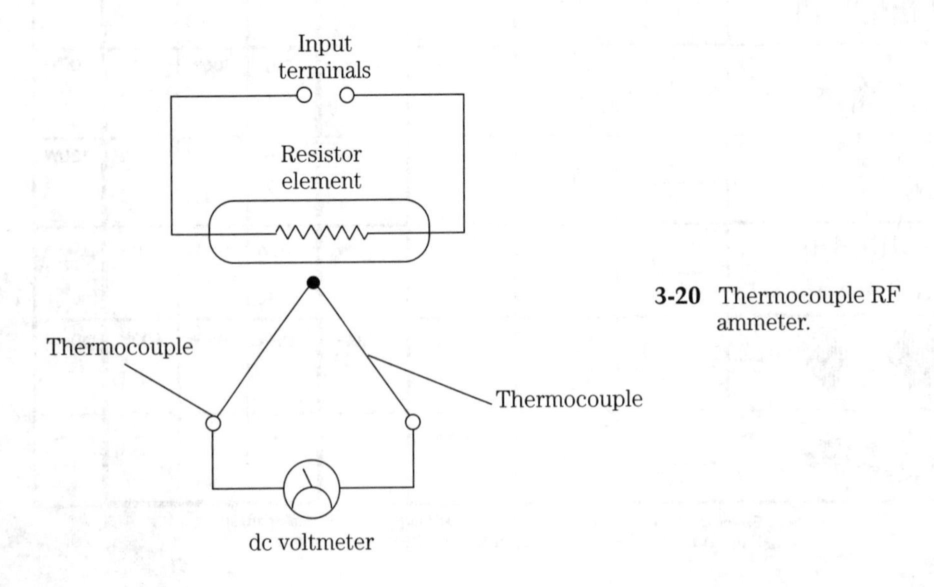

3-20 Thermocouple RF ammeter.

Because it works on the basis of power-dissipated heating a resistance, a thermocouple RF ammeter is inherently an RMS-reading device. Because of this feature it is very useful for making average power measurements. If we know the RMS current and the resistive component of the load impedance, and if the reactive component is zero or very low, then we can determine RF power from the familiar expression

$$P = I^2R$$

There is, however, a significant problem that keeps thermocouple RF ammeters from being universally used in RF power measurement: those instruments are highly frequency-dependent. Even at low frequencies, it is recommended that the meters be mounted on insulating material with at least a ⅝-inch spacing between the meter and its metal cabinet. Even with that precaution, however, there is a strong frequency dependence that renders the meter less useful at higher frequencies. Some meters are advertised to operate into the low-VHF region, but a note of caution is necessary. That recommendation requires a copy of the calibrated frequency response curve for that specific meter so that a correction factor can be added or subtracted from the reading. At 10 MHz and higher, the readings of the thermocouple RF ammeter must be taken with a certain amount of skepticism unless the original calibration chart is available.

We can also measure RF power by measuring the voltage across the load resistance (see Fig. 3-21). In the circuit of Fig. 3-21 the RF voltage appearing across the load is scaled downward to a level compatible with the voltmeter by resistor voltage divider R_2/R_3. The output of this divider is rectified by CR_1, and filtered to dc by the action of capacitor C_2.

The method of measuring the voltage in a simple diode voltmeter is valid only if the RF signal is unmodulated and has a sinusoidal waveshape. While these criteria are met in many transmitters, they are not universal. If the voltmeter circuit is peak reading, as in Fig. 3-21, then the peak power is

$$P = \frac{V_o^2}{R_1}$$

The average power is then found by multiplying the peak power by 0.707. Some meter circuits include voltage dividers that precede the meter and thereby convert the reading to RMS, thus also convert the power to average power. Again, it must be stressed that terms like RMS, average, and peak have meaning only when the input RF signal is both unmodulated and sinusoidal. Otherwise, the readings are meaningless unless calibrated against some other source.

It is also possible to use various bridge methods for measurement of RF power. In Fig. 3-22, we see a bridge set up to measure both forward and reverse power. This circuit was once popular for VSWR meters. There are four elements in this quasi-Wheatstone bridge circuit: R_1, R_2, R_3, and the antenna impedance (connected to the bridge at J_2). If R_{ant} is the antenna resistance, then we know that the bridge is in balance (i.e., the null condition) when the ratios R_1/R_2 and R_3/R_{ant} are equal. In an ideal situation, resistor R_3 will have a resistance equal to R_{ant}, but that may overly limit the usefulness of the bridge. In some cases, therefore, the bridge will use a compromise

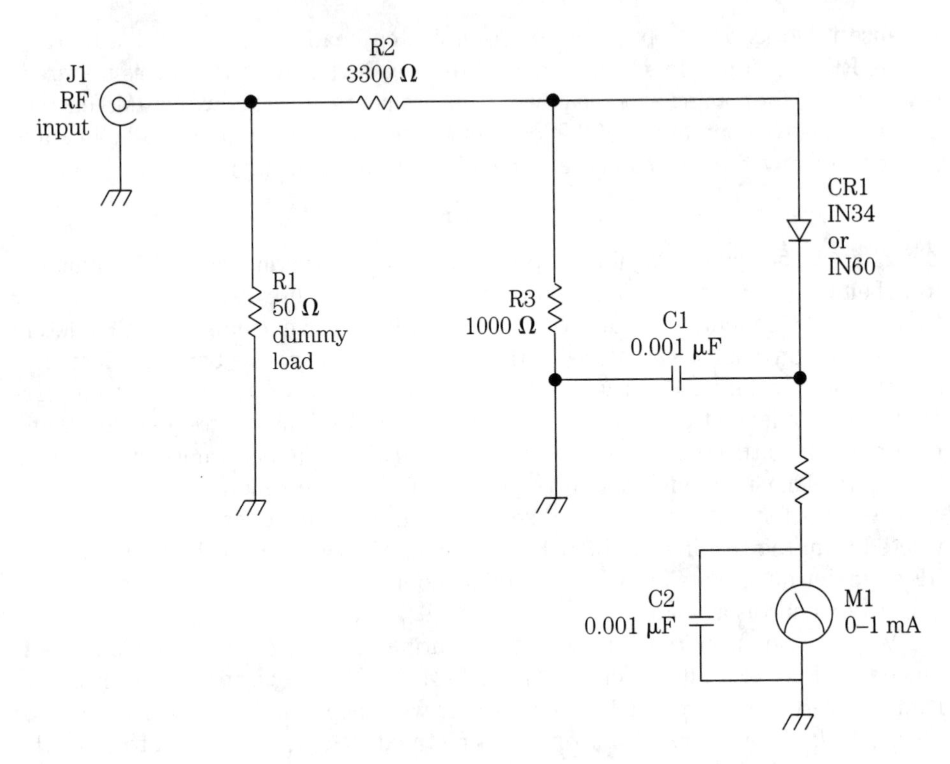

3-21 RF wattmeter based on resistor voltage divider.

value such as 68 ohms for R_3. Such a resistor will be usable on both 50-ohm and 75-ohm antenna systems with but small error. Typically, these meters are designed to read relative power level rather than the actual power.

An advantage of this type of meter is that we can get an accurate measurement of VSWR by proper calibration. With the switch in the FORWARD position, and RF power applied to J_1 (XMTR), potentiometer R_6 is adjusted to produce a full-scale deflection on meter M_1. When the switch is then set to the REVERSE position, the meter will read reverse power relative to the VSWR. An appropriate "VSWR" scale is provided.

A significant problem with the bridge of Fig. 3-22 is that it cannot be left in the circuit while transmitting because it dissipates a considerable amount of RF power in the internal resistances. These meters, during the time when they were popular, were provided with switches that bypassed the bridge when transmitting. The bridge was only in the circuit when making a measurement.

An improved bridge circuit is the capacitor/resistor bridge in Fig. 3-23; this circuit is called the *micromatch bridge*. Immediately we see that the micromatch is improved over the conventional bridge because it uses only 1 ohm in series with the line (R_i). This resistor dissipates considerably less power than the resistance used in the previous example. Because of this low value resistance we can leave the micromatch in the line while transmitting. Recall that the ratios of the bridge arms must be equal for the null condition to occur. In this case, the capacitive reactance ratio of

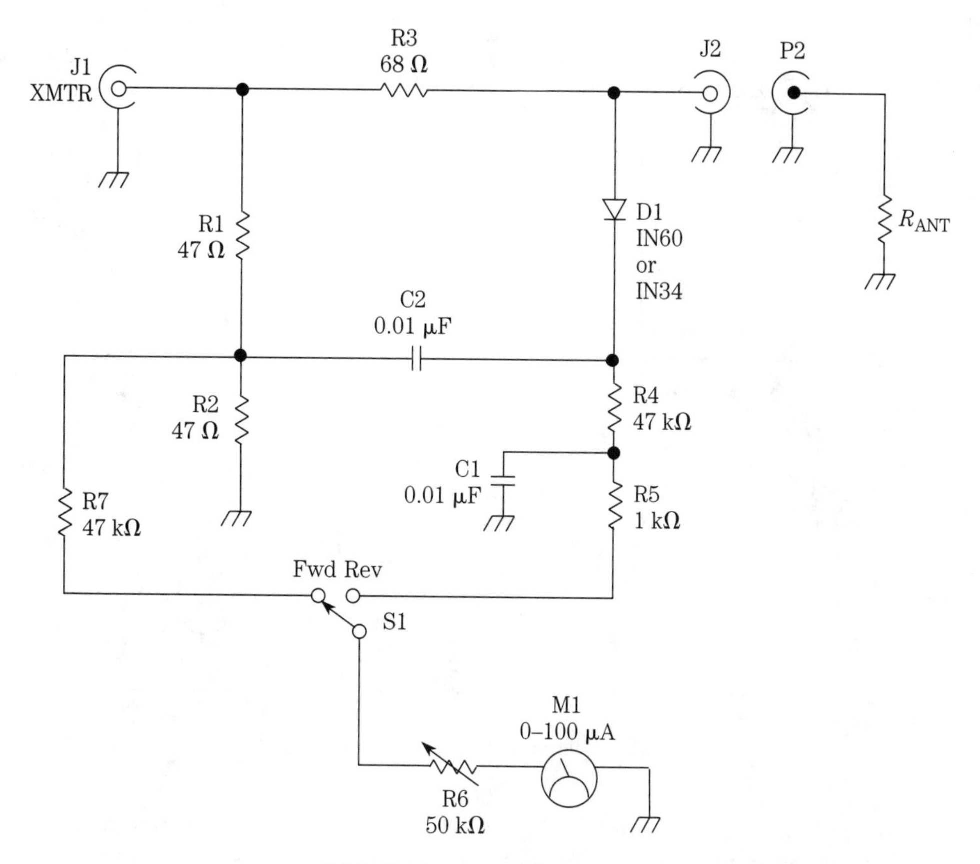

3-22 Bridge-type RF wattmeter.

C_1/C_2 must match the resistance ratio R_1/R_{ant}. For a 50-ohm antenna, the ratio is 1/50, and for 75-ohm antennas it is 1/75 (or, for the compromise situation, 1/68). The small-value trimmer capacitor (C_2) must be adjusted for a reactance ratio with C_1 of 1/50, 1/75, or 1/68, depending upon how the bridge is set up.

The sensitivity control can be used to calibrate the meter. In one version of the micromatch there are three power ranges (10 watts, 100 watts, and 1000 watts). Each range has its own sensitivity control, and they can be switched in and out of the circuit as needed.

The *monomatch bridge* circuit in Fig. 3-24 is the instrument of choice for HF and low-VHF applications. In the monomatch design, the transmission line is segment *B*, while RF sampling elements are formed by segments *A* and *C*. Although the original designs were based on a coaxial cable sensor, later versions used either printed circuit foil transmission line segments or parallel brass rods for *A*, *B*, and *C*.

The sensor unit is basically a directional coupler with a detector element for both forward and reverse directions. For best accuracy, diodes CR_1 and CR_2 should be matched, as should R_1 and R_2. The resistance of R_1 and R_2 should match the transmission line surge impedance, although in many instruments a 68-ohm compromise resistance is used.

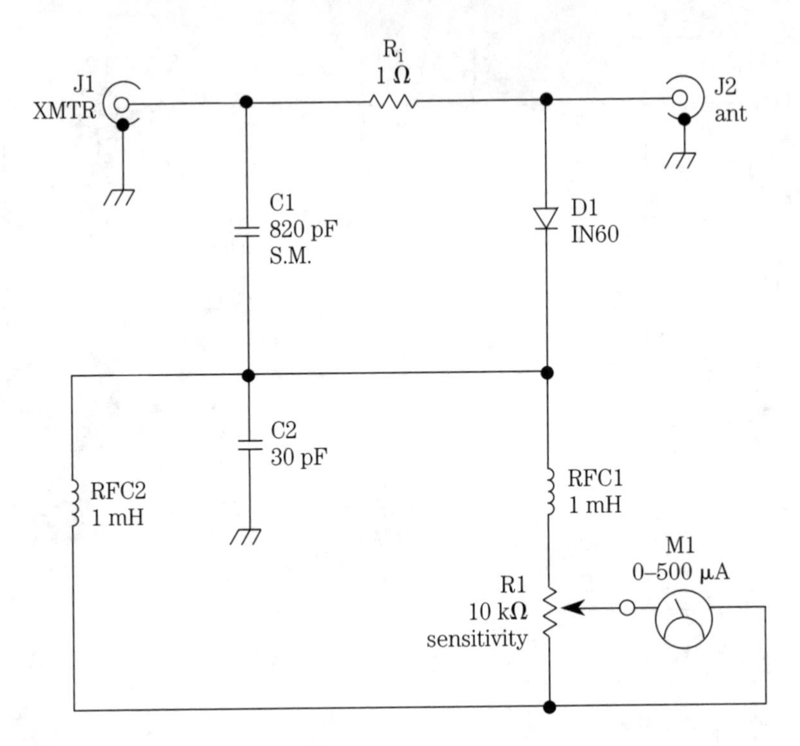

3-23 Micromatch RF wattmeter.

The particular circuit shown in Fig. 3-24 uses a single dc meter movement to monitor the output power. Many modern designs use two meters (one each for forward and reverse power).

One of the latest designs in VSWR meter sensors is the current transformer assembly shown in Fig. 3-25. In this instrument a single-turn ferrite toroid transformer is used as the directional sensor. The transmission line passing through the hole in the toroid "doughnut" forms the primary winding of a broadband RF transformer. The secondary, which consists of 10 to 40 turns of small enamel wire, is connected to a measurement bridge circuit ($C_1 + C_2$ + load) with a rectified dc output.

Figures 3-26 and 3-27 show instruments based on the current transformer technique. In Fig. 3-26, we see the Heath Model HM-102 high-frequency VSWR/power meter. The sensor is a variant on the current transformer method, with L1 being the toroid transformer. This instrument measures both forward and reflected power, and can be calibrated to measure VSWR. The instrument shown in Fig. 3-27 is based on the monomatch design.

The Bird Model 43 Thruline RF wattmeter shown in Fig. 3-28 has for years been one of the industry standards in communications service work. Although it is slightly more expensive than lesser instruments, it is also versatile, and is accurate and rugged. The Thruline meter can be inserted into the transmission line of an antenna system with so little loss that it may be left permanently in the line during normal operations. The Model 43 Thruline is popular with land mobile and marine radio technicians.

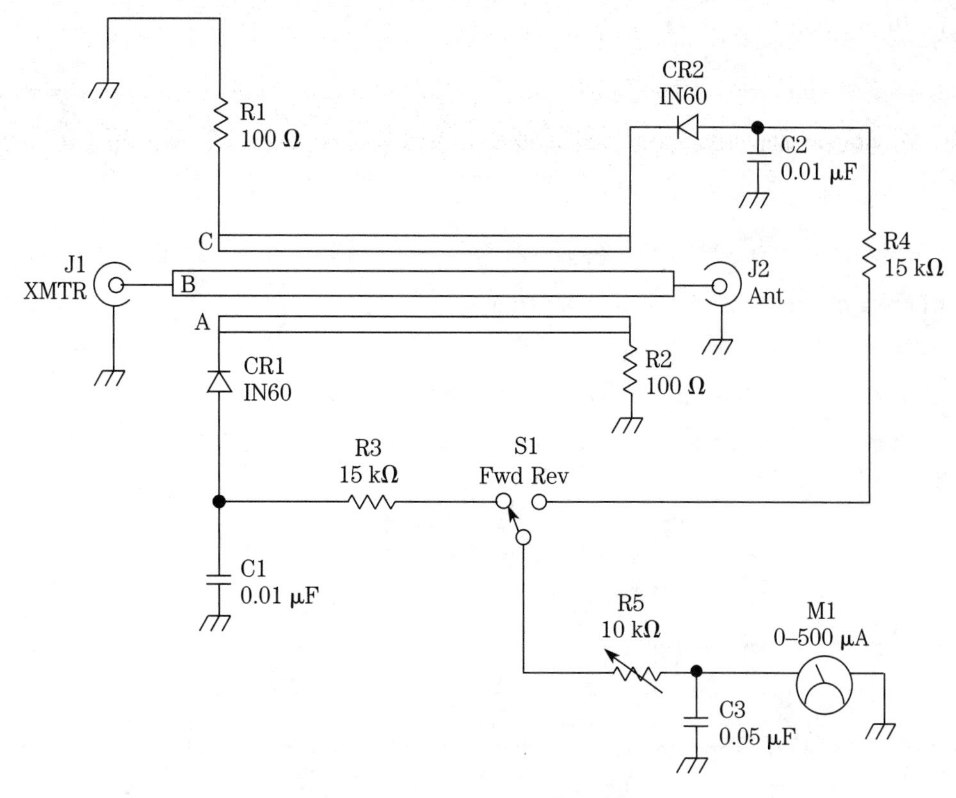

3-24 Monomatch RF wattmeter.

The heart of the Thruline meter is the directional coupler transmission line assembly shown in Fig. 3-29; it is connected in series with the antenna or dummy load transmission line. The plug-in directional element can be rotated 180 degrees to measure both forward and reverse power levels. A sampling loop and diode detector are contained within each plug-in element. The main RF barrel is actually a special coaxial line segment with a 50-ohm characteristic impedance. The Thruline sensor works due to the mutual inductance between the sample loop and center conductor of the coaxial element. Figure 3-30 shows an equivalent circuit. The output voltage from the sampler (e) is the sum of two voltages, e_r and e_m. Voltage e_r is created by the voltage divider action of R and C on transmission line voltage E. If R is much less than X_c, then we may write the expression for e_r as:

$$e_r = \frac{RE}{X_c} = REj\varpi C$$

Voltage e_m, on the other hand, is due to mutual inductance, and is expressed by:

$$e_m = Ij\varpi \pm m$$

We now have the expression for both factors that contribute to the total voltage e. We know that

$$e = e_r + e_m$$

so, by substitution

$$e = j\varpi M\left(\frac{E}{Z_o}\right) \pm I$$

By recognizing that, at any given point in a transmission line, E is the sum of the forward (E_f) and reflected (E_r) voltages, and that the line current is equal to:

$$\frac{E_f}{Z_o} = \frac{E_r}{Z_o}$$

where Z_o is the transmission line impedance.

We may specify e in the forms

$$e = \frac{j\varpi M(2E_f)}{Z_o}$$

and

$$e = \frac{j\varpi M(2E_r)}{Z_o}$$

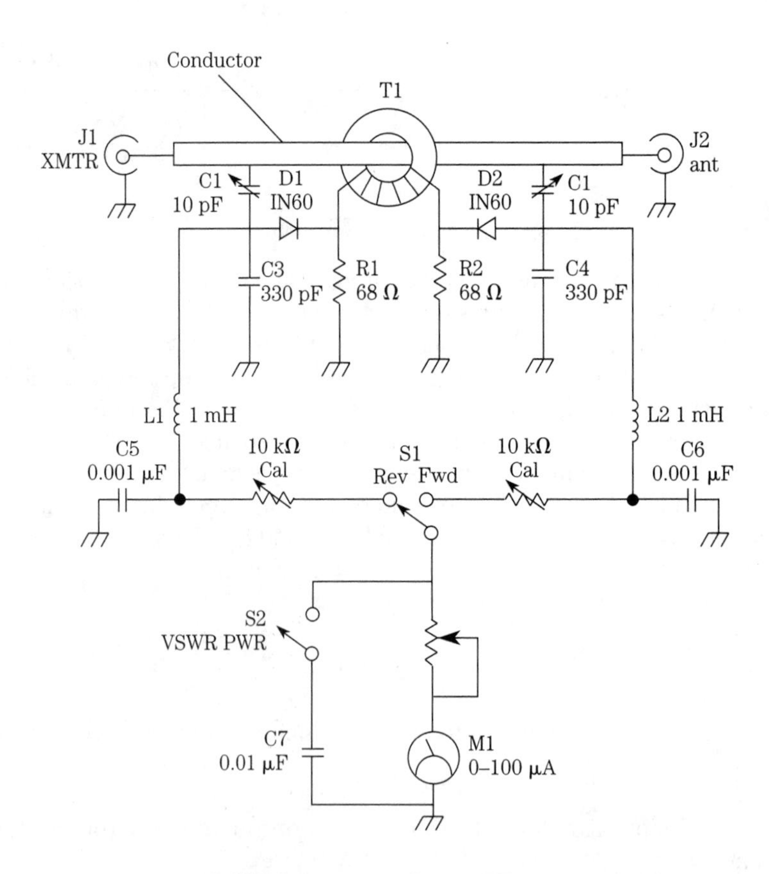

3-25 Current transformer RF wattmeter.

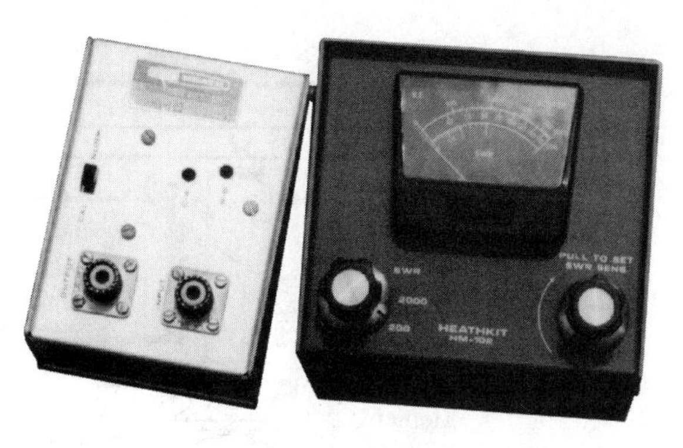

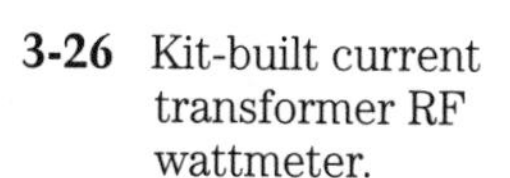
3-26 Kit-built current transformer RF wattmeter.

3-27 Monomatch RF power/VSWR meter.

3-28 Bird Model 43 in-line wattmeter.

The output voltage e of the coupler then, is proportional to the mutual inductance and frequency (by virtue of $j\varpi M$). But the manufacturer terminates R in a capacitive reactance, so the frequency dependence is lessened (see Fig. 3-31). Each element is custom-calibrated, therefore, for a specific frequency and power range. Beyond the specified range for any given element, however, performance is not guaranteed. There are a large number of elements available that cover most commercial applications.

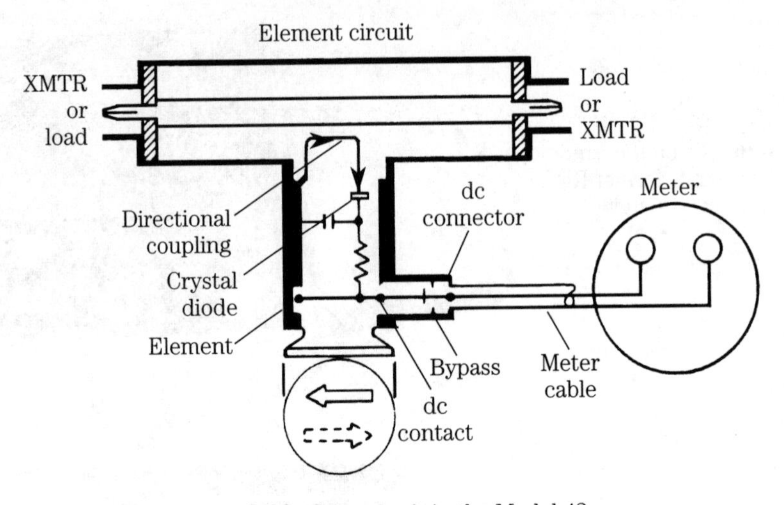

3-29 Sensor unit in the Model 43.

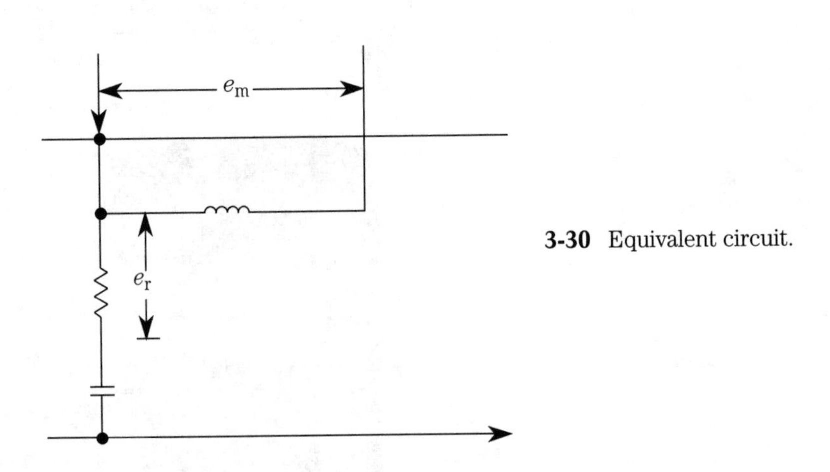

3-30 Equivalent circuit.

The Thruline meter is not a VSWR meter, but rather a power meter. VSWR can be determined from the formula, or by using the nomograph in Fig. 3-32. Some of the Thruline series intended for very high power (Fig. 3-33) applications use an inline coaxial cable coupler (for broadcast-style hardline) and a remote indicator.

Building and using the RF noise bridge

In this section, we will explore a device that has applications in general RF electronics as well as antenna work: the *RF noise bridge*. It is one of the most useful, low-cost, and often overlooked test instruments in the servicer's armamentarium.

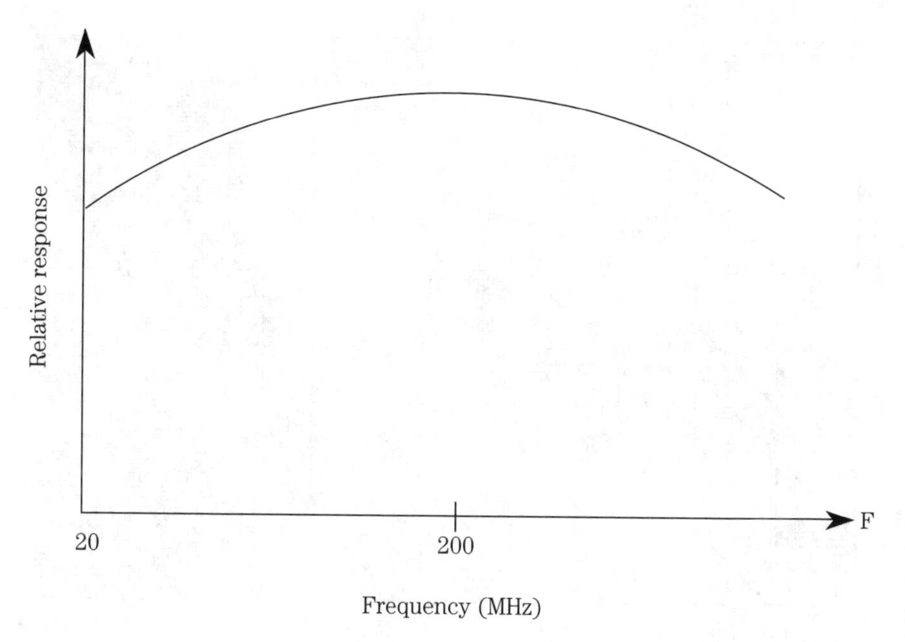

3-31 Typical frequency dependence in VHF region.

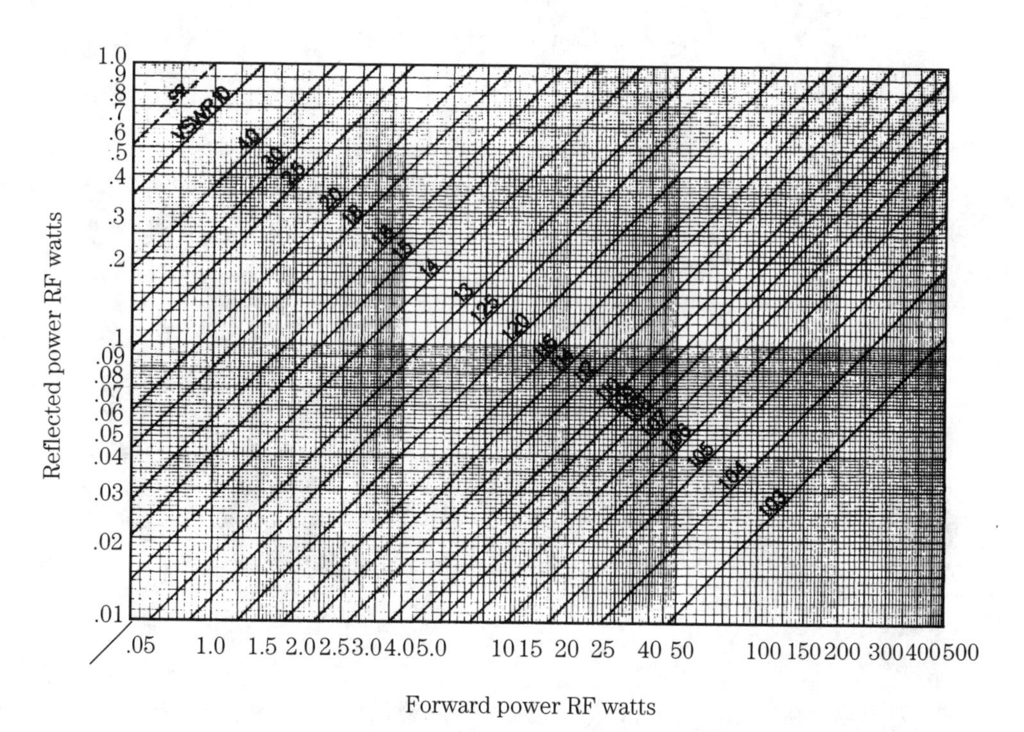

3-32 A. High-power RF wattmeters.

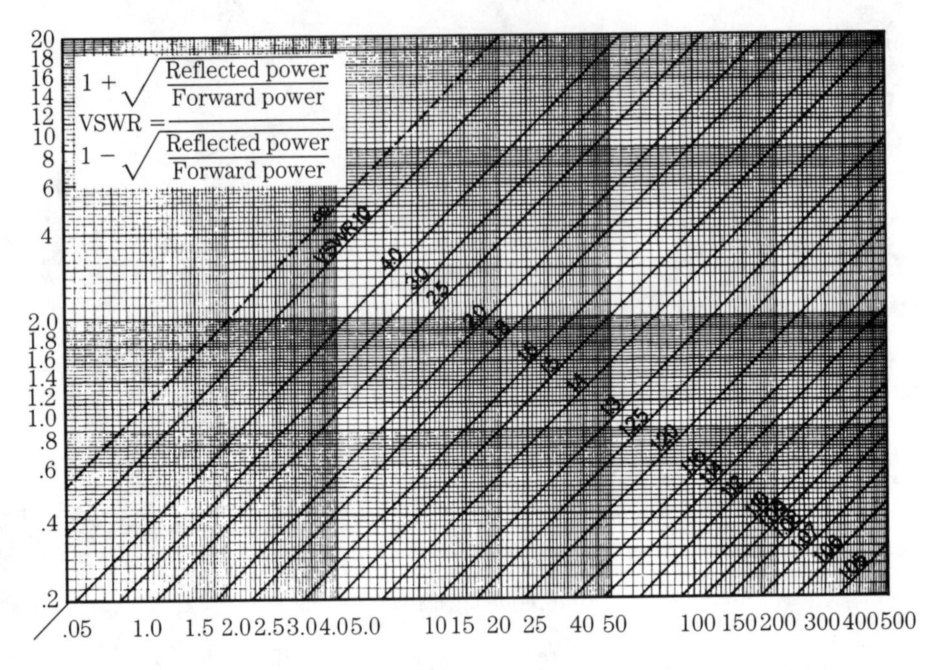

3-32 B. High-power RF wattmeters.

3-33 Nomograph for finding VSWR from forward and reflected RF power.

Several companies have produced low-cost noise bridges: Omega-T, Palomar Engineers, and most recently, The Heath Company. The Omega-T device (Fig. 3-34A) is a small cube with minimal dials, and a pair of BNC coax connectors (ANTENNA and RECEIVER). The dial is calibrated in ohms, and measures only the resistive component of impedance. The Palomar Engineers device (Fig. 3-34B) is a little less eye-appealing, but it does everything the Omega-T does, plus, it allows you to make a rough measurement of the reactive component of impedance.

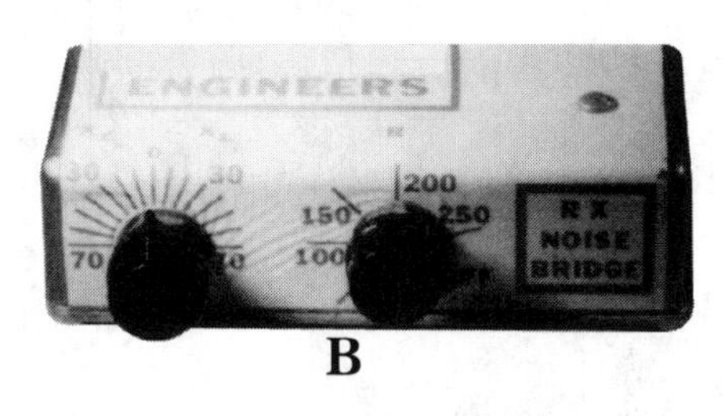

3-34 A. Omega-T noise bridge. B. Palomar noise bridge.

Over the years, I have found the noise bridge to be very useful for a variety of test and measurement applications, especially in the HF and low VHF regions—and those applications are not limited to the testing of antennas (which is the main job of the noise bridge). In fact, while the two-way technician (including CB) will measure antennas with the device, consumer technicians will find other applications.

Figure 3-35 shows a block diagram of this instrument. The bridge consists of four arms. The inductive arms (L 1b and L 1c) form a trifilar-wound transformer over a ferrite core with L 1a, so signal applied to L 1a is injected into the bridge circuit. The measurement consists of a series circuit of a 200-ohm potentiometer and a 120 pF variable capacitor. The potentiometer sets the range (0 to 200 ohms) of the resistive component of measured impedance, while the capacitor sets the reactance component. Capacitor C_2 in the UNKNOWN arm of the bridge is used to balance the measurement capacitor. With C_2 in the circuit, the bridge is balanced when C is approximately in the center of its range. This arrangement accommodates both inductive and capacitive reactances, which appear on either side of the "zero" point, i.e., the mid-range capacitance of C. When the bridge is in balance, the settings of R and C reveal the impedance across the UNKNOWN terminal.

A reverse-biased zener diode (zeners normally operate in the reverse-bias mode) produces a large amount of noise because of the avalanche process inherent in zener operation. While this noise is a problem in many applications, in a noise bridge it is highly desirable: the richer the noise spectrum the better. The spectrum is enhanced somewhat in some models because of a 1-kHz square wave modulator that chops the noise signal. An amplifier boosts the noise signal to the level needed in the bridge circuit.

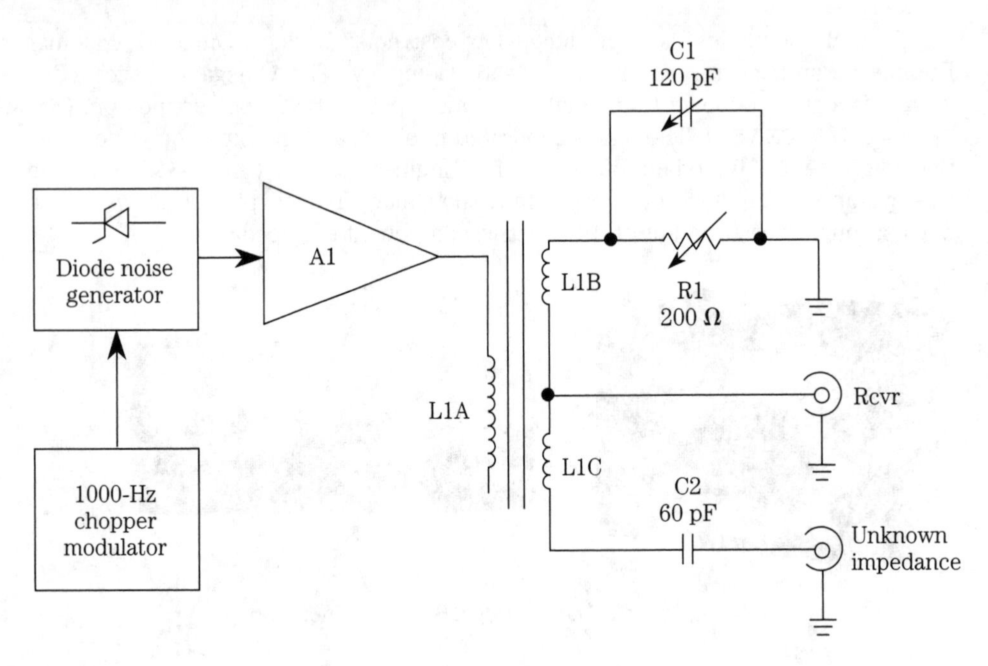

3-35 Block diagram of a noise bridge.

The detector used in the noise bridge is an HF receiver. The preferable receiver uses an AM demodulator, but both CW (Morse code) and SSB receivers will do in a pinch. The quality of the receiver depends entirely on the precision with which you need to know the operating frequency of the device under test.

Adjusting antennas

Perhaps the most common use for the antenna noise bridge is finding the impedance and resonant points of an HF antenna. Connect the RECEIVER terminal of the bridge to the ANTENNA input of the HF receiver through a short length of coaxial cable as shown in Fig. 3-36. The length should be as short as possible, and the characteristic impedance should match that of the antenna feedline. Next, connect the coaxial feedline from the antenna to the ANTENNA terminals on the bridge. You are now ready to test the antenna.

Finding impedance Set the noise bridge resistance control to the antenna feedline impedance (usually 50 or 75 ohms for most amateur antennas). Set the reactance control to mid-range (zero). Next, tune the receiver to the expected resonant frequency (F_{exp}) of the antenna. Turn the noise bridge on, and look for a noise signal of about S9 (will vary on different receivers, and if—in the unlikely event—that the antenna is resonant on the expected frequency).

Adjust the RESISTANCE control (R) on the bridge for a null, i.e., minimum noise as indicated by the S meter. Next, adjust the REACTANCE control (C) for a null. Repeat the adjustments of the R and C controls for the deepest possible null, as indicated by the lowest noise output on the S meter (there is some interaction between the two controls).

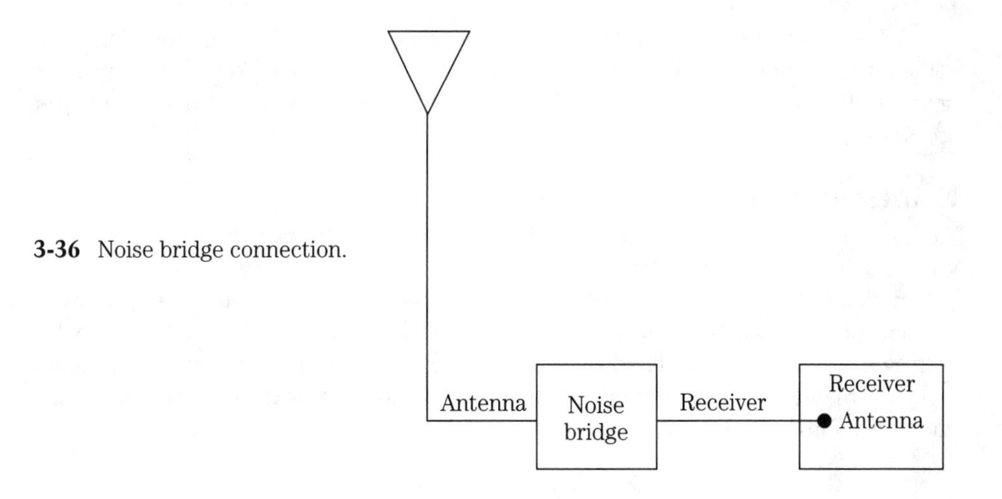

3-36 Noise bridge connection.

A perfectly resonant antenna will have a reactance reading of zero ohms, and a resistance of 50 to 75 ohms. Real antennas may have some reactance (the less the better), and a resistance that is different from 50 or 75 ohms. Impedance-matching methods can be used to transform the actual resistive component to the 50- or 75-ohm characteristic impedance of the transmission line.

If the resistance is close to zero, then suspect that there is a short circuit on the transmission line, and an open circuit if the resistance is close to 200 ohms.

A reactance reading on the X_L side of zero indicates that the antenna is too long, while a reading on the X_c side of zero indicates an antenna that is too short.

An antenna that is too long or too short should be adjusted to the correct length. To determine the correct length, we must find the actual resonant frequency, F_r. To do this, reset the REACTANCE control to zero, and then ***SLOWLY*** tune the receiver in the proper direction—downband for too long and upband for too short—until the null is found. On a high-Q antenna the null is easy to miss if you tune too fast. Don't be surprised if that null is out of band by quite a bit. The percentage of change is given by dividing the expected resonant frequency (F_{exp}) by the actual resonant frequency (F_r), and multiplying by 100:

$$Change = \frac{(F_{exp})(100\%)}{F_r}$$

Resonant frequency Connect the antenna, noise bridge and the receiver in the same manner as above. Set the receiver to the expected resonant frequency: i.e., $468/F$ for half-wavelength types and $234/F$ for quarter-wavelength types. Set the resistance control to 50 ohms or 75 ohms, as appropriate for the normal antenna impedance and the transmission line impedance. Set the reactance control to zero. Turn the bridge on and listen for the noise signal.

Slowly rock the reactance control back and forth to find on which side of zero the null appears. Once the direction of the null is determined set the reactance control to zero, and then tune the receiver towards the null direction (downband if null is on X_L side and upband if on the X_c side of zero).

A less than ideal antenna will not have exactly 50 or 75 ohms impedance, so some adjustment of R and C to find the deepest null is in order. You will be surprised how far off some dipoles and other forms of antennas can be if they are not in "free space," i.e., if they are close to the Earth's surface.

Nonresonant antenna adjustment

We can operate antennas on frequencies other than their resonant frequency if we know the impedance. In order for the antenna to radiate properly, however, it is necessary to match the impedance of the antenna feedpoint to the source (e.g., a transmission line from a transmitter). We can find the feedpoint resistance from the setting of the potentiometer in the noise bridge. The reactances can be calculated from the reactance measurement on the bridge by looking at the capacitor setting—and using a little arithmetic.

$$X_c = X = \left(\frac{55}{68 - C}\right) - 2340$$

or

$$X_L = X = 2340 - \left(\frac{159{,}155}{68 + C}\right)$$

Now, plug "X" calculated from one of the above into $X_f = X/F$ where F is the desired frequency in MHz.

Other RF jobs for the noise bridge

The noise bridge can be used for a variety of jobs. We can find the values of capacitors and inductors, the characteristics of series and parallel tuned resonant circuits, and for adjusting transmission lines.

Transmission line length Some antennas and (non-noise) measurements require antenna feedlines that are either quarter wavelength or half wavelength at some specific frequency. In other cases, a piece of coaxial cable of specified length is required for other purposes: for instance, the dummy load used to service depth sounders is nothing but a long piece of shorted coax that returns the echo at a time interval that corresponds to a specific depth. We can use the bridge to find these lengths as follows:

1. Connect a short-circuit across the UNKNOWN terminals and adjust R and X for the best null at the frequency of interest (note: both will be near zero).
2. Remove the short-circuit.
3. Connect the length of transmission line to the UNKNOWN terminal—it should be longer than the expected length.
4. For quarter-wavelength lines, shorten the line until the null is very close to the desired frequency. For half-wavelength lines, do the same thing—except that the line must be shorted at the far end for each trial length.

Transmission line velocity factor The velocity factor of a transmission line (usually designated by the letter "V" in equations) is a decimal fraction that tells us how fast the radio wave propagates along the line relative to the speed of light in free space. For example, foam dielectric coaxial cable is said to have a velocity factor of $V = 0.80$. This number means that the signals in the line travel at a speed 0.80 (or 80 percent) of the speed of light.

Since all radio wavelength formulas are based on the velocity of light, you need the V value to calculate the physical length needed to equal any given electrical length. For example, a half-wavelength piece of coax has a physical length of $[(492)(V)/F_{MHz}]$ feet. Unfortunately, the real value of V is often a bit different from the published value. You can use the noise bridge to find the actual value of V for any sample of coaxial cable as follows:

1. Select a convenient length of the coax more than 12 feet in length and install a PL-259 RF connector (or other connector compatible with your instrument) on one end, and short-circuit the other end.
2. Accurately measure the physical length of the coax in feet; convert the "remainder" inches to a decimal fraction of one foot by dividing by 12 (e.g., 32'8" = 32.67' because 8"/12" = 0.67). Alternatively, cut off the cable to the nearest foot and reconnect the short circuit.
3. Set the RESISTANCE and REACTANCE controls to zero.
4. Adjust the monitor receiver for the deepest null. Use the null frequency to find the velocity factor $V = FL/492$, where V is the velocity factor (a decimal fraction); F is the frequency in MHz; and L is the cable length in feet.

Tuned circuit measurements

An inductor/capacitor (LC) tuned tank circuit is the circuit equivalent of a resonant antenna, so there is some similarity between the two measurements. You can measure resonant frequency with the noise bridge to within ±20 percent or better if care is taken. This accuracy may seem poor, but it is better than you can usually get with low-cost signal generators, dip meters, absorption wavemeters and the like.

Series-tuned circuits A series-tuned circuit exhibits a low impedance at the resonant frequency, and a high impedance at all other frequencies. Start the measurement by connecting the series-tuned circuit under test across the UNKNOWN terminals of the bridge. Set the RESISTANCE control to a low resistance value, close to zero ohms. Set the REACTANCE control at mid-scale (zero mark). Next, tune the receiver to the expected null frequency, and then tune for the null. Make sure that the null is at its deepest point by rocking the R and X controls for best null. At this point, the receiver frequency is the resonant frequency of the tank circuit.

Parallel-resonant tuned circuits A parallel-resonant circuit exhibits a high impedance at resonance, and a low impedance at all other frequencies. The measurement is made in exactly the same manner as for the series-resonant circuits, except that the connection is different. Figure 3-37 shows a two-turn link coupling that is needed to inject the noise signal into the parallel-resonant tank circuit. If the inductor is the toroidal type, then the link must go through the hole in the doughnut-shaped core and then connects to the UNKNOWN terminals on the bridge. After this, do exactly as you would for the series-tuned tank measurement.

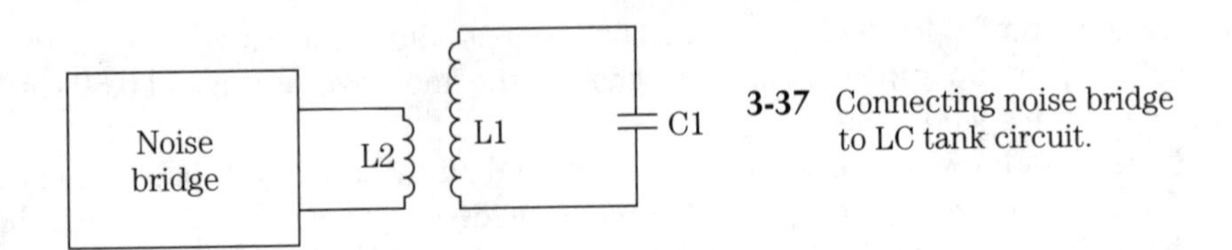

3-37 Connecting noise bridge to LC tank circuit.

Capacitance and inductance measurements

The Heathkit Model HD-1422 (a noise bridge similar to those discussed in this chapter) comes with a calibrated 100-pF silver mica test capacitor (called CTEST in the Heath literature), and a calibrated 4.7 µH test inductor (called LTEST), which are used to measure inductance and capacitance, respectively. The idea is to use the test components to form a series-tuned resonant circuit with an unknown component. If you find the resonant frequency, you can calculate the unknown value. In both cases, the series-tuned circuit is connected across the UNKNOWN terminals of the HD-1422, and the series-tuned procedure above is followed.

Inductance To measure inductance, connect the 100-pF CTEST in series with the unknown coil across the UNKNOWN terminals of the HD-1422. When the null frequency is found, find the inductance from: $L = 253/F^2$; L is the inductance in microhenrys (µH) and F is the frequency in megahertz (MHz).

Capacitance Connect LTEST across the UNKNOWN terminals in series with the unknown capacitance. Set the RESISTANCE control to zero, tune the receiver to 2 MHz, and readjust the REACTANCE control for null. Without readjusting the noise bridge control, connect LTEST in series with the unknown capacitance and retune the receiver for a null. Capacitance can now be calculated from $C = 5389/F^2$; C is in picofarads (pF), F is in megahertz (MHz).

4

Capacitance and capacitors for RF circuits

Capacitors are devices that store electrical energy in an internal electrical field in an insulating dielectric material. They are one of the two components used in RF tuning circuits (the other being inductors—see chapter 6). Like the inductor, the capacitor is an energy storage device. While the inductor stores electrical energy in a magnetic field, the capacitor stores energy in an *electrical* (or *electrostatic*) field; electrical charge (Q) is stored in the capacitor. But more about that shortly.

The basic capacitor consists of a pair of metallic plates facing each other, separated by an insulating material called a *dielectric*. This arrangement is shown schematically in Fig. 4-1A, and in a more physical sense in Fig. 4-1B. The fixed capacitor shown in Fig. 4-1B consists of a pair of square metal plates separated by a dielectric (i.e., an insulator). Although this type of capacitor is not very practical, it was used quite a bit in early radio transmitters. Spark gap transmitters of the 1920s often used a glass-and-tin-foil capacitor fashioned very much like Fig. 4-1B. Layers of glass and foil are sandwiched together to form a high-voltage capacitor. A 1-foot-square capacitor made of ⅛-inch-thick glass and foil has a capacitance up to about 2000 pF, depending on the specific glass material used. You will see this effect in a later experiment.

Units of capacitance

The *capacitance* (C) of the capacitor is a measure of its ability to store current, or more properly, electrical charge. The principal unit of capacitance is the *farad* (named after physicist Michael Faraday). One farad is the capacitance that will store one coulomb of electrical charge (6.28×10^{18} electrons) at an electrical potential of one volt. Or, in math form:

$$C_{\text{farads}} = \frac{Q_{\text{coulombs}}}{V_{\text{volts}}} \tag{4-1}$$

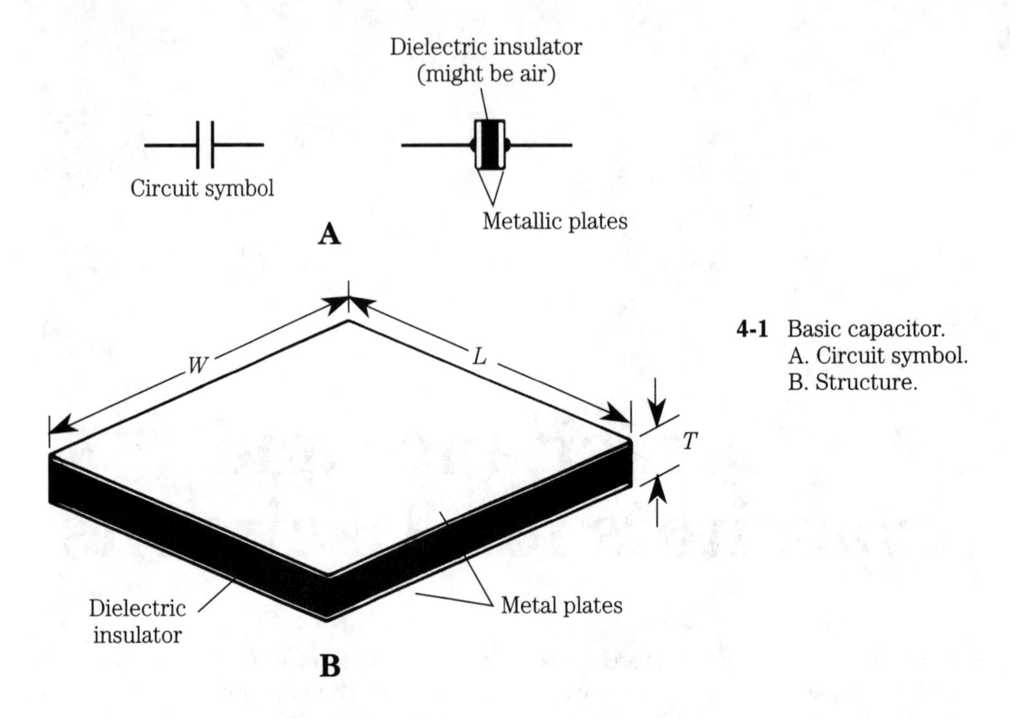

4-1 Basic capacitor.
A. Circuit symbol.
B. Structure.

The farad is far too large a unit for practical RF electronics work, so subunits are typically used instead. The *microfarad* (μF) is 0.000001 farad (1 F = 10^6 μF). The *picofarad* (pF) is 0.000001 μF, which is 0.000000000001 F, or 10^{-12} farads. In older radio texts and schematics the picofarad was called the *micromicrofarad* (μμF or mmF), but never fear: 1 μμF = 1 pF.

The capacitance of the capacitor is directly proportional to the area of the plates (in terms of Fig. 4-1B, $L \times W$), inversely proportional to the thickness (T) of the dielectric (or the spacing between the plates, if you prefer), and directly proportional to the dielectric constant (K) of the dielectric.

Dielectric constant is a property of the insulator material used for the dielectric. The dielectric constant is a measure of the material's ability to support electric flux, and is thus analogous to the permeability of a magnetic material. The standard of reference for dielectric constant is a perfect vacuum, which by definition has a value of $K = 1.000$. Other materials are compared with the vacuum. The values of K for some common materials are:

Material	K
Vacuum	1.0000
Dry air	1.0006
Paraffin (wax) paper	3.5
Glass	5 to 10
Mica	3 to 6
Rubber	2.5 to 35
Dry wood	2.5 to 8
Pure (distilled) water	81

The value of capacitance in any given capacitor is found from:

$$C = \frac{0.0224\, KA\, (N-1)}{T} \qquad \textbf{[4-2]}$$

where

C is the capacitance in picofarads (pF)
K is the dielectric constant
A is the area of one of the plates ($L \times W$), assuming that the two plates are identical, in square inches
N is the number of identical plates
T is the thickness of the dielectric

Experiment 4-1

Obtain some 36- or 40-gauge copper foil, and an ordinary 8.5 × 11 office page protector. The copper foil can be obtained at hobby and craft shops, especially stores that deal with doll house builders. The page protectors are made of cellophane or some other synthetic material, with a piece of black paper inside.

1. Remove the black paper from inside the page protector.
2. Cut two 8 × 10 inch pieces of copper foil. Glue or tape one foil section inside the page protector, and fasten the other copper foil section outside of the page protector.
3. Solder a 4-inch length of #22 solid hook-up wire to each copper foil section at one end of the assembly (they should be separated by about 1 inch).
4. Connect the wires to a digital capacitance meter and measure the capacitance. Write down the capacitance measured.
5. Repeat the experiment several times, using different dimensions for the copper foil sheets.

In several different attempts at this experiment, my results ranged from 415 to 490 pF.

In the experiment above, an RF variable frequency oscillator, which can be used for measuring either L or C, can be used. The C value is measured by comparing the oscillator frequency with the experiment capacitance connected into the circuit and without. The ratio of the frequency change is the square root of the difference in capacitance.

Breakdown voltage

The capacitor works by supporting an electrical field between two metal plates. This potential, however, can become too large. When the electrical potential, i.e., the voltage, gets too large, free electrons in the dielectric material (there are a few, but not many, in any insulator) are able to flow. If a stream of electrons gets started, the dielectric can break down and short the capacitor, allowing a current to pass between the plates. This is why the maximum breakdown voltage of the capacitor must not be exceeded. However, for practical purposes, a smaller voltage called the *dc working voltage* (WVdc) rating defines the *maximum safe voltage* that can be applied to the capacitor. Typical values are found in common electronic circuits from 8 WVdc to 1000 WVdc, although multi-kilovolt WVdc rated capacitors are available.

Circuit symbols for capacitors

The circuit symbols used to designate fixed-value capacitors are shown in Fig. 4-2A, and for variable capacitors in Fig. 4-2B. Both types of symbol are common. In certain types of capacitor, the curved plate shown on the left in Fig. 4-2A is usually the outer plate, i.e., the one closest to the outside package of the capacitor. This end of the capacitor is often indicated with a color band next to the lead attached to that plate.

The symbol for the variable capacitor is shown in Fig. 4-2B. This symbol is the fixed value symbol with an arrow through the plates. Small trimmer and padder capacitors are often denoted by the symbol of Fig. 4-2C. The variable set of plates is designated by the arrow.

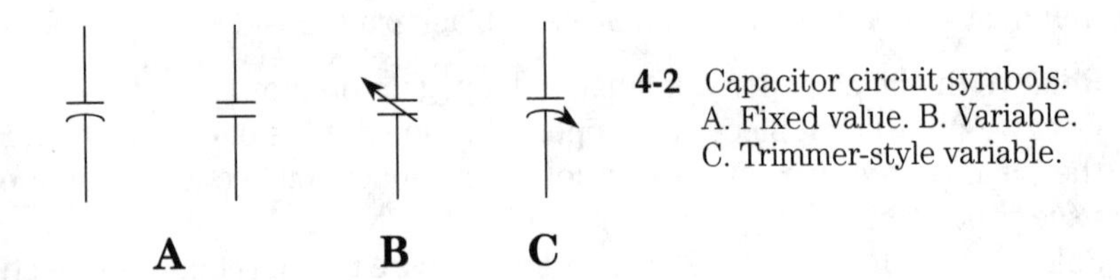

4-2 Capacitor circuit symbols. A. Fixed value. B. Variable. C. Trimmer-style variable.

Fixed capacitors

There are several types of fixed capacitors found in typical electronic circuits. They are classified by dielectric type: paper, mylar, ceramic, mica, polyester and others.

Paper dielectric capacitors

The construction of old-fashioned *paper capacitors* is shown in Fig. 4-3. It consists of two strips of metal foil, sandwiched on both sides of a strip of paraffin wax paper. The strip sandwich is then rolled up into a tight cylinder. This rolled up cylinder is then packaged in either a hard plastic, bakelite, or paper-and-wax case. When the case is cracked, or the wax end plugs are loose, replace the capacitor. Even though it might test good, it won't last for long. Paper capacitors come in values from about 300 pF to about 4 μF. The breakdown voltages range from 100 WVdc to 600 WVdc.

Paper capacitors are used for a number of different applications in older circuits—such as bypassing, coupling, and dc blocking. Unfortunately, no component is perfect. The long rolls of foil used in the paper capacitor exhibit a significant amount of stray inductance. As a result, paper capacitors are not used for high frequencies. Although they are found in some early shortwave receiver circuits, they are not used at all in VHF circuits.

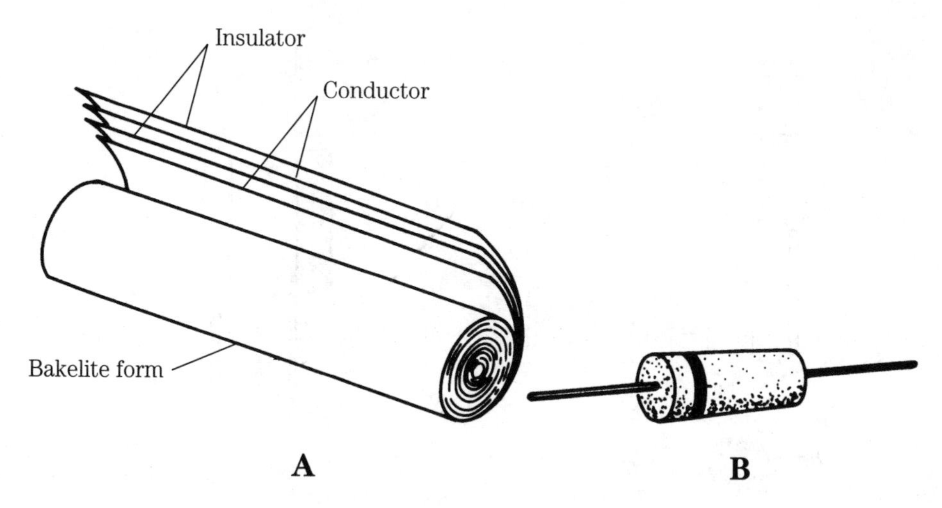

4-3 A. Paper capacitor construction. B. Finished capacitor.

Mylar dielectric capacitors

In modern applications, or when older equipment that used paper capacitors, use a *mylar dielectric capacitor* in place of the paper capacitor. These capacitors use a sheet of stable mylar synthetic material for the dielectric, and are of dipped construction (Fig. 4-4). Select a unit with exactly the same capacitance rating, and a WVdc rating that is equal to or greater than the original WVdc rating.

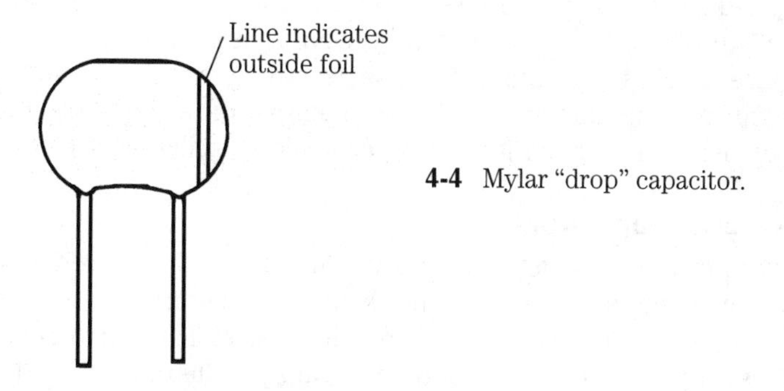

4-4 Mylar "drop" capacitor.

Ceramic dielectric capacitors

Several forms of *ceramic capacitors* are shown in Fig. 4-5. These capacitors come in values from a few picofarads up to 0.5 μF. The working voltages range from 400 WVdc to more than 30,000 WVdc. The common garden variety disk ceramic capacitors are usually rated at either 600 WVdc or 1000 WVdc. The tubular ceramic capacitors are typically much smaller in value than disk or flat capacitors, and are used extensively in VHF and UHF circuits for blocking, decoupling, bypassing, coupling, and tuning.

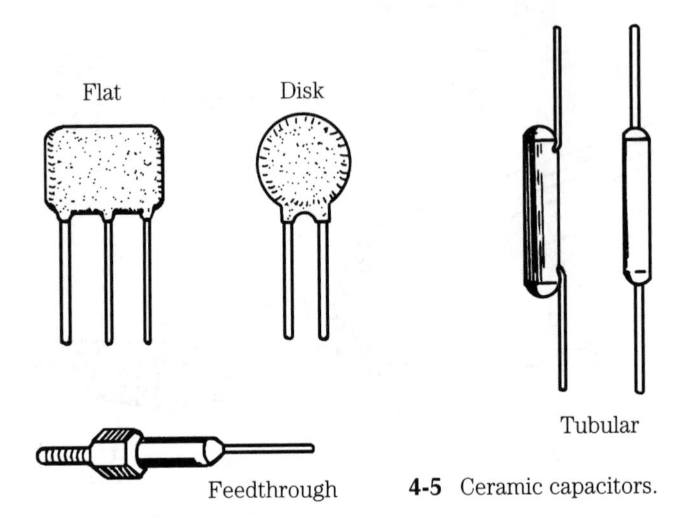

4-5 Ceramic capacitors.

The feedthrough type of ceramic capacitor is used to pass dc and low-frequency ac lines through a shielded panel. These capacitors are often used to filter or decouple lines that run between circuits that are separated by the shield for purposes of electromagnetic interference (EMI) reduction.

Ceramic capacitors are often rated as to *temperature coefficient*. This specification is the change of capacitance per change of temperature in degrees Celsius. A "P" prefix indicates a positive temperature coefficient, an "N" indicates a negative temperature coefficient, and the letters "NPO" indicate a zero temperature coefficient (NPO stands for "negative positive zero"). Do not ad-lib on these ratings when servicing a piece of electronic equipment. Use exactly the same temperature coefficient as the original manufacturer used. Non-zero temperature coefficients are often used in oscillator circuits to temperature-compensate the oscillator's frequency drift.

Mica dielectric capacitors

Several types of mica capacitor are shown in Fig. 4-6. The *fixed mica capacitor* consists of either metal plates on either side of a sheet of mica, or a sheet of mica that is silvered with a deposit of metal on either side. The range of values for mica capacitors is normally 50 pF to 0.02 μF at voltages in the range of 400 WVdc to 1000 WVdc.

The mica capacitor shown in Fig. 4-6C is called a *silver-mica* capacitor. These capacitors are low-temperature coefficient, although for most applications an NPO disk ceramic will serve better than all but the best silver-mica units. Mica capacitors are typically used for tuning and other higher frequency applications.

Other capacitors

Today, equipment designers can select from a number of different dielectric capacitors that were not commonly available (or available at all) a few years ago. The polycarbonate, polyester and polyethylene capacitors are used in a wide variety of

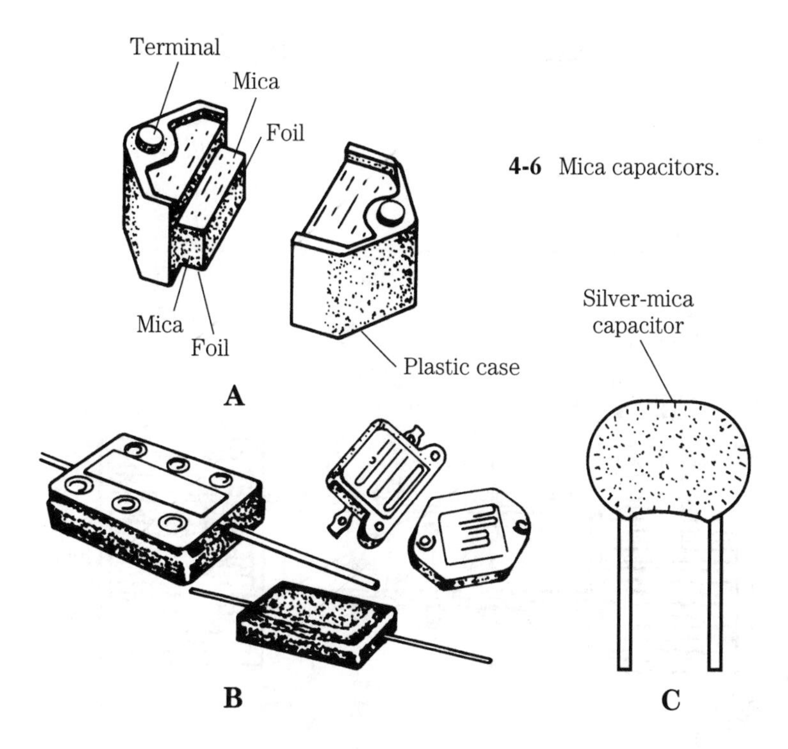

4-6 Mica capacitors.

applications where the above discussed capacitors once ruled supreme. In digital circuits, we find tiny 100 WVdc capacitors that carry ratings of 0.01 µF to 0.1 µF. They are used for decoupling the noise on the +5 Vdc power supply line. In circuits such as timers and op-amp Miller integrators, where the leakage resistance across the capacitor is very important, you might want to use a polyethylene capacitor. Check current catalogues for various old and new style capacitors—the applications paragraph in the catalog will tell you in which applications that they will serve, and the type of antique capacitor they will replace.

Capacitors in ac circuits

When an electrical potential is applied across a capacitor, current flows as charge is stored in the capacitor. As the charge in the capacitor increases, the voltage across the capacitor plate rises until it equals the applied potential. At this point the capacitor is fully charged, and no further current will flow.

Figure 4-7 shows an analogy for the capacitor in an ac circuit. The actual circuit is shown in Fig. 4-7A, and consists of an ac source connected in parallel across the capacitor (C). The mechanical analogy is shown in Fig. 4-7B. The capacitor (C) consists of a two-chamber cylinder in which the upper and lower chambers are separated by a flexible membrane or diaphragm. The "wires" are pipes to the "ac source" (which is a pump). As the pump moves up and down, pressure is applied to first one side of the diaphragm then the other, alternately forcing fluid to flow into and out of the two chambers of the "capacitor."

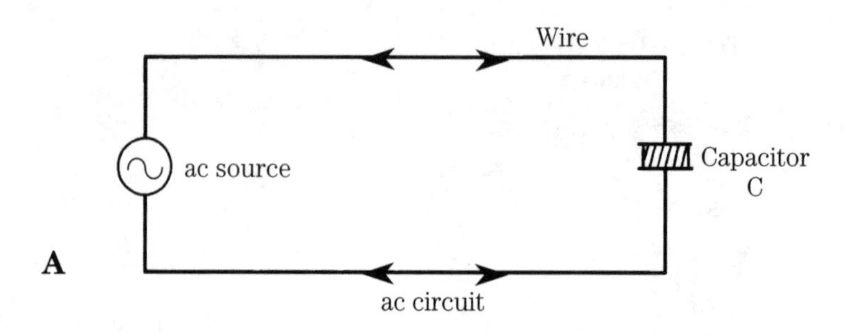

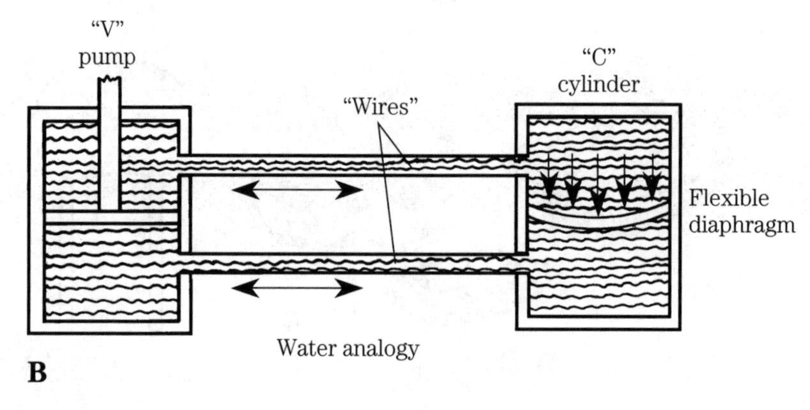

4-7 Capacitor circuit analogy. A. Circuit. B. Plumbing analogy.

The ac circuit mechanical analogy is not perfect, but it works for our purposes. Now let's apply these ideas to the electrical case. In Fig. 4-8, we see a capacitor connected across an ac (sine wave) source. In Fig. 4-8A, the ac source is positive, so negatively charged electrons are attracted from Plate A to the ac source, and electrons from the negative terminal of the source are repelled toward Plate B of the capacitor. On the alternate half-cycle (Fig. 4-8B), the polarity is reversed, so electrons from the new negative pole of the source are repelled toward Plate A of the capacitor, and electrons from Plate B are attracted toward the source. Thus, current will flow in and out of the capacitor on alternating half-cycles of the ac source.

Voltage and current in capacitor circuits

Consider the circuit in Fig. 4-9: an ac source (V) connected in parallel with the capacitor (*C*). It is the nature of a capacitor to oppose these changes in the applied voltage (the inverse of the action of an inductor). As a result, the voltage (*V*) lags behind the current (*I*) by 90 degrees. These relationships are shown in terms of sine waves in Fig. 4-9B, and in vector form in Fig. 4-9C.

Want to remember the difference between the action of inductors (L) and capacitors (C) on the voltage and current? In earlier texts, they used the letter E to denote voltage, so they could make a little mnemonic:

ELI the ICE man

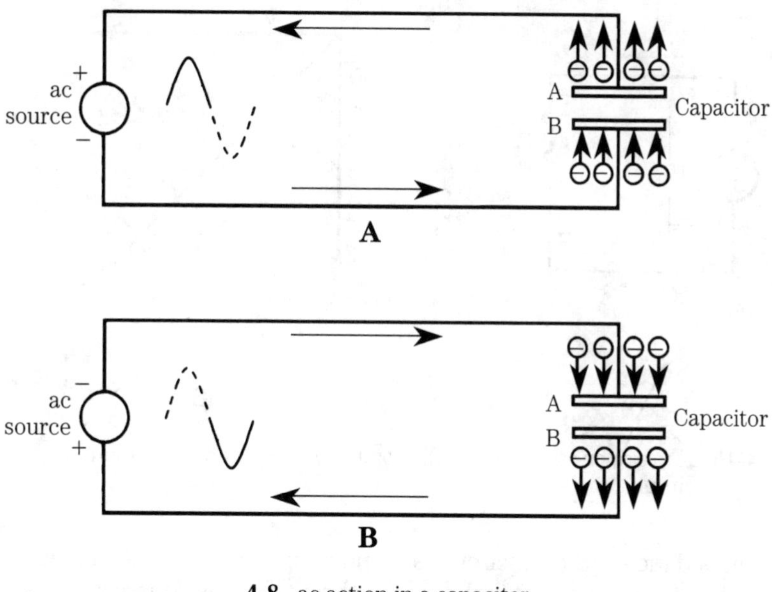

4-8 ac action in a capacitor.

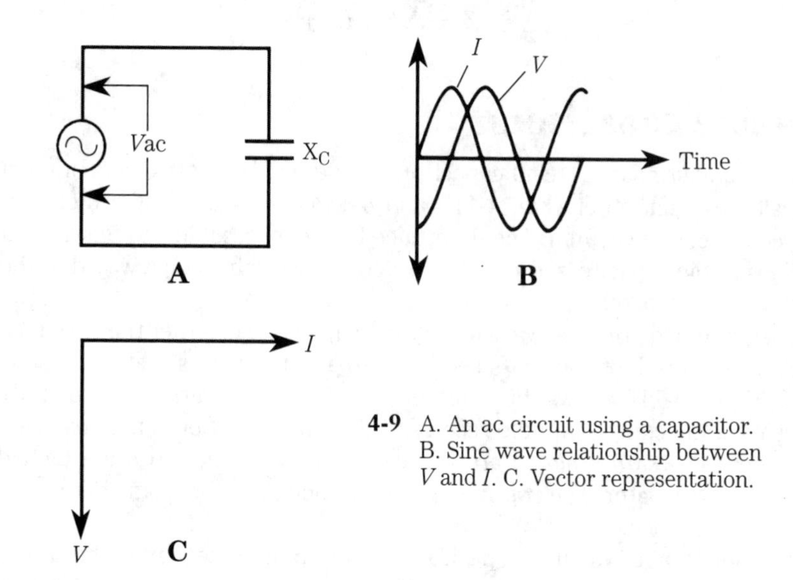

4-9 A. An ac circuit using a capacitor. B. Sine wave relationship between *V* and *I*. C. Vector representation.

"ELI the ICE man" suggests that in the inductive (L) circuit, the voltage (*E*) comes before the current (*I*)—ELI, and in a capacitive (C) circuit the current comes before the voltage—ICE.

The action of a circuit containing a resistance and capacitance is shown in Fig. 4-10A. As in the case of the inductive circuit, there is no phase shift across the resistor, so the *R* vector points in the "east" direction (Fig. 4-10B). The voltage across the capacitor, however, is phase shifted –90 degrees, so its vector points "south." The total resultant phase shift (θ) is found using the Pythagorean rule to calculate the angle between V_r and V_t.

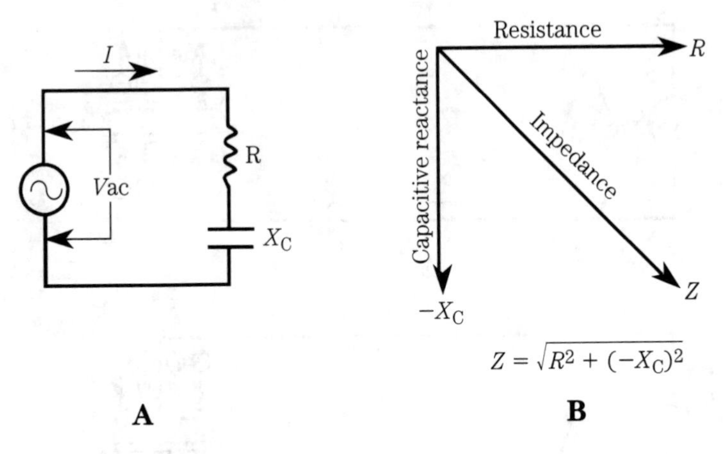

4-10 A. Resistor-capacitor (RC) circuit. B. Vector representation of circuit action.

The impedance of the RF circuit is found in exactly the same manner as the impedance of an RL circuit, that is the root of the sum of the squares:

$$Z = \sqrt{R^2 + (X_c)^2} \qquad \textbf{[4-3]}$$

Variable capacitors

Variable capacitors are, like all capacitors, made by placing two sets of metal plates in parallel to each other (Fig. 4-11B and 4-11A), separated by a dielectric of air, mica, ceramic, or a vacuum. The difference between variable and fixed capacitors is that, in variable capacitors, the plates are constructed in such a way that the capacitance can be changed.

There are two principal ways to vary the capacitance: either the spacing between the plates is varied, or the cross-sectional area of the plates that face each other is varied. Figure 4-11B shows the construction of a typical variable capacitor used for the main tuning control in radio receivers. The capacitor consists of two sets of parallel plates. The *stator plates* are fixed in their position, and are attached to the frame of the capacitor. The *rotor plates* are attached to the shaft that is used to adjust the capacitance.

Another form of variable capacitor found in radio receivers is the *compression capacitor* shown in Fig. 4-11C. It consists of metal plates separated by sheets of mica dielectric. In order to increase the capacitance, the manufacturer may increase the area of the plates and the mica, or the number of layers (alternating mica/metal) in the assembly. The entire capacitor will be mounted on a ceramic or other form of holder. If mounting screws or holes are provided, they are part of the holder assembly.

Still another form of variable capacitor is the *piston capacitor* shown in Fig. 4-11D. This type of capacitor consists of an inner cylinder of metal coaxial to, and inside of, an outer cylinder of metal. An air, vacuum or (as shown) ceramic dielectric separates the two cylinders. The capacitance is increased by inserting the inner cylinder further into the outer cylinder.

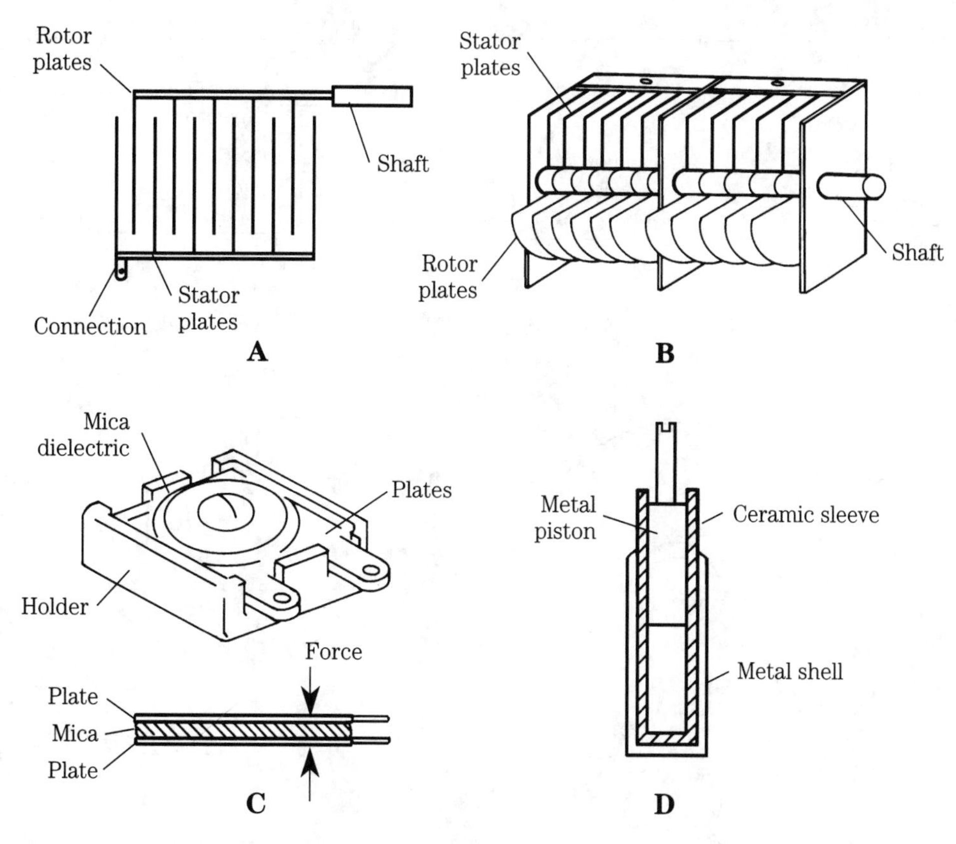

4-11 Variable capacitor. A. Air-variable in schematic form. B. Typical large air-variable. C. Mica compression trimmer. D. Piston trimmer.

The small compression or piston-style variable capacitors are sometimes combined with air-variable capacitors. Although not exactly correct usage, the smaller capacitor used in conjunction with the larger air-variable is called a *trimmer capacitor*. These capacitors are often mounted directly on the air-variable frame, or very close by in the circuit. In many radios the "trimmer" is actually part of the air-variable capacitor.

There are actually two uses for small variable capacitors in conjunction with the main tuning capacitor in radios. First, there is the true "trimmer," i.e., a small-valued variable capacitor in parallel with the main capacitor (Fig. 4-12A). These capacitors are used to trim the exact value of the main capacitor. The other form of small capacitor is the *padder* capacitor (Fig. 4-12B), which is connected in series with the main capacitor. The error in terminology referred to above is calling both series and parallel capacitors "trimmers," when only the parallel connected capacitor is properly so-called.

Air-variable main tuning capacitors

The capacitance of an air-variable capacitor at any given setting is a function of how much of the rotor plate set is shaded by the stator plates. In Fig. 4-13A, the rotor

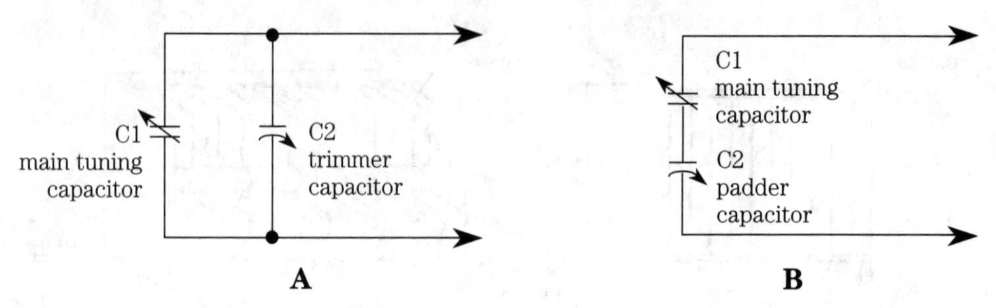

4-12 A. Parallel connection indicates a trimmer (C2). B. Series connection indicates a padder.

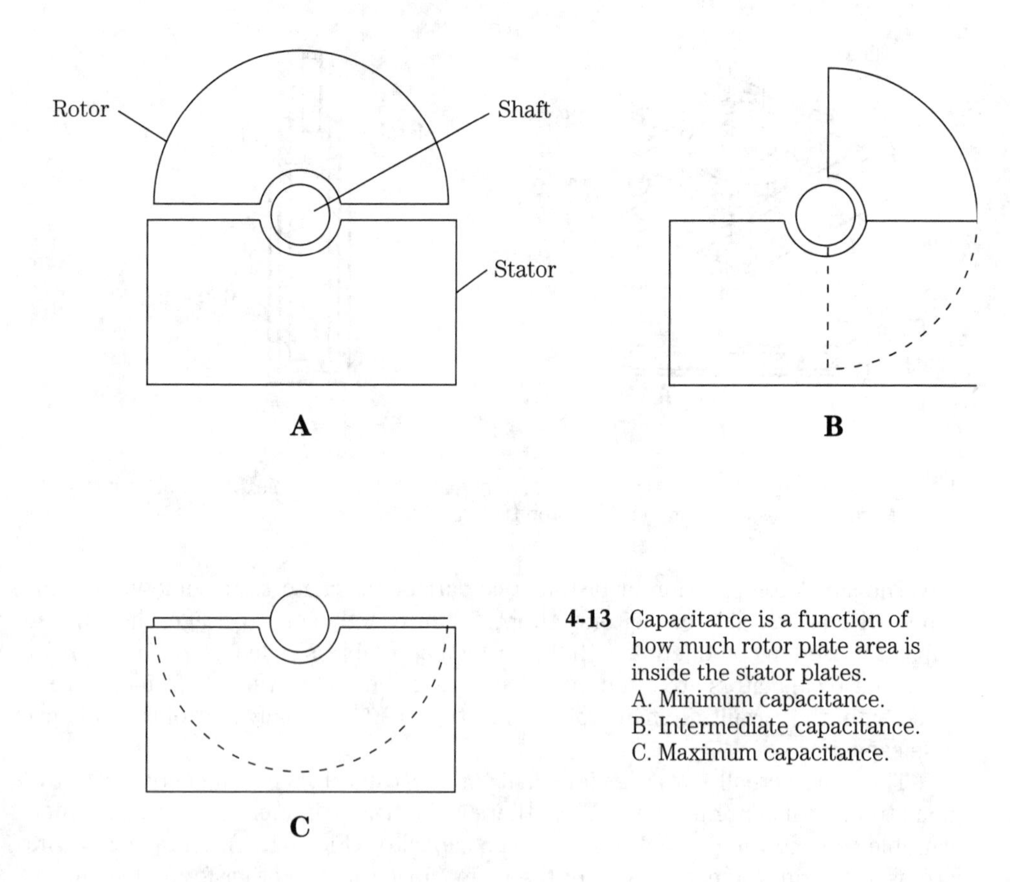

4-13 Capacitance is a function of how much rotor plate area is inside the stator plates.
A. Minimum capacitance.
B. Intermediate capacitance.
C. Maximum capacitance.

plates are completely outside of the stator plate area. Because the shading is zero, the capacitance is minimum. In Fig. 4-13B, however, the rotor plate set has been slightly meshed with the stator plate, so some of its area is shaded by the stator. The capacitance in this position is at an intermediate value. Finally, in Fig. 4-13C the rotor is completely meshed with the stator so the cross-sectional area of the rotor that is shaded by the stator is maximum. Therefore, the capacitance is also maximum. Remember these two rules:

1. Minimum capacitance is found when the rotor plates are completely unmeshed with the stator plates; and
2. Maximum capacitance is found when the rotor plates are completely meshed with the stator plates.

Figure 4-14 shows a typical single-section variable capacitor. The stator plates are attached to the frame of the capacitor, which is grounded in most radio circuits. Front and rear plates have bearing surfaces to ease the rotor's action. These capacitors were often used in early multi-tuning knob TRF radio receivers (the kind where each RF tuned circuit had its own selector knob). But that design was not very good, so the *ganged-variable capacitor* (Fig. 4-15) was invented. These capacitors are basically two or (in the case of Fig. 4-15) three variable capacitors mechanically ganged on the same rotor shaft.

4-14 Typical air-variable capacitor.

In Fig. 4-15 all three sections of the variable capacitor have the same capacitance, so they are identical to each other. If this capacitor is used in a superheterodyne radio, the section used for the local oscillator (LO) tuning must be padded with a series capacitance in order to reduce the overall capacitance. This trick is done to permit the higher frequency LO to track with the RF amplifiers on the dial.

In many superheterodyne radios you will find variable tuning capacitors in which one section (usually the front section) has fewer plates than the RF amplifier section (an example shown in Fig. 4-16). These capacitors are sometimes called *cut-plate capacitors* because the LO section plates are cut to permit tracking of the LO with the RF.

4-15 Three-section ganged air variable capacitor.

4-16 Cut-plate variable has two sections of different capacitance.

Straight-line capacitance vs. straight-line frequency capacitors

The variable capacitor shown in Fig. 4-13 has the rotor shaft in the geometric center of the rotor plate half-circle. The capacitance of this type of variable capacitor varies directly with the rotor shaft angle. As a result, this type of capacitor is called a *straight-line capacitance* model. Unfortunately, as you will see in chapter 7, the frequency of a tuned circuit based on inductors and capacitors is not a linear (straight-line) function of capacitance. If a straight-line capacitance unit is used for the tuner, then the frequency units on the dial will be cramped at one end and spread out at the other (you've probably seen such radios). But some capacitors have an offset rotor shaft (Fig. 4-17A) that compensates for the nonlinearity of the tuning circuit. The shape of the plates, and the location of the rotor shaft, are designed to produce a linear relationship between the shaft angle and the resonant frequency of the tuned circuit in which the capacitor is used. A comparison between straight-line capacitance and straight-line frequency capacitors is shown in Fig. 4-17B.

4-17 A. Constant frequency capacitance (note rotor plate cut).

Special variable capacitors

In the sections above, the standard forms of variable capacitor were discussed. These capacitors are largely used for tuning radio receivers, oscillators, signal generators and other variable frequency LC oscillators. In this section, we will take a look at some special forms of variable capacitor.

4-17 B. Comparison of a constant frequency (left) to a constant capacitance (right) capacitor (note difference in plate cuts).

Split-stator capacitors The split-stator capacitor is one in which two variable capacitors are mounted on the same shaft. The symbol for the split-stator capacitor is shown in Fig. 4-18. This particular example also illustrates the main use of these capacitors: antenna tuners—especially balanced types of tuner circuit. The split-stator capacitor normally uses a pair of identical capacitors, each the same value, turned by the same shaft. The rotor is common to both capacitors. Thus, the capacitor will tune two tuned circuits at the same time.

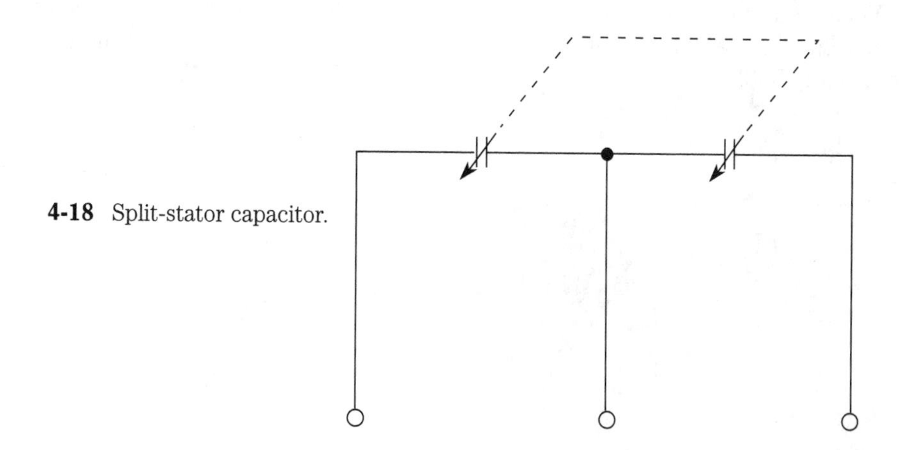

4-18 Split-stator capacitor.

Differential capacitors Although some differential capacitors are often mistaken for split-stator capacitors, they are actually quite different. The split-stator capacitor is tuned in tandem, i.e., both capacitor sections have the same value at any given shaft setting. The differential capacitor, on the other hand, is arranged so that one capacitor section increases in capacitance, while the other section decreases in exactly the same proportion.

Figure 4-19A shows both the mechanical construction and circuit symbol for a differential capacitor, while Fig. 4-19B shows a typical example. Note that the rotor plate is set to equally shade both stator-A and stator-B. If the shaft is moved clockwise, it will shade more of stator-B, and less of stator-A, so C_a will decrease and C_b will increase by exactly the same amount. Note: the total capacitance (C_t) is constant no matter what position the rotor shaft takes, only the proportion between C_a and C_b changes.

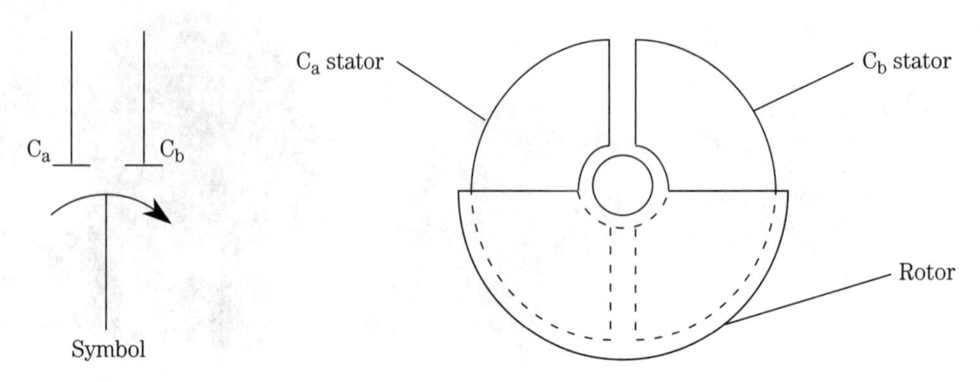

4-19 Differential capacitor. A. Circuit symbol and structure.

Differential capacitors are used in impedance bridges, RF resistance bridges, and other such instruments. If you buy or build a high-quality RF impedance bridge for antenna measurements, for example, it is likely that it will have a differential capacitor as the main adjustment control. The two capacitors are used in two arms of a Wheatstone bridge circuit. It is expensive to build such a bridge, however. I recently bought the differential capacitor in Fig. 4-19B for such an instrument, and it cost nearly $50!

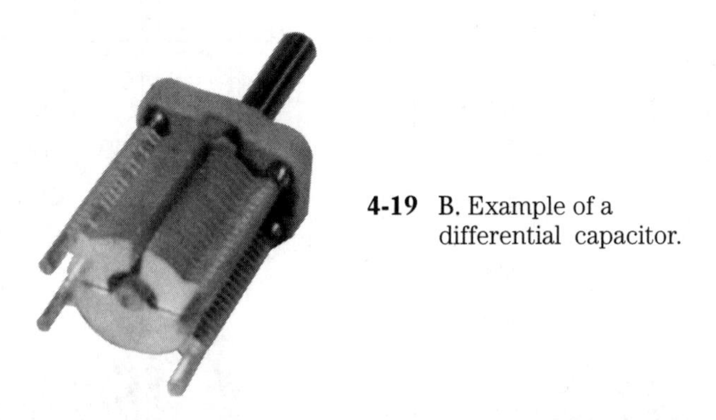

4-19 B. Example of a differential capacitor.

"Transmitting" variable capacitors The one requirement of transmitting variable capacitors (and certain antenna tuner capacitors) is the ability to withstand high voltages. The high-power ham radio or AM broadcast transmitter will have a dc potential of 1500 volts to 7500 volts on the RF amplifier anode, depending upon the type of tube used. If amplitude modulated, the potential can double. Also, if certain antenna defects arise, then the RF voltages in the circuit can rise quite high. As a result, the variable capacitor used in the final amplifier plate circuit must be able to withstand these potentials.

Two forms of transmitting-variable capacitors are typically found in RF power amplifiers and antenna tuners. Figure 4-20 shows a transmitting *air-variable* ca-

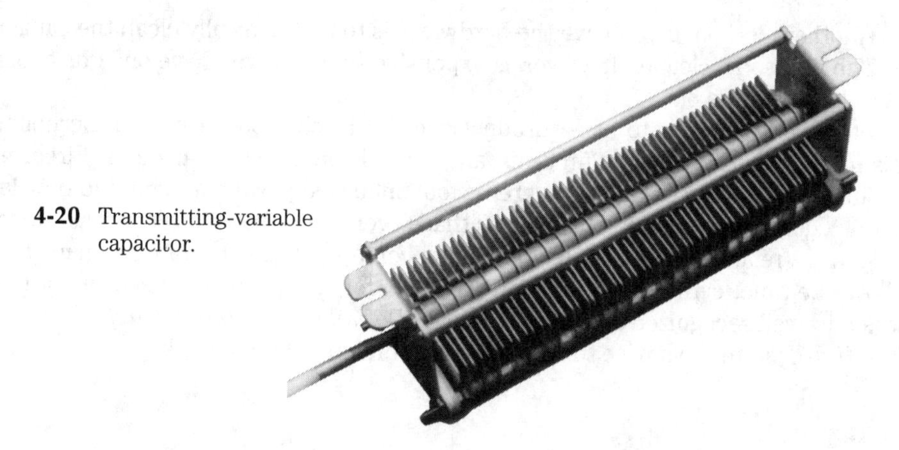

4-20 Transmitting-variable capacitor.

pacitor. The shaft of this particular capacitor is nylon, so it can be mounted either with the frame grounded or with the frame floating at high voltage. The other form of transmitting variable is the *vacuum-variable*. This type of capacitor is a variation of the piston capacitor, but it has a vacuum dielectric (K-factor = 1.0000). The model shown in Fig. 4-21 is an 18-pF to 1000-pF model that is driven from a 12-Vdc electric motor. Other vacuum-variables are manually driven.

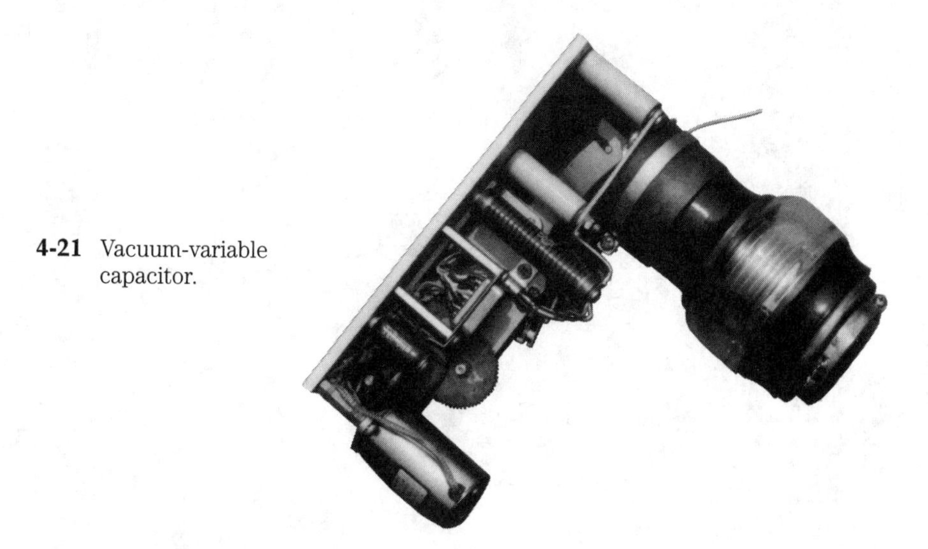

4-21 Vacuum-variable capacitor.

Variable capacitor cleaning note

Antique radio buffs often find that the main tuning capacitors in their radios are full of crud, grease, and dust. Similarly, ham radio operators working the hamfest circuit looking for linear amplifier and antenna tuner parts often find just what they need, but the thing is all gooped up with scum, crud, grease and other stuff. There are several things that can be done about it. First, try using dry compressed air. It will remove dust, but not grease. Aerosol cans of compressed air can be bought from a lot of sources, including automobile parts stores and photography stores.

Another method, if you have the hardware, is to ultrasonically clean the capacitor. The ultrasonic cleaner, however, is expensive so unless you have one don't rush out to lay down the bucks.

Still another way is to use a product such as Birchwood Casey Gun Scrubber. This product is used to clean firearms, and is available in most gun shops. Firearms become all gooped up because gun grease, oil, unburned powder and burned powder residue combine to create a crusty mess that's every bit as hard to remove as capacitor gunk. A related product is the degunking compound used by auto mechanics.

At one time, carbon tetrachloride was used for this purpose. However, carbon tet is now well recognized as a health hazard. *Do not use carbon tetrachloride for cleaning*, despite the advice to the contrary found in old radio books.

5 Using and stabilizing varactor diodes

Have you tried to buy an air-variable capacitor for a receiver project recently? They are very rare these days. I've seen them advertised in some amateur radio parts catalogs, in British electronics catalogs, and in antique radio supplies catalogs in the USA, but otherwise it's catch as catch can. So what to do? Well, it seems that commercial radio manufacturers today use *voltage-variable capacitance diodes*, commonly called *varactors*, for the radio tuning function. These special semiconductor diodes exhibit a capacitance across the pn junction that is a function of the reverse bias potential (see Fig. 5-1).

The diode representations of Figs. 5-1A and 5-1B are in the form of pn junction diode block diagrams. In the *n-type region* negative charge carriers (electrons) predominate, while in the *p-type region* positive charge carriers (holes) predominate. When a reverse bias potential is applied, as in Fig. 5-1A, the charge carriers are pulled away from the junction region to form a *depletion zone* that is depleted of charge carriers (hence acts like an insulator or "dielectric"). The situation is the same as in a charged capacitor: an insulator separating two electrically conductive regions. Thus, a capacitance is formed across the junction that is a function of the width of the depletion zone. And because the size of the depletion zone is a function of applied voltage (compared Figs. 5-1A and 1B), the capacitance of the junction is also a function of applied voltage. A varactor is a diode in which this function is enhanced and stabilized.

Figures 5-2A and 5-2B show two common circuit symbols for a varactor diode. In both cases, the normal diode "arrow" symbol is combined with a pair of parallel lines representing a capacitor. In some cases, I've seen a variant on Fig. 5-2A in which an arrow is drawn through the parallel plates by extending one side of the arrow symbol. I suppose that's used to indicate the property of "variableness."

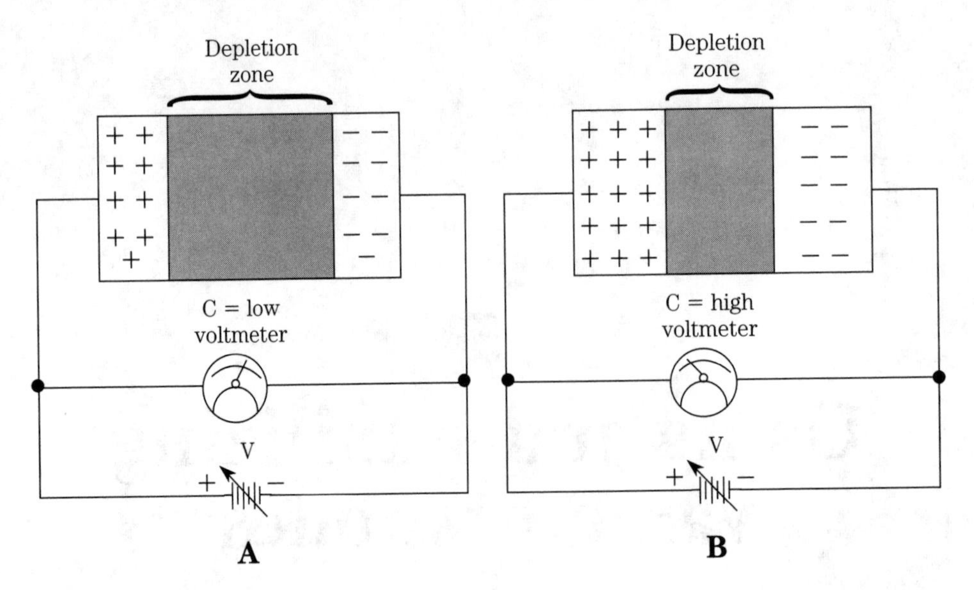

5-1 Varactor diode under two different reverse-bias conditions.

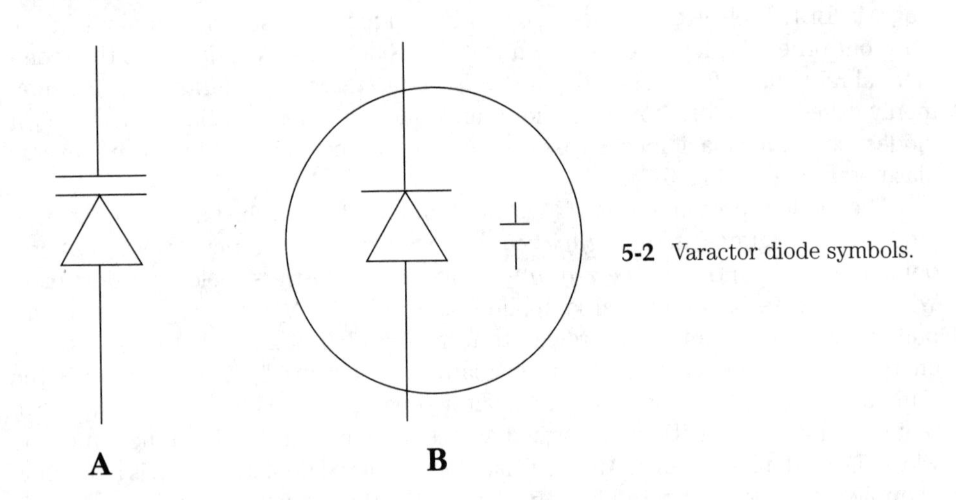

5-2 Varactor diode symbols.

Several types of varactors are listed in Table 5-1. Some of them can be easily obtained through the ECG and NTE replacement transistor lines, which are sold by parts houses that normally deal with radio/TV repair shops. Look up the specs for NTE-611 TO NTE-618, or ECG-611 to ECG-618 to see if they are appropriate for your application. Alternatively, look up the replacements for those diodes in the table from the ECG or NTE crossover directories.

Table 5-1

Type No.	Capacitance range (pF)	Tuning ratio	Frequency ratio
1N5139	6.8 - 47	2.7 - 3.4	1.6 - 1.8
MV2101	6.8 - 100	1.6 - 3.3	1.6 - 1.8
MMBV105G	120 - 550	10 - 14	3.2 - 3.7
MV209	30	5 - 6.5	2.2 - 2.5
BB212	10 -550	—	7.4

Varactor tuning circuits

The varactor diode wants to see a voltage that is proportional to the desired capacitance. Several different circuits are used to provide this function, some of which are shown in Figs. 5-3 and 5-4. In all cases, the tuning voltage must be supplied from a very stable reference voltage source. It is normally considered good engineering practice to provide $+V_{ref}$ from a separate voltage regulator that serves only the varactor, even when the maximum value of the voltage is the same as the rest of the circuit (e.g., +12 volts). Therefore, always use a voltage regulator to provide the tuning voltage source potential. Most varactors use a maximum voltage of around +30 to +40 volts, while many intended for car radio applications are rated only to +12 or +18 volts (check!).

The simplest, and probably the most popular varactor tuning circuit is shown in Fig. 5-3A. In this circuit, a potentiometer (R_1) is connected across the V_{ref} supply, so the tuning voltage (V_t) is a function of the potentiometer wiper position. In many cases, a 0.001-µF to 0.01-µF capacitor is connected from the wiper of the potentiometer to ground in order to snuff any noise pulses so they don't alter the tuning (they are, as far as the diode is concerned, valid tuning voltage signals!). A series

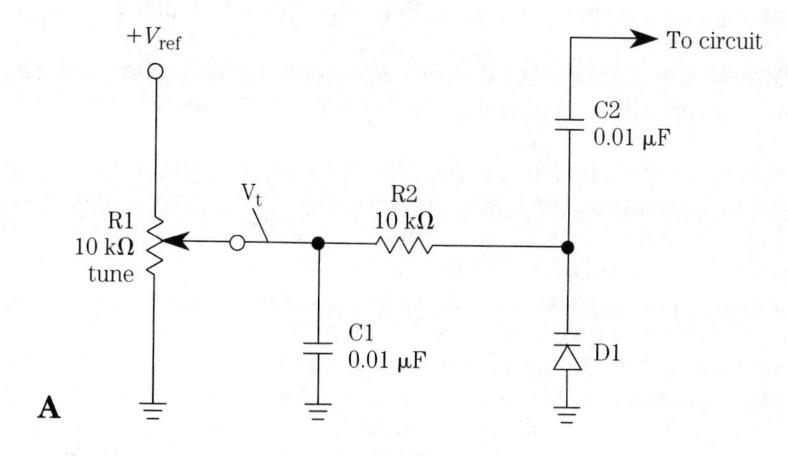

5-3 Varactor diode tuning voltage circuits.

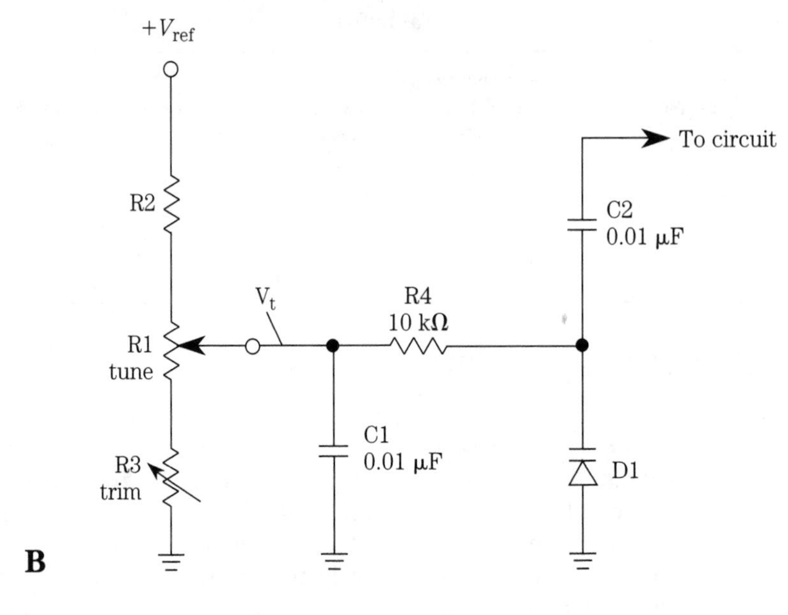

5-3 Continued.

current-limiting resistor (R1), usually of a value between 4.7 kilohms and 100 kilohms, is used to protect the diode in case the voltage gets to the breakover point, as well as to isolate its capacitance from the tuning circuit (otherwise, C1 would always predominate). In many cases, a dc-blocking capacitor (C2) is needed to prevent the tuning voltage from affecting following circuits, or other circuit voltages from affecting the varactor diode tuning voltage. From the point in Fig. 5-3A marked "TO CIRCUIT" the varactor network acts like a variable capacitor.

A variant circuit is shown in Fig. 5-3B. In this circuit, the tuning voltage is only a small portion of the reference voltage. Thus, the tuning voltage is produced by a voltage divider made up of three resistors: R_1, R_2, and R_3. In some cases, one or more of the other resistors will be a trimmer potentiometer used to set the "fine" or "vernier" frequency of the overall circuit.

Regardless of which tuning circuit is used, the resistors, including the potentiometer, should be low-temperature coefficient types in order to reduce thermal drift. Ordinary carbon composition resistors are probably not suitable for most applications.

If you wish to sweep a band of frequencies, i.e., in a sweep generator or swept receiver (e.g., panadaptor or spectrum analyzer), replace the $+V_{ref}$ potential with a sawtooth waveform. The sawtooth waveform is a linear ramp that rises to a specified maximum voltage, then drops back to zero abruptly. Unfortunately, it is rarely the case that the sawtooth voltage range, the desired swept frequency range, and the varactor voltage characteristic, are in sync with each other. For those situations, we need to be able to provide a sawtooth of variable amplitude to set the *sweep width* and a dc offset tuning voltage to provide the *center frequency* function. Figure 5-4 shows how this might be done.

The circuit of Fig. 5-4 uses three operational amplifiers to provide the combination tuning voltage. Op-amp A_1 provides a variable amplitude sweep width control to change the sawtooth amplitude. If feedback resistor R_5 is made 10 kilohms, then the output sawtooth will have the same amplitude as the input sawtooth. If higher or lower amplitude is needed, adjust the gain of A_1 by selecting a different R5 value: Gain = $R_5/R_6 = -R_5/10$ kilohms (the "–" indicates that the circuit is an inverter). For tuning voltages to 18 volts, ordinary 741s can be used for A_1 through A_3.

Digital frequency control can be accomplished by supplying the reference voltage ($+V_{ref}$) from a digital-to-analog converter (DAC) that has a voltage output. The binary number applied to the DAC binary inputs will set the tuning voltage, which in turn sets the capacitance of the diode. Those who wish to experiment with low-cost components will find that the eight-bit National DAC0800 series devices (available in most local parts stores in the Jameco Jim-Pak display) will provide 256 steps of voltage (hence also of capacitance and frequency). An op-amp is recommended to convert the current output of the DAC080x to a voltage (the *National Linear Data Book* gives example circuits, as well as specs for the different devices in the series).

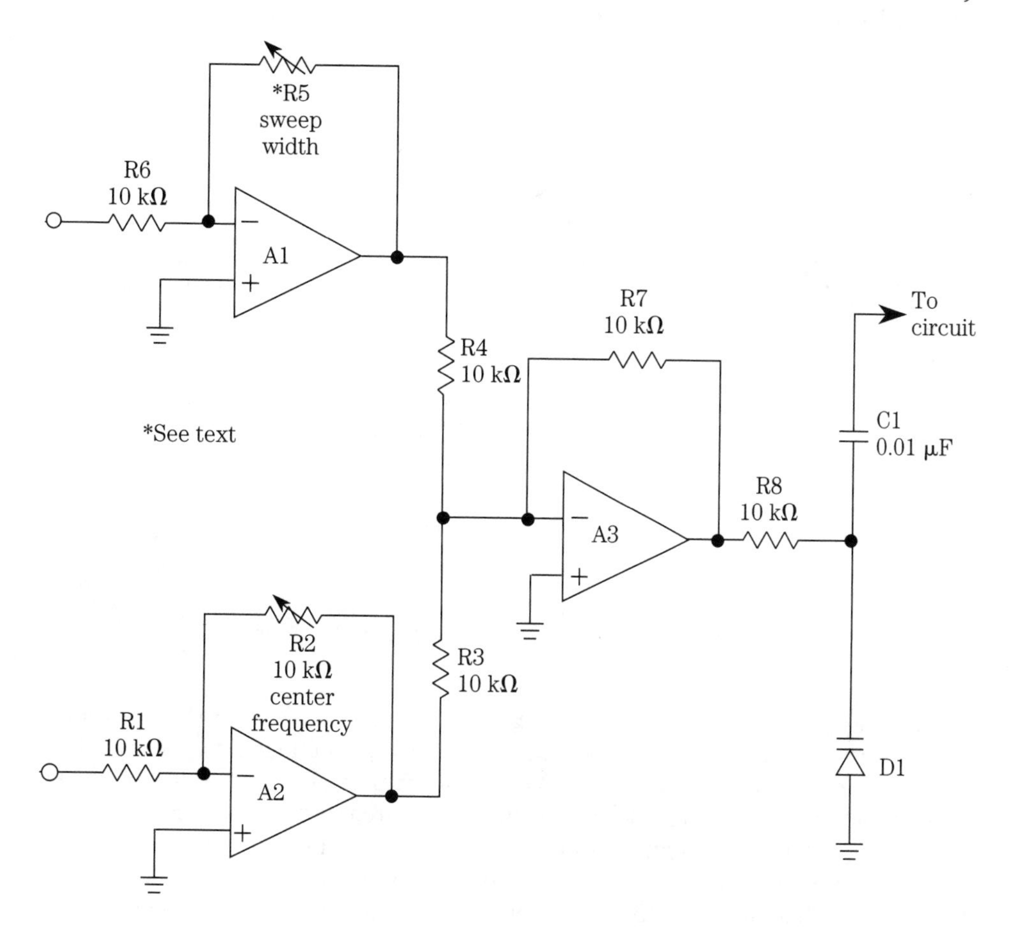

5-4 Sawtooth/fixed voltage combiner circuit.

Temperature compensation

There is one nasty little problem with the varactor tuning circuit—the thermal drift can be horrible! According to one source, the temperature coefficient of capacitance (ppm/°C) varied from about 30 ppm/°C at $+V_{ref}$ = 30 volts to 587 ppm/°C at $+V_{ref}$ = 1 volt. Ouch! There are three approaches to this problem:

- Ignore it
- Use the circuit in Fig. 5-5
- Use the circuit in Fig. 5-6

The circuit of Fig. 5-5 uses a fixed, regulated voltage for $+V_{ref}$, but passes it through an ordinary silicon diode (D_2) that is in close thermal proximity to the varactor diode (so they see the same temperature environment). When resistor R_1 is set to draw a current through D_2 sufficient to get the voltage drop into the 0.6-volt region, then the output voltage $+V_{ref}$ will track the thermal changes to counteract the change of capacitance. In practice, R_1 can be the tuning potentiometer when diodes such as 1N4148 or 1N914 are used.

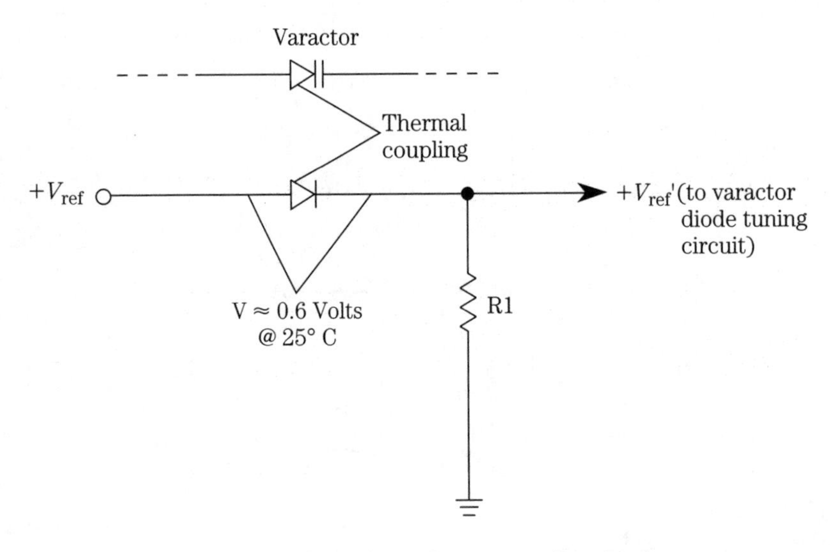

5-5 Simple diode thermal compensation circuit.

Figure 5-6 shows a circuit using a special zener diode voltage regulator that is sold in Europe under both MVS-460-2 and ZTK33B-type numbers. It appears to be a +33-volt zener diode that has a –2.3 mV/°C temperature coefficient. It will provide a nominal +33-volt output for all input voltages (V) greater than 34 Vdc. Again, the temperature stabilizer (which looks like a diode) is placed in close thermal proximity to the varactor diode being protected. The MVS-460-2 part is in a TO-92-like plastic package, while the ZTK33B is in the normal glass diode package (similar to 1N60 devices).

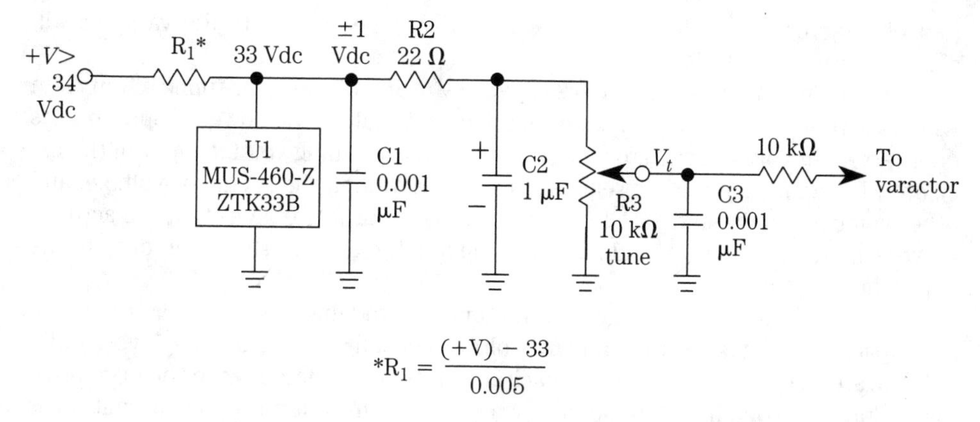

$$*R_1 = \frac{(+V) - 33}{0.005}$$

5-6 Using a special varactor thermal regulator IC device.

Unfortunately, the MVS-460-2 and ZTK33B are hard to find in the USA. I bought some from Maplins Professional Supplies in England (P.O. Box 777, Rayleigh, Essex, SS6 8LU, England) for £0.382[1] each in lots of 25 or more. Unfortunately, with a minimum practical order of several pounds sterling, plus a shipping charge of £8 for USA and Canada, it is best to order 25 or more. This translates to $27.38 or so—if the price still holds as of publication date. Ordering from UK is reasonably easy. You can get an international money order denominated in pounds sterling at many banks, but the fee might make you puke (my bank gets $15, which is why I opened a UK checking account). Alternatively, they will accept Visa, Mastercard or American Express cards. The bank card company will make the currency conversion for you, and they use the rate in effect on the day they make the conversion. I've used all three types of cards to make purchases from UK electronics and old book dealers (my other passion), and experienced no problems. Just give them the card number, the expiration date, and your signature to authorize the charge.

Well, that's that for varactors. If you want to know more theoretical smoke about the subject, I recommend Motorola Semiconductor's application note AN847 *Tuning Diode Design Techniques* (Motorola Technical Literature Distribution Center, POB 20912, Phoenix, AZ 85036).

Varactor tuning voltage sources

The capacitance of a varactor diode is a function of the applied reverse bias potential. Because of this, it is essential that a stable, noise-free source of bias is provided. If the diode is used to tune an oscillator, for example, frequency drift will result if the dc potential is not stable. Besides causing ordinary dc drift, noise affects the opera-

[1]As of this writing £1 = $1.57, but the rate changes daily so check before sending money order denominated in £ sterling.

tion of varactors. Anything that varies the dc voltage applied to the varactor will cause a capacitance shift.

Electronic servicers should be especially wary of varactor-tuned circuits in which the tuning voltage is derived from the main regulated dc power supply and has no intervening voltage regulator that serves *only* the tuning voltage input of the oscillator. Dynamic shifts in the regulator load, variation in the regulator voltage, and other complications can create local oscillator (LO) drift problems that are actually power supply problems . . . and have nothing at all to do with the tuner, despite the apparent symptoms.

The specifications for any given varactor are listed in two ways. First, the nominal capacitance is taken at a standard voltage (usually 4 volts dc, but 1 volt and 2 volts dc are also used). Second, a capacitance ratio is expected when the dc reverse bias voltage is varied from 2 to 30 Vdc (whatever the maximum voltage permitted as the applied potential for a particular diode). The NTE and ECG replacement semiconductor line (Type 614) is typical. According to the *NTE Service Replacement Guide and Cross Reference*, the 614 device has a nominal capacitance of 33 pF at 4 volts reverse-bias potential, and a C2-C30 ratio of 3:1.

Varactor applications

Varactors are electronically variable capacitors. In other words, they exhibit a variable capacitance that is a function of the applied reverse bias potential. This phenomenon leads to several common applications in which capacitance is a consideration. Figure 5-7 shows a typical varactor-tuned LC tank circuit. The link-

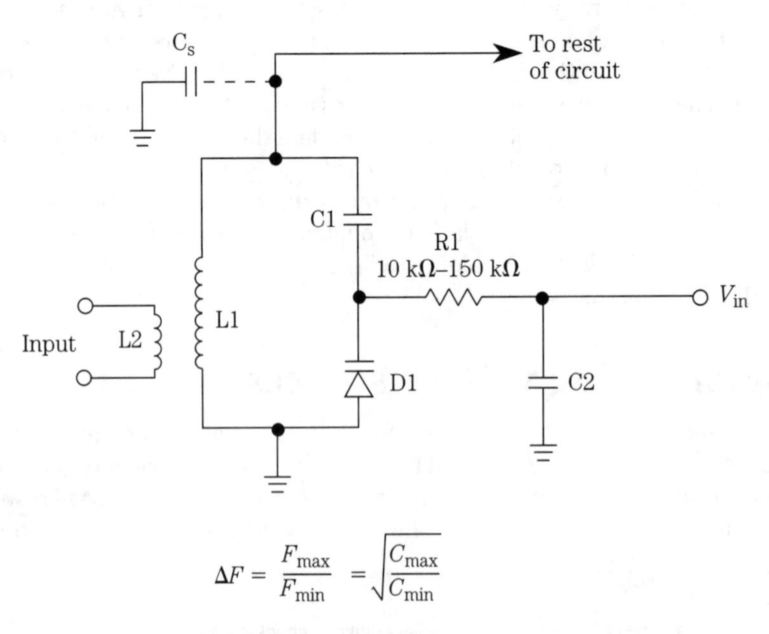

$$\Delta F = \frac{F_{max}}{F_{min}} = \sqrt{\frac{C_{max}}{C_{min}}}$$

5-7 Typical varactor-tuned LC tank circuit.

coupled inductor (L_2) is used to input RF signal to the tank when the circuit is used for RF amplifiers, and in many oscillator circuits, it serves as the output to take signal energy to other circuits. The principal LC tank circuit consists of a main inductor (L_1) and a capacitance made up from the series equivalent of C_1 and varactor D_1, or:

$$C_t = \frac{C_1 C_{D1}}{C_1 + C_{D1}} \qquad \textbf{[5-1]}$$

In addition, you must also take into account the stray capacitance (C_S) that exists in all electronic circuits. (The blocking capacitor and series resistor functions were discussed previously.) Capacitor C_2 is used to filter the tuning voltage, V_{in}.

Because the resonant frequency of an LC-tuned tank circuit is a function of the square root of the inductance/capacitance product, the maximum/minimum frequency of the varactor-tuned tank circuit varies as the square root of the capacitance ratio of the varactor diode. This value is the ratio of the capacitance at minimum reverse bias over the capacitance at maximum allowable reverse bias. A consequence of this fact is that the tuning characteristic curve (voltage versus frequency) is basically a parabolic function.

6
Inductors and inductance

Inductors form a very large part of RF electronic circuitry with applications ranging from radio tuning, to filters, to RFI/EMI suppression, to impedance matching. In this chapter we will take a look at the subjects of inductance, inductors (coils), and how to build the inductors that you need in practical RF circuits. I will also discuss the related topic of RF transformers.

Because inductor circuits involve phase relationships between voltages and currents when ac is applied, it is necessary to understand the use of vectors. If you do not understand basic vectors, I recommend that you see Appendix A in the back of this book.

Inductor circuit symbols

Figure 6-1 shows various circuit symbols used in schematic diagrams to represent inductors. Figures 6-1A and 6-1B represent alternate but equivalent forms of the same thing; i.e., a fixed value, air-core inductor ("coil" in the vernacular). The other forms of inductor symbol shown in Fig. 6-1 are based on Fig. 6-1A, but are just as valid if the "open-loop" form of Fig. 6-1B is used instead.

The form shown in Fig. 6-1C is a *tapped fixed-value air-core inductor*. By providing a tap on the coil, different values of fixed inductance are achieved. The inductance from one end of the coil to the tap is a fraction of the inductance available across the entire coil. By providing one or more taps, several different fixed values of inductance can be selected. Radio receivers and transmitters sometimes use the tap method, along with a bandswitch, to select different tuning ranges or "bands."

Variable inductors are shown in Figs. 6-1D and 6-1E. Both forms are used in schematic diagrams, although in some countries, Fig. 6-1D implies a form of construction whereby a wiper or sliding electrical contact rides on the uninsulated turns of the coil. Figure 6-1E implies a construction where variable inductance is achieved by moving a magnetic core inside of the coil.

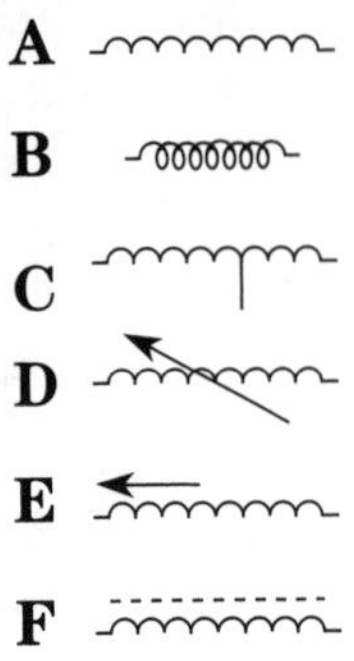

A

B

C

D

E

F

6-1 Forms of inductor symbol.
A. Fixed (open loop style).
B. Fixed (closed loop style).
C.Tapped. D. Variable (style 1).
E. Variable (style 2).
F. Powdered iron or ferrite
core inductor.

Figure 6-1F indicates a fixed value (or tapped, if desired) inductor with a powdered iron, ferrite or nonferrous (e.g., brass) core. The core will increase (ferrite or powdered iron) or decrease (brass) the inductance value relative to the same number of turns on an air core coil.

Inductance and inductors

Inductance (L) is a property of electrical circuits that opposes changes in the flow of current. Note that word "changes," it is important. Inductance is somewhat analogous to the concept of inertia in mechanics. An inductor stores energy in a magnetic field (a fact which we will see is quite important). In order to understand the concept of inductance we must understand these physical facts:

- When an electrical conductor moves relative to a magnetic field, a current is generated (or induced) in the conductor. An *electromotive force* (EMF or voltage) appears across the ends of the conductor.
- When a conductor is in a magnetic field that is changing, a current is induced in the conductor. As in the first case, an EMF is generated across the conductor.
- When an electrical current moves in a conductor, a magnetic field is set up around the conductor.

According to *Lenz's law*, the EMF induced into a circuit is ". . . in a direction that opposes the effect that produced it." From this fact we can see the following effects:

- A current induced by either the relative motion of a conductor and a magnetic field, or changes in the magnetic field, always flows in the direction that sets up a magnetic field that opposes the original magnetic field.
- When a current flowing in a conductor changes, the magnetic field that it generates changes in a direction that induces a further current into the conductor that opposes the current change that caused the magnetic field to change.
- The EMF generated by a change in current will have a polarity that is opposite the polarity of the potential that created the original current.

The unit of inductance (L) is the *henry* (H). The accepted definition of the henry is the inductance that creates an EMF of 1 volt when the current in the inductor is changing at a rate of 1 ampere per second, or mathematically:

$$V = L\left(\frac{\Delta I}{\Delta t}\right) \qquad \textbf{[6-1]}$$

where

V is the created EMF in volts (V)
L is the inductance in henrys (H)
I is the current in amperes (A)
t is the time in seconds (s)
Δ indicates a "small change in . . .."

The *henry* (H) is the appropriate unit for large inductors such as the smoothing filter chokes used in dc power supplies, but is far too large for RF circuits. In those circuits the subunits of millihenrys (mH) and microhenrys (μH) are used. These are related to the henry by: 1 henry = 1000 millihenrys (mH) = 1,000,000 microhenrys (μH). Thus, 1 mH = 10^{-3} H and 1 μH = 10^{-6} H.

The phenomenon listed that we are concerned with here is called *self-inductance*: when the current in a circuit changes, the magnetic field generated by that current change also changes. This changing magnetic field induces a counter current in the direction that opposes the original current change. This induced current also produces an EMF (discussed above), which is called the *counter electromotive force* (CEMF). As with other forms of inductance, self-inductance is measured in henrys and its subunits.

Although the term *inductance* refers to several phenomena, when used alone it typically means self-inductance, and will be so used in this chapter unless otherwise specified (e.g., mutual inductance). However, keep in mind that the generic term can have more meanings than is commonly attributed to it.

Inductance of a single straight wire

Although it is commonly assumed that inductors are "coils," and therefore consist of at least one, usually more, turns of wire around a cylindrical form, it is also true that a single, straight piece of wire possesses inductance. This inductance of a wire in which the length is at least 1000 times its diameter is given by:

$$L_{\mu H} = 0.00508\, b\left(Ln\left(\frac{4a}{d}\right) - 0.75\right) \qquad \textbf{[6-2]}$$

(Ratios less than $1/d > 1000$ are more difficult to calculate.)

The inductance values of representative small wires is very small in absolute numbers, but at higher frequencies becomes a very appreciable portion of the whole inductance needed. Consider a 12-inch length of #30 wire (dia. = 0.010 inches). Plugging these values into Eq. 6-2 yields:

$$L_{\mu H} = 0.00508\ (12\ \text{in})\left(Ln\left(\frac{4\times 12\ \text{in}}{0.010\ \text{in}}\right) - 0.75\right) \qquad \textbf{[6-3]}$$

$$L_{\mu H} = 0.471\ \mu H \qquad \textbf{[6-4]}$$

An inductance of 0.471-µH seems terribly small, and at 1 MHz it is small compared with inductances typically used as that frequency. But at 100 MHz, 0.471 µH could easily be more than the entire required circuit inductance. RF circuits have been created in which the inductance of a straight piece of wire is the total inductance. But when the inductance is an unintended consequence of the circuit wiring, then it can become a disaster at higher frequencies. Such unintended inductance is called *stray inductance*, and can be reduced by using broad, flat conductors to wind the coils. An example is the "printed circuit" coils wound on cylindrical forms in the FM radio receiver tuner in Fig. 6-2.

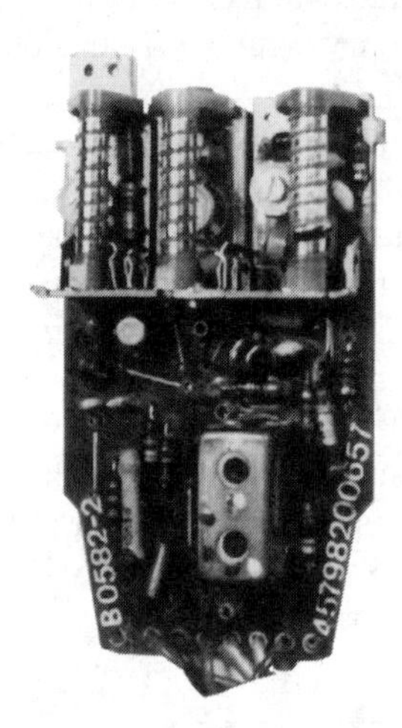

6-2 Printed circuit inductors.

Self-inductance can be increased by forming the conductor into a multi-turn coil (Fig. 6-3) in such a manner that the magnetic field in adjacent turns reinforces each other. This requirement means that the turns of the coil must be insulated from each other. A coil wound in this manner is usually called an *inductor*, or simply a *coil*, in RF/IF circuits, but to be correct, the inductor pictured in Fig. 6-3 is called a *solenoid-wound coil* if the length (a) is greater than the diameter (d).

Several factors affect the inductance of a coil. Perhaps the most obvious are the length, the diameter and the number of turns in the coil. Also affecting the inductance is the nature of the *core* material and its cross-sectional area. In the example of Fig. 6-3 the core is air and the cross-sectional area is directly related to the diameter, but in many radio circuits the core is made of powdered iron or ferrite materials (about which, more later).

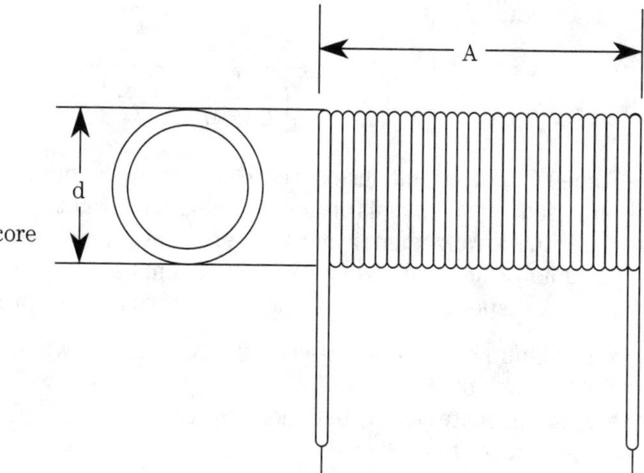

6-3 Solenoid-wound air-core inductor (A>B).

Combining two or more inductors

When inductors are connected together in a circuit, their inductances combine similar to the resistances of several resistors in parallel or series. For inductors in which their respective magnetic fields do not interact:

A. Series connected inductors:

$$L_{total} = L_1 + L_2 + L_3 + \ldots + L_n \quad \textbf{[6-5]}$$

B. Parallel connected inductors:

$$L_{total} = \frac{1}{\left(\frac{1}{L_1} + \frac{1}{L_2} + \frac{1}{L_3} + \ldots + \frac{1}{L_n}\right)} \quad \textbf{[6-6]}$$

Or, in the special case of two inductors in parallel:

$$L_{total} = \frac{L_1 \times L_2}{L_1 + L_2} \quad \textbf{[6-7]}$$

If the magnetic fields of the inductors in the circuit interact, the total inductance becomes somewhat more complicated to express. For the simple case of two inductors in series, the expression would be:

A. Series inductors:

$$L_{total} = L_1 + L_2 \pm 2M \quad \textbf{[6-8]}$$

Where M is the *mutual inductance* caused by the interaction of the two magnetic fields (note: +M is used when the fields aid each other, and –M is used when the fields are opposing).

B. Parallel inductors:

$$L_{total} = \frac{1}{\left(\frac{1}{L_1 \pm M}\right) + \left(\frac{1}{L_2 \pm M}\right)} \quad \textbf{[6-9]}$$

Some LC tank circuits use air-core coils in their tuning circuits. Where multiple coils are used, adjacent coils are usually aligned at right angles to the other one. The reason for this arrangement is not for mere convenience, but rather it is a tactic used by the radio designer in order to prevent unintended interaction of the magnetic fields of the respective coils. In general, for coils in close proximity to each other:

- Maximum interaction between the coils occurs when the coils' axes are parallel to each other.
- Minimum interaction between the coils occurs when the coils' axes are at right angles to each other.

For the case where the coil axes are along the same line, the interaction depends on the distance between the coils.

Inductors in ac circuits

Impedance (Z) is the total opposition to the flow of alternating current (ac) in a circuit, and as such it is analogous to resistance in dc circuits. The impedance is made up of a resistance component (R) and a component called *reactance* (X). Like resistance, reactance is measured in ohms. If the reactance is produced by an inductor, then it is called *inductive reactance* (X_L), and if by a capacitor it is called *capacitive reactance* (X_C). Inductive reactance is a function of the inductance and the frequency of the ac source:

$$X_L = 2\,\pi f L \quad \textbf{[6-10]}$$

where

X_L is the inductive reactance in ohms (Ω)
f is the ac frequency in hertz (Hz)
L is the inductance in henrys (H)

In a purely resistive ac circuit (Fig. 6-4) the current (I) and voltage (V) are said to be *in-phase* with each other; i.e., they rise and fall at exactly the same times in the ac cycle (Fig. 6-5). In vector notation (Fig. 6-6), the current and voltage vectors are along the same axis, which is an indication of the zero degree phase difference between the two.

In an ac circuit that contains only an inductor (Fig. 6-7), and is excited by a sine wave ac source, the change in current is opposed by the inductance. As a result, the current (I) in an inductive circuit lags behind the voltage (V) by 90 degrees. This is shown vectorially in Fig. 6-8A, and as a pair of sine waves in Fig. 6-8B.

The ac circuit that contains a resistance and an inductance (Fig. 6-9A) shows a phase shift (θ), shown vectorially in Fig. 6-9B, that is other than the 90 degrees seen in purely inductive circuits. The phase shift is proportional to the voltage across the

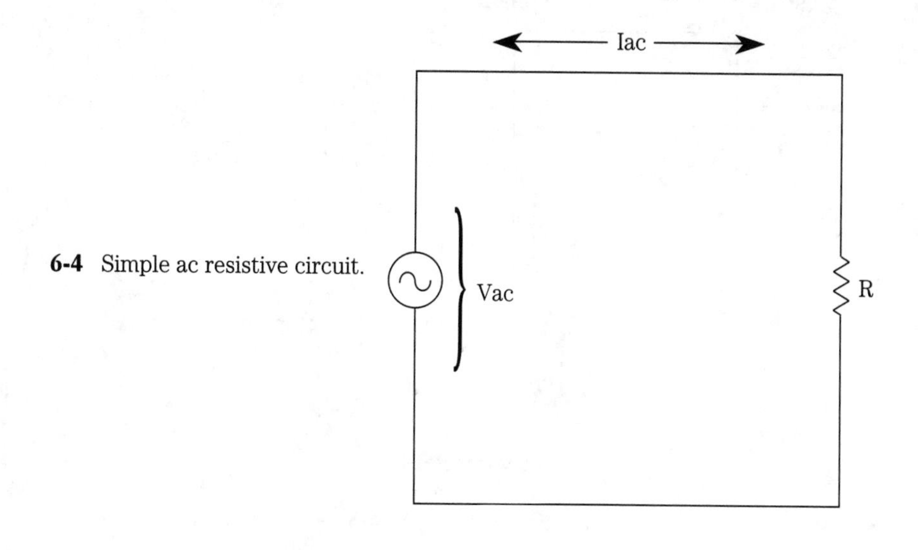

6-4 Simple ac resistive circuit.

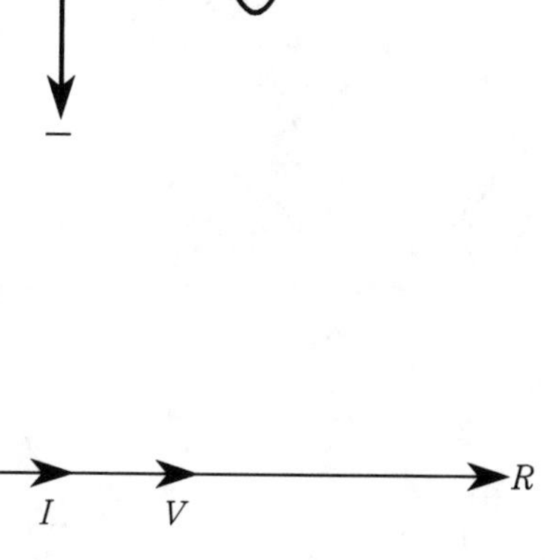

6-5 I and V phase relationship in resistive circuit.

6-6 Vector relationships in resistive circuit.

inductor and the current flowing through it. The *impedance* (Z) of this circuit is found by the *Pythagorean rule*, also called the root of the sum of the squares method (see Fig. 6-9C):

$$Z = \sqrt{R^2 + X_L^2} \qquad \textbf{[6-11]}$$

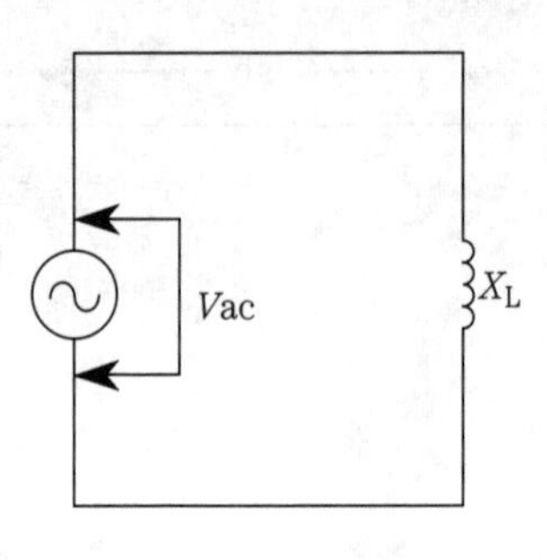

6-7 Inductive ac circuit.

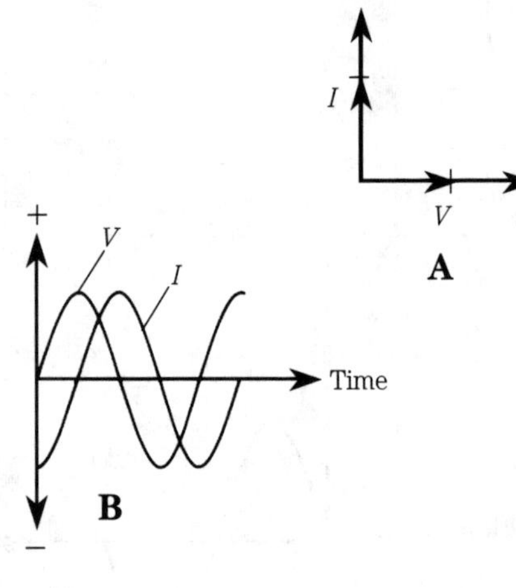

6-8 A. Vector relationships in ac inductive circuit. B. Phase relationships in inductive circuit.

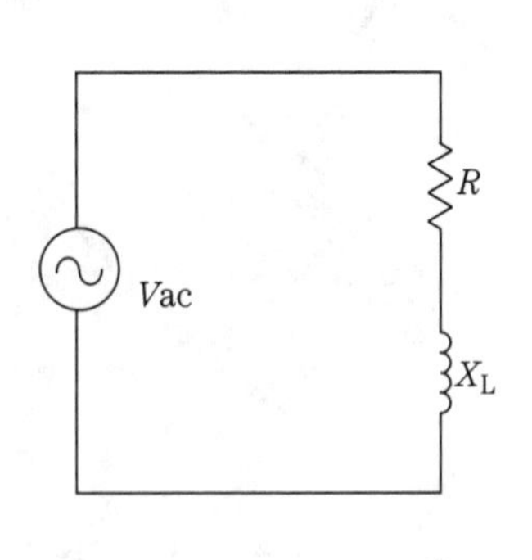

+X
X
Z
Inductive reactance
Impedance (Z)
+R
R
Resistance

B

Z
X_L
R

$$Z = \sqrt{R^2 + (X_L)^2}$$

C

6-9 A. ac resistor-inductor circuit.
B. Vector phase relationships.
C. "Triangle rule" for impedance.

Air-core inductors

An *air-core inductor* actually has no core, so might also be called a *coreless coil.* Although it can be argued that the performance of an inductor differs between air and vacuum cores, the degree of difference is so negligible as to fall into the "decimal dust" category.

Three forms of air core inductor can be recognized. If the length (b) of a cylindrical coil is greater than, or equal to, the diameter (d), then the coil is said to be *solenoid-wound.* But if the length is much shorter than the diameter, then the coil is said to be *loop-wound.* There is a gray area around the breaking point between these inductors where the loop-wound coil seems to work somewhat like a solenoid-wound coil, but in the main most loop- wound coils are such that $b << d$. The principal uses of loop-wound coils is in making loop antennas for interference-nulling and radio direction-finding (RDF) applications.

Solenoid-wound air-core inductors

An example of the solenoid-wound air-core inductor is shown in Fig. 6-3. This form of coil is longer than its own diameter. The inductance of the solenoid-wound coil is given by:

$$L_{\mu H} = \frac{a^2N^2}{9a + 10b} \qquad \textbf{[6-12]}$$

where

$L_{\mu H}$ is the inductance in microhenrys (μH)
a is the coil radius in inches (in)
b is the coil length in inches (in)
N is the number of turns in the coil

The above equation will allow calculation of the inductance of a known coil, but usually we need to know the number of turns (N) required to achieve some specific inductance value determined by the application. For this purpose, we rearrange the equation in the form:

$$N = \sqrt{\frac{L(9a + 10b)}{a}} \qquad \textbf{[6-13]}$$

Several forms of solenoid-wound air-core inductor are shown in Fig. 6-10. The version shown in Fig. 6-10A is homemade on a form of 25 mm PVC plumbing pipe, sawn to a length of about 80 mm and fitted at one end with solder lugs and secured with small machine screws and hex nuts. The main inductor is shown in the dark-colored #24 AWG enamel coated wire, while a small winding is made from #26 AWG insulated hook-up wire. The use of two coaxial inductors makes this assembly a transformer. The small primary winding can be connected between ground and an aerial, while the larger secondary winding can be resonated with a variable capacitor.

The air-core coil shown in Fig. 6-10B is part of an older radio transmitter, where it forms the anode tuning inductance. (Amateur radio operators may recognize this

6-10 A. Solenoid wound inductor with a transformer "coupling" link winding.

B. Inductor in transmitter final amplifier.

C. Tapped air-core inductor.

D. Tapped inductor on mobile HF antenna.

E. Adjustable "transmitting" inductor.

unit as the Heathkit DX-60B transmitter.) Because several different frequency bands must be accommodated, several coils are wound on the same form. The required sections are switch-selected according to the band of operation. Another method for achieving tapped inductance is shown in Fig. 6-10C. This coil is a commercial air-core coil made by: Barker & Williamson, 10 Canal Street, Bristol, PA 19007, USA; (215) 788-5581. Some models of this coil stock come with alternate windings indented to facilitate the connection of a tap, but in any case it is easy to press in the windings adjacent to the connection point. The variant shown in Fig. 6-10D was found, oddly enough, on a 1000-watt HF motorcycle mobile. The whip antenna is resonated by using an alligator clip to select the number of turns from the coil.

The air-core coil shown in Fig. 6-10E is a rotary inductor of the type found in some HF transmitters, antenna tuning units and other applications where continuous control of frequency is required. The inductor coil is mounted on a ceramic form that can be rotated using a shaft protruding from one end. As the form rotates, a movable shorting element rides along the turns of the coil to select the required inductance. Note in Fig. 6-10E that the pitch (number of turns per unit length) is not constant along the length of the coil. This "pitch-winding" method is used to provide a nearly constant change of inductance for each revolution of the adjustment shaft.

Adjustable coils

There are several practical problems with using the standard fixed coil discussed above. For one thing, the inductance cannot easily be adjusted either to tune the radio or to trim the tuning circuits to account for the tolerances in the circuit.

Air-core coils are difficult to adjust. They can be lengthened or shortened; the number of turns can be changed; or a tap or series of taps can be established on the coil in order to allow an external switch to select the number of turns that are allowed to be effective. None of these methods are very elegant, even though all have been used in one application or another.

The solution to the adjustable inductor problem, which was developed relatively early in the history of mass produced radios (and is still used today), is to insert a powdered iron or ferrite core (or "slug") inside of the coil form (Figs. 6-11A and 6-11B). The permeability of the core will increase or decrease the inductance according to how much of the core is inside the coil. If the core is made with either a hexagonal hole or screwdriver slot, the inductance of the coil can be adjusted by moving the core in or out of the coil. These coils are called *slug-tuned inductors*. An example is shown in Fig. 6-11C.

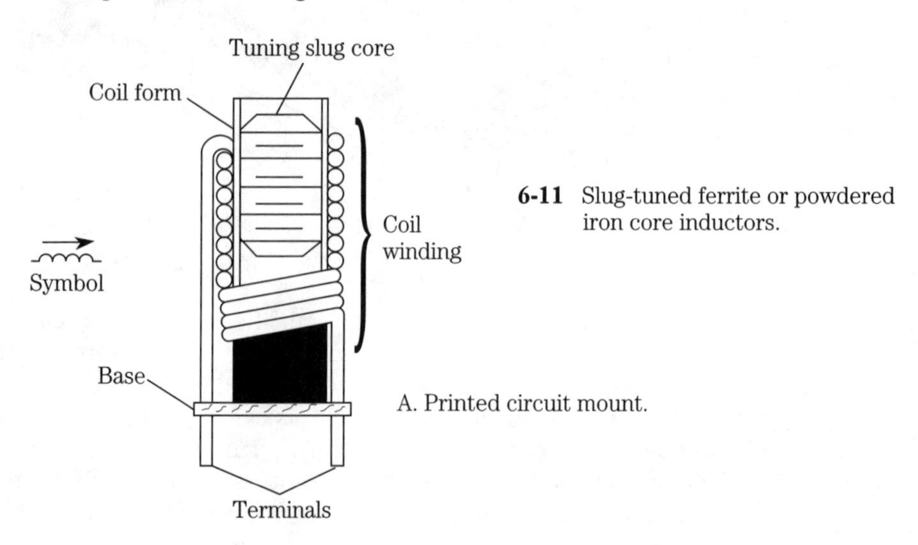

6-11 Slug-tuned ferrite or powdered iron core inductors.

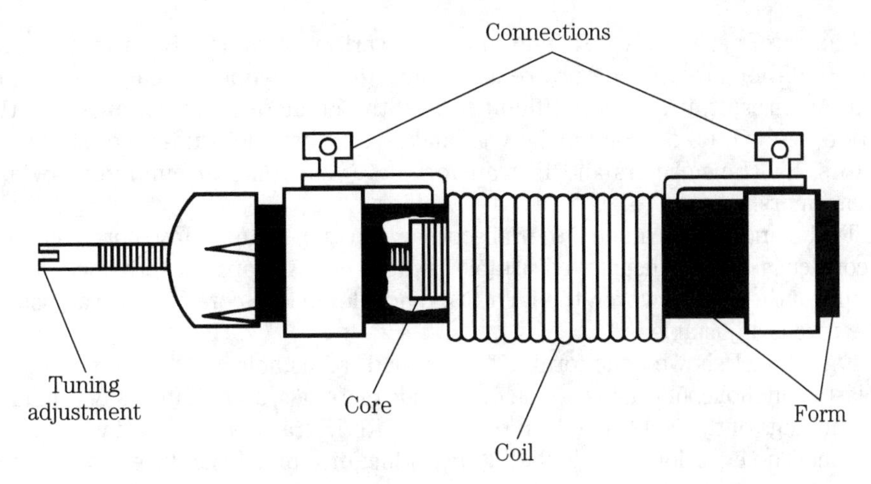

B. Flange mount.

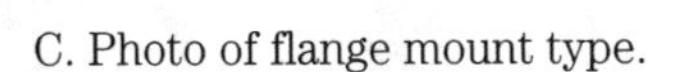
C. Photo of flange mount type.

6-11 Continued.

Winding your own coils

Inductors (L) and capacitors (C) are the principal components used in RF tuning circuits (also called *resonant circuits* and *LC tank circuits*. The *resonant frequency* of a tank circuit is the frequency to which the LC combination is tuned to, and is found from:

$$F = \frac{1}{2\pi \sqrt{LC}} \qquad \textbf{[6-14]}$$

or, if either the inductance (L) or capacitance (C) is either known or preselected, then the other can be found by solving Eq. 6-14 for the unknown, or:

$$C = \frac{1}{4\pi^2 F^2 L} \qquad \textbf{[6-15]}$$

and

$$L = \frac{1}{4\pi^2 F^2 C} \qquad \textbf{[6-16]}$$

In all three equations, L is in henrys, C is in farads, and the frequency is in hertz (don't forget to convert values to microhenrys and picofarads after calculations are made).

Capacitors are easily obtained in a wide variety of values. But tuning inductors are either unavailable, or are available in other people's ideas of what you need. As a result, it is often difficult to find the kinds of parts you need. In this section, we will take a look at how to make your own slug-tuned adjustable inductors, RF transformers and IF transformers (yes, you *can* build your own IF transformers!).

Tuning inductors can be either air-core or ferrite/powdered iron core coils. The air-core coils are not usually adjustable unless clumsy taps are provided on the winding of the coil. However, the ferrite and powdered iron core coils are adjustable if the core is adjustable.

Figure 6-11 showed one form of "slug tuned" adjustable coil. The form is made of plastic, phenolic, fiberglass, nylon or ceramic materials, and is internally threaded. The windings of the coil (or coils in the case of RF/IF transformers) are wound onto the form. The equation for calculating the inductance of a single-layer coil is found in any good radio book, but is not needed for our purposes. We have a simpler way. The *tuning slug* is a ferrite or powdered iron coil core that mates with the internal threads in the coil form. A screwdriver slot or hex hole in either (or both) ends allows adjustment. The inductance of the coil depends on how much of the core is inside the coil windings.

Amidon Associates coil system

It was once difficult to obtain coil forms to make your own project inductors. But Amidon Associates, Inc., 2216 East Gladwick, Dominguez Hills, CA 90220, USA; (310) 763-5770 (voice), (310) 763-2250 (fax) sells a series of slug-tuned inductor forms that can be used to make any value coil you are likely to need. Figure 6-12A shows a sectioned view of the Amidon forms, while Fig. 6-12B shows an exploded view.

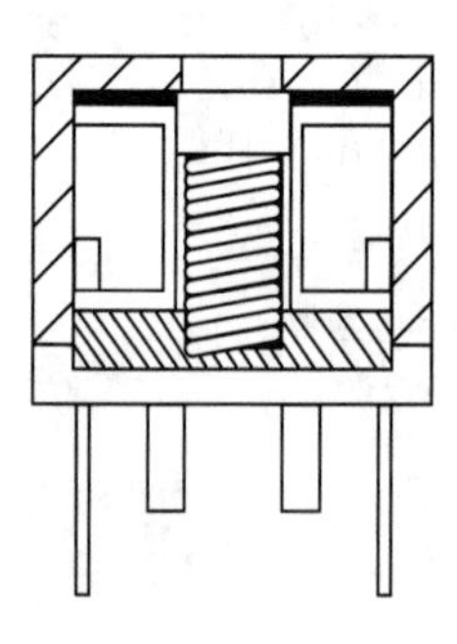

6-12 A. Sectioned view of shielded slug-tuned core.

Table 6-1 shows the type numbers, frequency ranges (in MHz), and other specifications for the coil forms made by Amidon. Three sizes of coil forms are offered. The L-33-X are 0.31-inch square and 0.40-inch high; the L-43-X are 0.44-inch square and 0.50-inch high; and the L-57-X are 0.56-inch square and 0.50-inch high. The "X" in each type number indicates the type of material, which in turn translates to the operating frequency range (see Table 6-1). Now, let's see how the coil forms are used.

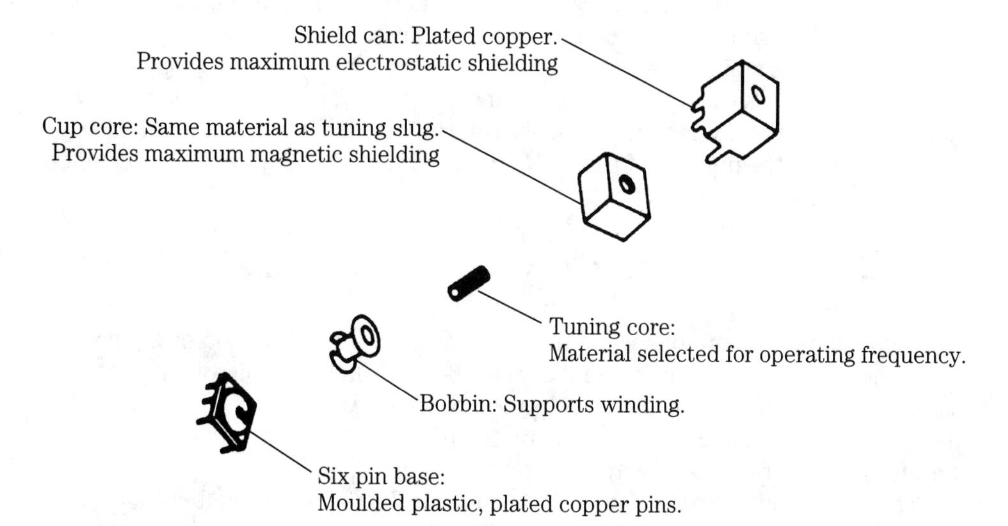

6-12 B. Exploded view showing sub-components.

Table 6-1. Amidon coil form specifications

Part number	Frequency range (MHz)	A_L Value	Ratio	Q_{max}
L-33-1	0.30–1.0	76	1.7:1	80
L-33-2	1.00–10	68	1.5:1	90
L-33-3	0.01–0.5	80	1.8:1	70
L-33-6	10–50	60	1.5:1	100
L-33-10	25–100	54	1.4:1	120
L-33-17	50–200	48	1.3:1	130
L-43-1	0.30–1.00	115	1.6:1	110
L-43-2	1.00–10	98	1.6:1	120
L-43-3	0.01–0.5	133	1.8:1	90
L-43-6	10–50	85	1.4:1	130
L-43-10	25–100	72	1.3:1	150
L-43-17	50–200	56	1.2:1	200
L-57-1	0.30–1.0	175	3:1	*
L-57-2	1.00–10	125	2:1	*
L-57-3	0.01–0.5	204	3:1	*
L-57-6	10–50	115	2:1	*
L-57-10	25–100	100	2:1	*
L-57-17	50–200	67	1.5:1	*

Determine the required inductance from Eq. 6-16. In my experiment to test these coils I decided to build a 15 MHz WWV converter that reduced the WWV frequency to an 80/75-meter band frequency. Thus, I needed a tuned circuit that would tune 15 MHz. It is generally a good idea to have a high capacitance to inductance ratio in order to maintain a high "Q" factor. I selected a 56-pF/NPO capacitor for the tuned circuit because a) it is in the right range for "high," and b) a dozen or so were in my junk box. According to Eq. 6-16, therefore, I needed a 2-µH inductor.

To calculate the number of turns (N) required to make any specific inductance, use the following equation:

$$N = 100\sqrt{\frac{L_{\mu H}}{0.9\,A_L}} \qquad \textbf{[6-17]}$$

The inductance is in microhenrys (µH). The "A_L" factor is a function of the properties of the core material, and is found in Table 6-1; the units are microhenrys per 100 turns (µH/100 turns). In my case, I selected an L-57-6 (which covers the correct frequency range), which has an A_L value of 115 µH/100 turns. According to Eq. 6-17, therefore, I need 14 turns of wire.

The coil is wound from #26 to #32 wire. Ideally, Litz wire is used, but that is both hard to find and difficult to solder. For most projects ordinary enamel-coated "magnet wire" will suffice. A razor knife (such as X-acto) and soldering iron tip can be used to remove the enamel from the ends of the wire. Because the forms are so small, I recommend using the #32 size.

Winding the coil can be a bit tricky if your vision needs augmentation as much as mine. But using tweezers, needle nose pliers, and a magnifying glass on a stand made it relatively easy. Figure 6-13 shows the method for winding a coil with a tapped winding. Anchor one end of the wire with solder on one of the end posts, and use this as the reference point. In my case, I wanted a 3-turn tap on the 14-turn coil, so I wound three turns and then looped the wire around the center post. After this point was soldered, the rest of the coil was wound and then anchored at the remaining end post. A dab of glue, clear fingernail polish or Q-Dope will keep the coil windings from moving.

If you make an RF/IF transformer, there will be two or more windings. Try to separate the primary and secondary windings if both are tuned. If one winding is not tuned, then simply wind it over the "cold" (i.e., ground) end of the tuned winding—no separation is desired.

The Amidon coil forms are tight, but they do have sufficient space for very small

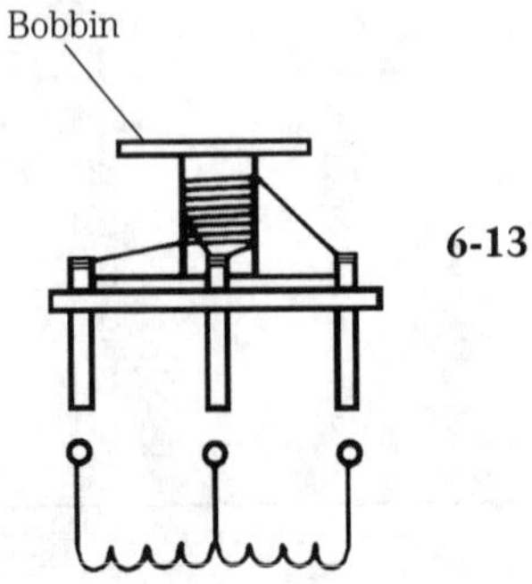

6-13 Tapped inductor on a bobbin form.

disk ceramic capacitors inside. The 56-pF capacitors that I selected fit nicely inside the shielded can of the coil, so I elected to place it there. Thus, I've basically made a 15-MHz IF transformer.

After constructing the 15-MHz RF coil, I tested it and found that I could adjust the slug-tuned coil to 15 MHz with a nice tolerance on either side of the design resonant frequency. It worked!

Although slug-tuned inductors are sometimes considered a bit beyond the hobbyist or ham, that is not entirely true. The Amidon Associates, Inc. L-series coil forms are easily used to make almost any inductor that you are likely to need.

Toroid core inductors and transformers

Many electronic construction projects intended for hobbyists and amateur radio operators call for inductors or radio frequency (RF) transformers wound on *toroidal cores*. A toroid is a doughnut shaped object, i.e., a short, flat cylinder (often with rounded edges) that has a hole in the center (see Fig. 6-14). The toroidal shape is desirable for inductors because it permits a relatively high inductance value with few turns of wire by virtue of the core's *permeability* (μ), and, perhaps most important, the geometry of the core makes the coil self-shielding. That latter attribute makes the toroid inductor easier to use in practical RF circuits.

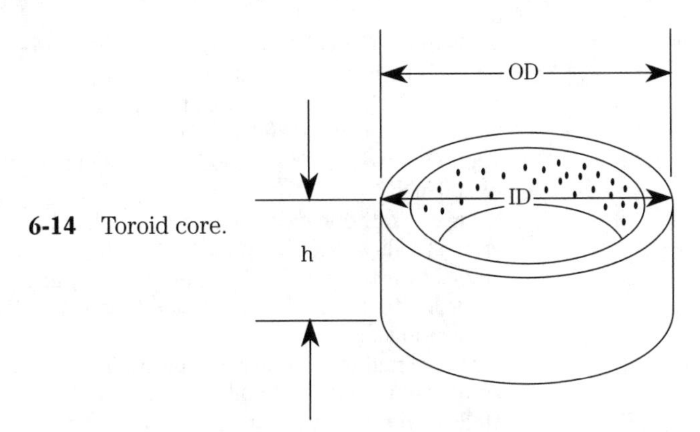

6-14 Toroid core.

Regular solenoid-wound cylindrical inductors have a magnetic field that goes outside of the immediate vicinity of the windings, and can thus intersect nearby inductors and other objects. Unintentional inductive coupling can cause a lot of serious problems in RF electronic circuits, so it should be avoided wherever possible. The use of a toroidal shape factor, with its limited external magnetic field, makes it possible to mount the inductor close to other inductors (and other components) without too much undesired interaction.

Materials used in toroidal cores

Toroidal cores come in a variety of materials that are usually grouped into two general classes: powdered iron and ferrites. These groups are further subdivided as discussed below.

Powdered iron materials The powdered iron cores come in two basic formulations: Carbonyl irons and hydrogen reduced irons. The *carbonyl materials* are well regarded for their temperature stability; they have permeability (μ) values that range from 1 mu to about 35 mu. The carbonyls offer very good "Q" values to frequencies of 200 MHz. Carbonyls are used in high-power applications, as well as in variable-frequency oscillators and other applications where temperature stability becomes important. However, note that no powdered iron material or ferrite is totally free of temperature variation, so oscillators using these cores must be temperature compensated for proper operation. The *hydrogen-reduced iron* devices offer permeabilities up to 90 mu, but are lower "Q" than carbonyl devices. They find their main usage in electromagnetic interference (EMI) filters. The powdered iron materials are the subject of Table 6-2.

Table 6-2. Powdered iron core materials

Material	Permeability (μ)	Comments
0	1	Used up to 200 MHz. Inductance varies with method of winding.
1	20	Made of Carbonyl C. Similar to Mixture No. 3 but is more stable, and has a higher volume resistivity.
2	10	Made of Carbonyl E. High Q and good volume resistivity over range of 1 to 30 MHz.
3	35	Made of Carbonyl HP. Very good stability and good Q over range of 0.05 to 0.50 MHz.
6	8	Made of Carbonyl SF. Is similar to Mixture No. 2, but has higher Q over range 20 to 50 MHz.
10	6	Type W powdered iron. Good Q and high stability from 40 to 100 MHz.
12	3	Made of synthetic oxide material. Good Q, but only moderate stability over the range 50 to 100 MHz.
15	25	Made of Carbonyl GS6. Excellent stability and good Q over range 0.1 to 2 MHz. Recommended for AM BCB and VLF applications.
17	3	Carbonyl material similar to Mixture No. 12, but has greater temperature stability but lower Q than No. 12.
26	75	Made of Hydrogen Reduced Iron. Has very high permeability. Used in EMI filters and dc chokes.

Ferrite materials The name "ferrite" implies that the material is iron-based, but that is not the case; ferrite materials are actually grouped into nickel-zinc and manganese-zinc types. The *nickel-zinc material* has a high volume resistivity and high Q over the range 0.50 to 100 MHz. The temperature stability is only moderate, however. The permeabilities of nickel-zinc materials are found in the range 125 to 850 mu. The *manganese-zinc materials* have higher permeabilities than nickel-zinc, and are on the order of 850 to 5000. Manganese-zinc materials offer high Q over the range 0.001 to 1 MHz. They have low volume resistivity and moderate saturation flux density. These materials are used in switching power supplies from 20 to 100 kHz, and for EMI attenuation in the range of 20 to 400 MHz. See Table 6-3 for information on ferrite materials.

Table 6-3. Ferrite materials

N–Z: Nickel-zinc
M–Z: Manganese-zinc

Material	Permeability (μ)	Remarks
33	850	M–Z. Used over 0.001 to 1 MHz for loopstick antenna rods. Low-volume resistivity.
43	850	N–Z. Medium wave inductors and wideband transformers to 50 MHz. High attenuation over 30 to 400 MHz. High-volume resistivity.
61	125	N–Z. High Q over 0.2 to 15 MHz. Moderate temperature stability. Used for wideband transformers to 200 MHz.
63	40	High Q over 15 to 25 MHz. Low permeability and high-volume resistivity.
67	40	N–Z. High Q operation over 10 to 80 MHz. Relatively high flux density and good temperature stability. Is similar to Type 63, but has lower volume resistivity. Used in wideband transformers to 200 MHz.
68	20	N–Z. Excellent temperature stability and high Q over 80 to 180 MHz. High-volume resistivity.
72	2000	High Q to 0.50 MHz, but used in EMI filters from 0.50 to 50 MHz. Low-volume resistivity.
J/75	5000	Used in pulse and wideband transformers from 0.001 to 1 MHz, and in EMI filters from 0.50 to 20 MHz. Low-volume resistivity and low core losses.
77	2000	0.001 to 1 MHz. Used in wideband transformers and power converters, and in EMI and noise filters from 0.5 to 50 MHz.
F	3000	Is similar to Type 77 above, but offers a higher volume resistivity, higher initial permeability, and higher flux saturation density. Used for power converters and in EMI/noise filters from 0.50 to 50 MHz.

Toroid core nomenclature

There are several different ways to designate toroidal cores, but the one used by Amidon Associates is perhaps that most commonly found in projects published for electronic hobbyists and amateur radio operators. Although the units of measure are the English system used in the USA, Canada (and formerly in the UK), rather than SI units, their use with respect to toroids seems widespread. The type number for any given core will consist of three elements: xx-yy-zz. The *xx* is a one or two letter designation of the general class of material, i.e., powdered iron (xx = "T") or ferrite (xx = "TF"). The *yy* is a rounded off approximation of the outside diameter (OD in Fig. 6-14) of the core in inches; "37" indicates a 0.375-inch (9.53 mm) core, while "50" indicates a 0.50-inch (12.7 mm) core. Some standard core sizes are shown in Table 6-4. The *zz* indicates the type (mixture) of material. A mixture No. 2 powdered iron core of 0.50-inch diameter would be listed as a T-50-2 core. The cores are color-coded to assist in identification.

Table 6-4. Standard toroid core sizes ("xx")

Core size	OD (in)	ID (in)	Thickness (in)
23	0.230	0.120	0.060
37	0.375	0.187	0.125
50	0.500	0.281	0.188
50A	0.500	0.312	0.250
50B	0.500	0.312	0.500
82	0.825	0.520	0.250
87A	0.870	0.540	0.500
114	1.142	0.750	0.295
114A	1.142	0.750	0.545
130	1.300	0.780	0.437
150	1.500	0.750	0.250
150A	1.500	0.750	0.500
193	1.930	1.250	0.750
200	2.000	1.250	0.550
240	2.400	1.400	0.500

Inductance of toroidal coils

The inductance of the toroidal core inductor is a function of the permeability of the core material, the number of turns, the inside diameter (ID) of the core, the outside diameter (OD) of the core, and the height (h) (see Fig. 6-14), and can be approximated by:

$$L_{\mu H} = 0.011684\, hn^2 \mu\, \text{LOG}_{10}\left(\frac{\text{OD}}{\text{ID}}\right) \qquad \textbf{[6-18]}$$

This equation is rarely used directly, however, because toroid manufacturers provide a parameter called the A_L value which relates inductance per 100 or 1000 turns of wire. Tables 6-5 and 6-6 show the A_L values of common ferrite and powdered iron cores, respectively. Table 6-7 shows some of the other properties of powdered iron cores.

Table 6-5. Common ferrite core A_L values

	Core type no. prefix: TF-yy-zz					
Core	Material type					
size	43	61	63	72	75	77
23	188	24.8	7.9	396	990	356
37	420	55.3	17.7	884	2210	796
50	523	68	22	1100	2750	990
50A	570	75	24	1200	2990	1080
50B	1140	150	48	2400	—	2160
82	557	73.3	22.8	1170	3020	1060
114	603	79.3	25.4	1270	3170	1140
114A	—	146	—	2340	—	—
240	1249	173	53	3130	6845	3130

Table 6-6. Common powdered iron A_L values

Core size	26	3	15	1	2	6	10	12	0
	Core material type (mix)								
12	—	60	50	48	20	17	12	7	3
16	—	61	55	44	22	19	13	8	3
20	—	90	65	52	27	22	16	10	3.5
37	275	120	90	80	40	30	25	15	4.9
50	320	175	135	100	49	40	31	18	6.4
68	420	195	180	115	57	47	32	21	7.5
94	590	248	200	160	84	70	58	32	10.6
130	785	350	250	200	110	96	—	—	15
200	895	425	—	250	120	100	—	—	—

Table 6-7. Properties of powdered iron core types

Material type	Color code	Mu (μ)	Frequency (MHz)
41	Green	75	—
3	Gray	35	0.05–0.5
15	Red/white	25	0.1–2
1	Blue	20	0.5–5
2	Red	10	1–30
6	Yellow	8	10–90
10	Black	6	60–150
12	Green/white	3	100–200
0	Tan	1	150–300

Winding toroid inductors

There are two basic ways to wind a toroidal core inductor: close spaced winding and distributed winding. In *distributed winding* toroidal inductors, the turns of wire that are wound on the toroidal core are spaced evenly around the circumference of the core, with the exception of a gap of at least 30° between the ends (see Fig. 6-15A). The gap ensures that stray capacitance is kept to a minimum. The winding covers only 270° of the core circumference. In close winding toroids (Fig. 6-15B) the turns are made such that adjacent turns of wire touch each other, or nearly so. This practice raises the stray capacitance of the winding, which affects the resonant frequency, but can be done in many cases with little or no ill effect (especially where the capacitance and resonant point shift are negligible).

In general, close winding is used for inductors in narrow band tuned circuits, while distributed winding is used for broadband situations like conventional and BALUN RF transformers. The method of winding has a small effect on the final inductance of the coil. While this fact makes calculating the final inductance less predictable, it also provides a means of final adjustment of actual inductance in the circuit as-built.

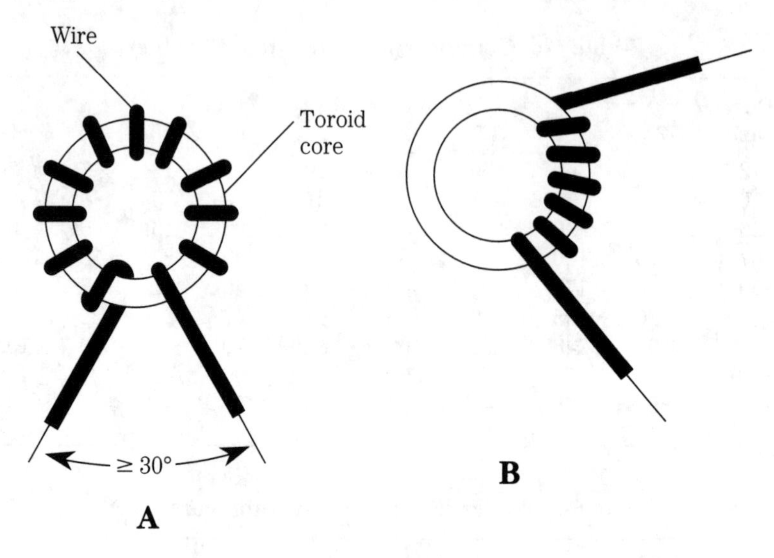

6-15 A. Distributed winding style (leave at least 30-degree gap at ends to reduce capacitance). B. Close winding style.

Calculating the number of turns needed

As in all inductors, the number of turns of wire determines the inductance of the finished coil. In powdered iron cores the A_L rating of the core is used with fair confidence to predict the number of turns needed.

For powdered iron cores

$$N = 100\sqrt{\frac{L_{\mu H}}{A_L}} \quad [6\text{-}19]$$

where

N is the number of turns
$L_{\mu H}$ is the inductance required in microhenrys (μH)
A_L is an attribute of the core material and size (μH/100 turns)

Example

Find the number of turns of wire required to make a 6 μH inductance from a T-50-2 (RED) powdered iron core ($A_L = 49$).

$$N = 100\sqrt{\frac{L_{\mu H}}{A_L}}$$

$$N = 100\sqrt{\frac{6_{\mu H}}{49}} = (100)\ (0.35) = 35 \text{ turns}$$

For ferrite cores

$$N = 1000\sqrt{\frac{L_{\mu H}}{A_L}} \quad [6\text{-}20]$$

where

L_{mH} is the inductance required in millihenrys (mH)
A_L is an attribute of the core material and size (mH/1000 turns)

Example

How many turns are needed to wind a 200 μH inductor on a ferrite FT-50A-43 core (A_L = 570 mH/1000 turns)? Note: 200 μH = 0.200 mH.

$$N = 1000 \sqrt{\frac{0.200}{570}}$$

$$N = (1000)\ (0.0187) = 18.7 \text{ turns}$$

The number of turns calculation often comes out to a fraction of a turn. With the possible exception of 0.5 turns, the actual turns count should be rounded off to the nearest turn. It is possible to round off to the nearest half turn, but it is not as easy to implement in practice.

Building the toroidal device

The toroid core or transformer is usually wound with enameled or formvar-insulated wire. For low-power applications (receivers, variable frequency oscillators, etc.) the wire will usually be #22 through #36 (with #26 being very common) SWG. For high-power applications, such as transmitters and RF power amplifiers, a heavier grade of wire is needed. For amateur radio high-power transmitter applications, #14 or #12 wire is usually specified, although wire as large as #6 has been used in some commercial applications. Again, the wire is enameled or formvar covered insulated wire.

In the high-power case, it is likely that high voltages will exist. In high powered RF amplifiers, such as those used by amateur radio operators in many countries, the potentials present across a 50-ohm circuit can reach hundreds of volts. In those cases, it is common practice to wrap the core with a glass-based tape such as Scotch 27.

High-powered applications also require a large area toroid, rather than the small toroids that are practical at lower power levels. Cores in the FT-150-zz to FT-240-zz, or T-130-zz to T-500-zz are typically used. In some high-powered cases, several identical toroids are stacked together and wrapped with tape to increase the power-handling capacity. This method is used quite commonly in RF power amplifier and antenna-tuning unit projects.

Binding the wires

Sometimes, the wires making up the toroidal inductor or transformer become loose. Some builders prefer to fasten the wire to the core by using one of the two methods shown in Fig. 6-16. In Fig. 6-16A, we see the use of a dab of glue, silicone adhesive, or the high-voltage sealant Glyptol (sometimes used in television receiver high voltage circuits) to anchor the end of the wire to the toroid core.

Other builders prefer the method shown in Fig. 6-16B. In this method, the end of the wire is looped underneath the first full turn and pulled taut. This method ef-

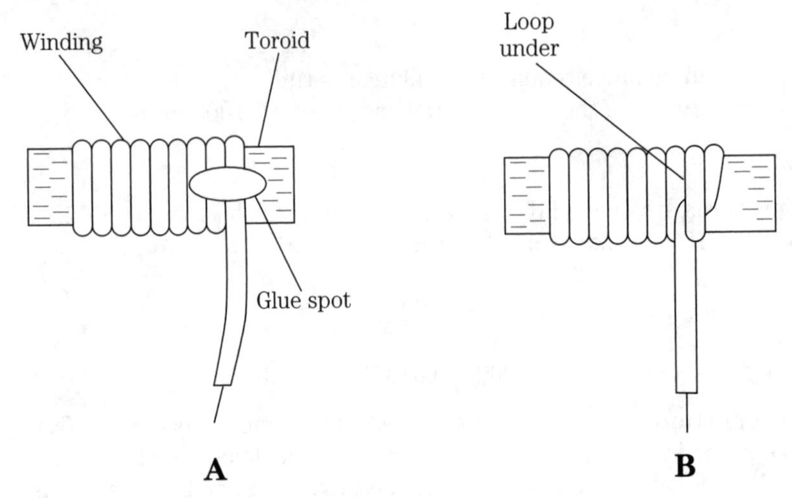

6-16 Securing ends of wire in toroid winding. A. Use of glue or RTV sealant. B. Use of wire overlap.

fectively anchors the wire, but some say it creates an anomaly in the magnetic situation that might provoke interactions with nearby components. In my experience, that situation is not very likely, and I have used this method regularly with no observed problems thus far.

When the final coil is ready, and both the turns count and spacing are adjusted to yield the required inductance, the turns can be anchored and the coil placed in service. A final sealing method is to coat the coil with a thin layer of clear lacquer, or "Q-dope" (intended by its manufacturer as an inductor sealant).

Mounting the toroidal core device

Toroids are sometimes a bit more difficult to mount than solenoid-wound coils, but the rules that you must follow are not as strict. The toroid, when built correctly, is essentially self-shielding—so less attention (not NO attention!) can be paid to components surrounding the inductor. In the solenoid wound coil, for example, the distance between adjacent coils and their orientation is important. Adjacent coils, unless well-shielded, must be placed at right angles to each other to lessen the mutual coupling between the coils. However, toroidal inductors can be closer together and either coplanar or adjacent planar with respect to each other. While some spacing must be maintained between toroidal cores (the winding and core manufacture not being perfect), the required average distance can be less than for solenoid-wound cores.

Mechanical stability of the mounting is always a consideration for any coil (indeed, any electronic component). For most benign environments, the core can be mounted directly to a printed wiring board (PWB) in the manner of Figs. 6-17A and 6-17B. In Fig. 6-17A, the toroidal inductor is mounted flat against the board; its leads are passed through holes in the board to solder pads underneath. The method of Fig. 6-17B places the toroid at right angles to the board, but still uses the leads soldered to copper pads on the PWB to anchor the coil. It is wise to use a small amount of RTV silicone sealant or glue to hold the coil to the board once it is found to work satisfactorily.

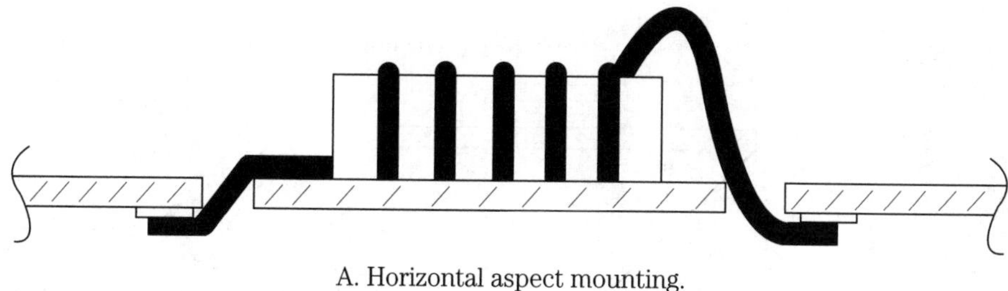

A. Horizontal aspect mounting.

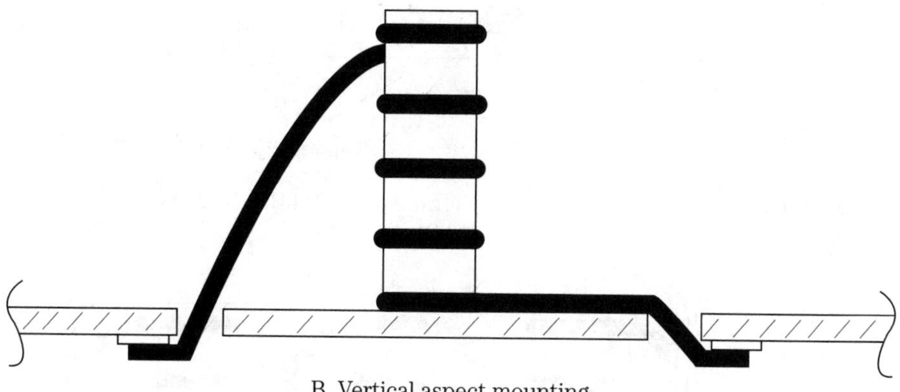

B. Vertical aspect mounting.

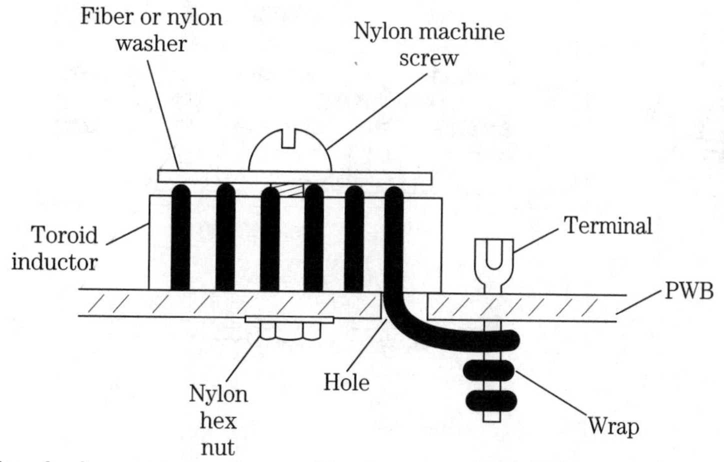

C. Use of nylon or fiber washers and hardware to secure PCB mounted toroid.

6-17 Printed circuit board mounting of toroid inductors.

If the environment is less benign with respect to vibration levels, then a method similar to Fig. 6-17C may be employed. Here the toroid is fastened to the PWB with a set of nylon machine screws and nut hardware, and a nylon or fiber washer. In high-powered antenna tuning units, it is common to see an arrangement similar to Fig. 6-17D. In this configuration, several toroidal cores are individually wrapped in glass tape, then the entire assembly is wrapped as a unit with the same

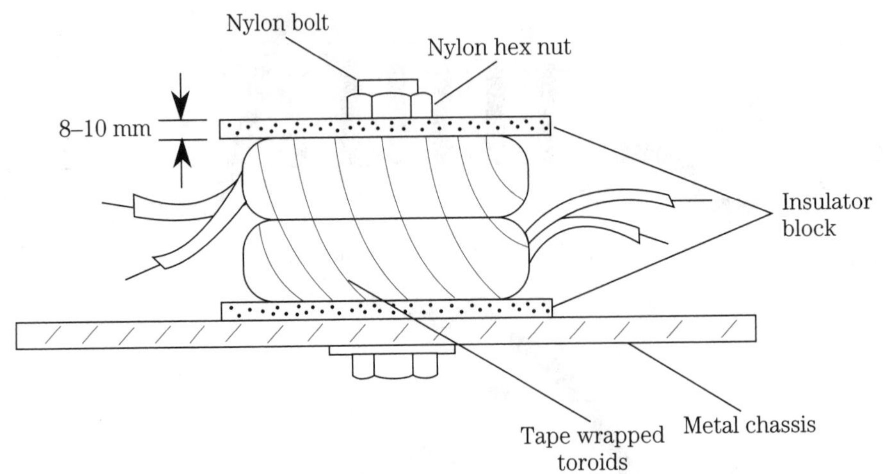

D. Mounting of high-power toroid transformer (e.g., BALUN) in transmitters and antenna tuning units.

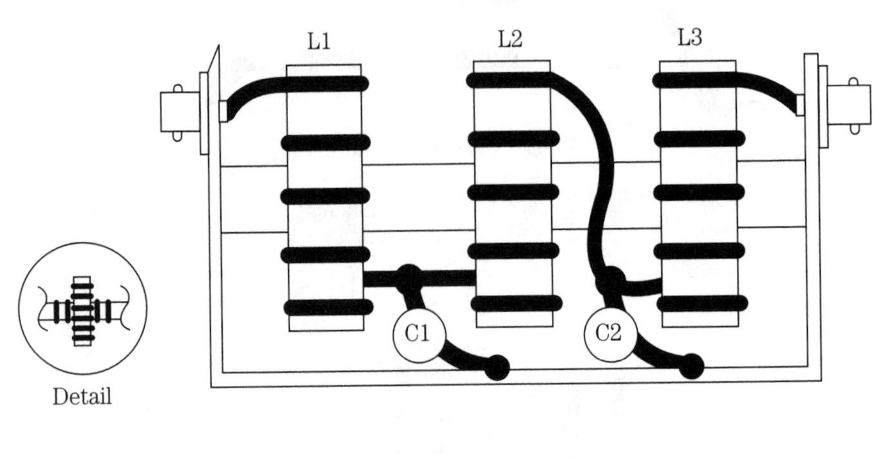

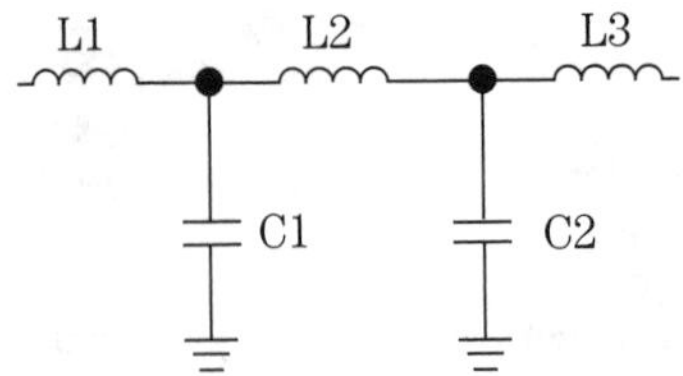

E. Mounting of multiple toroidal inductors on a rod or dowel.

6-17 Continued.

tape. This assembly is mounted between two insulators such as plastics, ceramic, or fiberboard, which are held together as a "sandwich" by a nylon bolt and hex nut.

Figure 6-17E shows a method for suspending toroidal cores in a shielded enclosure. I've used this method to make five-element low-pass filters (see inset) for use on my basement laboratory workbench. The toroidal inductors are mounted on a dowel,

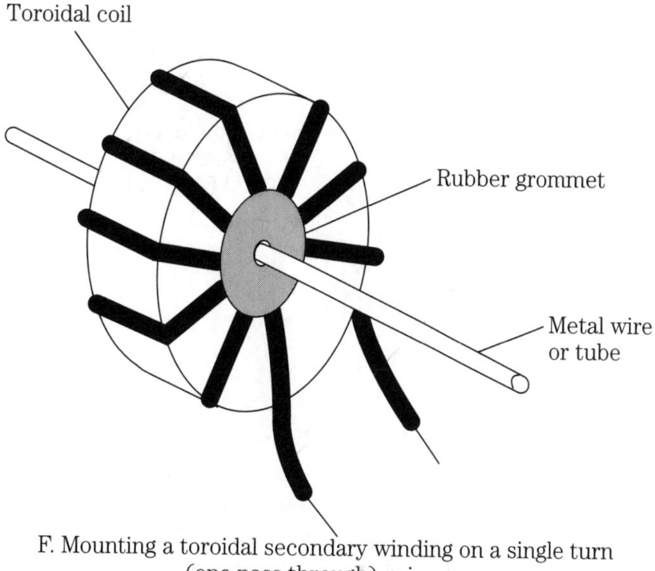

F. Mounting a toroidal secondary winding on a single turn (one pass through) primary.

6-17 Continued.

which is made of some insulating material such as wood, plastic, plexiglass, Lexan or other synthetic material. If the dowel is sized correctly, the inductors will be a tight slip fit, and need no further anchoring. Otherwise, a small amount of glue or RTV silicone sealant can serve to stabilize the position of the inductor. Care must be observed against force fitting, however, in order to avoid fracturing the toroid core.

Some people use a pair of undersized rubber grommets over the dowel, one pressed against either side of the inductor (see inset to Fig. 6-17E). If the grommets are taut enough, no further action is needed. Otherwise, they can be glued to the rod.

A related mounting method is used to make current transformers in homemade RF power meters (Fig. 6-17F). In this case a rubber grommet is fitted into the center of the toroid, and a small brass or copper rod is passed through the center hole of the grommet. The metal rod serves as a one-turn primary winding. A sample of the RF current flowing in the metal rod is magnetically coupled to the secondary winding on the toroid, where it can be either fed to an oscilloscope for display, or it can be rectified, filtered and displayed on a dc current meter that is calibrated in watts or VSWR units.

Toroidal RF transformers

Both narrow band tuned and broadband RF transformers can be accommodated by toroidal powdered iron and ferrite cores. The schematic symbols used for transformers are shown in Fig. 6-18. These symbols are largely interchangeable, and are all seen from time to time. In Fig. 6-18A the two windings are shown adjacent to each other, but the core is shown along only one of them. This is done to keep the drawing simple, and does not imply in any way that the core does not affect one of the windings. The core may be represented either by one or more straight lines, as shown, or by dotted lines. The method shown in Fig. 6-18B is like the conventional

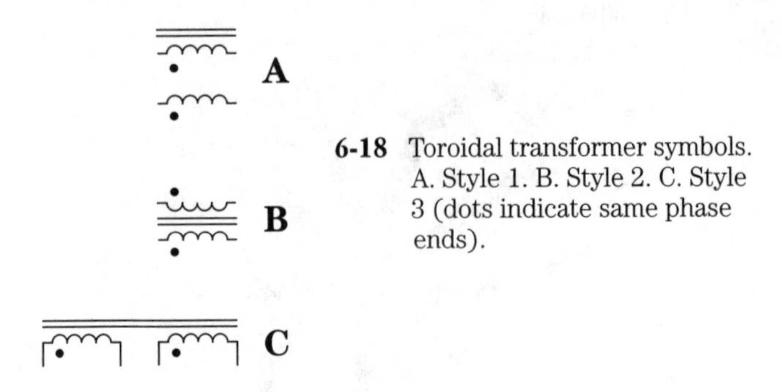

6-18 Toroidal transformer symbols. A. Style 1. B. Style 2. C. Style 3 (dots indicate same phase ends).

transformer representation in which the windings are juxtaposed opposite each other with the core between them. In Fig. 6-18C, the core is extended and the two windings are shown along one side of the core bars.

In each of the transformer representations of Fig. 6-18, dots are shown on the windings. These dots tell us the "sense" of the winding, and represent the same end of the coils. Thus, the wires from two dotted ends are brought to the same location, and the two coils are wound in the same direction. Another way of looking at it is that if a third winding is used to excite the core from an RF source, the phase of the signals at the dot ends will be the same; the phase of the signal at the undotted ends will also be the same, but will be opposite that of the dotted ends.

The windings of the toroidal transformer can be spaced at different locations around the circumference of the toroid when the device is narrow band, but for wideband operation a *bifilar winding* scheme is used (Fig. 6-19A). In this type of winding scheme, the wires, A and B, are held closely parallel to each other as they are wound around the core. When the job is finished, ends A1 and B1 will be at the same location, while A2 and B2 will be at another location on the toroid core. A trifilar wound transformer is shown in Fig. 6-19B.

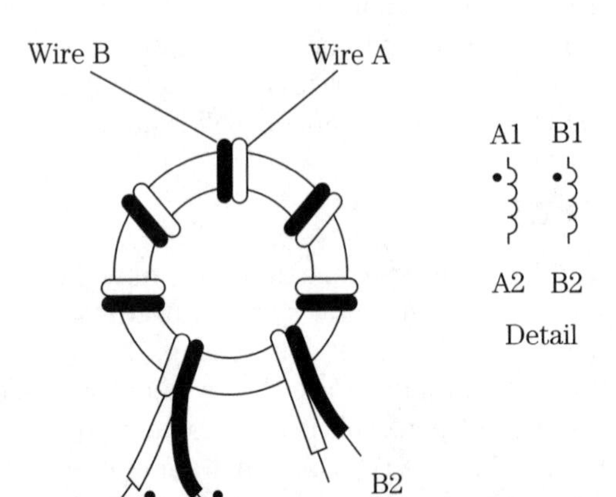

6-19 A. Bifilar winding of toroid transformer (inset schematic shows key to windings).

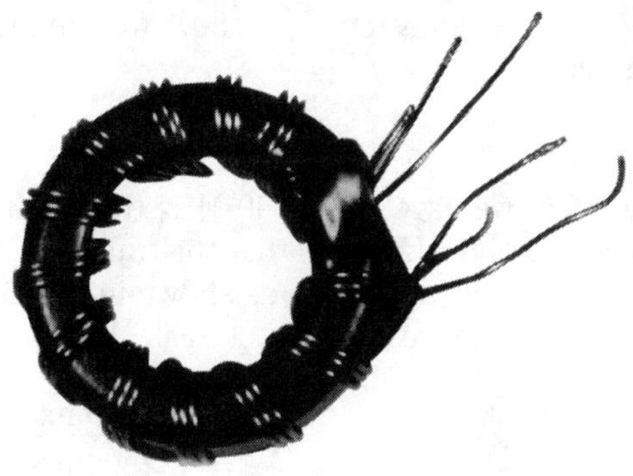

6-19 B. Actual toroid wound BALUN transformer.

Conventional transformers

One of the principal uses of transformers in RF circuits is for impedance transformation. When the secondary winding of a transformer is connected to a load impedance, the impedance seen "looking into" the primary will be a function of the load impedance and the turns ratio of the transformer (see Fig. 6-20A). The relationship is:

$$\frac{N_p}{N_s} = \sqrt{\frac{Z_p}{Z_s}} \quad [6\text{-}21]$$

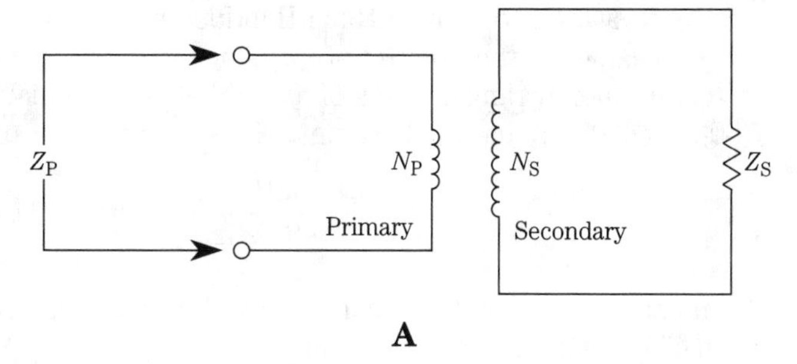

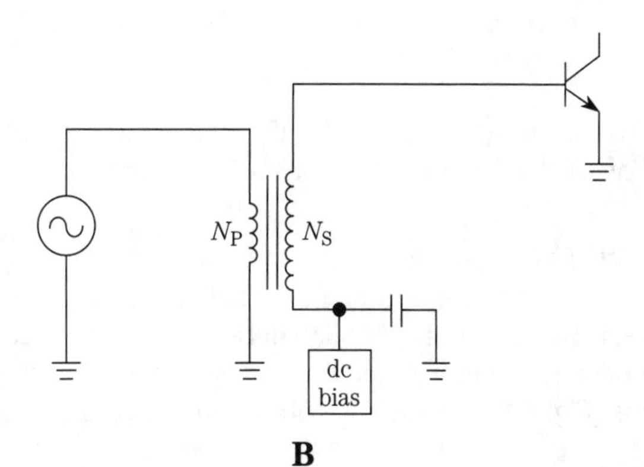

6-20 A. Transformer circuit showing source and load resistances. B. Typical example in solid state circuit.

With the relationship of Eq. 6-21 we can match source and load impedances in RF circuits.

Example

Assume that we have a 3- to 30-MHz transistor RF amplifier with a base input impedance of 4 ohms (Z_s), and that the transistor amplifier has to be matched to a 50-ohm source impedance (Z_p), as shown in Fig. 6-20B. What turns ratio is needed to effect the impedance match? Let's calculate:

$$\frac{N_p}{N_s} = \sqrt{\frac{50 \text{ ohms}}{4 \text{ ohms}}} = 3.53{:}1$$

A general design rule for the value of inductance used in transformers is that the inductive reactance at the lowest frequency must be four times (4X) the impedance connected to that winding. In the case of the 50-ohm primary of the transformer above, then, the inductive reactance of the primary winding should be 4 × 50 ohms, or 200 ohms. The inductance should be:

$$L_{\mu H} = \frac{(200 \text{ ohms})\ (10^6)}{(2\pi)\ (F)}$$

$$L_{\mu H} = \frac{(200 \text{ ohms})\ (10^6)}{(2\pi)\ (3{,}000{,}000)} = 10.6\ \mu\text{H}$$

Now that we know that a 10.6 µH inductance is needed, we can select a toroidal core and calculate the number of turns needed. The T-50-2 (RED) core covers the correct frequency range, and is of a size that is congenial to easy construction. The T-50-2 (RED) core has an A_L value of 49, so the number of turns required:

$$N = 100\sqrt{\frac{10.6\mu\text{H}}{49}} = 46.5 \text{ turns} \approx 47 \text{ turns}$$

The number of turns in the secondary must be such that the 3.53:1 ratio is preserved when 47 turns are used in the primary:

$$N_s = \frac{47 \text{ turns}}{3.53} = 13.3 \text{ turns} \approx 13 \text{ turns}$$

If we wind the primary with 47 turns, and the secondary with 13 turns, then we will convert the 4-ohm transistor base impedance to the 50-ohm systems impedance.

Example

A Beverage wave antenna is constructed for the AM broadcast band (530 to 1700 kHz). By virtue of its construction and installation, it exhibits a characteristic impedance Z_o of 600 ohms. What is the turns ratio required of a transformer at the feed end (Fig. 6-21) to match a 50-ohm receiver input impedance?

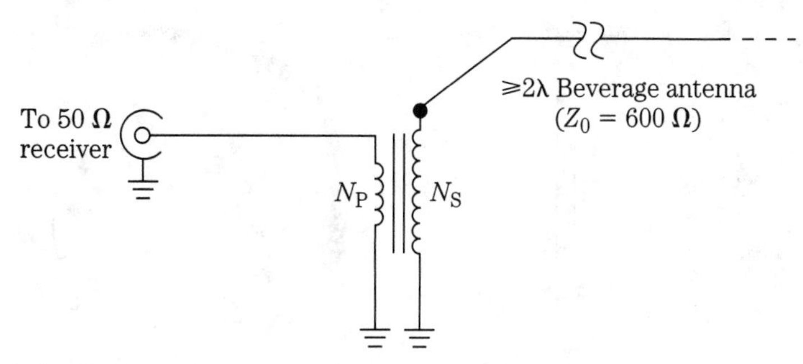

6-21 Use of a coupling transformer to impedance match a 600-ohm Beverage or long-wire antenna to a 50-ohm receiver.

$$\frac{N_s}{N_p} = \sqrt{\frac{600 \text{ ohms}}{50 \text{ ohms}}} = 3.46{:}1$$

The secondary requires an inductive reactance of 4 × 600 ohms, or 2400 ohms. To obtain this inductive reactance at the lowest frequency of operation requires an inductance of:

$$L_{\mu H} = \frac{2400 \text{ ohms} \times 10^6}{(2\pi)\ (530{,}000)} = 721\ \mu H$$

Checking a table of powdered iron toroid cores, it is found that the –15 (RED/WHT) mixture will operate over the 0.1 to 2 MHz region. Selecting a T-106-15 (RED/WHT) core gives us an A_L value of 345. The number of turns required to create an inductance of 721 μH is:

$$L_{\mu H} = 100\sqrt{\frac{721}{345}} = 145 \text{ turns}$$

the primary winding must have

$$N_p = \frac{145 \text{ turns}}{3.46} = 42 \text{ turns}$$

Winding the conventional RF transformer

When the windings of the conventional transformer are equal, i.e., where the turns ratio is 1:1, it is universal practice to wind the two coils in the bifilar manner discussed above. A special case of RF transformers called BALUN transformers (discussed below) uses this manner of winding exclusively. In cases where the windings are not equal, as is often the case in conventional transformers, there are three approaches to winding the coils. Figure 6-22A shows an RF transformer in which high impedance (hi-Z) and low impedance (low-Z) windings are used. The two different styles of winding the coils are shown in Figs. 6-22B and 6-22C. The method shown in Fig. 6-22B keeps the primary and secondary separated on the core. This method is

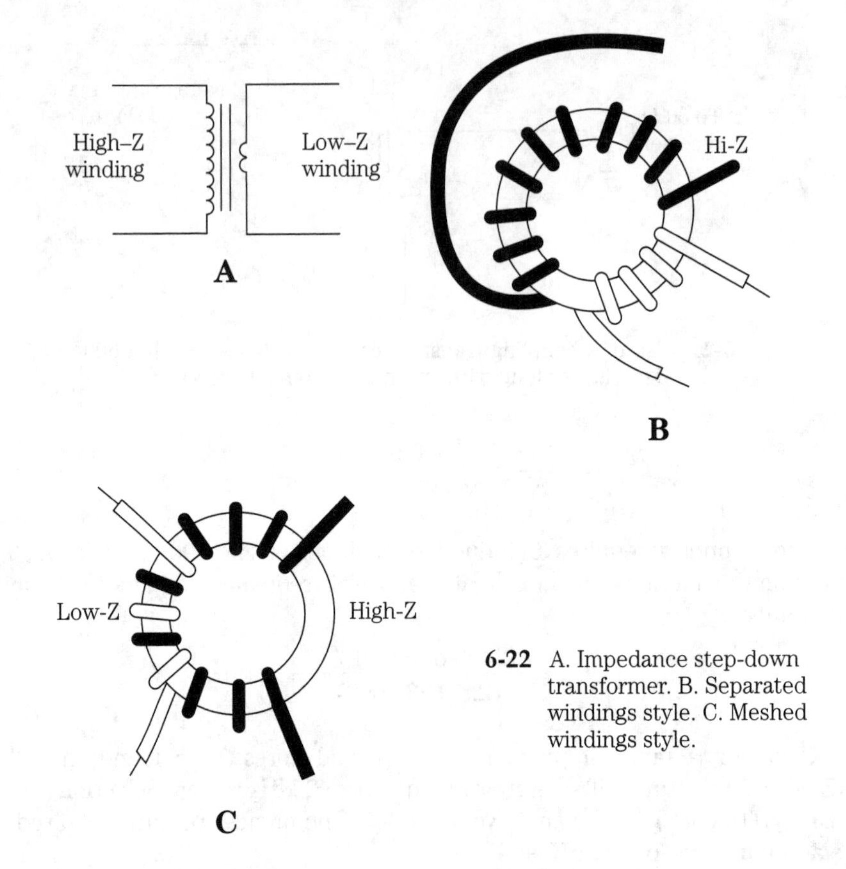

6-22 A. Impedance step-down transformer. B. Separated windings style. C. Meshed windings style.

suitable for use in narrow bandwidth applications, for example in the tuning circuit of a radio receiver.

The method in Fig. 6-22C intersperses the turns of the low-Z winding over or among the windings of the hi-Z winding. This method can be used for narrow band or relatively wideband applications. But if a transformer must be truly wideband, the best winding method is to wind the low-Z and hi-Z coils in the bifilar manner as far as is needed to accommodate the low-Z winding. Starting from one point on the core, the wires are kept bifilar until the low-Z coil is completed, and then monofilar the rest of the way until the hi-Z part is completed.

Connecting the conventional transformer in the circuit

A conventional RF transformer schematic symbol might include small dots, or some other device, to indicate the sense of the windings. They can also be used to determine the phasing of the signal transmitted through the transformer. In Fig. 6-23A the same ends of both windings are grounded, so the output signal is 180 degrees out of phase with the input signal. In Fig. 6-23B, on the other hand, the opposite ends of the two windings are grounded, so the output signal is in phase with the input signal.

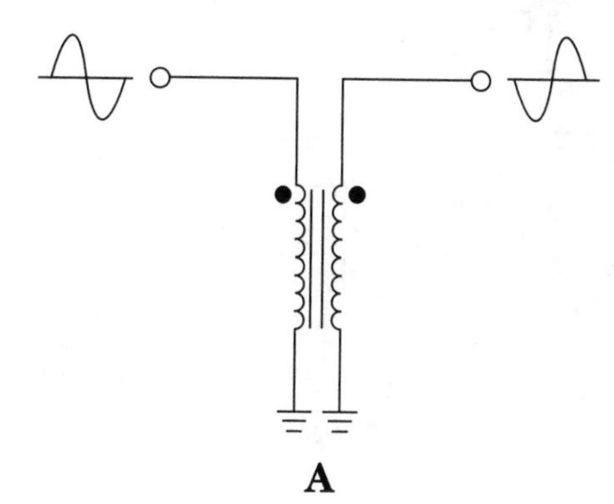

6-23 A. Phase reversal connection. B. In-phase connection.

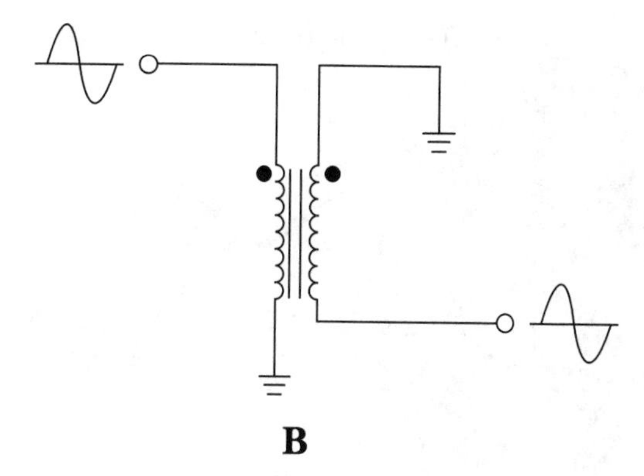

Autotransformers

An *autotransformer* differs from conventional transformers in that there is only one winding, which is tapped to provide the two impedance levels needed. Figure 6-24 shows the autotransformer in two different connection schemes. The connection scheme in Fig. 6-24A results in an in-phase output signal, while that of Fig. 6-24B produces an out-of-phase signal across the load.

Winding the autotransformer proceeds along the same lines as for a straight coil, except that the two sections of the winding are broken at a point to create the tap. Two methods can be used in doing this job. In one method, the entire winding is one continuous piece of wire. A small loop is made at the tap, and is made available to the rest of the circuit. The enameled insulation can be scraped away and the wire tinned with solder. The other method, as shown in Fig. 6-25, breaks the two sections into two discrete windings, A-B and B-C. The connection at the junction is soldered for electrical and mechanical integrity. It is very important that the two windings maintain the same sense. The A-B winding and B-C windings must be wound in the same direction. The starting turns of both sections in Fig. 6-25 start in the same direction, as is needed to maintain the sense of the coils.

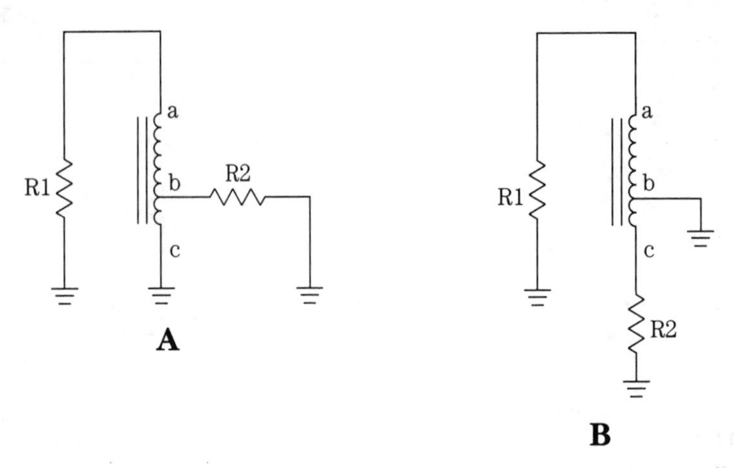

6-24 Autotransformer connections. A. In-phase. B. Out of phase.

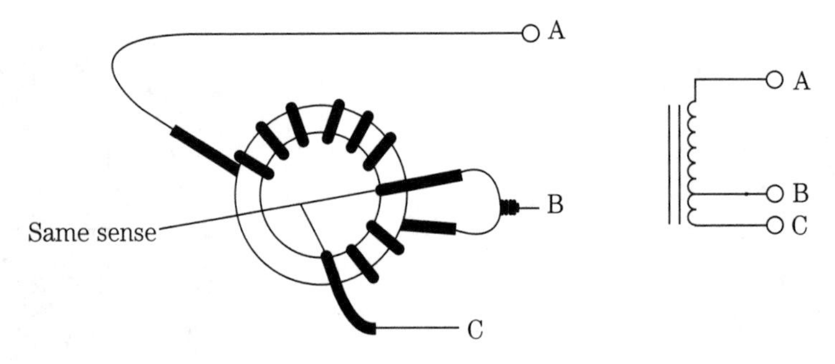

6-25 Creating a tap on a toroid winding.

BALUNs, BAL-BALs, and UN-UNs

A special category of RF transformers, sometimes called *transmission line transformers*, are available in several different configurations—depending on the type of load at each winding and the impedance ratio. The *BALUN* transformer gets its name from BALanced-UNbalanced, which describes the relationship between the source and load types. In the BALUN, one load will be unbalanced with respect to ground (e.g., a coaxial cable from a standard 50-ohm transmitter output), while the other will be balanced with respect to ground (e.g., a dipole antenna). Amateur radio operators and SWLs often use 1:1 impedance ratio BALUN transformers at the feedpoint of dipole and other balanced antennas because it ensures that the pattern is a more nearly ideal bidirectional "figure-8." Other common BALUN devices are available in 4:1 impedance ratios. These devices can be used to match the feedpoint impedances of high impedance antennas such as the G5RV, the folded dipole, or the long wire.

Figure 6-26 shows the two most common forms of voltage BALUN transformer. In the 1:1 impedance ratio version shown in Fig. 6-26A, there are three bifilar windings on the same core, while in the 4:1 impedance ratio version of Fig. 6-26B there are two bifilar windings. In both cases, the sense of the windings is very important, and must be scrupulously observed.

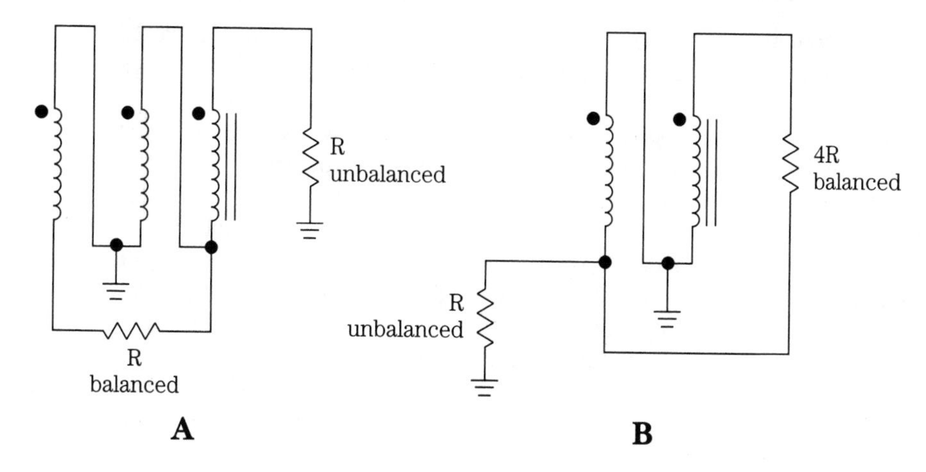

6-26 BALUN transformers. A. 1:1 impedance ratio. B. 4:1 impedance ratio.

A pair of RF transformers are shown in Fig. 6-27. Although the transformer of Fig. 6-27A is usually called a 1:1 BALUN transformer in the literature, it is not technically in that category. Instead, it is an RF isolation transformer. It serves the function of converting the balanced load to an unbalanced form that is compatible with the unbalanced input.

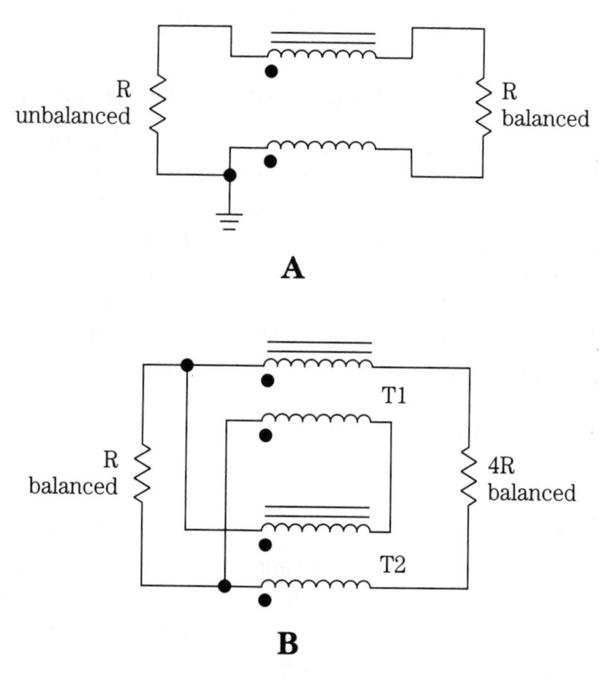

6-27 A. Alternate 1:1 BALUN. B. 4:1 BAL-BAL transformer.

The transformer shown in Fig. 6-27B is a *BAL-BAL* in that it has a balanced load at both ends. The impedance ratio of this transformer is 4:1. It can be used to convert high impedance antenna feedpoints to a lower impedance while retaining the balanced feature. It is also occasionally used in RF power amplifier circuits. This circuit actually consists of two transformers connected together.

The circuit shown in Fig. 6-28 is an *UN-UN* transformer, i.e., it has an unbalanced load at both ends. This device is actually a pair of 4:1 transformers in cascade, resulting in a 16:1 impedance ratio. One use for this transformer is to convert extremely low impedances to 50 ohms, as might be seen in RF power amplifiers or in vertical antennas in some installation situations. An example might be the 3- to 4-ohm base impedance in a bipolar transistor RF power amplifier circuit. In order to match the 50-ohm input impedance of the system, the 16:1 UN-UN transformer of Fig. 6-28 can be used.

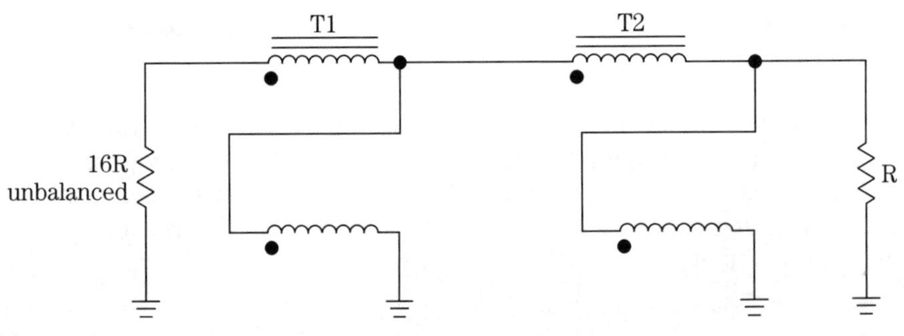

6-28 16:1 ratio UN-UN transformer.

Binocular core inductors and transformers

Recently I have been working with AM broadcast band (540 to 1700 kHz), LF, and VLF RF circuits—so needed some pretty large inductance values. For example, when the standard 365 pF capacitor is used to tune the AM BCB, an inductance of about 225 µH is used. At lower frequencies (LF and VLF), even higher inductances are needed. To achieve this inductance with Type 15 powdered iron material in the 0.44-inch toroid core (i.e., the T-44-15 RED/WHT core) the A_L is 160. This core would require 117 turns to achieve this inductance. That's a lot of winding. In fact, it may not even be possible to fit that many turns on a 0.44 inch core. When switching to ferrite, which tends to have higher A_L values than powdered iron, it is possible to get away with fewer turns. The 225-µH coil done on an FT-50-43 the A_L is 523, so the required number of turns is 21 . . . which is considerably easier to wind than 117 turns!

Binocular cores

The toroidal core has a certain charm because it is easy to use, is predictable, and is inherently self-shielding because of its geometry. But there is another core shape that offers very high inductance values in a small volume. The *binocular core* (Fig. 6-29) offers very high A_L values in small packages, so can create very high inductance values without being excessively large. A binocular core that uses Type 43 fer-

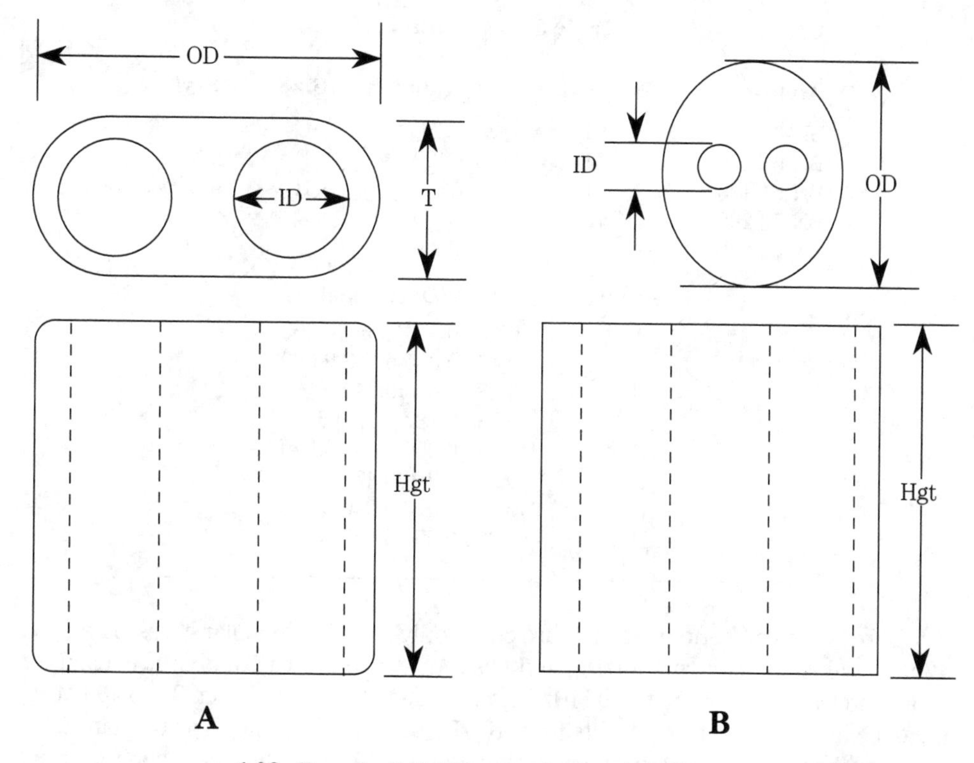

6-29 Bazooka BALUN cores. A. Type 1. B. Type 2.

rite, and is about the same weight and size as the T-50-43 (A_L = 523) has an A_L value of 2890. Only 8.8 turns are required to achieve 225 μH on this core.

Figure 6-29 shows two types of binocular core. The Type 1 binocular core is shown in Fig. 6-29A. It is larger than the Type 2 (Fig. 6-29B), and has larger holes. It can therefore be used for larger value inductors and transformers. The Type 2 core can be considered as a 2-hole ferrite bead.

Table 6-8 shows several popular-sized binocular cores and their associated A_L values. The center two digits of each part number is the type of ferrite material used to make the core (e.g., BN-xx-202), while the last digits refer to the size and style of the core.

Table 6-8

Part No.	Material	A_L value	Size	Style
BN-43-202	43	2890	A	1
BN-43-2302	43	680	B	1
BN-43-2402	43	1277	C	1
BN-43-3312	43	5400	D	1
BN-43-7051	43	6000	E	1
BN-61-202	61	425	A	1
BN-61-2302	61	100	B	1

Table 6-8 Continued

Part No.	Material	A_L value	Size	Style
BN-61-2402	61	280	C	1
BN-61-1702	61	420	F	2
BN-61-1802	61	310	H	2
BN-73-202	73	8500	A	1
BN-73-2402	73	3750	G	1

Size codes	OD/ID/hgt/thick
A	0.525/0.150/0.550/0.295
B	0.136/0.035/0.093/0.080
C	0.280/0.070/0.240/0.160
D	0.765/0.187/1.00/0.375
E	1.130/0.250/1.130/0.560
F	0.250/0.050/0.470/—
G	0.275/0.070/0.240/0.160
H	0.250/0.050/0.250/—

Three types of ferrite materials are commonly used in binocular cores. The Type 43 material is a nickel-zinc ferrite, and has a permeability (μ) of 850. It is used for wideband transformers up to 50 MHz, and has high attenuation from 30 to 400 MHz. It can be used in tuned RF circuits from 10 kHz to 1000 kHz. The Type 61 material is also nickel-zinc, and has a permeability of 125. It offers moderate-to-good thermal stability, and a high "Q" over the range of 200 kHz to 15 MHz. It can be used for wideband transformers up to 200 MHz. The Type 73 material has a permeability of 2500, and offers high attenuation from 500 kHz to 50 MHz.

The binocular core can be used for a variety of RF inductor devices. Besides the single, fixed inductor, it is also possible to wind conventional transformers and BALUN transformers of various types on the core.

Figure 6-30 shows Type 1 binocular cores wound in various ways. The normal manner of winding the turns of the inductor is shown in Fig. 6-30A; the wire is passed from hole-to-hole around the central wall between the holes. The published A_L values for each core are based on this style of winding, and it is the most commonly used.

An edge-wound coil is shown in Fig. 6-30B. In this coil, the turns are wound around the outside of the binocular core. To check the difference, I wound a pair of BN-43-202 cores with ten turns of #26 wire; one in the center (Fig. 6-30A) and one around the edge (Fig. 6-30B). The center-wound version produced 326 μH of inductance, while the edge wound produced 276 μH with the same number of turns.

Counting the turns on a binocular core is a little different than you might expect. A single "U" shaped loop that enters and exits the core on the same side (Fig. 6-30C) counts as one turn. When the wire is looped back through a second time (Fig. 6-30D) there are two turns.

Winding the binocular core

Some people think that it is easier to use these cores than toroids, and after spending a rainy weekend winding LF and AM BCB coils (after pumping ground water out

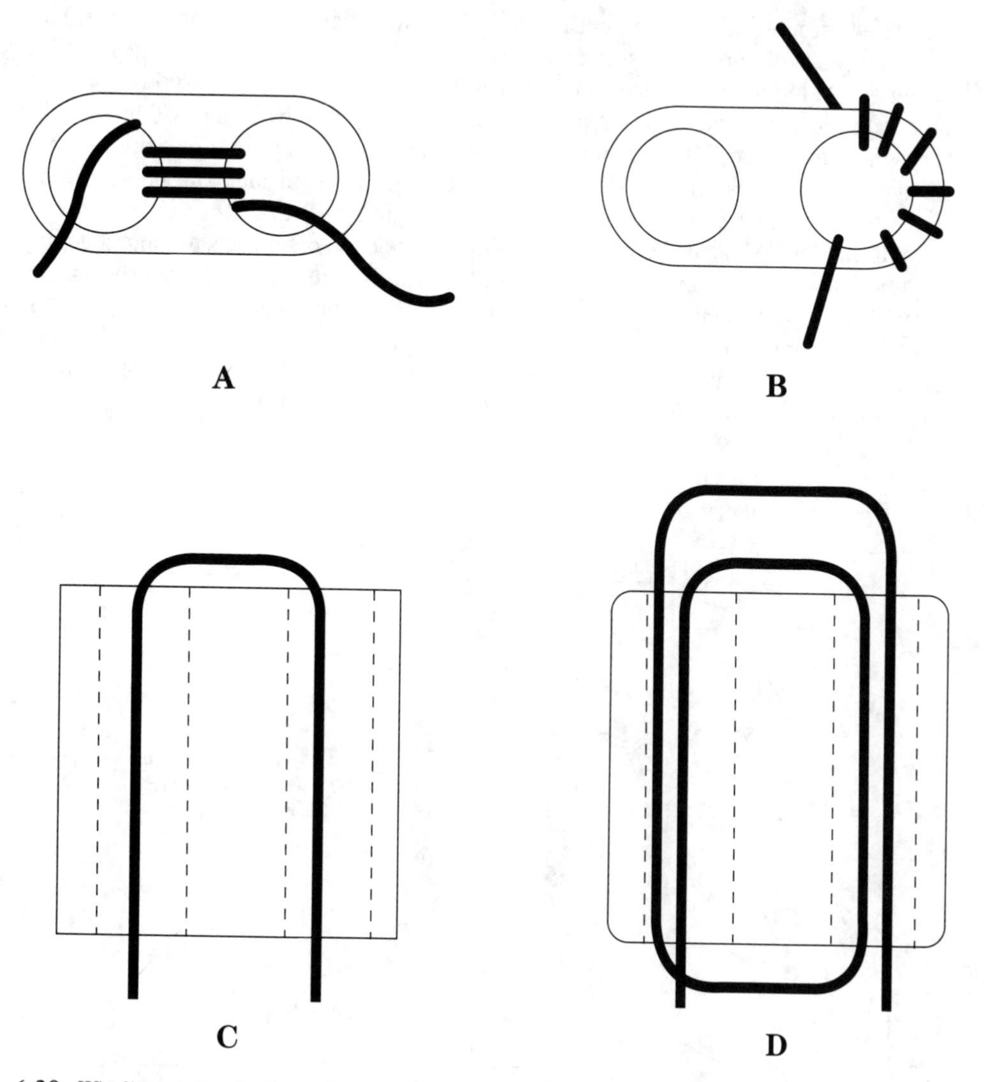

6-30 Winding styles for bazooka transformers. A. Through the center. B. Around the edge (less predictable inductance). C. Single-turn winding (no doubling back). D. Two-turn winding.

of the basement workshop!) I am inclined to agree—partially. The "partially" means that they are easier to work than toroids if you do it correctly. It took me some experimenting to figure out a better way than holding the core in one hand, the existing wires already on the core in another hand, and then wind the remaining coils with a third hand. Not being a Martian, I don't have three hands, and my "Third Hand" bench tool didn't seem to offer much help. Its alligator clip jaws were too coarse for the #36 enameled wire that I was using for the windings. So enter a little "mother of invention" ingenuity (it's amazing how breaking a few wires can focus one's attention on the problem). In fact, I came up with two related methods between gurgles of my portable sump pump.

The first method is shown in Fig. 6-31. The binocular core is temporarily affixed to a stiff piece of cardboard stock such as a 5 × 7 inch card, or a piece cut from the stiffener used in men's shirts at the laundry. The cardboard is taped to the work surface, and the core is taped to the cardboard. One end of the wire that will be used for the winding is taped to the cardboard with enough leader to permit working the end of the coil once it is finished (2-3 inches). Pass the wire through the holes enough times to make the coil needed, and then anchor the free end to the cardboard with tape. If the device has more than one winding, make each one in this manner, keeping the ends taped down as you go. Once all of the windings are in place, seal the assembly with Q-Dope or some other sealant (RTV silicone, rubber cement, etc.). Q-Dope is intended for inductors, and can be purchased from G-C dealers, or by mail from: Ocean State Electronics, P.O. Box 1458, 6 Industrial Drive, Westerly, RI 02891; Phone (800) 866-6626 (orders only), (401) 596-3080 (voice) or (401) 596-3590 (fax).

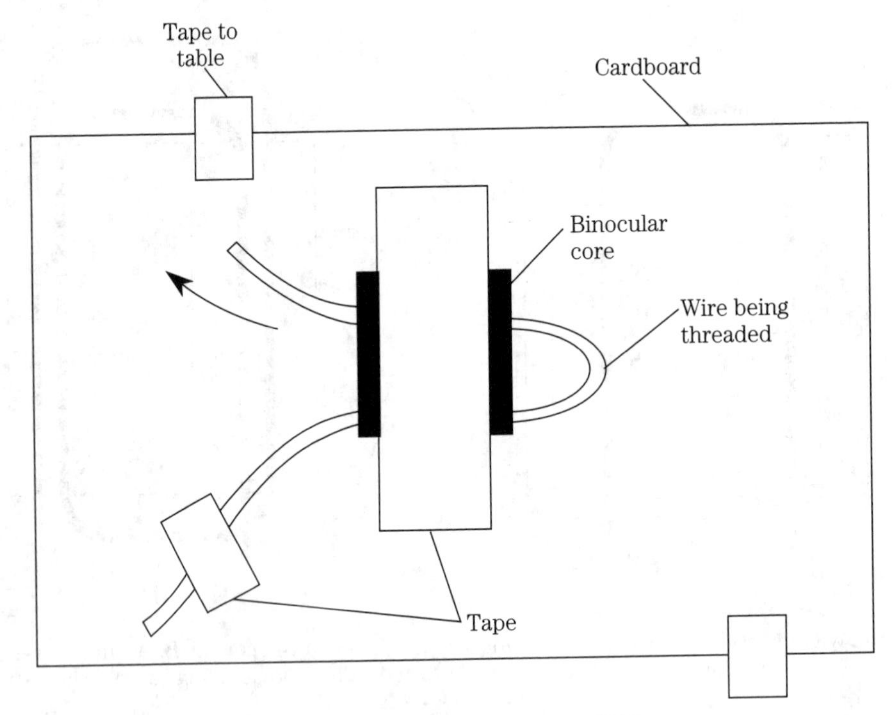

6-31 Winding "jig" for the bazooka BALUN core.

The second method involves making a header for the binocular core. This header can be permanent, and can be installed into the circuit just like any other coil with a header. When built correctly, the header will be spaced on 0.100-inch centers, so is compatible with DIP printed circuit boards and perforated wiring board. I used perforated wiring board of the sort that has printed circuit pads (none of which connected to each other) at each hole.

Figure 6-32 shows the basic configuration for my homebrew header (a DIP header can also be used). These connectors are intended to connect wiring or other

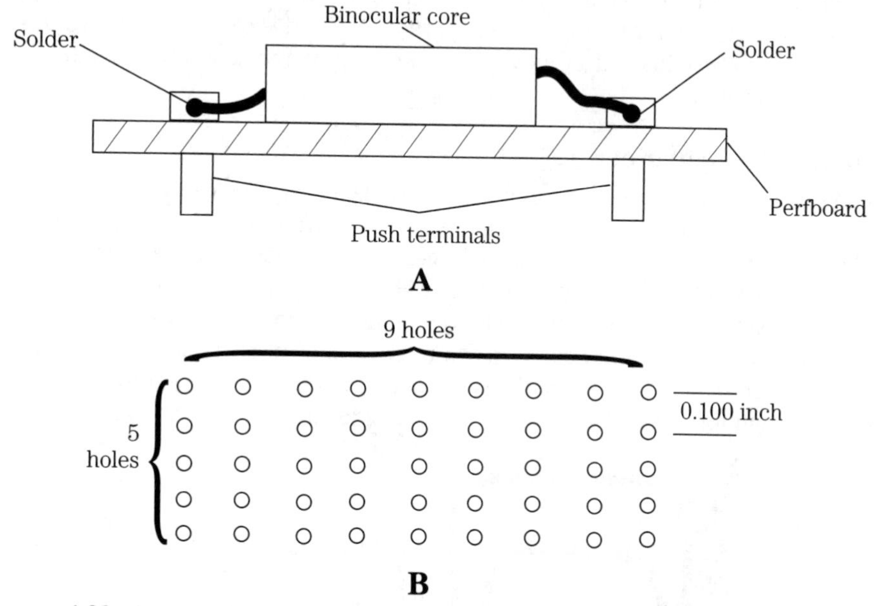

6-32 Construction of a perfboard header or carrier for the bazooka BALUN.

components to a DIP printed circuit board designed for digital integrated circuits. I found that a small segment of printed perfboard, 0.100 inch centers on the holes, that contained five rows by nine columns of holes (see Fig. 6-32B) was sufficient for the 0.525 × 0.550 inch BN-xx-202. Larger or smaller hole matrices can be cut for larger or smaller binocular cores.

The connections to the header are perfboard push terminals (available any place that perfboard and printed circuit making supplies are sold). I used the type of perfboard that has solder terminals so that the push terminals can be held to the board with solder. Otherwise, they have a distinct tendency to back out of the board with handling.

When the header is finished, the binocular core is fastened to the top surface of the header with tape, and then the pins of the header are pushed into a large piece of perfboard. This step is done to stabilize the assembly on the work surface. It might be a good idea to stabilize the perfboard to the table with tape to keep it from moving about as you wind the coils.

Once the header and core are prepared, then it is time to make the windings. Scrape the insulation off one end of the wire for about ¼ inch. An X-acto knife, scalpel or similar tool can be used to do this job. Turn the wire over several times to make sure that the enamel insulation is scraped away around the entire circumference. Some people prefer to burn the insulation off with a soldering iron, which also serves to tin the end of the wire as it burns the insulation away. I've found that method to be successful when the smaller gauges are used, but when good-quality #26 or larger wire is used, the scraping method seems to work better. If the scraping method is used, follow the scraping by tinning the exposed end of the wire with solder. Each winding of the transformer can be made by threading the wire through the core as needed. As each winding is finished, the loose end is cleaned, tinned and soldered to its push terminal. After all windings are completed, seal the assembly with Q-Dope or equivalent.

Homebrew binocular cores

You can build custom "binocular cores" from toroidal cores. The toroids are easily obtained from many sources, and are available in many different mixtures of both powdered iron and ferrite. You can also make larger binocular cores using toroids because of the wide range of toroid sizes. Binocular cores are available in a limited range of mixtures and sizes. Figure 6-33 shows the common way to make your own binocular core: stack a number of toroid cores in the manner shown. It is common practice to wrap each stack in tape, then place the two stacks together and wrap the assembly together. Although four toroids are shown on each side, any number can be used.

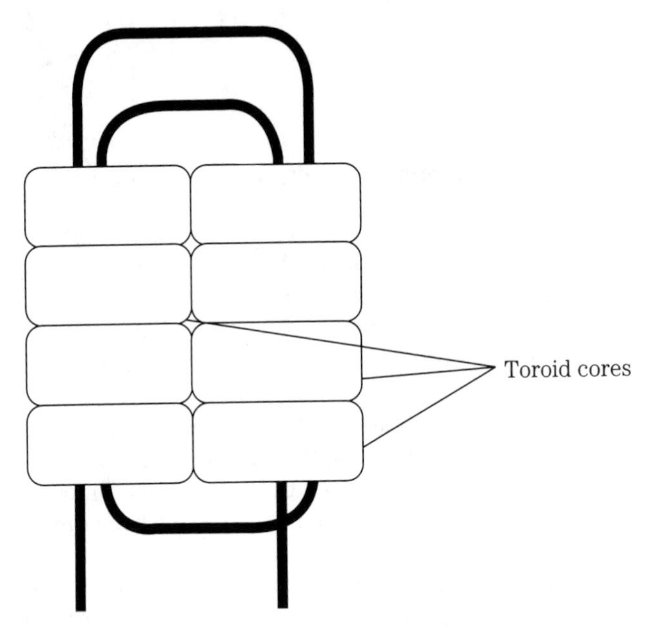

6-33 Stacking toroid cores to make a higher-power bazooka core.

A variation on the theme is shown in Fig.6-34. This binocular core is designed to have a single-turn winding consisting of a pair of brass tubes passed through the center holes of the toroid stacks. The ends of the stacks are held together with a pair of copper-clad printed circuit boards. The rear panel has no copper removed, while the front panel is etched to isolate the two brass tubes. The pads around the brass tubes at the front end are used to make connections to the tubing (which serves as a single turn winding).

The other winding of the transformer is made of ordinary insulated wire, which is passed through the brass tubes the correct number of turns to achieve the desired turns ratio. This type of binocular core was once popular with ham op-

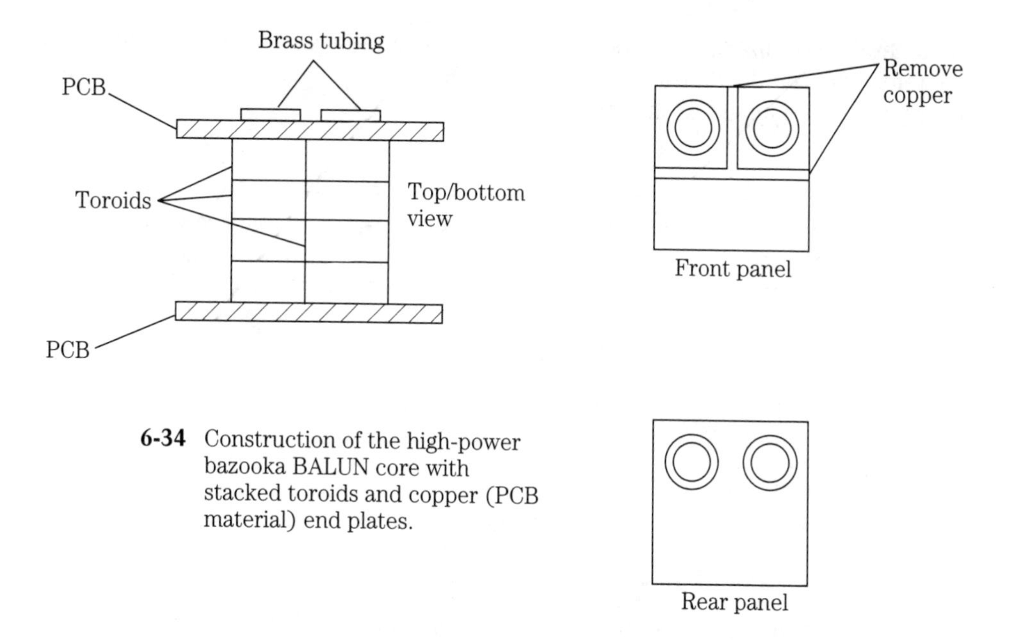

6-34 Construction of the high-power bazooka BALUN core with stacked toroids and copper (PCB material) end plates.

erators who built their own solid-state RF power amplifiers. The high-power transformers needed to match the impedances of the base and collector terminals of the RF transistors were not easily available on the market, so they had to "roll their own."

The binocular core is not as well-known as the toroid core, but for many applications, it is the core of choice. This is especially true when low frequencies are used, or when large inductances are needed in a small package . . . and you don't want to work your arm off hand winding the large number of turns that would be required on a solenoid-wound or toroidal core.

Ferrite and powdered iron rods

Ferrite rods are used in low frequency medium wave, AM broadcast band, and LF/VLF receivers to form "loopstick" antennas. These antennas are also used in radio direction finders because they possess a *figure-8* reception pattern that counterpoises two deep nulls with a pair of main lobes. Ferrite rod inductors can also be used in any application, up to about 10 MHz, where a high inductance is needed.

Two permeability figures are associated with the ferrite rod, but only one is easily available (see Table 6-9): the *initial permeability*. It is this figure that is used in the equations for inductance. The *effective permeability* is a little harder to pin down, and is dependent on such factors as:

- The length/diameter ratio of the rod
- Location of the coil on the rod (centered is most predictable)
- The spacing between turns of wire
- The amount of air space between the wire and the rod

In general, maximizing effective A_L and inductance requires placing the coil in the center of the rod. The best Q, on the other hand, is achieved when the coil runs nearly the entire length of the rod.

Table 6-9

Part No.	Permeability	Approx. A_L[1]	Ampere-turns
R61-025-400	125	26	110
R61-033-400	125	32	185
R61-050-400	125	43	575
R61-050-750	125	49	260
R33-037-400	800	62	290
R33-050-200	800	51	465
R33-050-400	800	59	300
R33-050-750	800	70	200

[1] Approximate value for coil centered on rod, covering nearly the entire length, made of #22 wire. Actual A_L may vary with situation.

Several common ferrite rods available from *Amidon Associates* are shown in Table 6-9. The "R" indicates "rod," while the number associated with the "R" (e.g., R61) indicates the type of ferrite material used in the rod. The following numbers (e.g., -025-) denote the diameter (025 = 0.25 inch, 033 = 0.33 inch and 050 = 0.50 inch); the length of the rod is given by the last three digits (200 = 2 inches, 400 = 4 inches and 750 = 7.5 inches). Type 61 material is used from 0.2 to 10 MHz, while the Type 33 material is used in VLF applications.

In some cases, the A_L rating of the rod will be known, while in others only the permeability (μ) is known. If either is known we can calculate the inductance produced by any given number of turns.

For the case where the A_L is known:

$$L_{\mu H} = N_p^2 A_L \times 10^{-4} \qquad [6\text{-}22]$$

For the case where μ is known:

$$L_{\mu H} = (4 \times 10^{-9})\,\pi N_p^2 \mu \left(\frac{A_e(\mathrm{cm}^2)}{l_e(\mathrm{cm})} \right) \qquad [6\text{-}23]$$

where

N_p is the number of turns of wire
μ is the core permeability
A_e is the cross-sectional area of the core (cm^2)
l_e is the length of the rod's flux path (cm)

Example

Find the number of turns required on an R61-050-750 ferrite rod to make a 220 μH inductor for use in the AM BCB with a 10-365 pF variable capacitor.

Solution

Solving Eq. 6-22 for N:

$$N = \sqrt{\frac{L_{\mu H}}{A_L \times 10^{-4}}}$$

$$N = \sqrt{\frac{220_{\mu H}}{(43)\ (10^{-4})}} = 226$$

Figure 6-35 shows several popular ways for ferrite rod inductors to be wound for service as loopstick antennas in radio receivers. Figure 6-35A shows a transformer circuit in which the main tuned winding is broken into two halves, A and B. These windings are at the end regions of the rod, but are connected together in series at the center. The main coils are resonated by a dual capacitor, C1A and C1B. A coupling winding, of fewer turns, is placed at the center of the rod, and is connected to the coaxial cable going to the receiver. In some cases,the small coil is also tuned, but usually with a series capacitor (see Fig. 6-35B). Capacitor C2 is usually a much larger value than C1 because the coupling winding has so many fewer turns than the main tuning windings. The antenna in Fig. 6-35C is a little different: The two halves of the tuning winding are connected together at the center, and connected directly to the

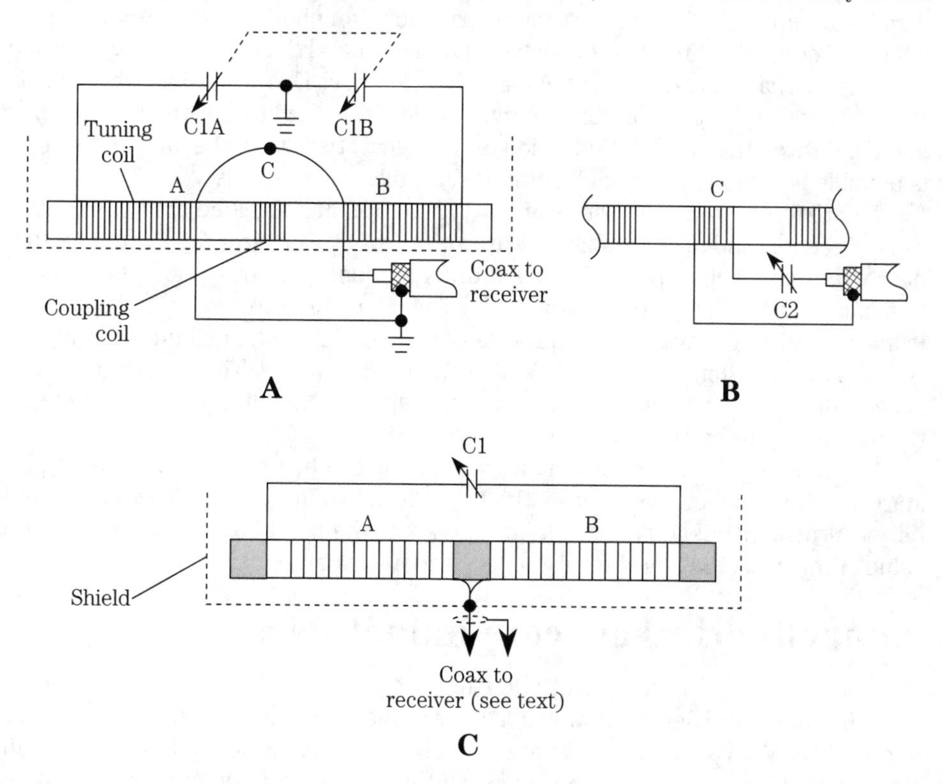

6-35 Loopstick antenna. A. Overall circuit for double-tuned winding. B. Tuning the coupling loop. C. Single-tuned variety.

center conductor of the coaxial cable, or to a single downlead. A single capacitor is used to resonate the entire winding (both halves).

An RF choke wound on a ferrite rod is shown in Fig. 6-36; the circuit is shown in Fig. 6-36A, while the actual choke is shown in Fig. 6-36B.

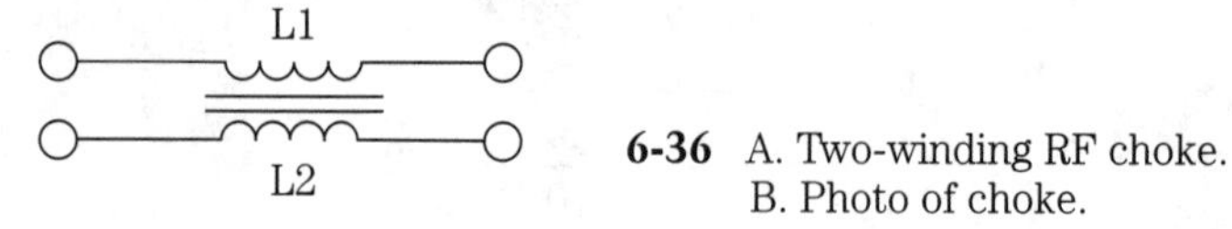

L1 and L2 are bifilar wound

A

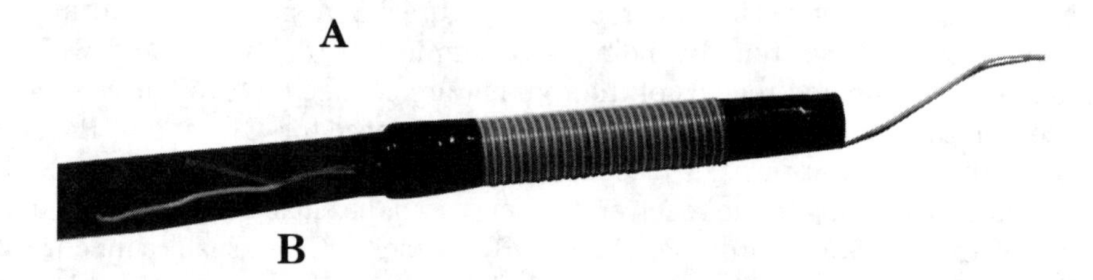

B

6-36 A. Two-winding RF choke.
B. Photo of choke.

Project 6-1

A radio direction-finding antenna can be used for a number of purposes, only one of which is finding the direction from which a radio signal arrives. Another use is in suppressing co-channel and adjacent channel interference. This is possible when the desired station is in a direction close to right angles from the line between the receiver and the desired transmitter. Reduction of the signal strength of the interfering signal is possible because the loopstick antenna has nulls off both ends.

Figure 6-37 shows a loopstick antenna mounted in a shielded compartment for radio direction finding. The shield is used to prevent electrical field coupling from nearby sources such as power lines and other stations, yet doesn't affect the reception of the magnetic field of radio stations. The aluminum can be one-half of an electronic hobbyist's utility box, of appropriate dimensions, or can be built custom from Harry & Harriet Homeowner Do-It-Yourself hardware stores. The loopstick antenna is mounted by nonmetallic cable ties to nylon spacers that are, in turn, fastened to the aluminum surface with nylon hardware.

The number of turns required for the winding can be found experimentally, but start with the number called for by the formula shown in Eq. 6-22. The actual number of turns depends in part on the frequency of the band being received, and the value of the capacitors used to resonate the loopstick antenna.

Noncylindrical air-core inductors

Most inductors used in radio and other RF circuits are either toroidal or solenoid-wound cylindrical. There is, however, a class of inductors that are neither solenoidal nor toroidal. Many *loop antennas* are actually inductors fashioned into either triangle, square, hexagon or octagon shapes. Of these, the most common is the square-wound loop coil (Fig. 6-38). Two basic forms are recognized: *flat-wound* (Fig.

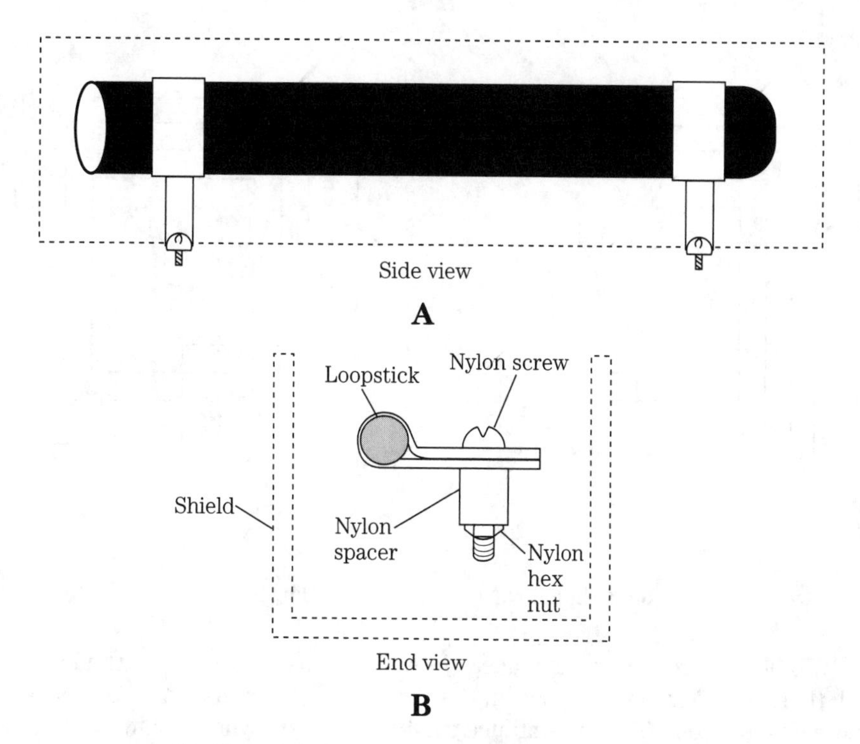

6-37 Mounting the loopstick in a shielded enclosure.

6-38A) and *depth-wound* (Fig. 6-38B). The equation for the inductance of these shaped coils is a bit difficult to calculate, but the equation provided in 1946 by F.W. Grover of the United States National Bureau of Standards (since renamed National Institute of Standards and Technology) is workable:[1]

$$L_{\mu H} = K_1 N^2 A \left(Ln\left(\frac{K_2 AN}{(N+1)B} \right) + K_3 + \left(\frac{K_4 (N+1)B}{AN} \right) \right) \qquad [6\text{-}24]$$

where

A is the length of each side in centimeters (cm)
B is the width of the coil in centimeters (cm)
N is the number of turns in the coil (close wound)
K_1, K_2, K_3 and K_4 are found in Table 6-10

Table 6-10. "K" values for loop antennas

Shape	K1	K2	K3	K4
Triangle	0.006	1.1547	0.6553	0.1348
Square	0.006	1.4142	0.3794	0.3333
Hexagon	0.012	2.00	0.6553	0.1348
Octagon	0.016	2.613	0.7517	0.0715

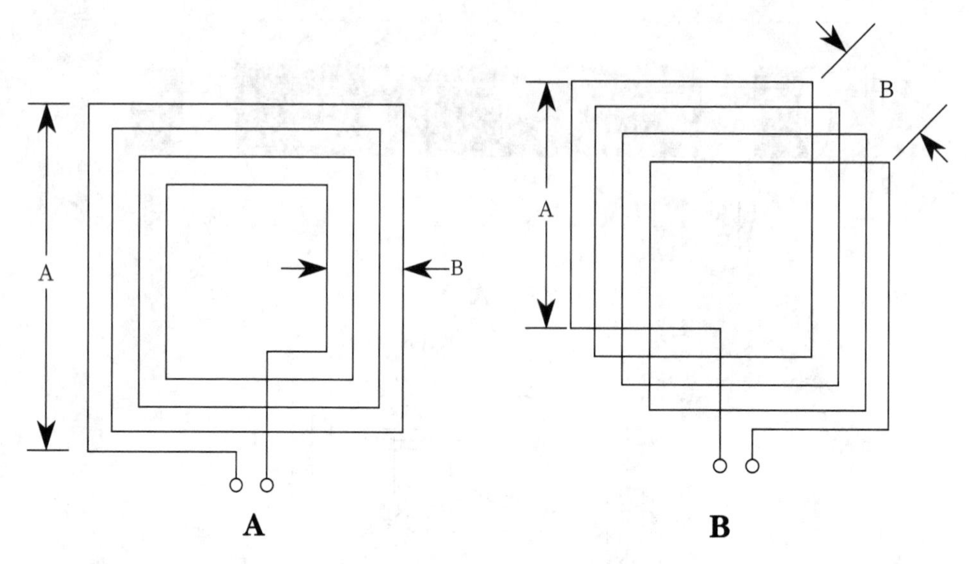

6-38 Inductive loop direction-finding antenna.

Whenever conductors are placed side-by-side, as in the case of the loop-wound coil, there is a capacitance between the conductors (even if formed by a single loop of wire, as in a coil). This capacitance can be significant when dealing with radio circuits. The estimate of distributed loop capacitance (in pF) for square loops is given by Bramslev as about 60 A, where A is expressed in meters (m).[2] The distributed capacitance must be accounted for in making calculations of resonance. Subtract the distributed capacitance from the total capacitance required in order to find the value of the capacitor required to resonate the loop-wound coil.

References

1. F.W. Grover, *Inductance Calculation—Working Formulas and Tables*, D. Van-Nostrand Co., Inc. (New York, 1946).
2. G. Bramslev, "Loop Aerial Reception," *Wireless World*, Nov. 1952, pp 469–472.

7

Inductor-capacitor (LC) resonant tank circuits

When you use an inductor (L) and a capacitor (C) together in the same circuit, the combination forms an *LC resonant circuit*, sometimes called a tank circuit or resonant tank circuit. These circuits are used to select one frequency, while rejecting all others (as in to tune a radio receiver). There are two basic forms of LC resonant-tank circuit: series (Fig. 7-1A) and parallel (Fig. 7-1B). These circuits have much in common, and much that makes them fundamentally different from each other.

The condition of *resonance* occurs when the capacitive reactance (X_c) and inductive reactance (X_L) are equal in magnitude ($|+X_L| = |-X_c|$). As a result, the resonant tank circuit shows up as purely resistive at the resonant frequency (Fig. 7-1C), and as a complex impedance at other frequencies. The LC resonant tank circuit operates by an oscillatory exchange of energy between the magnetic field of the inductor, and the electrostatic field of the capacitor, with a current between them carrying the charge.

Because the two reactances are both frequency-dependent, and because they are inverse to each other, the resonance occurs at only one frequency (f_r). We can calculate the standard resonance frequency by setting the two reactances equal to each other and solving for f. The result is:

$$F = \frac{1}{2\pi\sqrt{LC}} \qquad \textbf{[7-1]}$$

Series-resonant circuits

The *series-resonant circuit* (Fig. 7-1A), like other series circuits, is arranged so that the terminal current (I) from the source (V) flows equally in both components. The vector diagrams of Fig. 7-2A through 7-2C show the situation under three different conditions.

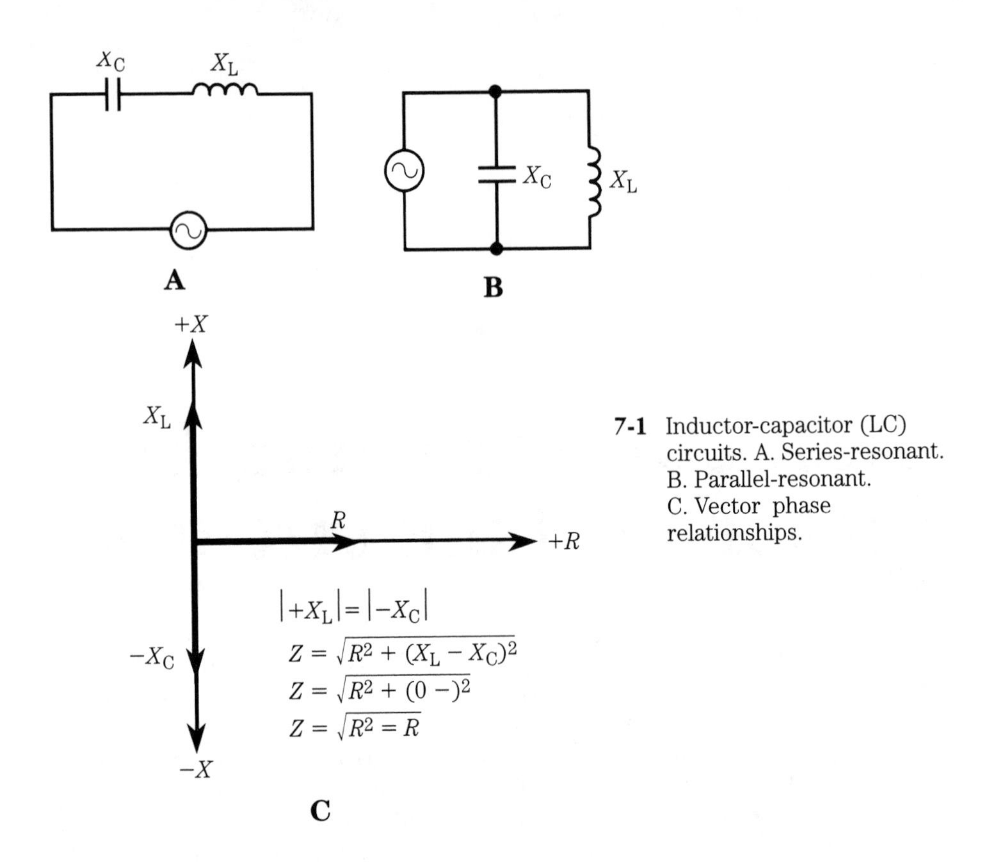

7-1 Inductor-capacitor (LC) circuits. A. Series-resonant. B. Parallel-resonant. C. Vector phase relationships.

In Fig. 7-2A, the inductive reactance is larger than the capacitive reactance, so the excitation frequency is greater than f_r. Note that the voltage drop across the inductor is greater than that across the capacitor, so the total circuit looks like it contains a small inductive reactance. In Fig. 7-2B, the situation is reversed: the excitation frequency is less than the resonant frequency, so the circuit looks slightly capacitive to the outside world. Finally, in Fig. 7-2C, the excitation frequency is at the resonant frequency, so $X_c = X_L$ and the voltage drops across the two components are equal but of opposite phase.

In a circuit that contains a resistance, an inductive reactance and a capacitive reactance, there are three vectors to consider (Fig. 7-3), plus a resultant vector. As in the other circuit, the north direction represents X_L, the south direction represents X_c, and the east direction represents *R*. Using the parallelogram method, we first construct a resultant for the *R* and X_c, which is shown as vector *A*. Next, we construct the same kind of vector (*B*) for *R* and X_c. The resultant (*C*) is made using the parallelogram method on *A* and *B*. Vector *C* represents the impedance of the circuit: the magnitude is represented by the length, and the phase angle by the angle between *C* and *R*.

Figure 7-4A shows a series-resonant LC tank circuit, and Fig. 7-4B shows the current and impedance as a function of frequency. The series-resonant circuit has a low impedance at its resonant frequency, and a high impedance at all other frequencies. As a result, the line current (*I*) from the source is maximum at the resonant frequency and the voltage across the source is minimum.

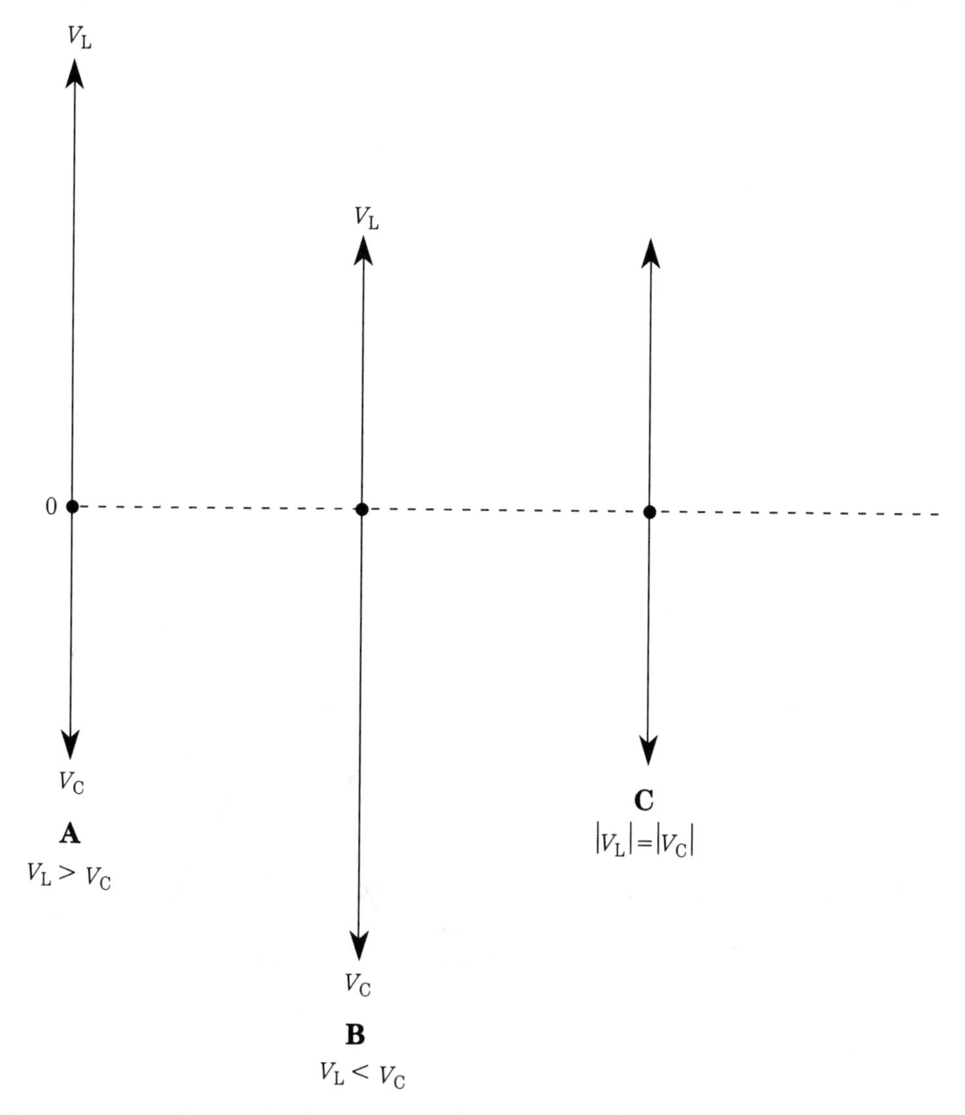

7-2 Phase relationships for: A. Inductive LC circuits. B. Capacitive LC circuits. C. Resonant LC circuits.

Parallel-resonant circuits

The *parallel-resonant tank circuit* (Fig. 7-5A) is the inverse of the series resonant circuit. The line current (I) from the source splits and flows in inductor and capacitor separately. The parallel-resonant circuit has its highest impedance at the resonant frequency, and a low impedance at all other frequencies (Fig. 7-5B). Thus, the line current from the source is minimum at the resonant frequency (Fig. 7-5B), and the voltage across the LC tank circuit is maximum. This fact is important in radio tuning circuits, as you will see in due course.

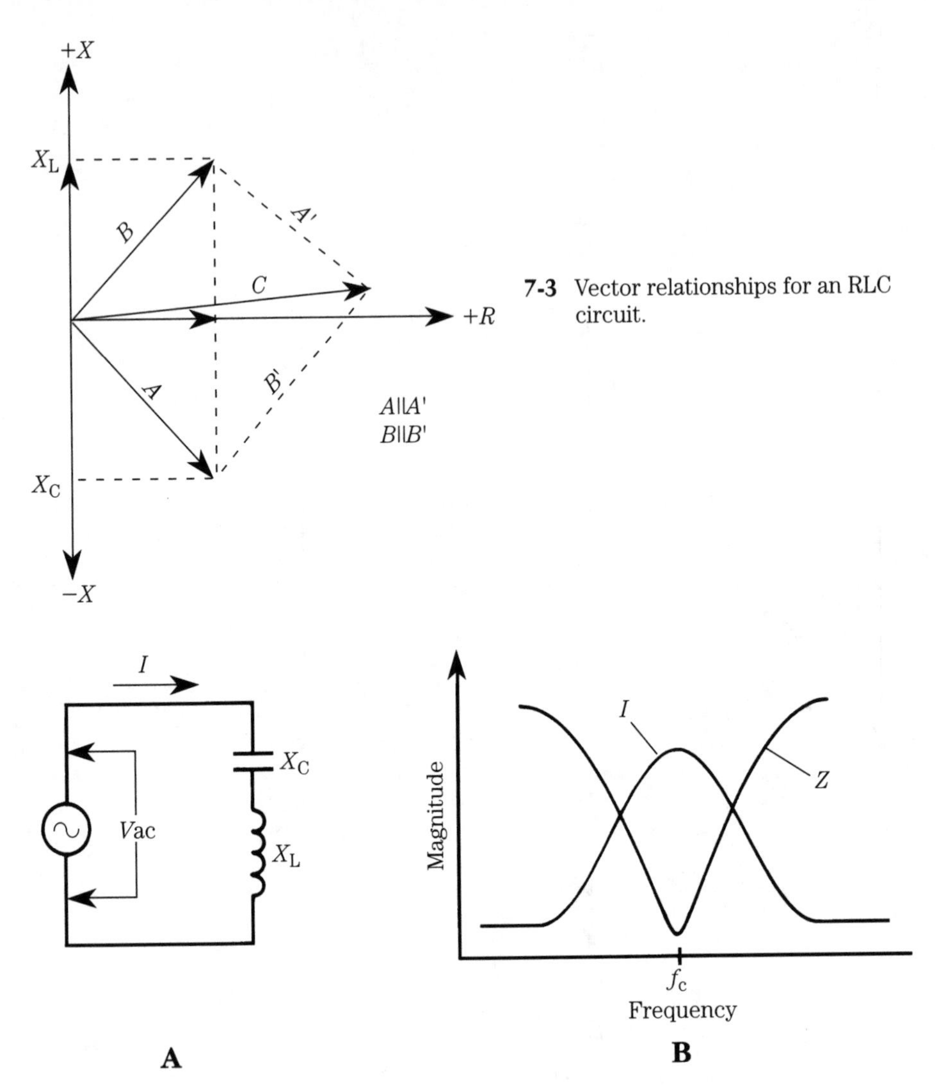

7-3 Vector relationships for an RLC circuit.

7-4 A. Series-resonant LC circuit. B. I and Z vs. frequency response.

Tuned RF/IF transformers

Many of the resonant circuits used in RF circuits, and especially in radio receivers, are actually transformers that couple signal from one stage to another. Figure 7-6 shows several popular forms of tuned, or *coupled*, RF/IF tank circuits. In Fig. 7-6A, one winding is tuned while the other is untuned. In the configurations shown, the untuned winding is the secondary of the transformer. This type of circuit is often used in transistor and other solid-state circuits, or when the transformer has to drive either a crystal or mechanical bandpass filter circuit. In the reverse configuration (L1 = output, L2 = input), the same circuit is used for the antenna-coupling network, or as the interstage transformer between RF amplifiers in TRF radios.

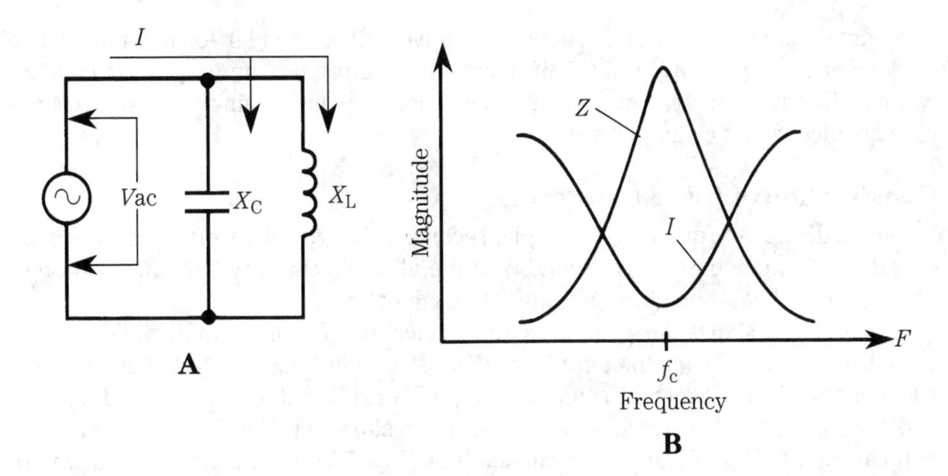

7-5 A. Parallel-resonant LC circuit. B. I and Z vs. frequency response.

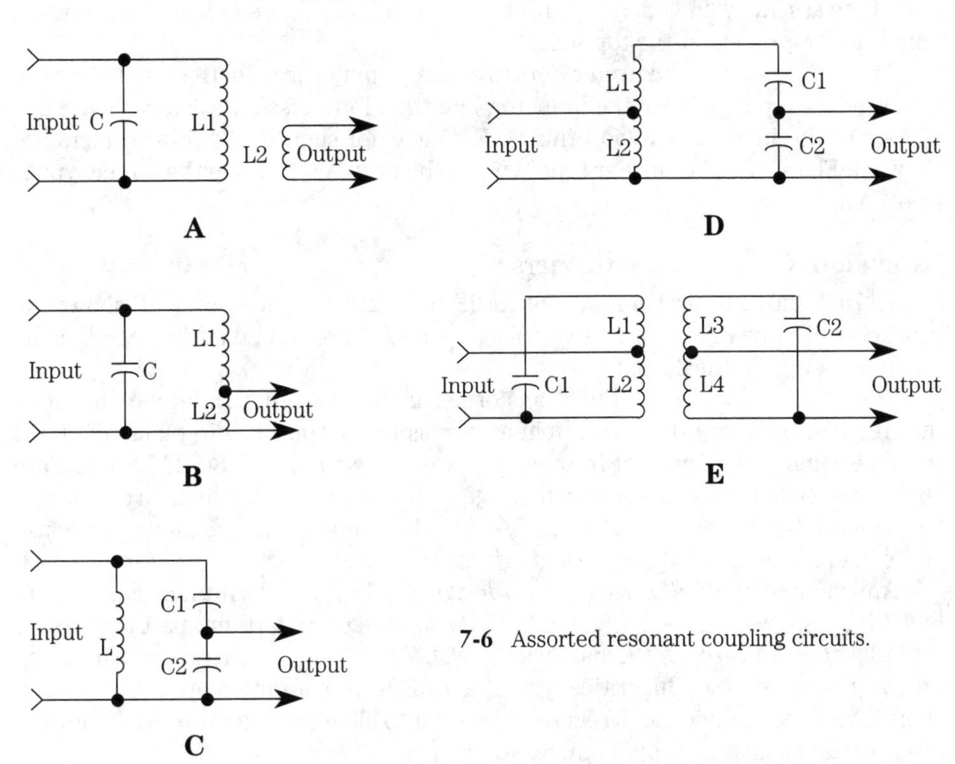

7-6 Assorted resonant coupling circuits.

The circuit in Fig. 7-6B is a parallel-resonant LC tank circuit that is equipped with a low-impedance tap on the inductor. This type of circuit is often used to drive a crystal detector or other low impedance load. Another circuit for driving a low-impedance load is shown in Fig. 7-6C. This circuit splits the capacitance that resonates the coil into two series capacitors. As a result, we have a capacitive voltage divider. The circuit in Fig. 7-6D uses a tapped inductor for matching low-impedance sources (e.g., antenna circuits), and a tapped capacitive voltage divider for low-impedance loads. Finally, the circuit in Fig. 7-6E uses a tapped primary and tapped secondary winding in order to match two low-impedance loads, while retaining the sharp band-pass characteristics of the tank circuit.

Construction of RF/IF transformers

Tuned RF/IF transformers built for radio receivers are typically wound on a common cylindrical form, and are surrounded by a metal shield that prevents interaction of fields of coils that are in close proximity to each other.

Figure 7-7A shows the schematic for a typical RF/IF transformer, while the sectioned view (Fig. 7-7B) shows one form of construction. This method of building the transformers was common at the beginning of World War II, and continued into the early transistor era. The methods of construction shown in Figs. 7-7C and 7-7D were popular prior to World War II. The capacitors in Fig. 7-7B were built into the base of the transformer, while the tuning slugs were accessed from holes in the top and bottom of the assembly. In general, you can expect to find the secondary at the bottom hole, and the primary at the top hole.

The term *universal wound* refers to a cross-winding system that minimizes the interwinding capacitance of the inductor, and therefore raises the self-resonant frequency of the inductor (a good thing). Examples of such RF/IF transformers are shown in Fig. 7-7E. The smaller type tends to be post-WWII, while the larger type is pre-WWII.

Bandwidth of RF/IF transformers

Figure 7-8A shows a parallel-resonant RF/IF transformer, while Fig. 7-8B shows the usual construction in which the two coils (L_1 and L_2) are wound at distance D apart on a common cylindrical form.

The *bandwidth* of the RF/IF transformer is the frequency difference between the frequencies where the signal voltage across the output winding falls off –6 dB from the value at the resonant frequency (f_r), as shown in Fig. 7-8C. If F_1 and F_2 are –6 dB (also called the –3 dB point when signal *power* is measured instead of voltage) frequencies, the bandwidth (BW) is $F_2 - F_1$. The shape of the frequency response curve in Fig. 7-8C is said to represent critical coupling.

An example of a *subcritical* or *undercoupled* RF/IF transformer is shown in Fig. 7-9. As shown in Figs. 7-9A and 7-9B, the windings are farther apart than in the critically coupled case, so the bandwidth (Fig. 7-9C) is much narrower than in the critically coupled case. Subcritically coupled RF/IF transformers are often used in shortwave or communications receivers in order to allow the narrower bandwidth to discriminate against adjacent channel stations.

The *overcritically coupled* RF/IF transformer is shown in Fig. 7-10. Here we note in Figs. 7-10A and 7-10B that the windings are closer together, so the bandwidth (Fig. 7-10C) is much broader. In some radio schematics and service manuals (not to mention early textbooks), this form of coupling was sometimes called "high fidelity" coupling because it allowed more of the sidebands of the signal (which carry the audio modulation) to pass with less distortion of frequency response.

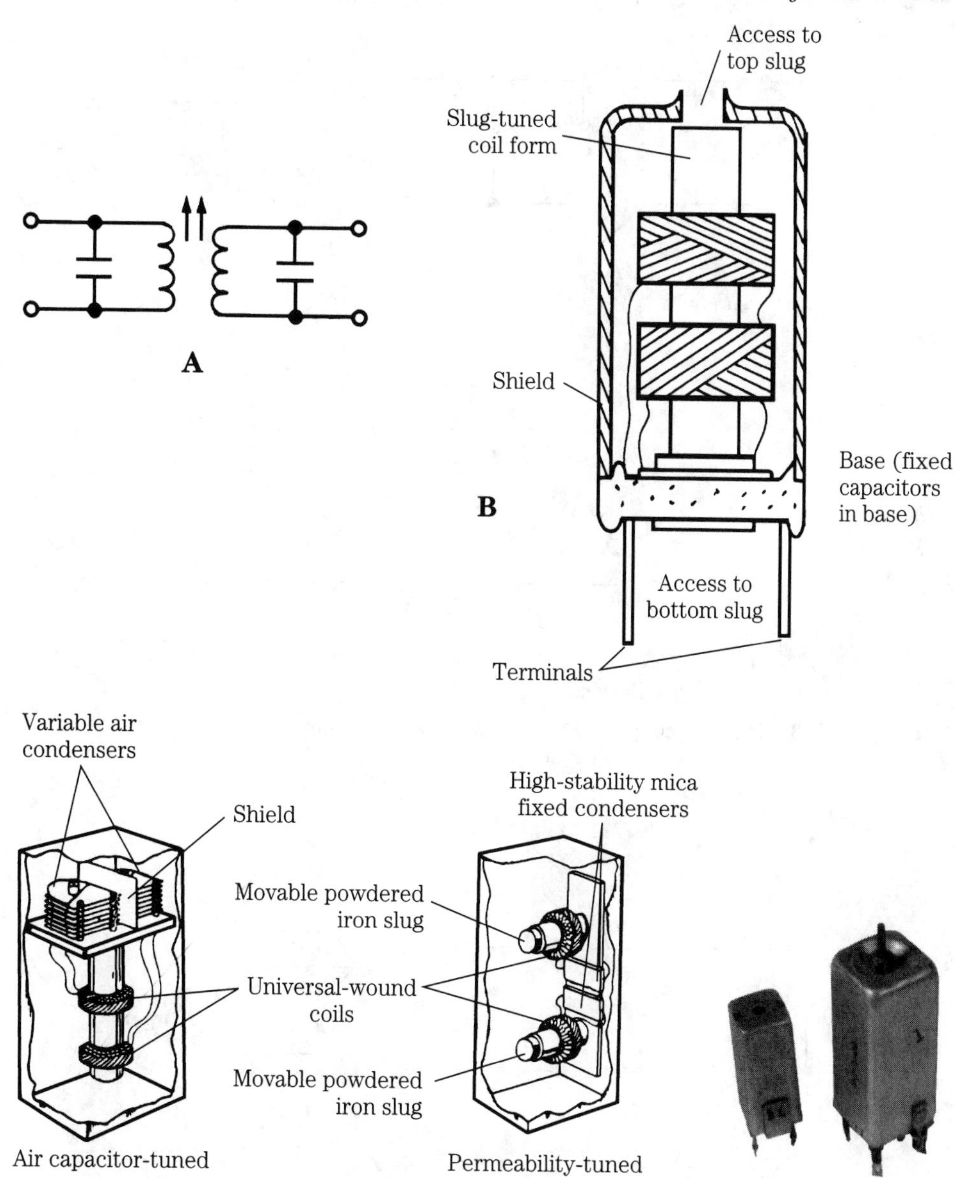

7-7 Slug-tuned transformers. A.Circuit symbol. B. Sectioned view of shielded transformer. C Air capacitor-tuned transformer. D. Permeability-tuned transformer. E. Actual transformers.

The bandwidth of the resonant tank circuit, or the RF/IF transformer, can be summarized in a *figure of merit* called Q. The Q of the circuit is the ratio of the bandwidth to the resonant frequency:

$$Q = \frac{BW}{F_r} \qquad \textbf{[7-2]}$$

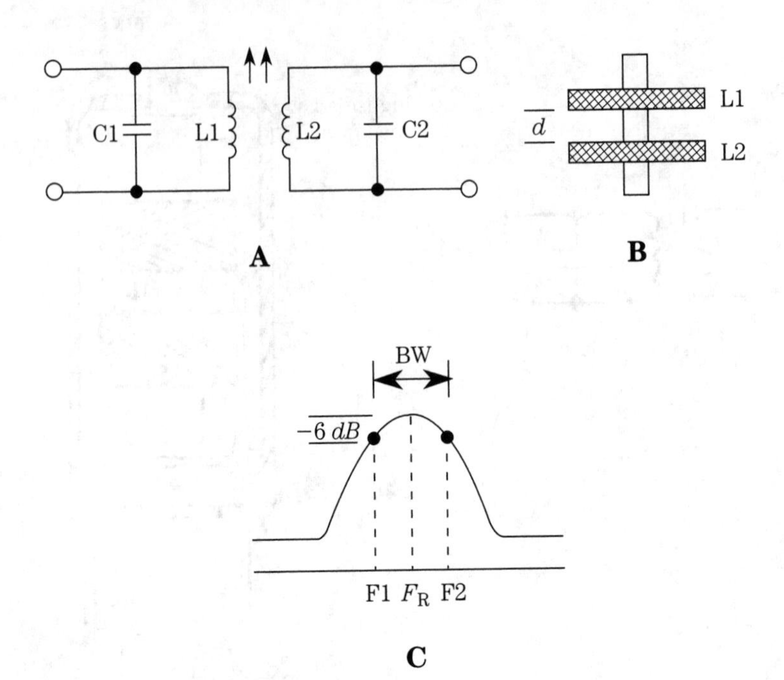

7-8 Critically coupled IF/RF transformer. A. Circuit. B. Place of coils on coil form. C. Frequency response.

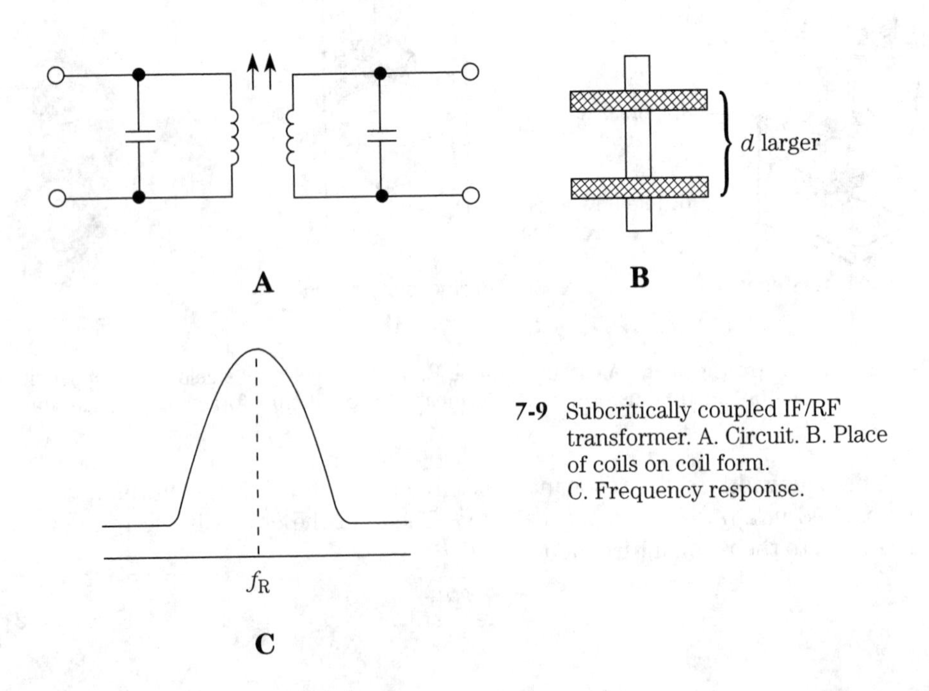

7-9 Subcritically coupled IF/RF transformer. A. Circuit. B. Place of coils on coil form. C. Frequency response.

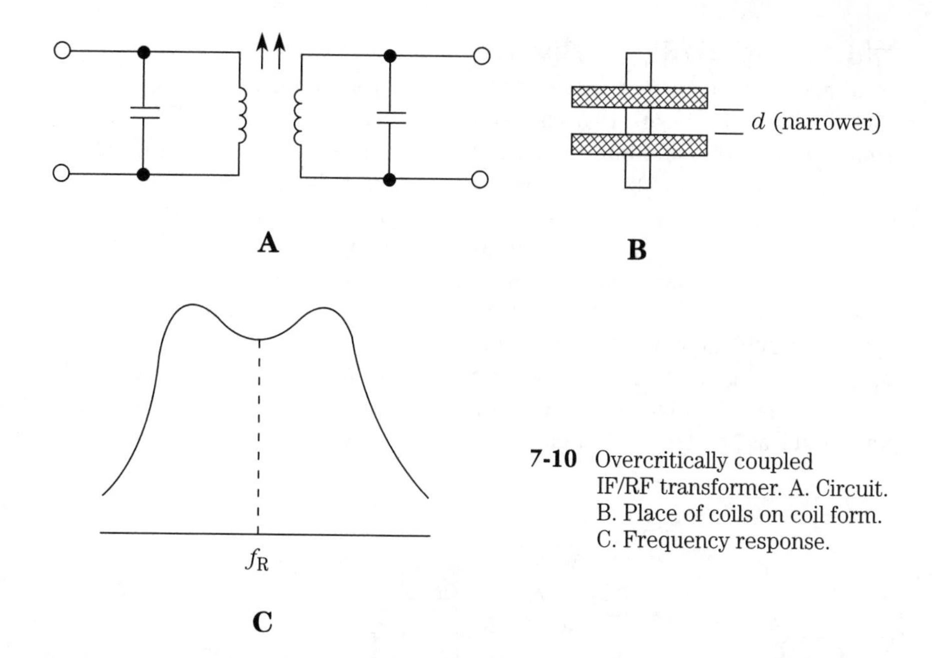

7-10 Overcritically coupled IF/RF transformer. A. Circuit. B. Place of coils on coil form. C. Frequency response.

An overcritically coupled circuit has a low Q, while a narrow bandwidth subcritically coupled circuit has a high Q.

A resistance in the LC tank circuit will cause it to broaden, that is, to lower its Q. The resistor is sometimes called a "de-Qing resistor." The "loaded Q" (i.e., Q when a resistance is present, as in Fig. 7-11A) is always less than the unloaded Q. In some radios, a switched resistor (Fig. 7-11B) is used to allow the user to broaden or narrow the bandwidth. This switch might be labelled "fidelity" or "tone" or something similar.

7-11 Resistors used to broaden the frequency response of an IF/RF transformer.
A. Parallel version.
B. Series version (with "fidelity switch").

Problems with IF/RF transformers

The IF and RF transformer represents a high potential for intermittent problems in radio receivers. There are two basic forms of problem with the IF transformer: intermittent operation and intermittent noise. The best cure for a bad IF or RF transformer is replacement, but because old IF and RF transformers are not always available today, we must place more emphasis on repair of the transformer.

Figure 7-12A shows the basic circuit for a single-tuned IF transformer (others might have a tuned secondary winding). If one of the very fine wires making up the coil breaks (Fig. 7-12B), operation is interrupted. The transformer can usually be diagnosed by light tapping on the shield can of the transformer, or by using a signal tracer or signal generator to find the point where the signal is interrupted. In some cases, the plate voltage of the IF amplifier is interrupted when the transformer opens, and this can be spotted using a dc voltmeter.

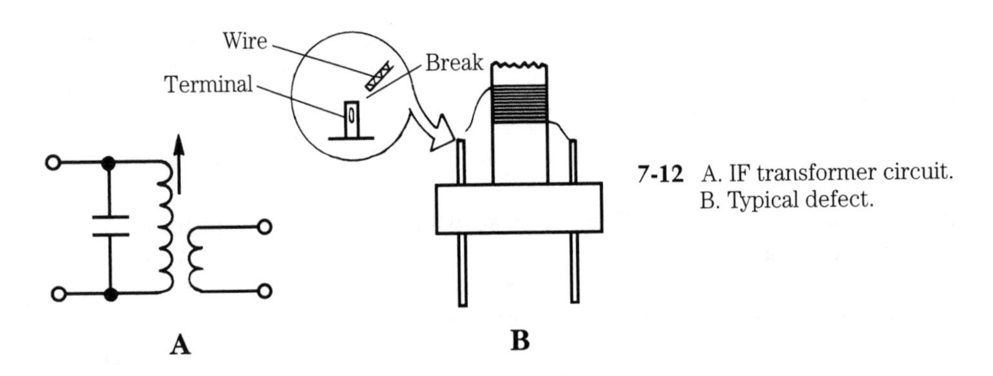

7-12 A. IF transformer circuit. B. Typical defect.

Repairing the IF transformer is a delicate operation. Examine the shield to determine how it is assembled. If it is secured with a screw or nut, merely remove the screw (or nut). If the IF transformer is sealed by metal tabs, you must very carefully pry the tabs open (don't break them, or bend them too far!). Slide the base and coil assembly out of the shield. If there is enough slack on the broken wire, simply solder the wire back onto the terminal. The heat of the soldering iron tip will burn away the enamel insulation. If the wire does not have enough slack, add a little length to the terminal by soldering a short piece of solid wire to the terminal, then solder the IF transformer wire to the added wire. In some cases, it is possible to remove a portion of one turn of the transformer winding in order to gain extra length, but this procedure will change the tuning.

The noisy IF transformer is more likely to need to be replaced than open types, but we are still faced with the lack of original or replacement components. The capacitor can be replaced on many IF transformers, and that can lead to a repair. If the capacitor is a discrete component soldered to the base, it is simple to replace it with another similar (if not identical) capacitor. But if the IF transformer uses an embedded mica capacitor (Fig. 7-13), the job becomes more complex.

If the capacitor element is visible (as in Fig. 7-13), clip one end of the capacitor where it is attached to the terminal. Then bridge a disk ceramic or mica capacitor be-

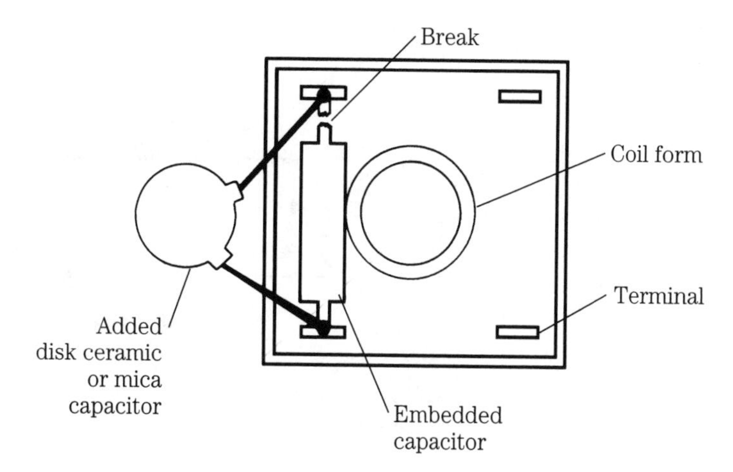

7-13 Replacement of embedded capacitor with a disk ceramic or silver mica fixed capacitor.

tween the two terminals. The value of the capacitor must be found experimentally . . . its value will resonate the coil to the IF frequency.

If the capacitor is embedded and not visible, it may become necessary to install a new terminal and solder the added capacitor to that terminal (with appropriate external circuit modifications).

Choosing component values for LC resonant tank circuits

Resonant LC tank circuits are used to tune radio receivers; it is these circuits that select the station to be received, while rejecting others. A superheterodyne radio receiver (the most common type) is shown in simplified form in Fig. 7-14. According to the superhet principle, the radio frequency being received (F_{RF}) is converted to another frequency, called the *intermediate frequency* (F_{IF}), by being mixed with a *local oscillator* signal (F_{LO}) in a nonlinear mixer stage. The output of the untamed mixer would be a collection of frequencies defined by:

$$F_{IF} = mF_{RF} \pm nF_{LO} \qquad \textbf{[7-3]}$$

where m and n are either integers or zero. For the simplified case in this article, only the first set of products ($m = n = 1$) are considered, so the output spectrum will consist of F_{RF}, F_{LO}, $F_{RF} - F_{LO}$ (difference frequency), and $F_{RF} + F_{LO}$ (sum frequency). In older radios, for practical reasons the difference frequency was selected for F_{IF}; today either sum or difference frequencies can be selected depending on the design of the radio.

Several LC tank circuits are used in this superhet radio. The antenna tank circuit (C_1/L_1) is found at the input of the RF amplifier stage, or if no RF amplifier is used it is placed at the input to the mixer stage. A second tank circuit (L_2/C_2), tuning the same range as L_1/C_1, is found at the output of the RF amplifier, or the input of the

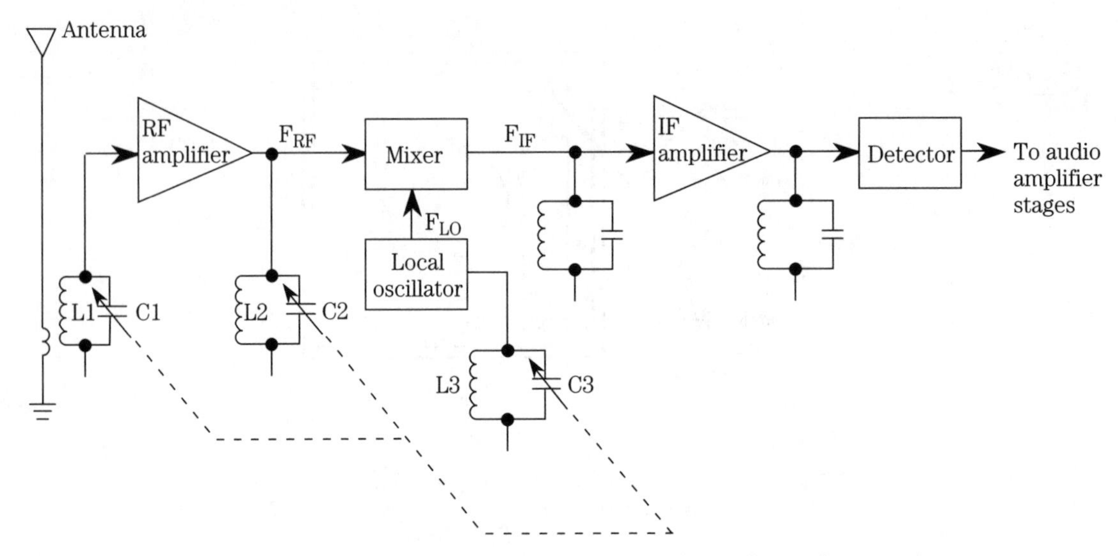

7-14 Use of tuned LC resonant circuits in a typical superheterodyne receiver.

mixer. Another LC tank circuit (L_3/C_3) is used to tune the local oscillator; this tank circuit sets the frequency that the radio will receive.

Additional tank circuits (only two shown) may be found in the IF amplifier section of the radio. These tank circuits will be fixed-tuned to the IF frequency, which in common AM broadcast band (BCB) radio receivers is typically 450 kHz, 455 kHz, 460 kHz, or 470 kHz, depending on the designer's choices (and sometimes the country of origin). Other IF frequencies are also seen, but these are most common. FM broadcast receivers typically use a 10.7 MHz IF, while shortwave receivers might use a 1.65 MHz, 8.83 MHz, 9 MHz or an IF frequency above 30 MHz.

Appendix B—BASIC program to calculate RF tank circuit component values—permits you to easily make these calculations on your computer.

The tracking problem

On a radio that tunes the front-end with a single knob (which is almost all receivers today), the three capacitors ($C_1 - C_3$ in Fig. 7-14) are typicall *ganged*, i.e., the capacitors are mounted on a single rotor shaft. These three tank circuits must *track* each other; i.e., when the RF amplifier is tuned to a certain radio signal frequency, the LO must produce a signal that is different from the RF frequency by the amount of the IF frequency. Perfect tracking is probably impossible, but the fact that your single-knob-tuned radio works is testimony to the fact that the tracking isn't too bad.

The issue of tracking LC tank circuits for the AM BCB receiver has not been a major problem for many years: the band limits are fixed over most of the world, and component manufacturers offer standard adjustable inductors and variable capacitors to tune the RF and LO frequencies. Indeed, some even offer three sets of coils: antenna, mixer input/RF amp output, and LO. The reason why the antenna and mixer/RF coils are not the same, despite tuning the same frequency range, is that these locations see different distributed or stray capacitances. In the USA, it is standard practice to use a 10 to 365 pF capacitor and a 200 μH inductor for the 540 to 1600 kHz AM BCB. In some other countries, slightly different combinations are sometimes used: 320 pF, 380 pF, 440 pF, 500 pF, and others are seen in catalogs.

Recently, however, two events coincided that caused me to examine the method of selecting capacitance and inductance values. First, I embarked on a design project to produce an AM DXers receiver that had outstanding performance characteristics. Second, the AM broadcast band was recently extended so that the upper limit is now 1700 kHz, rather than 1600 kHz. The new 540 to 1700 kHz band is not accommodated by the now-obsolete "standard" values of inductance and capacitance. So I calculated new candidate values. Shortly, we will see the result of this effort.

The RF amplifier/antenna tuner problem

In a typical RF tank circuit, the inductance is kept fixed (except for a small adjustment range that is used for overcoming tolerance deviations) and the capacitance is varied across the range. Figure 7-15 shows a typical tank circuit main tuning capacitor (C_1), trimmer capacitor (C_2) and a fixed capacitor (C_3) that is not always needed. The stray capacitances (C_S) includes the interwiring capacitance, the wiring to chassis capacitance, and the amplifier or oscillator device input capacitance. The frequency changes as the square root of the capacitance changes. If F_1 is the minimum frequency in the range, and F_2 is the maximum frequency, then the relationship is:

$$\frac{F_2}{F_1} = \sqrt{\frac{C_{max}}{C_{min}}} \quad [7\text{-}4]$$

or, in a rearranged form that some find more congenial

$$\left(\frac{F_2}{F_1}\right)^2 = \frac{C_{max}}{C_{min}} \quad [7\text{-}5]$$

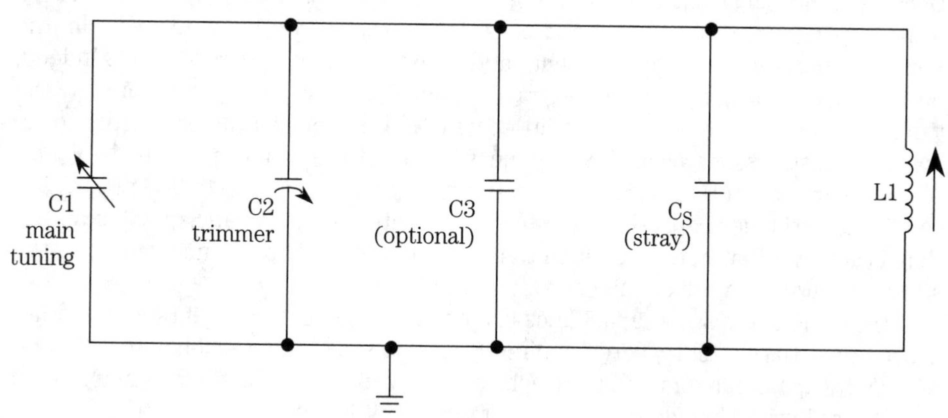

7-15 Typical line-up of components in an LC tuning circuit.

In the case of the new AM receiver, I wanted an overlap of about 15 kHz at the bottom end of the band, and 10 kHz at the upper end, so needed a resonant tank circuit that would tune from 525 kHz to 1710 kHz. In addition, because variable capacitors are widely available in certain values based on the old standards, I wanted to use a "standard" AM BCB variable capacitor. A 10 to 380 pF unit from a vendor was selected.

The minimum required capacitance, C_{min}, can be calculated from:

$$\left(\frac{F_2}{F_1}\right)^2 C_{min} = C_{min} + \Delta C \qquad \textbf{[7-6]}$$

where

F_1 is the minimum frequency tuned
F_2 is the maximum frequency tuned
C_{min} is the minimum required capacitance at F_2
ΔC is the difference between C_{max} and C_{min}

Example 7-1

Find the minimum capacitance needed to tune 1710 kHz when a 10 to 380 pF capacitor (ΔC = 380 – 10 pF = 370 pF) is used, and the minimum frequency is 525 kHz.

Solution

$$\left(\frac{F_2}{F_1}\right)^2 C_{min} = C_{min} + \Delta C$$

$$\left(\frac{1710\text{ kHz}}{525\text{ kHz}}\right)^2 C_{min} = C_{min} + 370\text{ pF}$$

$$10.609\, C_{min} = C_{min} + 370\text{ pF}$$

$$C_{min} = 38.51\text{ pF}$$

The maximum capacitance must be $C_{min} + \Delta C$, or 38.51 + 370 pF = 408.51 pF. Because the tuning capacitor (C_1 in Fig. 7-15) does not have exactly this range, external capacitors must be used, and because the required value is higher than the normal value, additional capacitors are added to the circuit in parallel to C_1. Indeed, because somewhat unpredictable stray capacitances also exist in the circuit, the tuning capacitor values should be a little less than the required values in order to accommodate strays, as well as the tolerances in the actual—versus published—values of the capacitors. In Fig. 7-15, the main tuning capacitor is C_1 (10 to 380 pF), C_2 is a small-value trimmer capacitor used to compensate for discrepancies, C_3 is an optional capacitor that may be needed to increase the total capacitance, and C_S is the stray capacitance in the circuit.

The value of the stray capacitance can be quite high, especially if there are other capacitors in the circuit that are not directly used to select the frequency (e.g., in Colpitts and Clapp oscillators, the feedback capacitors affect the LC tank circuit). In the circuit that I was using, however, the LC tank circuit is not affected by other capacitors. Only the wiring strays and the input capacitance of the RF amplifier or mixer stage need be accounted. From experience I apportioned 7 pF to C_S as a trial value.

The minimum capacitance calculated above was 38.51, there is a nominal 7 pF of stray capacitance, and the minimum available capacitance from C_1 is 10 pF. Therefore, the combined values of C_2 and C_3 must be 38.51 pF – 10 pF – 7 pF, or 21.5 pF. Because there is considerable reasonable doubt about the actual value of C_S, and because of tolerances in the manufacture of the main tuning variable capacitor (C_1), a wide range of capacitance for $C_2 + C_3$ is preferred. It is noted from several catalogs that 21.5 pF is near the center of the range of 45 pF and 50 pF trimmer capacitors. For example, one model lists its range as 6.8 pF to 50 pF, its center point is only slightly removed from the actual desired capacitance. Thus, a 6.8 to 50 pF trimmer was selected, and C_3 is not used.

Selecting the inductance value for L_1 (Fig. 7-15) is a matter of picking the frequency and associated required capacitance at one end of the range, and calculating from the standard resonance equation solved for L:

$$L_{\mu H} = \frac{10^6}{4\,\pi^2\,F^2_{low}\,C_{max}}$$

$$L_{\mu H} = \frac{10^6}{(4)\,(\pi^2)\,(525{,}000)^2\,(4.085 \times 10^{-10})} = 224.97 \approx 225\ \mu H$$

The RF amplifier input LC tank circuit and the RF amplifier output LC tank circuit are slightly different cases because the stray capacitances are somewhat different. In the example, I am assuming a JFET transistor RF amplifier, and it has an input capacitance of only a few picofarads. The output capacitance is not a critical issue in this specific case because I intend to use a 1-mH RF choke in order to prevent JFET oscillation. In the final receiver, the RF amplifier may be deleted altogether, and the LC tank circuit described above will drive a mixer input through a link coupling circuit.

The local oscillator (LO) problem

The local oscillator circuit must track the RF amplifier, and must also tune a frequency range that is different from the RF range by the amount of the IF frequency (455 kHz). In keeping with common practice, I decided to place the LO frequency 455 kHz *above* the RF frequency. Thus, the LO must tune the range 980 kHz to 2165 kHz.

There are three methods for making the local oscillator track with the RF amplifier frequency when single shaft tuning is desired: the trimmer capacitor method, the padder capacitor method, and the different-value cut-plate capacitor method.

Trimmer capacitor method

The *trimmer capacitor method* was shown in Fig. 7-15, and is the same as the RF LC tank circuit. Using exactly the same method as before, but with a frequency ratio of (2165/980), to yield a capacitance ratio of $(2165/980)^2 = 4.88{:}1$, solves this problem. The results were a minimum capacitance of 95.36 pF, and a maximum capacitance of 465.36 pF. An inductance of 56.7 μH is needed to resonate these capacitances to the LO frequency range.

There is always a problem associated with using the same identical capacitor for both RF and LO. It seems that there is just enough difference that tracking between them is always a bit off. Figure 7-16 shows the ideal LO frequency and the calculated LO frequency. The difference between these two curves is the degree of nontracking. The curves overlap at the ends, but are awful in the middle. There are two cures for this problem. First, use a *padder capacitor* in series with the main tuning capacitor (Fig. 7-17). Second, use a *different-value cut-plate capacitor*.

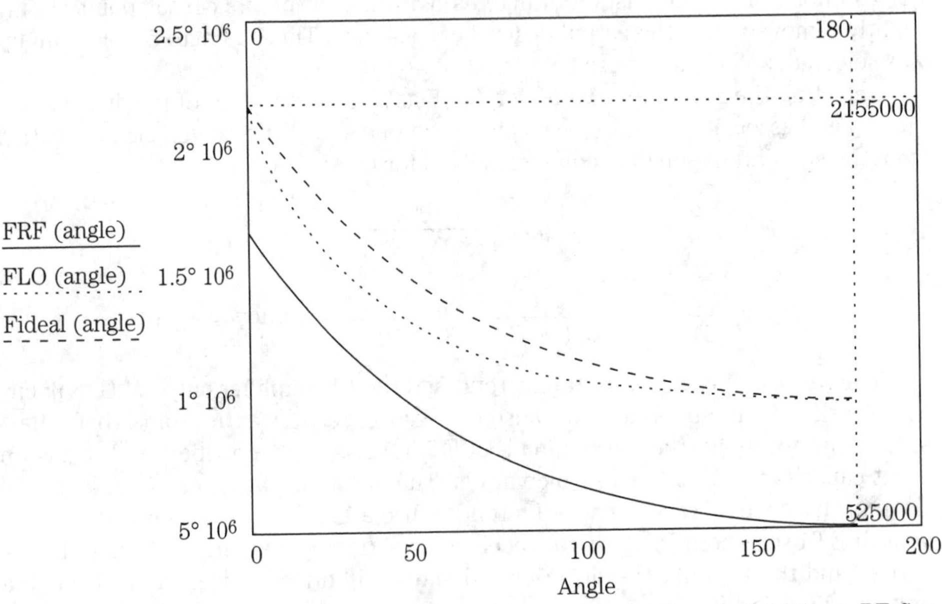

7-16 Tracking of LC tuned circuit when RF and LO signals are different. Solid line is RF, fine dotted (middle) line is LO, and coarse dotted line is ideal.

Padder capacitor method

Figure 7-17 shows the use of a padder capacitor (C_p) to change the range of the LO section of the variable capacitor. This method is used when both sections of the variable capacitor are identical. Once the reduced capacitance values of the C_1/C_p combination are determined the procedure is identical to above. But first, we have to calculate the value of the padder capacitor and the resultant range of the C_1/C_p combination. The padder value is found from:

$$\frac{C_{1max}\,C_p}{C_{1max} + C_p} = \left(\frac{F_2}{F_1}\right)^2 \left(\frac{C_{1min}\,C_p}{C_{1min} + C_p}\right) \quad \textbf{[7-7]}$$

and solving for C_p. For the values of the selected main tuning capacitor and LO:

$$\frac{(380 \text{ pF})\,(C_p)}{(380 + C_p) \text{ pF}} = (4.88)\left(\frac{(10 \text{ pF})\,(C_p)}{(10 + C_p \text{ pF})}\right) \quad \textbf{[7-8]}$$

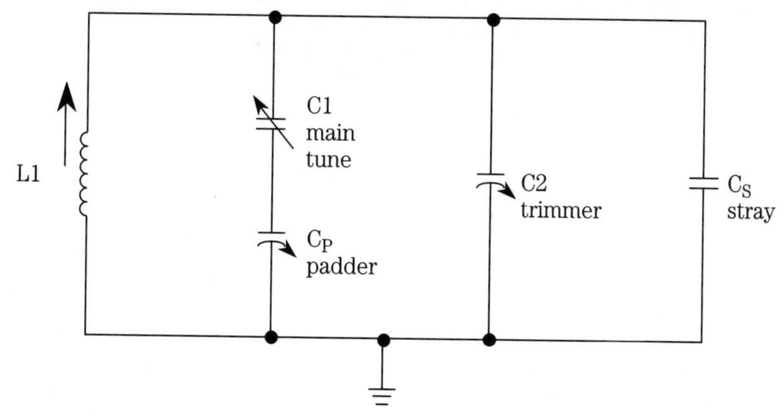

7-17 Typical LO tuning circuit uses a padder capacitor to compensate for different frequency range of LO and RF.

Solving for C_p by the least common denominator method (crude, but it works) yields a padder capacitance of 44.52 pF. The series combination of 44.52 pF and a 10- to 380-pF variable yields a range of 8.2 pF to 39.85 pF. An inductance of 661.85 μH is needed for this capacitance to resonate over 980 kHz to 2165 kHz.

Cut-plate capacitor method

A practical solution to the tracking problem that comes close to the ideal is to use a *cut-plate* capacitor. These variable capacitors have at least two sections, one each for RF and LO tuning. The shape of the capacitor plates are especially cut to a shape that permits a constant change of frequency for every degree of shaft rotation. With these capacitors, when well-done, it is possible to produce three-point tracking, or better.

8
Build simple radio receivers

Notes on designing simple very low frequency (VLF) receivers

The very low frequency (VLF) region is located between a few kilohertz up to around 300 kHz, depending upon whose designation system is used. For purposes of this chapter, VLF represents the 5 to 100 kHz region. The reason for this seemingly arbitrary designation is that many ham band and SWL communications receivers operate down to 100 kHz, and only a few operate below that limit.

There is quite a lot of radio activity in the region below 100 kHz. Perhaps the best known station is WWVB on 60 kHz. This station is operated from Colorado by the National Institutes for Standards and Technology (NIST). WWVB is a very accurate time and frequency station, and for many purposes, is preferred over the high-frequency WWV and WWVH transmissions. The U.S. Navy operates submarine communications stations in the VLF region: NSS on 21.4 kHz from Annapolis, MD (400 kW), and NAA on 24 kHz from Cutler, ME (1000 kW) being two most commonly heard. There are also Omega Navigational System stations in the range 10 to 14 kHz, with principal frequencies being 10.2, 11.05, 11.33, and 13.6 kHz. Other stations, in the USA and abroad, are found throughout the VLF region.

But DXing in the VLF band is not all that easy. Besides the fact that propagation doesn't support "skip" the way the 20-meter ham band does, there is also the huge amount of noise signals found in the VLF region. Two sources seem to predominate. First, the 60-Hz power lines are terrible offenders. While it may seem counterintuitive that a 60-Hz signal could be of much concern at, say, 30 kHz, it is nonetheless a fact. The high harmonics are due to the fact that the alternating current from the power lines is not pure, and contains harmonics. In addition, the large amounts of power carried by normal residential power lines makes even harmonics strong enough to interfere with sensitive receivers.

The second form of interference is neighborhood television sets. The horizontal oscillator in TV receivers operates at 15.734 kHz, and it is a pulse. As a result, harmonics from television sets are found up and down the VLF spectrum. And furthermore, the TV horizontal pulse produces its own "sidebands," so each harmonic actually wipes out a lot of spectrum space on either side of the integer multiple of 15.734 kHz. Listening to VLF allows one to identify the evenings when a popular TV presentation is on the air. As a result of TV interference, it is common to find VLFers listening during daylight hours and during the period between 2330 and daybreak.

Amateur scientists use VLF receivers in two different types of activity. Some are active in monitoring solar activity that affects radio propagation. *Sudden Ionospheric Disturbances* (SIDs) can be noted by sudden increases in VLF signal levels (Taylor and Stokes 1991; Taylor 1993). The SID monitoring activity takes place in the 20 to 30 kHz region, although some articles cite frequencies as high as 60 kHz.

The other VLF amateur science activity involves looking for naturally created radio signals called "whistlers." These signals are believed to be created by lightning storms, and are propagated over long distances. They occur in the 1 to 10 kHz region (Mideke 1992 and Eggleston 1993). At least one project under the sponsorship of NASA engaged amateurs to look at whistlers (Pine 1991).

Types of receiver

Virtually all common forms of receivers are used in VLF receiver designs, except possibly the crystal set. In this chapter, we will consider the superheterodyne, the direct-conversion, the tuned-radio-frequency, and the tuned-input-gain-block methods. We will also examine the use of a converter to translate the VLF bands to the HF bands so that an ordinary ham band or SWL receiver can be used.

Superheterodyne receivers

The "*superhet*" (Fig. 8-1) is the basic receiver design used in communications and broadcast receivers. It dates from the 1920s, and is the most successful form of receiver design. In the superhet receiver, the incoming RF signal (at frequency F_1) is filtered by a tuned RF resonant circuit or a bandpass filter, and then applied to a mixer circuit.

In most cases, the RF signal is amplified in an RF amplifier (as shown in Fig. 8-1), but that is not a necessary requirement. The mixer nonlinearly combines F_1 with the signal from a local oscillator (at frequency F_2), to produce an output spectrum of $F_3 = mF_1 \pm nF_2$. In our simplified case, $m = n = 1$, so the output will consist of the two original signals (F_1, F_2), the sum signal ($F_1 + F_2$) and the difference signal ($F_1 - F_2$).

A filter at the output of the mixer selects either sum or difference signal as F_3, and this is called the *intermediate frequency* (IF). Most of the receiver's gain and selectivity are provided in the IF amplifier. The output of the IF amplifier is fed to a detector that will demodulate the type of signal being received. For an AM signal, a simple envelope detector is used, while for CW and SSB, a product detector is used.

In older radios, only the difference signal was used for the IF frequency—but in modern receivers, either the sum or the difference is used. In a VLF receiver, it is possible to use a 10 to 100 kHz RF range, a local oscillator range of 465 kHz to 555 kHz, to produce an IF of 455 kHz (one of the common "traditional" frequencies).

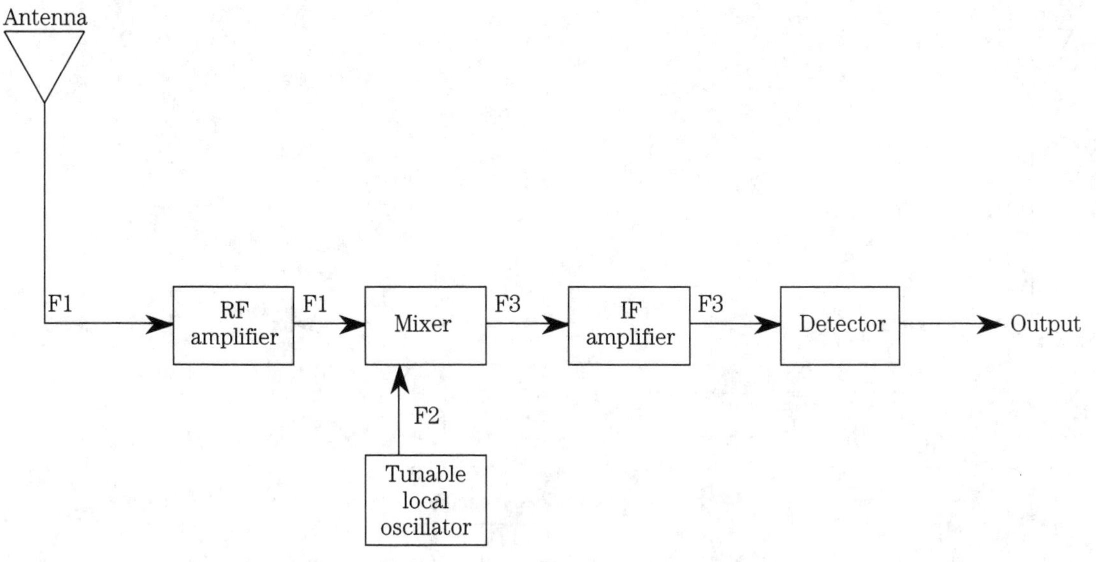

8-1 Block diagram for a superheterodyne receiver.

Converters

A sub-class of superheterodyne receivers is the *converter* technique (Fig. 8-2), in which the VLF band is frequency-translated to the high-frequency (HF) bands. The typical converter circuit is rather simple. An input filter (either bandpass or tuned to a specific frequency) feeds an optional RF amplifier, and then a mixer. A fixed-frequency crystal oscillator mixes with the RF signal to produce an output on an IF that can be heard on the ham band, or on some other shortwave band. For example, to receive 10 to 100 kHz, the input filter (Fig. 8-2) could be a bandpass filter with –3 dB points at 10 and 100 kHz. The local oscillator could be a 3600-kHz crystal oscillator. The output filter is a 3610 to 3710 kHz bandpass filter, the output of which is fed to an 80-meter ham band receiver. One problem that must be overcome is to reduce the feedthrough of the 3600-kHz crystal oscillator signal to the receiver. A combination of using double-balanced mixer and proper termination of the mixer, will generally cure this problem.

Tuned-radio-frequency (TRF) receivers

The TRF receiver (Fig. 8-3) uses a cascade chain of tuned RF amplifiers (A_1, A_2, and A_3) to amplify the radio signal. The TRF was the first really sensitive design in the early 1920s, and was eclipsed by the superhet in popular commercial receivers. But in the VLF range, the TRF is still popular, especially amongst homebrewers. A problem with the TRF receiver is the possibility of unwanted oscillations . . . which are common in tuned triode devices like npn bipolar transistors.

Peter Taylor's column in the Spring 1993 *Communications Quarterly* gave details of a 20- to 30-kHz TRF receiver for SID monitoring that was designed by Art Stokes (Taylor 1993). An improvement on the design is given in Fig. 8-3.

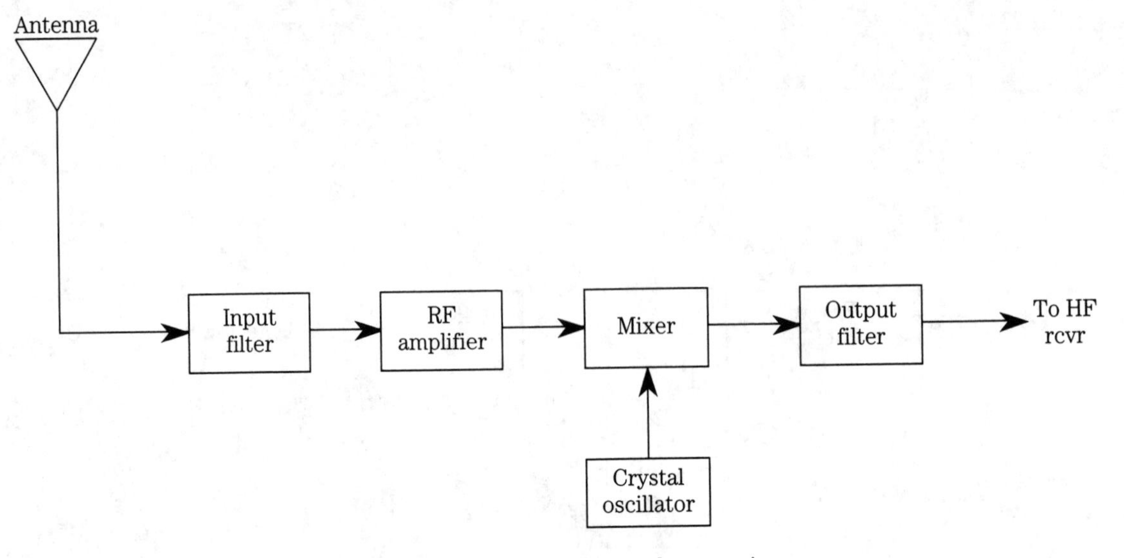

8-2 Front end of a superheterodyne receiver.

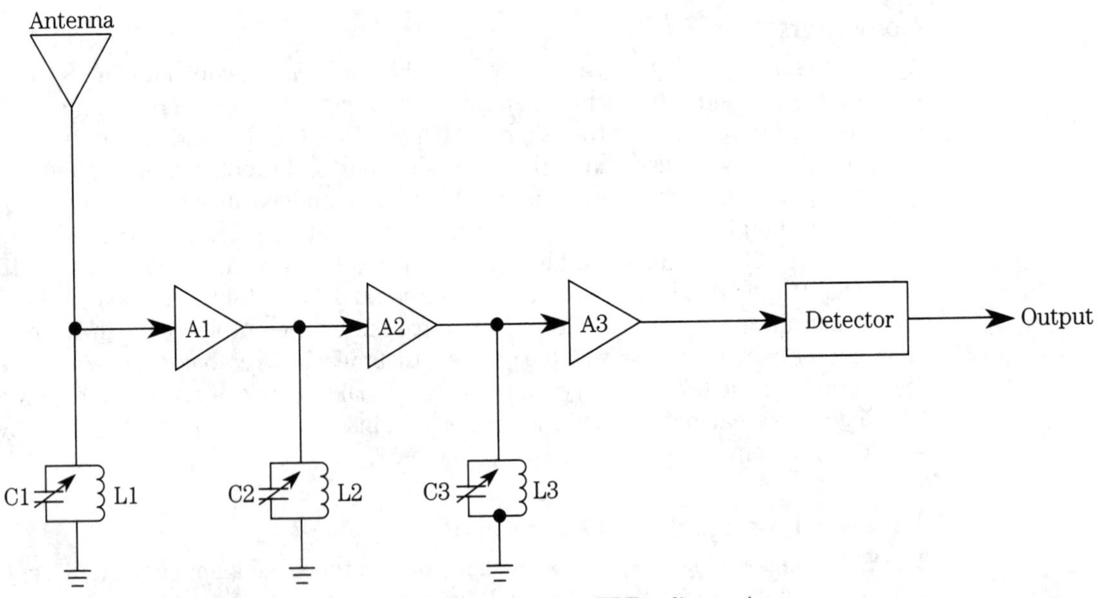

8-3 LC-tuned circuits in a TRF radio receiver.

Tuned-gain-block receivers (TGB)

The TGB receiver (Fig. 8-4) is basically a variant of the TRF receiver, except that the tuning circuits are all up front, ahead of the gain. Two benefits are realized. First, the oscillation problem of conventional TRF receivers is avoided because the gain block is untuned. Second, the tuning up front eradicates much of the unwanted noise prior to it being amplified. This argument was the basis for high-grade receivers of the

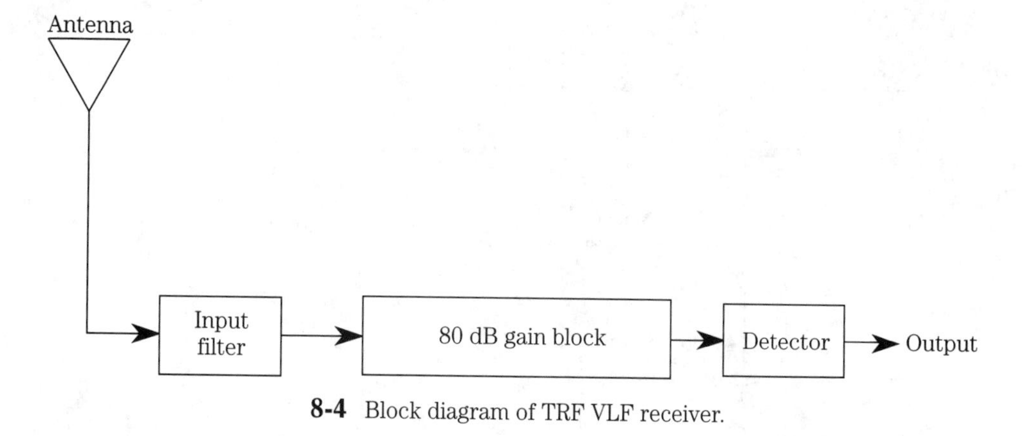

8-4 Block diagram of TRF VLF receiver.

1960s such as the Squires-Sanders. Art Stokes' design published in Taylor/Stokes (1991) was based on this concept. I prepared a printed circuit layout for this receiver, and provided it to Peter Taylor and Art Stokes (it may be published in an upcoming "Solar Spectrum" column).

Tuning circuit problems

The principal problem to solve in designing and building VLF receivers is the tuning circuits. The capacitance and inductance values tend to be rather large. If it is desired to use a standard "broadcast variable" capacitor (which is typically 10 to 365 pF in capacitance value), a 20- to 30-kHz receiver needs an inductor in the 88 mH range. If you calculate the value of capacitance needed to resonate 88 mH at 20 to 30 kHz, the result is 720 pF. A smaller variable capacitance is used because the distributed capacitance of large coils, such as 88 mH, is typically quite large.

The Stokes TGB receiver (1991 design) uses a tuned circuit such as Fig. 8-5. Originally, a J.W. Miller 6319 inductor was used, but these apparently are no longer available. The 1993 TRF design used 88 mH "telephone" toroid inductors. The TRF circuit (Fig. 8-5) uses the 88-mH inductor, paralleled by a 365-pF variable capacitor, and isolated for dc by a pair of 100-pF capacitors (C_2 and C_3 in Fig. 8-5).

A problem with the simple circuit of Fig. 8-5 is that it is not easy to adjust the tuning range; i.e., the receiver cannot be aligned. A variant on the theme, shown in Fig. 8-6, adds a small variable "trimmer" capacitor (C_4) shunted across the main tuning capacitor (C_1), and a trimmer inductor ($L1B$) in series with the main fixed inductor ($L1A$). It would be nice to obtain a single inductor with that tuning range, but they are hard to obtain these days. As shown, the circuit can be adjusted over about 10% of the inductance range. In one version, a 100-μH fixed inductor was connected in series with a 56-μH variable inductor.

When using the TGB design, all the tuning is done ahead of the gain block. This poses certain problems for the tuning circuit, especially if more than one tuned LC circuit is desired for selectivity purposes. Figure 8-7 shows two methods for combining LC-tuned circuits; both of them fall into the "mutual reactance" method, but with different approaches. In the version shown in Fig. 8-7A, a small capacitance (33 to 120

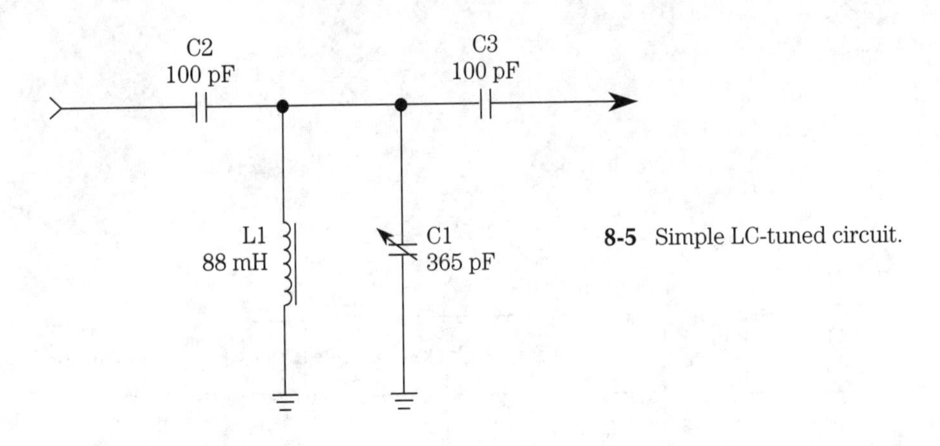

8-5 Simple LC-tuned circuit.

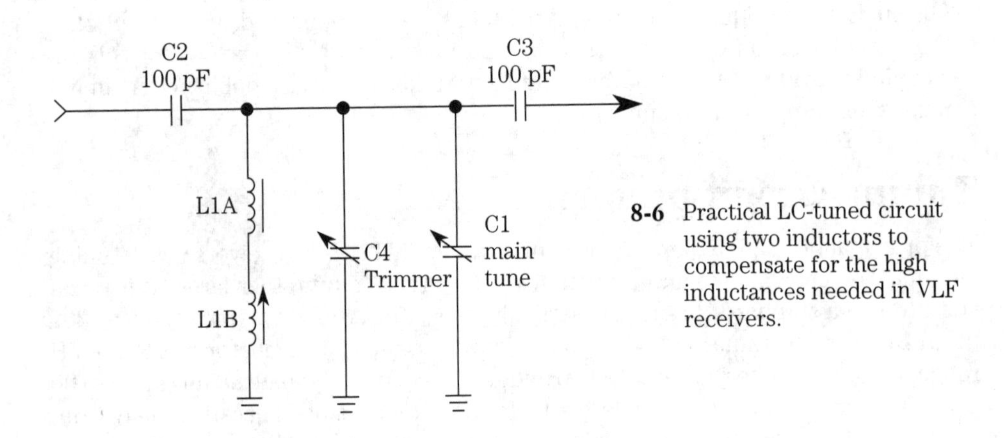

8-6 Practical LC-tuned circuit using two inductors to compensate for the high inductances needed in VLF receivers.

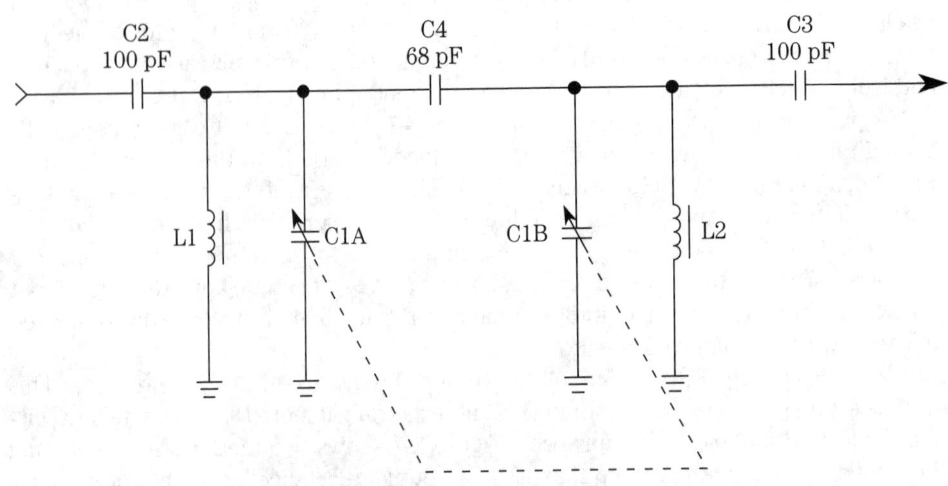

8-7 Mutual impedance coupling. A. Capacitive.

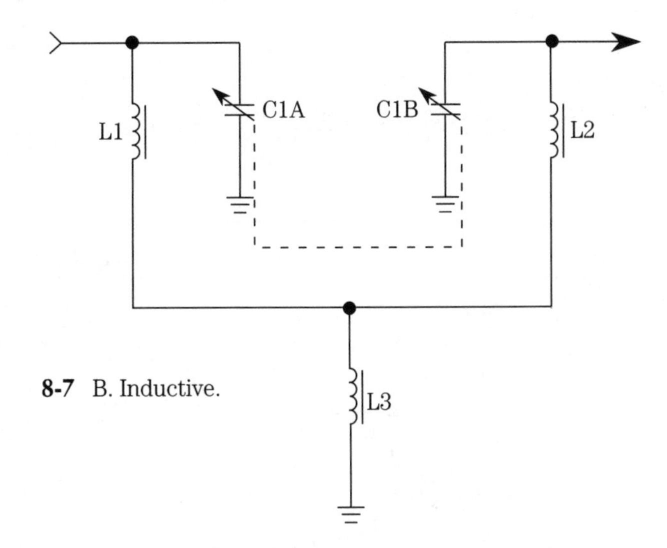

8-7 B. Inductive.

pF) is used to couple the two LC-resonant circuits (L_1/C_{1A} and L_2/C_{1B}). In Fig. 8-7B, a common inductor (L_3) in used for the same purpose. Experiments show that a value of 150 to 700 µH is needed for L_3. *The ARRL Handbook for Radio Amateurs* (all recent editions) provides details on selecting values for the components in these circuits.

A different approach to the design of receiver front ends is shown in Fig. 8-8. One of the problems in VLF receiver design is providing low-impedance link-coupling into and out of the LC tank circuit. Large inductance coils are not very often available with low impedance transformer windings. For example, in the Digi-Key catalog, the largest coil with such a winding is the 220-µH unit, which is intended for the AM broadcast band. A solution was found in the series-inductor method detailed above (Fig. 8-6). In this case, however, a 56-mH variable inductor was connected in series with a xenon flash tube trigger transformer (T1 and T2). The version selected

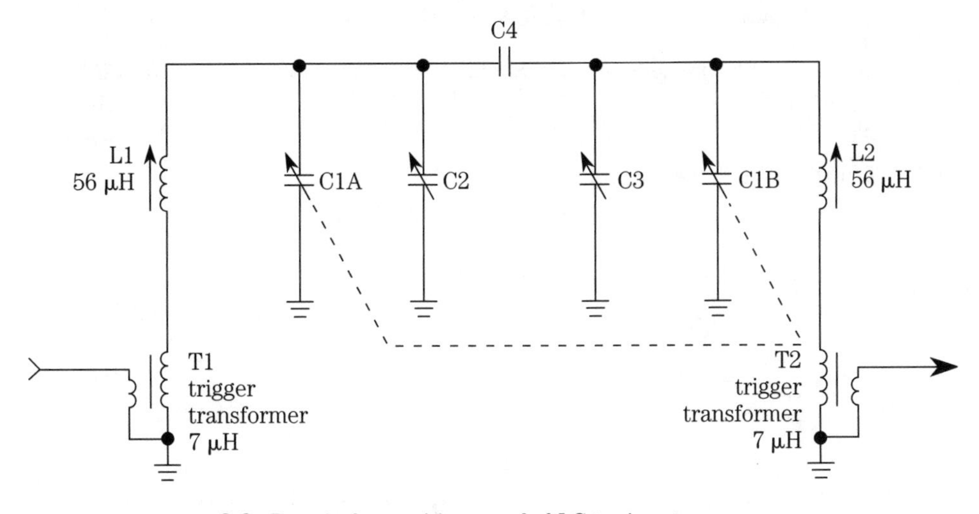

8-8 Practical capacitive coupled LC tuning stage.

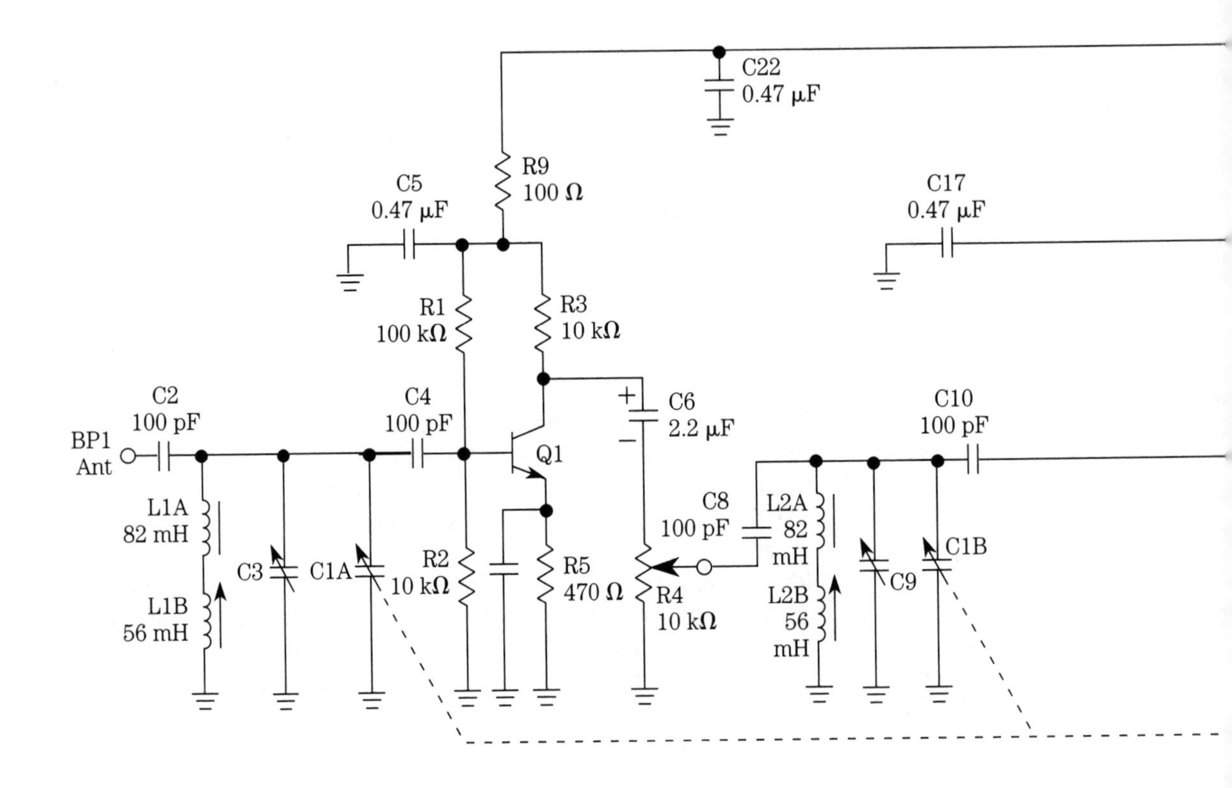

Q1, Q2, Q3: 2N4401
C1: 3 × 365 pF (see text)
C3, C9, C14: 8-80 pF trimmer capacitor (Sprague-Goodman GZC80000) Digi-key SG3010
L1A, L2A, L3A: 82 mH (Toko 181LY-823J) Digi-key TK4424
L1B, L2B, L3B: 56 mH (Toko CLNS-T1039Z) Digi-key TK1724

8-9 Schematic of VLF receiver project.

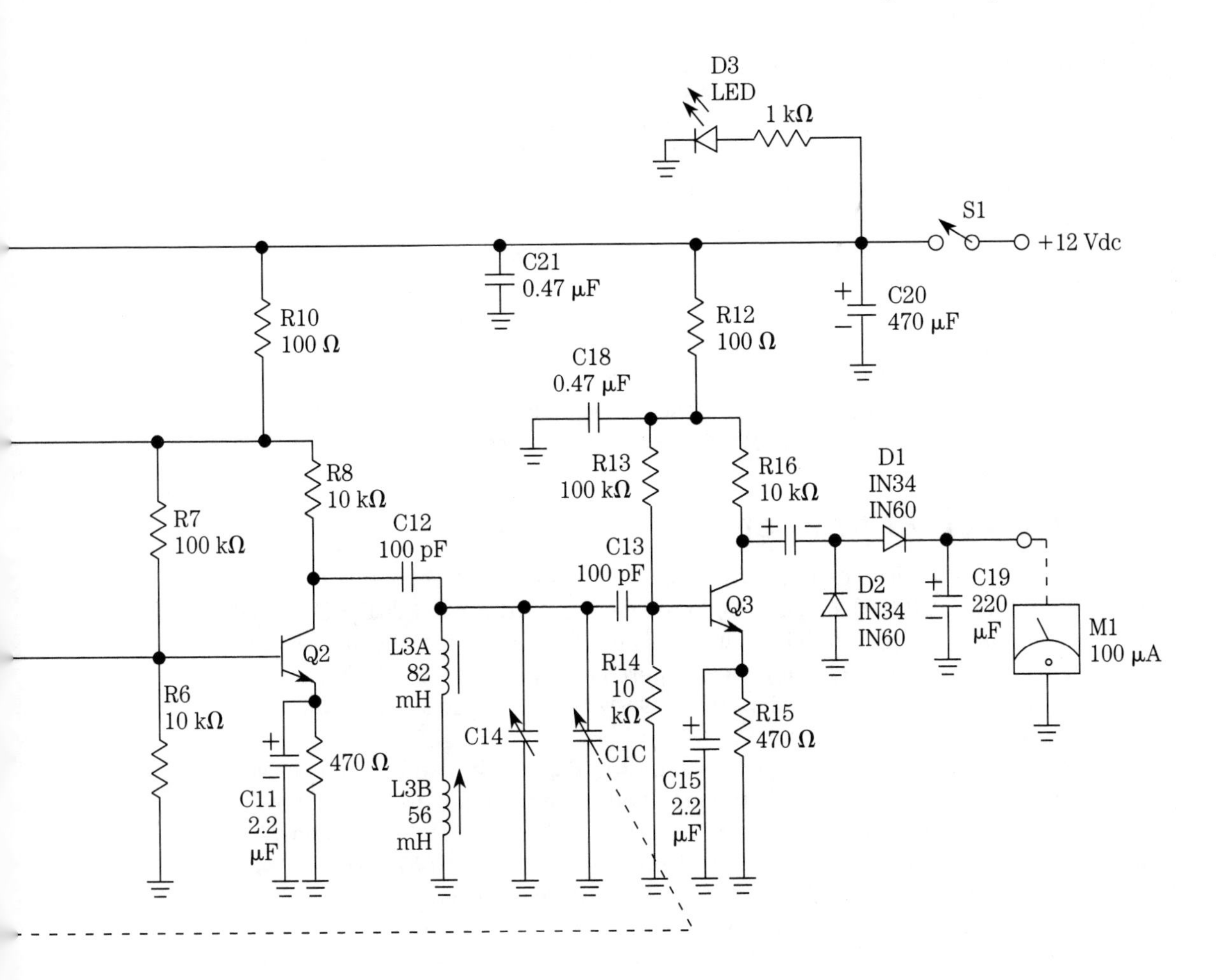
D3
LED
1 kΩ
S1
+12 Vdc
C21
0.47 μF
R10
100 Ω
R12
100 Ω
C20
470 μF
C18
0.47 μF
R8
10 kΩ
R7
100 kΩ
C12
100 pF
R13
100 kΩ
R16
10 kΩ
D1
IN34
IN60
C13
100 pF
Q3
D2
IN34
IN60
C19
220
μF
M1
100 μA
Q2
L3A
82
mH
R14
10
kΩ
R6
10 kΩ
C14
C1C
R15
470 Ω
470 Ω
C11
2.2
μF
L3B
56
mH
C15
2.2
μF

was intended to fire a 6000 volt fast rise time pulse into a xenon flash tube, and has a nominal inductance of 6.96 mH ± 20 percent. The measured inductance was 7.1 μH, and the self-resonance frequency was 140 kHz. Without the trimmer capacitors, the tuning range was 30 to 60 kHz when a 2 × 380-pF variable capacitor was used. Adding a 600-pF trimmer across each section of the main tuning capacitor reduced the tuning range to 20 to 30 kHz.

Trigger transformers are widely available from mail order sources. However, I ordered several from an English source: Maplin Electronics, P.O. Box 3, Rayleigh, Essex, SS6 8LR, England. They have variable capacitors, coil forms, and a number of other things of interest to amateur radio constructors. U.S. credit cards accepted include Visa and American Express, and the currency exchange is automatic.

Another alternative, although I've not tested it, is to use pulse transformers. Unfortunately, these transformers typically have limited turns ratios (2:1:1).

A VLF receiver project

After reading Peter Taylor's column in the Spring 1993 issue of *Communications Quarterly* (Taylor 1993), I decided to build the modified Art Stokes TRF SID monitoring receiver. It uses three npn bipolar transistor RF amplifier stages in cascade. The version shown in the article, however, required a different tuning shaft for each tuning capacitor. I wanted a single-shaft tuning system. Also added is a small amount of decoupling isolation between each stage.

The circuit for the modified receiver is shown in Fig. 8-9. It uses the same basic circuit as the Stokes design, but with the modified tuning circuits discussed above. The 82-mH fixed inductors (L_{1A}, L_{2A}, and L_{3A}) are Toko 181LY-823J, which are available from Digi-Key—P.O. Box 677, Thief River Falls, MN, 56701-0677, (800) 344-4539—under catalog number TK-4424. The 56-mH variable inductors are Toko CLNS-T1039Z, available under Digi-Key TK-1724. An additional degree of adjustment is provided by using an 8- to 80-pF trimmer capacitor across each section of the 3 × 365 pF variable main tuning capacitor. These capacitors are Sprague-Goodman GZC8000 units (Digi-Key SG3010).

The output circuit for this receiver reflects the fact that it is a SID monitor receiver. The detector is a voltage doubler (D_1/D_2) made from germanium diodes. The original diodes specified were 1N34, but 1N60 will work well also. If you cannot find these diodes (Radio Shack and Jim-Pak sells them), then try using replacements from the "universal" service shop replacement lines such as SK, NTE, and ECG. The NTE-109 and ECG-109 will work well. The output of the detector is heavily integrated by a 470-μF electrolytic capacitor. The output as shown is designed to feed a current-input recorder or a microammeter. If a voltage output is desired, connect a resistor (3.3 kΩ to 10 kΩ) across capacitor C_{19}.

Figure 8-10A shows a printed circuit board for use with this circuit, while Fig. 8-10B is the components placement "road map." The board is laid out for the specific components described above, and shown in Fig. 8-9. Variations on the theme can be accommodated by using different value inductors from the same Toko series (see Digi-Key catalog) (the L_1 series coils are Toko size 10RB, while the L_2 series of coils are size 10PA). Also, if you don't want to use two coils in each tuning circuit, short

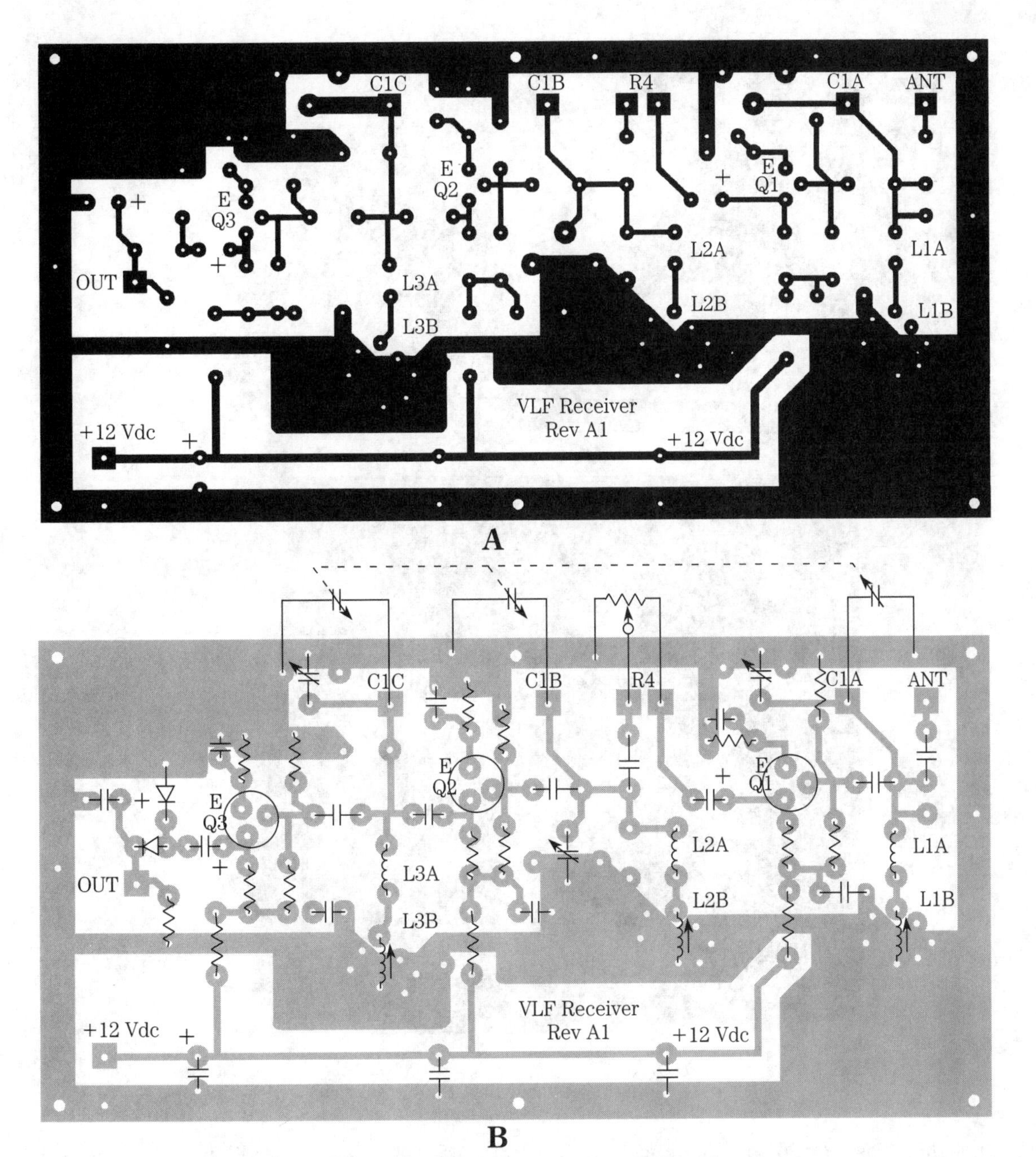

8-10 Printed circuit board. A. Pattern. B. Parts layout.

out the holes for L_1 positions and use a coil in the L_2 positions with the required inductance.

The final receiver is shown in Fig. 8-11. A front panel view is in Fig. 8-11A, while an internal view is shown in Fig. 8-11B. The tuning capacitor was a three-section model purchased from Antique Electronic Supply, Tempe, AZ. Although I first

8-11 Finished receiver. A. Front panel view. B. Internal view.

thought it was a 3 × 365 pF unit, it measured at 550 pF (which is better for VLF anyway). A 2-inch vernier dial from Ocean State Electronics—P.O. Box 1458, Westerly, RI 02891, (800) 866-6626—was used to drive the variable capacitor (note: Ocean State also sells multi-section variable capacitors). The final receiver tuned from 16.5 to 31 kHz.

The printed circuit board in Fig. 8-10A is available from the author. Contact him at P.O. Box 1099, Falls Church, VA 22041 for the price.

References

1. Eggleston, Gary (1993), "The Sky Chorus," *Popular Electronics*, July 1993, p. 46.
2. Mideke, Michael (1992), "Listening to Nature's Radio," *Science Probe!*, July 1992, p. 87.
3. Pine, Bill (1991), "High School Support for Space Physics Research." Unpublished paper sent to participants in Project INSPIRE.
4. Taylor, Peter and Arthur Stokes (1991), "Recording Solar Flares Indirectly," *Communications Quarterly*, Summer 1991, p. 29.
5. Taylor, Peter (1993), "The Solar Spectrum: Update on the VLF Receiver," *Communications Quarterly*, Spring 1993, p. 51.

9
RF amplifier and preselector circuits

In this chapter, we will take a look at small-signal radio frequency (RF) amplifiers. These devices can be used to preamplify radio signals from antennas prior to input to a receiver, at the output of a signal generator or oscillator circuit, or for a variety of other purposes.

Low-priced shortwave receivers often suffer from performance problems that are a direct result of trade-offs the manufacturers made in order to produce a low-cost model. In addition, older receivers often suffer the same problems, as do many homebrew radio receiver designs. Chief among these are sensitivity, selectivity, and image response.

Sensitivity is a measure of the receiver's ability to pick up weak signals. One cause of poor sensitivity is low gain in the front end of the radio receiver, although the IF amplifier contributes most of the gain.

Selectivity is a measure of the ability of the receiver to a) separate two closely spaced signals, and b) reject unwanted signals that are not on or very near the desired frequency being tuned. The selectivity provided by a preselector is minimal for very closely spaced signals (that is the job of the IF selectivity in a receiver), but it is used for reducing the effects (e.g., input overloading) of large local signals . . . so it fits the second half of the definition.

Image response affects only superheterodyne receivers (which most are), and receiving an image is an inappropriate response to a signal transmitted at a frequency of twice the receiver's IF frequency from the frequency that the receiver is tuned to. A superhet receiver converts the signal frequency (RF) to an intermediate frequency (IF) by mixing it with a local oscillator (LO) signal generated inside the receiver. The IF can be either the sum or difference between the LO and RF (i.e., LO+RF or LO–RF), but in most older receivers and nearly all low-cost receivers, it is the difference (LO–RF). The problem is that there are always two frequencies that meet "difference" criteria: LO–RF, and an *image frequency* (F_i) that is LO+IF. Thus, both F_i – LO and LO – RF are equal to the IF frequency. If the image frequency gets through the radio's front end tuning to the mixer, it will appear in the output as a valid signal.

A cure for all of these problems is a little circuit called the *active preselector*. A preselector can be either active or passive. In both cases, however, the preselector includes an inductor/capacitor (LC) resonant circuit that is tuned to the frequency that the receiver is tuned to. The preselector is connected between the antenna and the receiver antenna input connector (Fig. 9-1). Therefore, it adds a little more selectivity to the front end of the radio to help discriminate against unwanted signals. The difference between the active and passive designs is that the active design contains an RF amplifier stage, while the passive design does not. Thus, the active preselector also deals with the sensitivity problem of the receiver. In this section we will take a look at several active preselector circuits you can build and adapt to your own needs.

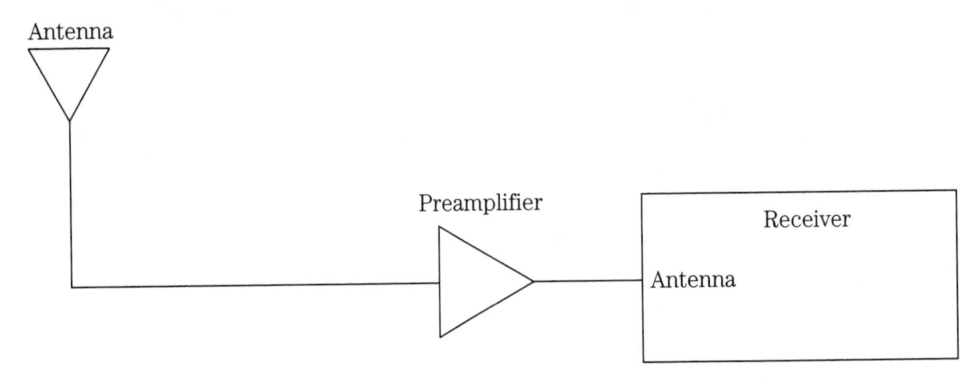

9-1 Position of a preamplifier with a receiver.

The preselector circuits in this section are based on either of two devices: the MPF-102 *junction field-effect transistor* (JFET), and the 40673 *metal-oxide semiconductor field-effect transistor* (MOSFET). Both of these devices are easily available from both mail order sources and from local distributor replacement lines (for example, the MPF-102 is the NTE-312 and the 40673 is the NTE-222). These transistors were selected because they are both easily obtained and are well-behaved into the VHF frequency region.

Preselectors should be built inside of shielded metal boxes to prevent RF leakage around the device. Select boxes that are either die-cast, or are made of sheet metal and have an overlapping lip. Do not use the lower-cost tab-fit sheet metal box.

Noise and preselectors

The weakest radio signal your radio can detect is determined mainly by the noise level in the receiver. Some noises arrive from outside sources, while other noises are generated inside the receiver. At the VHF/UHF range, the internal noise is predominant, so it is common to use a *low-noise preamplifier* ahead of the receiver. The preamplifier will reduce the noise figure for the entire receiver. If you select a commercial ready-built or kit VHF preamplifier, such as the units sold by Hamtronics, Inc. (NY)—65 Moul Rd., Hilton, NY 14468-9535, (716) 392-9430—make sure you specify the low-noise variety for the first amplifier in the system.

The low-noise amplifier (LNA) should be mounted on the antenna if it is wideband, and at the receiver if it is tunable (Note: the term *preselector* only applied to tuned versions, while *preamplifier* could denote either tuned or wideband models). Of course, if your receiver is used only for one frequency, it can also be mounted at the antenna. The reason for mounting the amplifier right at the antenna is to build up the signal and improve the signal-to-noise ratio (SNR) prior to feeding the signal into the transmission line where losses cause the signal to weaken somewhat.

A preselector can improve the performance of your receiver, whether you listen to VLF, AM broadcast band, shortwave, or the VHF/UHF bands. The circuits presented in this article allow you to "roll your own" and be successful at it.

Project 9-1: JFET preselector

Figure 9-2 shows the most basic form of JFET preselector. This circuit will work into the low-VHF frequency region. This circuit is in the common-source configuration, so the input signal is applied to the gate and the output signal is taken from the drain. Source bias is supplied by the voltage drop across resistor R2, and drain load by a series combination of a resistor (R3) and a radio frequency choke (RFC1). The RFC should be 1000 μH (1 mH) at the AM broadcast band and HF (shortwave) bands, and 100 μH in the low-VHF region (>30 MHz). At VLF frequencies below the broadcast band use 2.5 mH for RFC1, and increase all 0.01 μF capacitors to 0.1 μF. All capacitors are either disk ceramic, or one of the newer dielectric capacitors (if rated for VHF service . . . be careful, not all are!).

The input circuit is tuned to the RF frequency, but the output circuit is untuned. The reason for the lack of output tuning is that tuning both input and output permits the JFET to oscillate at the RF frequency . . . and that we don't want. Other possible causes of oscillation include layout, and a self-resonance frequency of the RFC that is too near the RF frequency (select another choke).

The input circuit consists of an RF transformer that has a tuned secondary (L_2/C_1). The variable capacitor (C_1) is the tuning control. Although the value shown is the standard 365 pF "AM broadcast variable," any form of variable can be used if the inductor is tailored to it. These components are related by:

$$f = \frac{1}{2\pi\sqrt{LC}} \qquad [9\text{-}1]$$

where

f is the frequency in hertz
L is the inductance in henrys
C is the capacitance in farads

Be sure to convert inductances from microhenrys to henrys, and picofarads to farads. Allow approximately 10 pF to account for stray capacitances, although keep

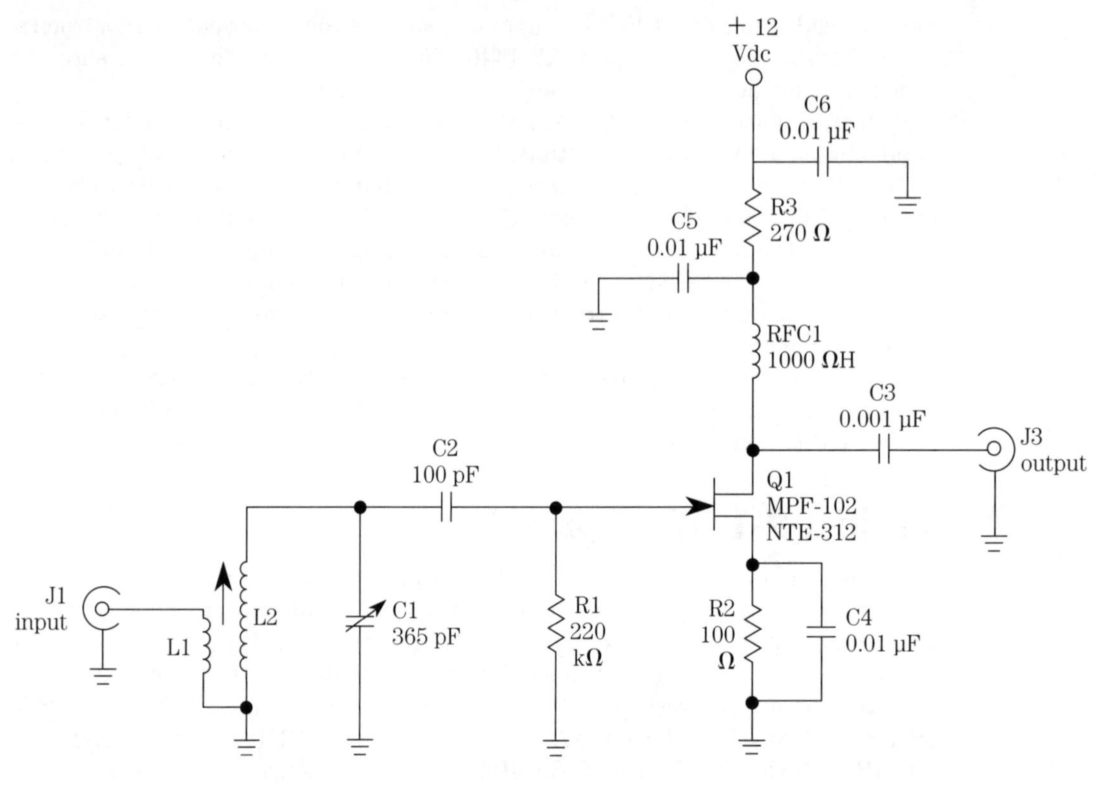

9-2 JFET preamplifier (common source configuration).

in mind that this number is a guess that may have to be adjusted (it is a function of your layout, among other things). We can also solve Eq. 9-1 for either L or C:

$$L = \frac{1}{39.5F^2C} \qquad \textbf{[9-2]}$$

Space does not warrant making a sample calculation, but we can report results for you to check for yourself. In a sample calculation, I wanted to know how much inductance is required to resonate 100 pF (90 pF capacitor plus 10 pF stray) to 10 MHz WWV. The solution, when all numbers are converted to hertz and farads, results in 0.00000253 H, or 2.53 µH. Keep in mind that the calculated numbers are close, but are nonetheless approximate . . . and the circuit may need tweaking on the bench.

The inductor (L_1/L_2) may be either a variable inductor (as shown) from a distributor such as Digi-Key—P.O. Box 677, Thief River Falls, MN 56701, (800) 344-4539—or "homebrewed" on a toroidal core. Most people will want to use the T-50-6 (RED) or T-68-6 (RED) toroids—Amidon Associates, 12033 Otsego Street, North Hollywood, CA 91607—for shortwave applications. The number of turns required for the toroid is calculated from:

$$N = 100\sqrt{\frac{L_{\mu H}}{A_L}} \qquad \textbf{[9-3]}$$

where

$L_{\mu H}$ is in microhenrys
A_L is 49 for T-50-2 (RED) and 57 for T-68-2 (RED)

Example, a 2.53 µH coil needed for L2 (Fig. 9-2) wound on a T-50-RED core requires 23 turns. Use #26 or #28 enameled wire for the winding. Make L_1 approximately 4–7 turns over the same form as L_2.

Be careful when making JFET or MOSFET RF amplifiers in which both input and output are tuned. If the circuit is a common source circuit (i.e., where the input signal is across the gate and source), and the output signal is between the drain and source, there is the possibility of accidentally turning the circuit into a dandy little oscillator. Sometimes, this problem is alleviated by tuning the input and output LC tank circuits to slightly different frequencies. In other cases, it is necessary to neutralize the stage. It is a common practice to make at least one end of the amplifier, usually the output, untuned in order to overcome this problem (although at the cost of some gain).

Figure 9-3 shows two methods for tuning both the input and output circuits of the JFET transistor. In both cases, the JFET is wired in the common gate configuration—so signal is applied to the source and output is taken from the drain. The dotted line indicates that the output and input tuning capacitors are ganged to the same shaft.

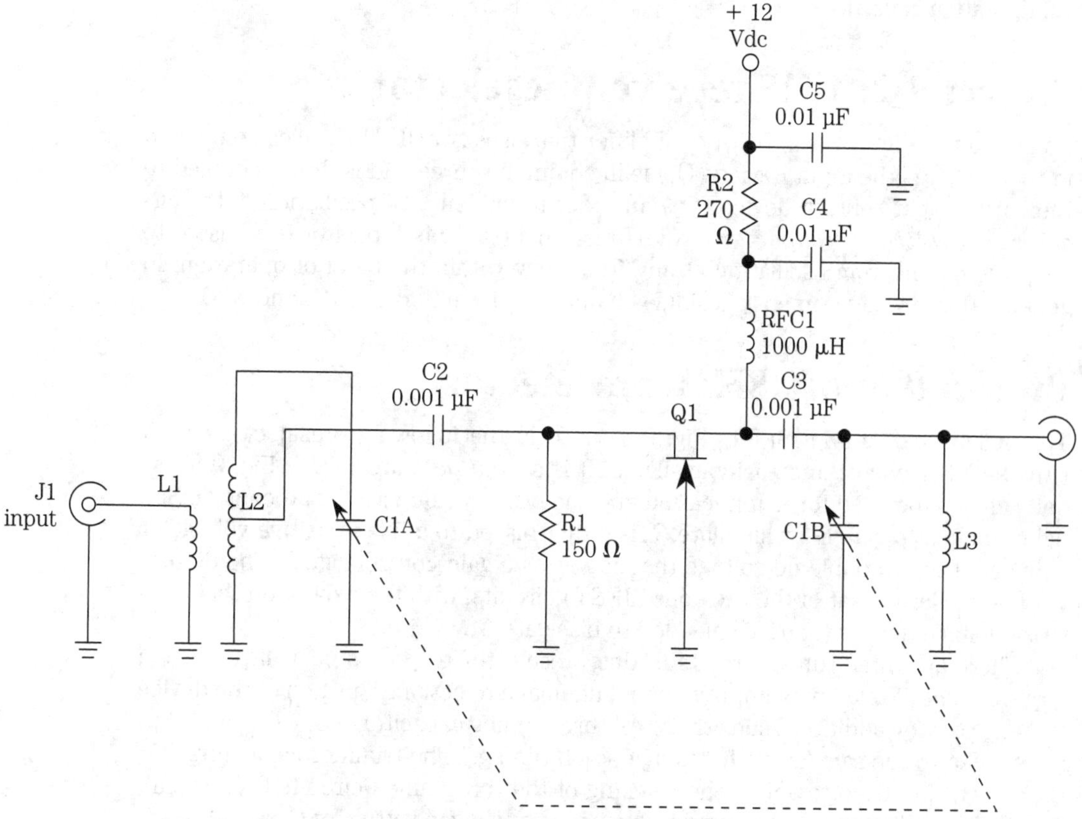

9-3 Common gate JFET preamplifier. A. Transformer coupling method 1.

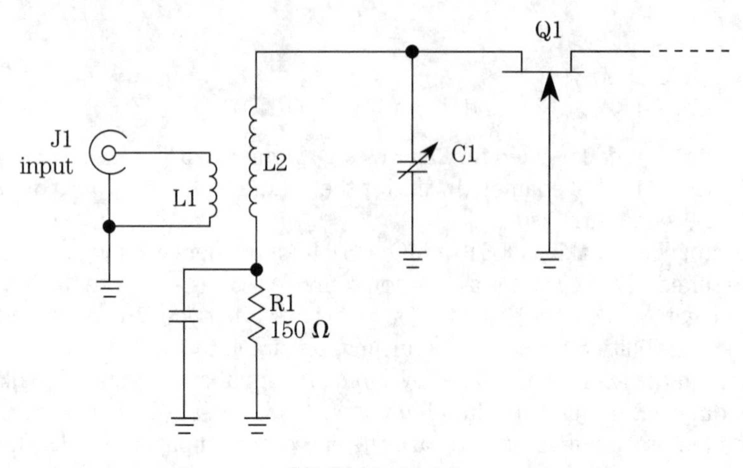

9-3 B. Method 2.

The source circuit of the JFET is low-impedance, so some means must be provided to match the circuit to the tuned circuit. In Fig. 9-3A a tapped inductor is used for L2 (tapped at ⅛ the coil winding), and in Fig. 9-3B a similar but slightly different configuration is used.

Project 9-2: VHF receiver preselector

The circuit in Fig. 9-4 is a VHF preamplifier that uses two JFET devices connected in *cascode*, i.e., the input device (Q_1) is in common source and is direct-coupled to the common gate output device (Q_2). In order to prevent self-oscillation of the circuit a *neutralization capacitor* (NEUT) is provided. This capacitor is adjusted to keep the circuit from oscillating at any frequency within the band of operation. In general, this circuit is tuned to a single channel by the action of L_2/C_1 and L_3/C_2.

Project 9-3: MOSFET preselector

The 40673 dual-gate MOSFET (Fig. 9-5) used in the following preselector circuit (Fig. 9-6) is low-cost and easily available. It is a *dual-gate MOSFET* (Fig. 9-6), so one gate can be used for amplification and the other for dc-based gain control. Signal is applied to gate G_1, while gate G_2 is either biased to a fixed positive voltage or connected to a variable dc voltage that serves as a gain control signal. The dc network is similar to that of the previous (JFET) circuits, with the exception that a resistor voltage divider (R_3/R_4) is needed to bias gate G_2.

There are three tuned circuits in this preselector project, so it will produce a large amount of selectivity improvement and image rejection. The gain of the device will also provide additional sensitivity. All three tuning capacitors (C_{1A}, C_{1B}, and C_{1C}) are ganged to the same shaft for "single-knob tuning." The trimmer capacitors (C_2, C_3, and C_4) are used to adjust the tracking of the three tuned circuits (i.e., ensure that they are all tuned to the same frequency at any given setting of C_{1A}–C_{1C}).

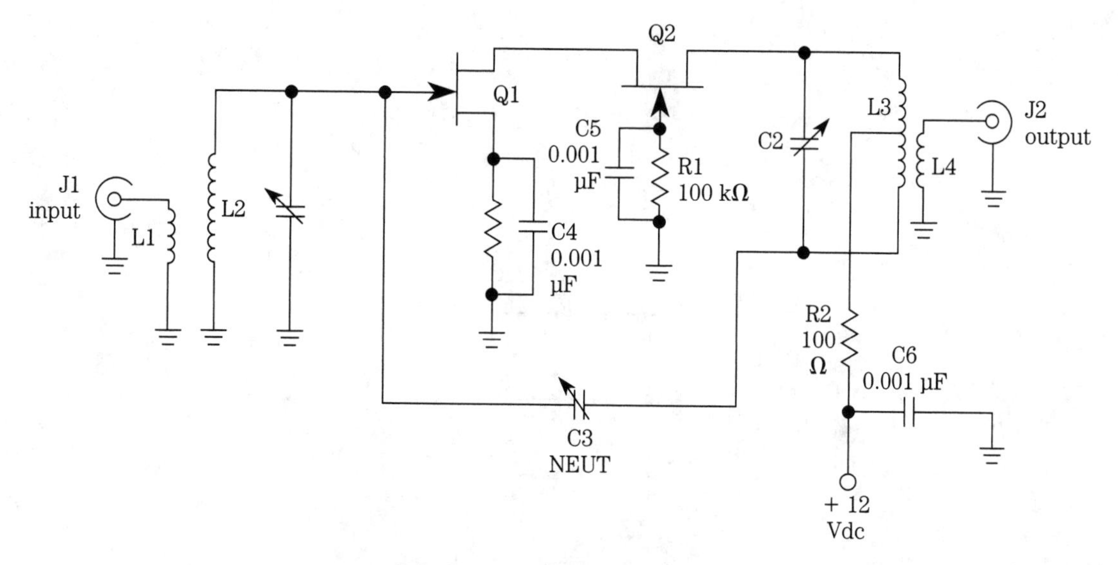

9-4 Cascode JFET preamplifier.

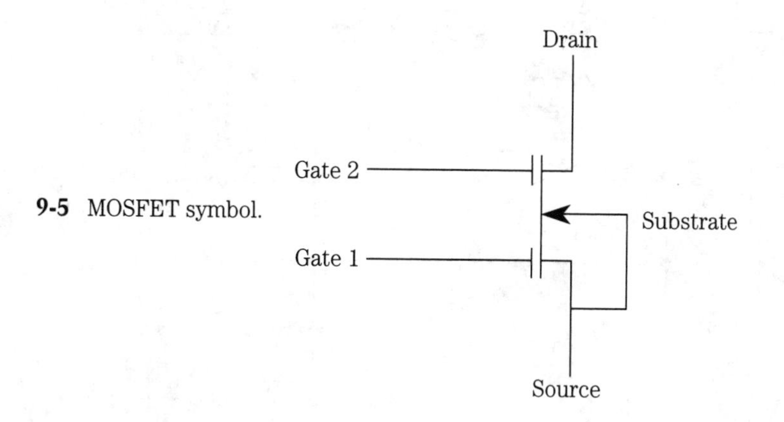

9-5 MOSFET symbol.

The inductors are of the same sort as described above. It is permissible to put L_1/L_2 and L_3 in close proximity to each other, but they should be separated from L_4 in order to prevent unwanted oscillation due to feedback arising from coil coupling.

Project 9-4: Voltage-tuned receiver preselector

The circuit in Fig. 9-7 is a little different. In addition to using only input tuning (which lessens the potential for oscillation), it also uses voltage tuning. The hard-to-find variable capacitors are replaced with *varactor diodes*, also called *voltage-variable capacitance diodes*. These pn-junction diodes exhibit a capacitance that is a function of the applied reverse bias potential, V_T. Although the original circuit was built and tested for the AM broadcast band (540 kHz to 1700 kHz), it can be changed to any band by correct selection of the inductor values. The designated varactor

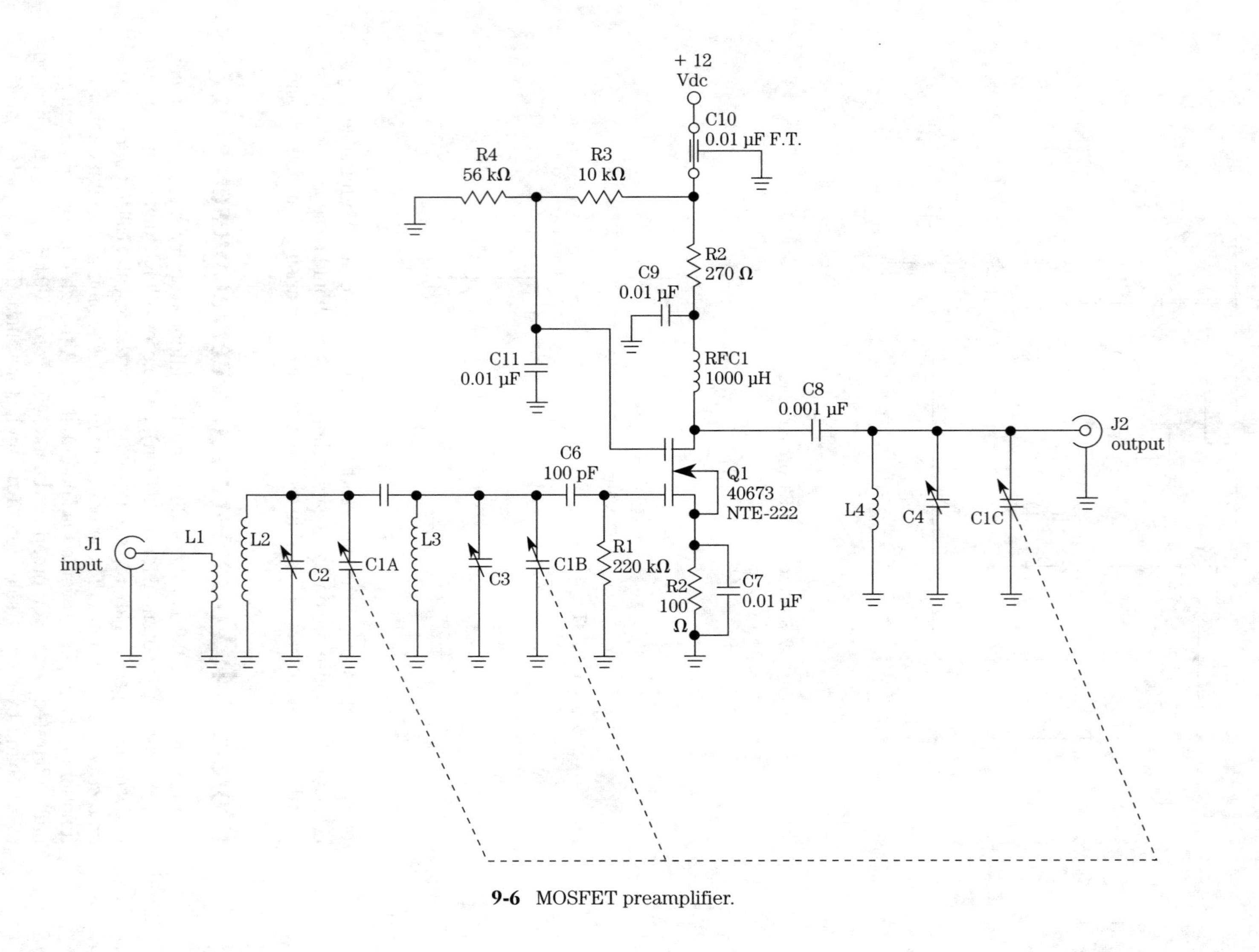

9-6 MOSFET preamplifier.

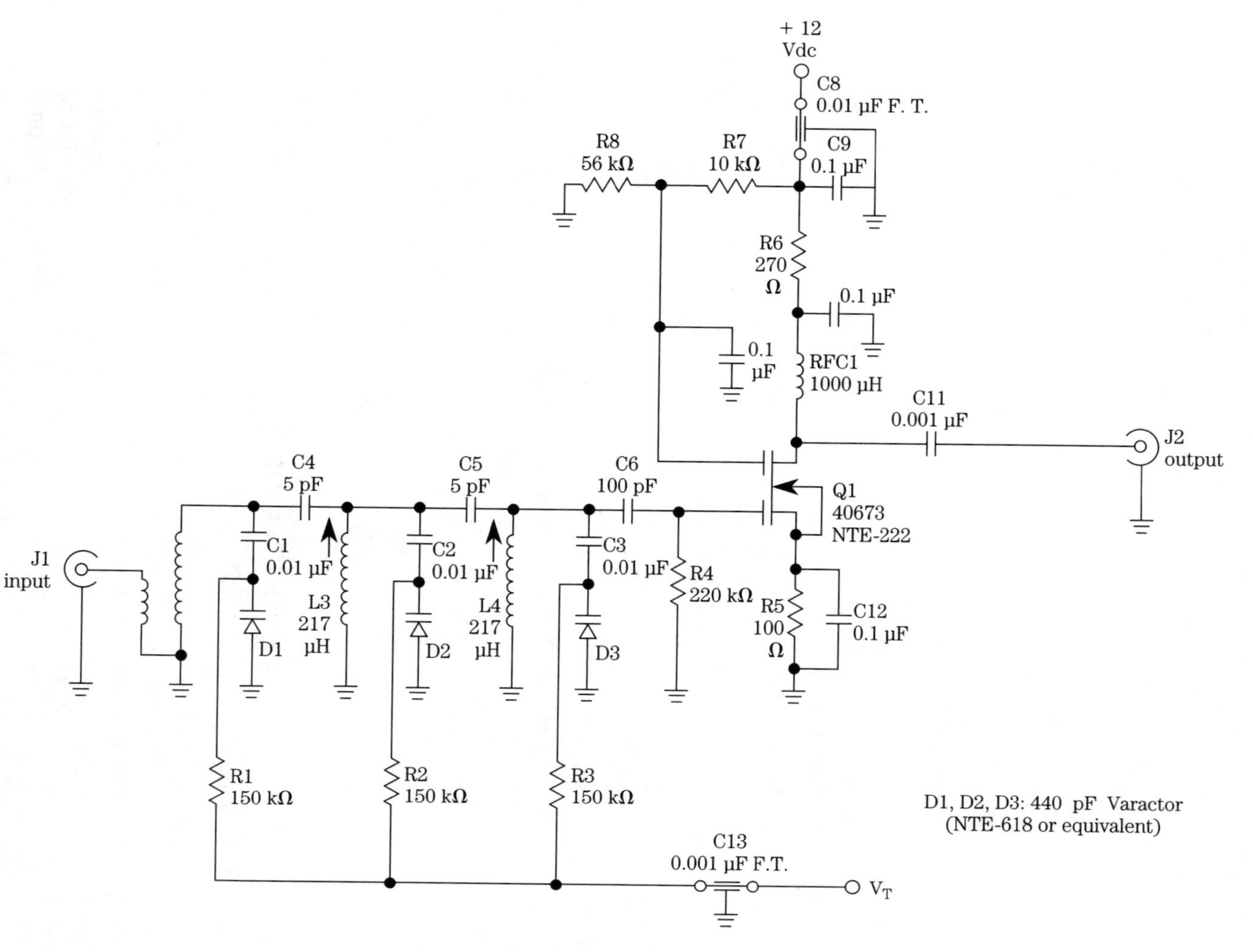

9-7 Voltage-tuned MOSFET preamplifier.

(NTE-618) offers a capacitance range of 440 pF down to 15 pF over the voltage range 0 to +18 Vdc.

The inductors may be either "store-bought" types or wound over toroidal cores. I used a toroid for L_1/L_2 (forming a fixed inductance for L_2) and "store-bought" adjustable inductors for L_3 and L_4. There is no reason, however, why these same inductors cannot be used for all three uses. Unfortunately, not all values are available in the form that has a low-impedance primary winding to permit antenna coupling.

An aluminum utility box was used for the shielded enclosure, and ordinary Vector or Radio Shack perfboard is used for constructing the circuit. The RF input and output connectors are SO-239 "UHF" coaxial connectors, although any other type can also be used. Select connectors that match your receiver and antenna system.

In both of the MOSFET circuits the fixed-bias network used to place gate G2 at a positive dc potential can be replaced with a variable-voltage circuit such as Fig. 9-8. The potentiometer in Fig. 9-8 can be used as an RF gain control to reduce gain on strong signals, and increase it on weak signals. This feature allows the active preselector to be custom set to prevent overload from strong signals.

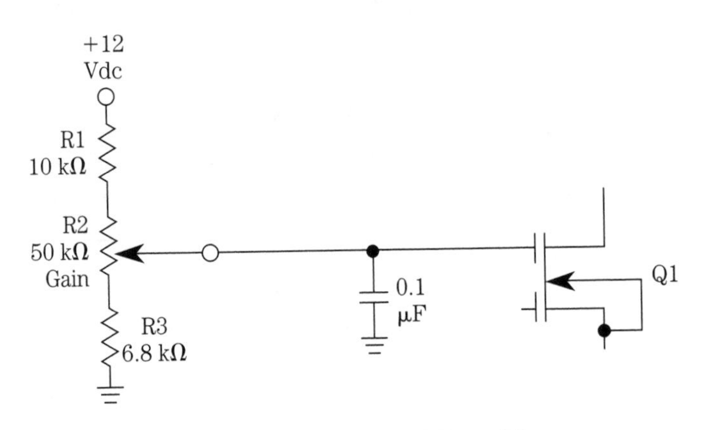

9-8 Gain control for MOSFET RF amplifier.

Project 9-5: Broadband RF preamplifier for VLF, LF, and AM BCB DXing

There are many situations where a broadband RF amplifier is needed. Typical applications include boosting the output of RF signal generators (which tend to be normally quite low level), antenna preamplification, loop antenna amplifier, and in the front ends of receivers. A number of different circuits are published (including some by me), but one failing I've noted on most of them is that they often lack response at the low end of the frequency range. Many designs offer –3 dB frequency response limits of 3 to 30 MHz, or 1 to 30 MHz, but rarely are the VLF, LF, or even the entire AM broadcast band (540 kHz to 1700 kHz)[1] covered.

[1]The AM BCB upper limit was extended to 1700 kHz recently.

The original purpose for building this amplifier was that I needed to boost AM BCB DX signals. Many otherwise fine communications or entertainment grade "general coverage" receivers operate from 100 kHz to 30 MHz, or so, and that range initially sounds real good to the VLF-through-AM BCB DXer. But when examined closer, the receiver lacks sensitivity on the bands below either 2 or 3 MHz, so its performance suffers in the lower end of the spectrum.

While most people only use the AM BCB to listen to powerful local stations (where receivers with no RF amplifier and a loopstick antenna will work nicely), those who are interested in DXing are not well-served. In addition to improving the receiver, I wanted to boost my signal generator's 50-ohm output to make it easier to develop some AM and VLF projects I am working on, and to provide a preamplifier for a square loop antenna that tunes the AM BCB.

Several requirements were developed for the RF amplifier. First, it had to retain the 50-ohm input and output impedances that are standard in RF systems. Second, it had to have a high dynamic range and third-order intercept point in order to cope with the bone-crunching signal levels on the AM BCB. One of the problems of the AM BCB is that those sought-after DX stations tend to be buried under multi-kilowatt local stations on adjacent channels. That's why high dynamic range, a high intercept point, and loop antennas tend to be required in these applications. I also wanted the amplifier to cover at least two octaves (4:1 frequency ratio), and in fact achieved a decade (10:1) response (250 kHz to 2500 kHz).

Furthermore, the amplifier circuit had to be easily modifiable to cover other frequency ranges up to 30 MHz. This last requirement would make the amplifier useful to a large number of readers, as well as extending its usefulness to me.

There are a number of issues to consider when designing an RF amplifier for the front end of a receiver. The dynamic range and intercept point requirements were mentioned above. Another issue is the amount of distortion products (related to the third-order intercept point) that are generated in the amplifier. It does no good to have a high-capability on the preamplifier, only to overload the receiver with a lot of extraneous RF energy it can't handle . . . energy that was generated by the preamplifier, not from the stations being received. These considerations point to the use of a push-pull RF amplifier design.

Push-pull RF amplifiers

The basic concept of a *push-pull amplifier* is demonstrated in Fig. 9-9. This type of circuit consists of two identical amplifiers that each process half of the input sine wave signal. In the circuit shown, this job is accomplished by using a center-tapped transformer at the input to split the signal, and another at the output to recombine the signals from the two transistors. The transformer splits the signal because its center tap is grounded, and thus serves as the common for the signals applied to the two transistors. Because of normal transformer action, the signal polarity at end "A" will be opposite that at end "B" when the center tap ("CT") is grounded. Thus, the two amplifiers are driven 180 degrees out of phase with each other; one will be turning on while the other is turning off, and vice versa.

The push-pull amplifier circuit is balanced, and as a result it has a very interesting property: even-order harmonics are cancelled in the output, so the amplifier out-

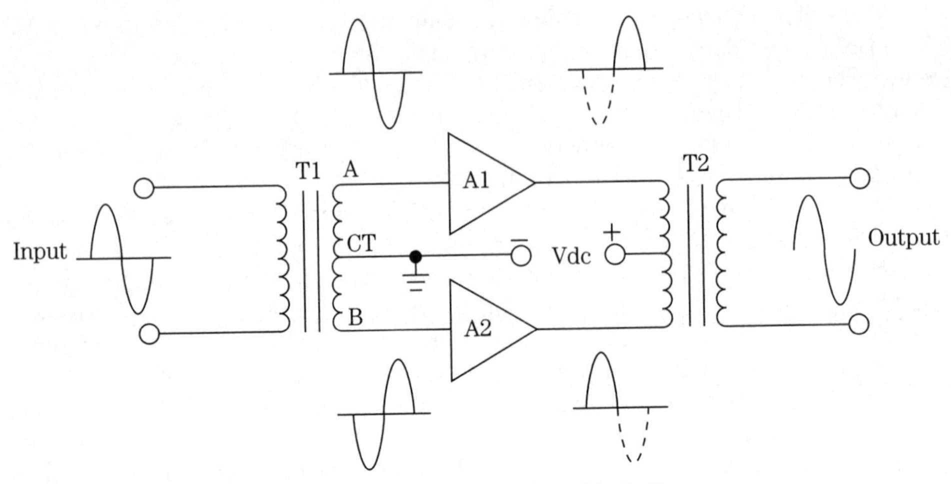

9-9 Push-pull broadband amplifier block diagram.

put signal will be cleaner than for a single-ended amplifier using the same active amplifier devices.

Project 9-6: The push-pull RF amplifier

The two general categories of push-pull RF amplifiers are tuned amplifiers and wideband amplifiers. The *tuned amplifier* has the inductance of the input and output transformers resonated to some specific frequency. In some circuits, the non-tapped winding may be tuned, but in others a configuration such as Fig. 9-10 might be used. In this circuit, both halves of the tapped side of the transformer are individually tuned to the desired resonant frequency. Where variable tuning is desired, a split-stator capacitor might be used to supply both capacitances.

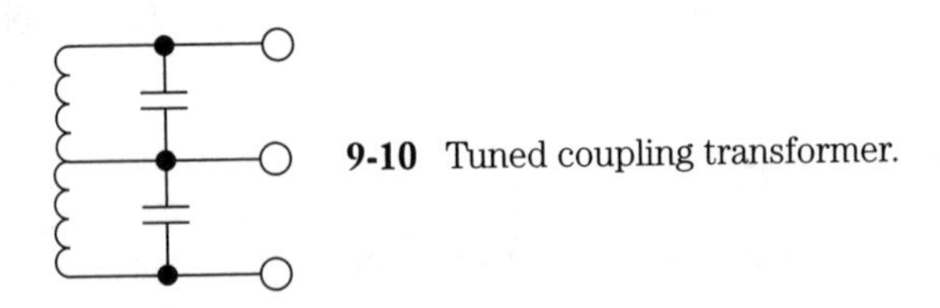

9-10 Tuned coupling transformer.

The *broadband* category of circuit is shown in Fig. 9-11A. In this type of circuit, a special transformer is usually needed. The transformer must be a broadband RF transformer, which means that it must be wound on a suitable core so that the windings are bifilar or trifilar. The transformer in Fig. 9-11A has three windings, of which one is much smaller than the others. These must be trifilar-wound for part of the way, and bifilar the rest of the way. This means that all three windings are kept parallel until no more turns are required of the coupling link, and then the remaining two windings are kept parallel until they are completed. Figure 9-11B shows an example for the case where the core of the transformer is a ferrite or powdered iron toroid.

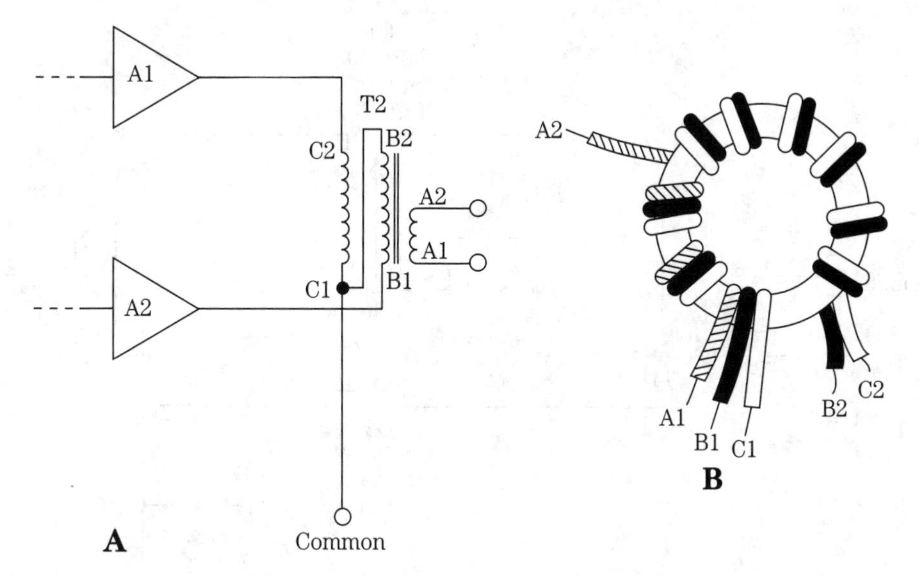

9-11 A. Untuned (broadband) coupling transformer. B. Winding the transformer on a toroid.

Circuit details

The actual RF circuit is shown in Fig. 9-12; it is derived from a similar circuit found in Doug DeMaw's excellent book *W1FB's ORP Notebook*—ARRL, 225 Main Street, Newington, CT 06111. The active amplifier devices are junction field-effect transistors (JFET) intended for service from dc to VHF. The device selected can be the ever-popular MPF-102, or its replacement equivalent from the SK, ECG, or NTE lines of devices. Also useful is the 2N4416 device. The device that I used was the NTE-451 JFET transistor. It offers a transconductance of 4000 microsiemens,[2] a drain current of 4 to 10 mA, and a power dissipation of 310 mW, with a noise figure of 4 dB maximum.

The JFET devices are connected to a pair of similar transformers, T_1 and T_2. The source bias resistor (R_1) for the JFETs, and its associated bypass capacitor (C_1), are connected to the center tap on the secondary winding of transformer T_1. Similarly, the +9 volt dc power supply voltage is applied through a limiting resistor (R_2) to the center tap on the primary of transformer T_2.

Take special note of those two transformers. These transformers are known generally as wideband transmission line transformers, and can be wound on either toroid or binocular ferrite or powdered iron cores. For the project at hand (because of the low frequencies involved), I selected a Type BN-43-202 binocular core. The Type 43 material used in this core is a good selection for the frequency range involved. The core can be obtained from either Amidon Associates—2216 East Gladwick, Dominguez Hills, CA 90220, (213) 763-5770 (voice) or (213) 763-2250 (fax)—or—Ocean State Electronics, P.O. Box 1458, 6 Industrial Drive, Westerly, RI 02891; (401) 596-3080 (voice), (401) 596-3590 (fax), or (800) 866-6626 (orders only).

---

[2]Note units: 1 μSiemen = 1 μMho.

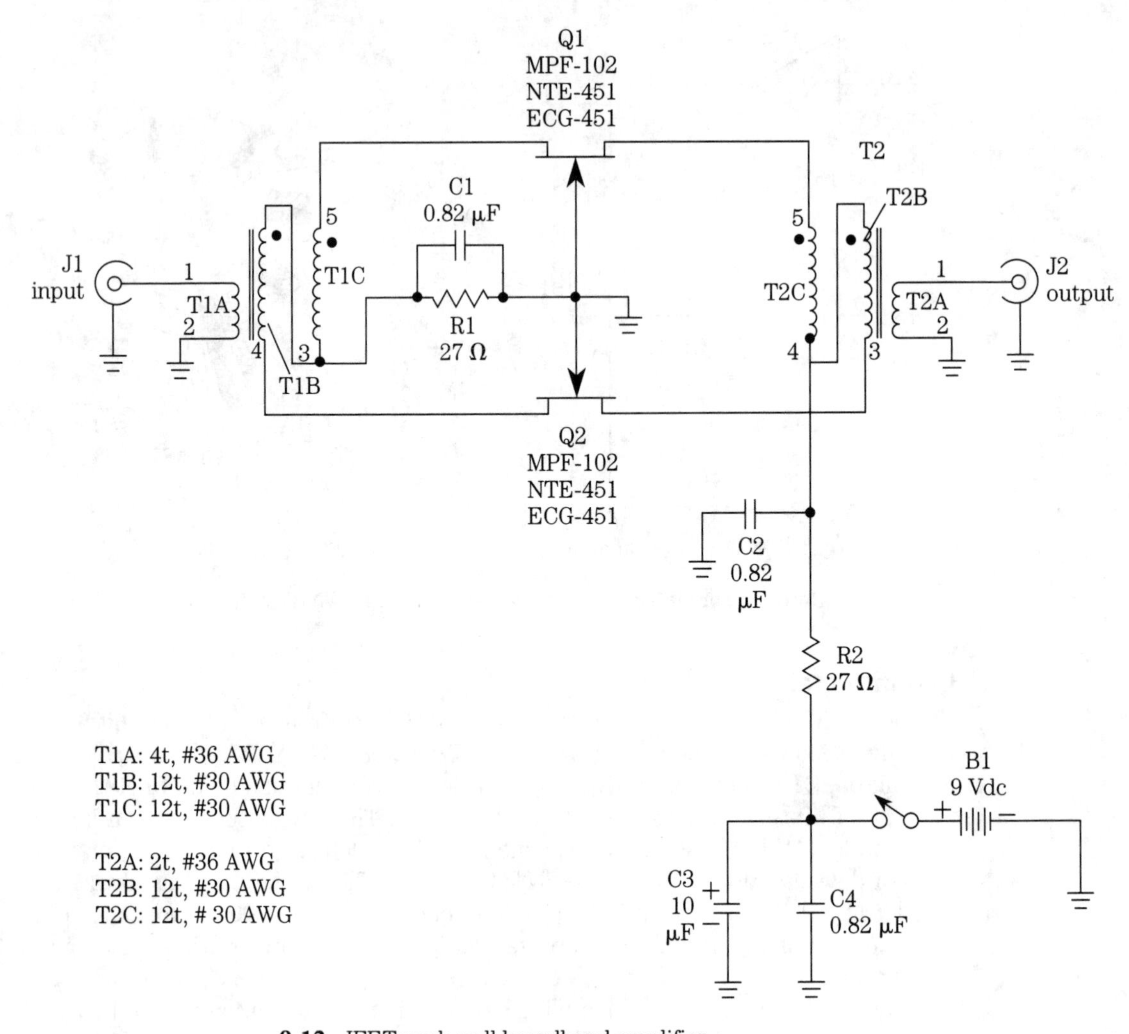

9-12 JFET push-pull broadband amplifier.

There are three windings on each transformer. In each case, the "B" and "C" windings are 12 turns of #30 AWG enameled wire wound in a bifilar manner. The coupling link in each is winding "A." The "A" winding on transformer T_1 consists of four turns of #36 AWG enameled wire—while on T_2, it consists of two turns of the same wire. The reason for the difference is that the number of turns in each is determined by the impedance matching job it must do (T_1 has a 1:9 pri/sec ratio, while T_2 has a 36:1 pri/sec ratio). Neither the source nor drain impedances of this circuit are 50 ohms (the system impedance), so there must be an impedance transformation function.

If the two amplifiers in the circuit were of the sort that had 50 ohm input and output impedances, such as the Mini-Circuits MAR-1 through MAR-8 devices, winding "A" in both transformers would be identical to windings "B" and "C." In that case, the impedance ratio of the transformers would be 1:1:1.

The details for transformers T_1 and T_2 are shown in Fig. 9-13. I elected to build a header of printed circuit perforated board for this part; the board holes are on 0.100 inch centers. The PC type of perfboard has a square or circular printed circuit soldering pad at each hole. A section of perfboard was cut with a matrix of five holes by nine holes. Vector Electronics push terminals are inserted from the unprinted side, and then soldered into place. These terminals serve as anchors for the wires that will form the windings of the transformer. Two terminals are placed at one end of the header, and three are placed at the opposite end.

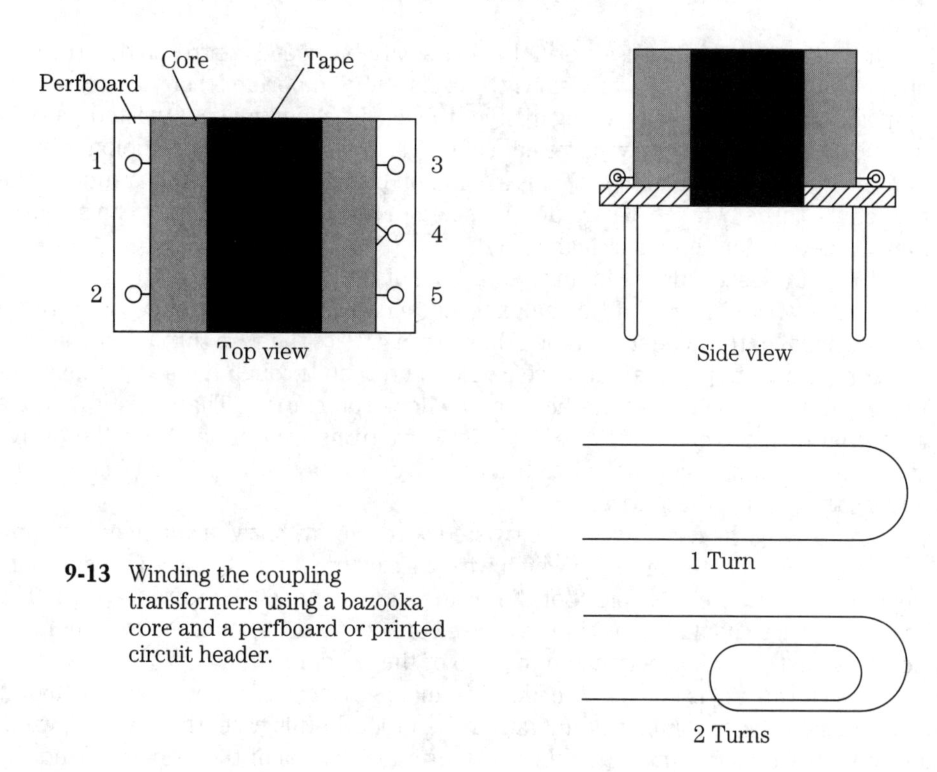

9-13 Winding the coupling transformers using a bazooka core and a perfboard or printed circuit header.

The coupling winding is connected to pins 1 and 2 of the header, and is wound first on each transformer. Strip the insulation from a length of #36 AWG enameled wire for about ¼ inch from one end. This can be done by scraping with a scalpel or X-acto knife, or by burning with the tip of a soldering pencil. Make sure that the exposed end is tinned with solder, and then wrap it around terminal 1 of the header. Pass the wire through the first hole of the binocular core, across the barrier between the two holes, and then through the second hole. This "U" shaped turn counts as one turn. To make transformer T_1 pass the wire through both sets of holes three more times (to make four turns). The wire should be back at the same end of the header as it started. Cut the wire to allow a short length to connect to pin 2. Clean the insulation off this free end, tin the exposed portion and then wrap it around pin 2 and solder. The primary of T_1 is now completed.

The two secondary windings are wound together in the bifilar manner, and consist of twelve turns each of #30 AWG enameled wire. The best approach seems to be twisting the two wires together. I use an electric drill to accomplish this job. Two pieces of wire, each 30 inches long, are joined together and chucked up in an electric drill. The other ends of the wire are joined together and anchored in a bench vise, or some other holding mechanism. I then back off, holding the drill in one hand, until the wire is nearly taut. Turning on the drill causes the two wires to twist together. Keep twisting them until you obtain a pitch of about eight to twelve twists-per-inch.

It is very important to use a drill that has a variable speed control so that the drill chuck can be made to turn very slowly. It is also very important that you follow certain safety rules, especially as regards your eyesight, when making twisted pairs of wire. Be absolutely sure to wear either safety glasses or goggles while doing this operation. If the wire breaks (and that is a common problem), it will whip around as the drill chuck turns. While #36 wire doesn't seem to be very substantial, it can severely injure an eye when thrown at high speed.

To start the secondary windings, scrape all of the insulation off both wires at one end of the twisted pair, and tin the exposed ends with solder. Solder one of these wires to pin 3 of the header, and the other to pin 4. Pass the wire through the hole of the core closest to pin 3, around the barrier, and then through the second hole, returning to the same end of the header as where you started. That constitutes one turn. Now do it eleven more times until all twelve turns are wound. When the twelve turns are completed, cut the twisted pair wires off to leave about ½ inch free. Scrape and tin the ends of these wires.

Connecting the free ends of the twisted wire is easy, but you will need an ohmmeter or continuity tester to see which wire goes where. Identify the end that is connected at its other end to pin 3 of the header, and connect this wire to pin 4. The remaining wire should be the one that was connected at its other end to pin 4 earlier; this wire should be connected to pin 5 of the header.

Transformer T_2 is made in the identical manner as transformer T_1, but with only two turns on the coupling winding rather than four. In this case, the coupling winding is the secondary, while the other two form two halves of the primary. Wind the two-turn secondary first, as was done with the four-turn primary on T_1.

The amplifier can be built on the same sort of perforated board as was used to make the headers for the transformers. Indeed, the headers and the board can be cut from the same stock. The size of the board will depend somewhat on the exact box you select to mount it in. For my purposes, the box was a Hammond 3 × 5.5 × 1.5-inch cabinet. Allowing room for the 9 Vdc battery at one end, and the input/output jacks and power switch at the other, left me with 2.5 × 3.5 inches of available space in which to build the circuit (Fig. 9-14).

I built the circuit from the output end backwards toward the input, so transformer T_2 was mounted first with pins 1 and 2 towards the end of the perfboard. Next, the two JFET devices were mounted, and then T_1 was soldered into place. After that, the two resistors and capacitors were added to the circuit. Connecting the elements together, and providing push terminals for the input, output, dc power supply ground and the +9 Vdc finished the board.

9-14 Completed preamplifier.

Because the input and output jacks are so close together, and because the dc power wire from the battery to the switch had to run the length of the box, I decided to use a shield partition to keep the input and output separated. This partition was made from one-inch brass stock. This material can be purchased at almost any hobby shop that caters to model builders. The RG-174/U coaxial cable between the input jack on the front panel and the input terminals on the perfboard run on the outside of the shield partition.

Variations on the theme

Three variations on the circuit extend its usefulness for many readers. First, there are those who want to use the amplifier at the output of a loop antenna that is remote-mounted. It isn't easy to go up to the roof or attic to turn on the amplifier any time you wish to use the loop antenna. Therefore, it is better to install the 9-Vdc power source at the receiver end, and pass the dc power up the coaxial cable to the amplifier and antenna. This method is shown in Fig. 9-15. At the receiver end RF is isolated from the dc power source by a 10-mH RF choke (RFC2), while the dc is kept from affecting the receiver input (which could short it to ground!) by using a blocking capacitor (C_4). All of these components should be mounted inside of a shielded box. At the amplifier end, lift the grounded side of the T_2 secondary, and connect it to RFC2, which is then connected to the +9-Vdc terminal on the perfboard. A decoupling capacitor (C_3) serves to keep the "cold" end of the T_2 secondary at ground potential for RF, while keeping it isolated from ground for dc.

A second variation is to build the amplifier for shortwave bands. This can be accomplished easily enough. First, reduce all capacitors to 0.1 μF. Second, build the transformers (T_1 and T_2) on a toroid core rather than the binocular core. In the original design (op-cit) a type TF-37-43 ferrite core was used with the same 12:12:2 and 12:12:4 turns scheme as used above.

Alternatively, select a powdered iron core such as T-50-2 (RED) or T-50-6 (YEL). I suspect that about twenty turns will be needed for the large windings, and four turns for the "A" winding on T_2 and seven turns for the "A" winding on T_1. You

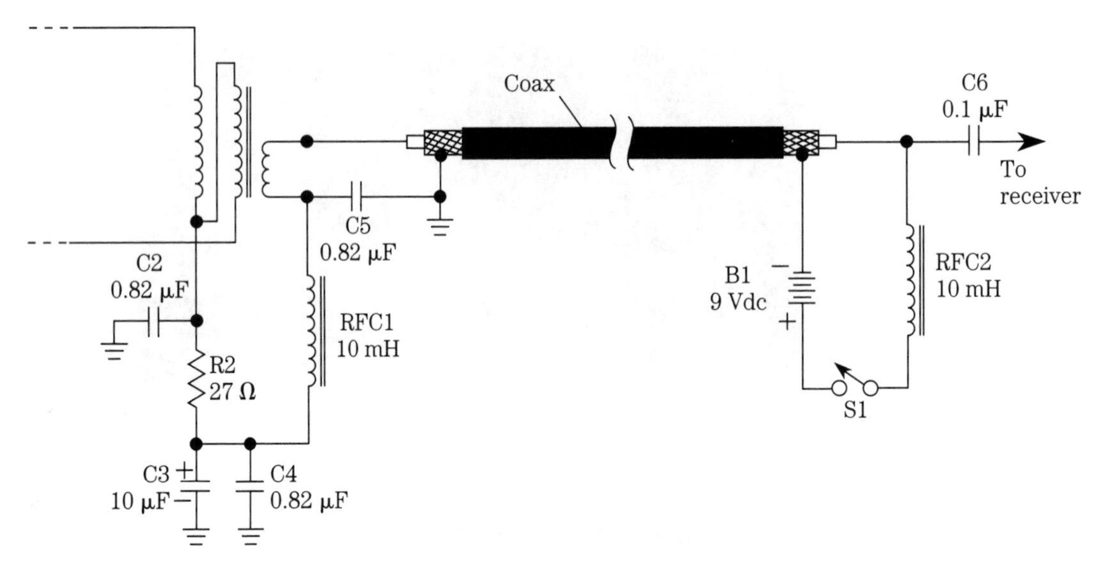

9-15 Powering a remote preamplifier.

can experiment with various cores and turns counts to optimize for the specific section of the shortwave spectrum that you wish to cover.

The third variation is to make the amplifier operate on a much lower frequency (e.g., well down into the VLF region). The principal changes are in the cores used for transformers T_1 and T_2, the number of turns of wire needed, and the capacitors needed. The Type 43 core will work down to 10 kHz, or so, but requires a lot more turns to work efficiently in that region. The Type 73 material, which is found in the BN-73-202 core, will provide an A_L value of 8500, as opposed to 2890 for the BN-43-202 device used in this chapter. Doubling the number of turns in each winding is a good starting point for amplifiers below 200 kHz. The Type 73 core works down to 1 kHz, so with a reasonable number of turns should work in the 20 to 100 kHz range as well.

Project 9-7: Broadband RF amplifier (50-ohm input and output)

This project is a highly useful RF amplifier that can be used in a variety of ways. It can be used as a preamplifier for receivers operating in the 3- to 30-MHz shortwave band. It can also be used as a post-amplifier following filters, mixers, and other devices that have an attenuation factor. It is common, for example, to find that mixers and crystal filters have a signal loss of 5 to 8 dB (this is called "insertion loss"). An amplifier following these devices will overcome that loss.

The amplifier can also be used to boost the output level of signal generator and oscillator circuits. In this service, it can be used either alone, in its own shielded container, or as part of another circuit containing an oscillator circuit.

The RF amplifier circuit is shown in Fig. 9-16. This circuit was originated by Hayward and used extensively by Doug DeMaw in various projects. The transistor (Q1) is a 2N5179 broadband RF transistor. It can be replaced by the NTE-316 or ECG-316 devices, if the original is not available to you. The NTE and ECG devices are intended for service and maintenance replacement applications, so they can often be found at local electronic parts distributors.

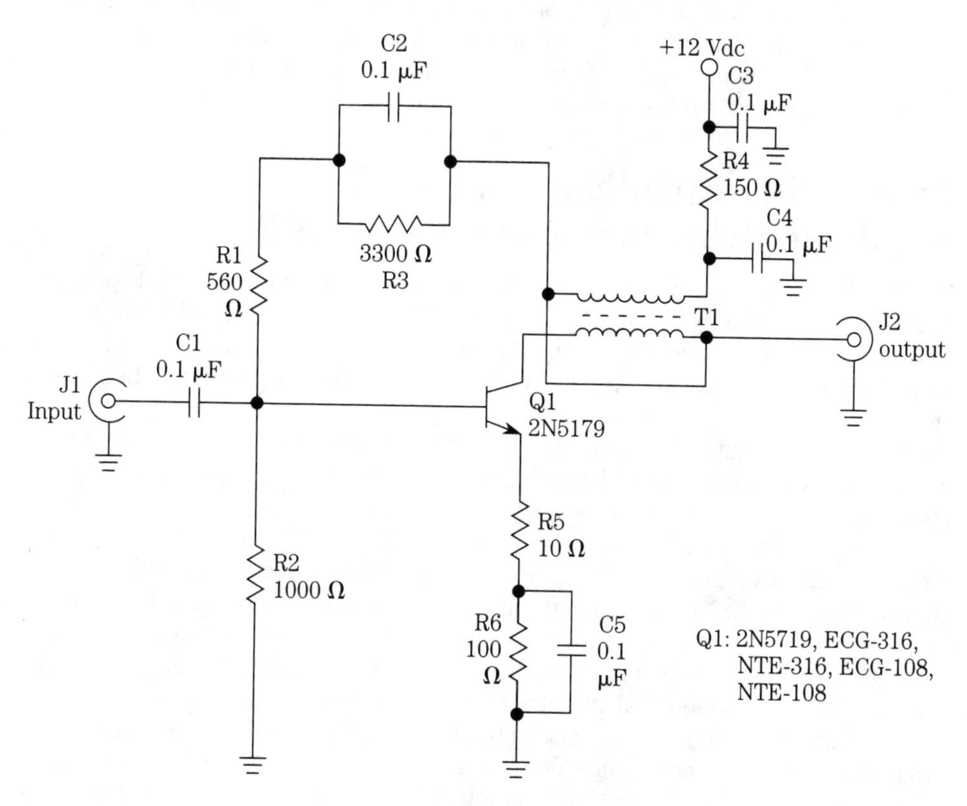

9-16 Feedback npn transistor preamplifier.

This amplifier has two main features: the degenerative feedback in the emitter circuit, and the feedback from collector to base. Degenerative, or negative, feedback is used in amplifiers to reduce distortion (i.e., make it more linear) and to stabilize the amplifier. One of the negative feedback mechanisms of this amplifier is seen in the emitter. The emitter resistance consists of two resistors, R5 is 10 ohms and R6 is 100 ohms. In most amplifier circuits, the emitter resistor is bypassed by a capacitor to set the emitter of the transistor at ground potential for RF signals, while keeping it at the dc level set by the resistance. In normal situations, the reactance of the capacitor should be no more than 1⁄10 of the resistance of the emitter resistor. The 10-ohm portion of the total resistance is left unbypassed, forming a small amount of negative feedback.

The collector-to-base feedback is accomplished by two means. First, a resistor-capacitor network (R1/R3/C2) is used; second, a 1:1 broadband RF transformer (T1) is used. This transformer can be homemade. Wind 15 bifilar turns of #26 enameled wire on a toroidal core such as the T-50-2 (RED) or T-50-6 (YEL); smaller cores can also be used.

The circuit can be built on perforated wire-board that has a grid of holes on 0.100 inch centers. Alternatively, you can use the printed circuit board pattern shown in Fig. 9-16B. In this version of the project, the PCB is designed for use with a Mini-Circuits 1:1 broadband RF transformer. Alternatively, use a homebrew transformer made on a small toroidal core. Use the size 37 core, with #36 enameled wire. As in the previous case, make the two windings bifilar.

Project 9-8: Broadband or tuned RF/IF amplifier using the MC-1350P

The MC-1350P is a variant of the MC-1590 device, but unlike the 1590, it is available in the popular and easy-to-use eight-pin mini-DIP package. It has sufficient gain to make a 30-dB amplifier, although it is a bit finicky and tends to oscillate if the circuit is not built correctly. Layout, in other words, can be a very critical factor because of the gain.

Readers who cannot find the MC-1350P are advised to seek the NTE-746 or ECG-746. These devices are MC-1350Ps, but are sold in the service and maintenance replacement lines, and are usually available locally.

Figure 9-17A shows the basic circuit for the MC-1350P amplifier. The signal is applied to the –IN input, pin 4, while the +IN input is decoupled to ground with a 0.1 µF capacitor. All capacitors in this circuit, except C6 and C7, should be disk ceramic, or one of the newer dielectrics that are competent at RF frequencies to 30 MHz. A capacitor in series with the input terminal, C1, is used to prevent dc riding on the signal from affecting the internal circuitry of the MC-1350P.

The output circuitry is connected to pin 1 of the MC-1350P. Because this circuit is broadband, the output impedance load is a radio frequency choke (L1). For most HF applications, L1 can be a 1-mH choke, although for the lower end of the shortwave region, the medium wave band and the AM broadcast band, use a 2.5-mH choke. The same circuit can be used for 455 kHz IF amplifier service if the coil (L1) is made 10 mH.

Pin 5 of the MC-1350P device is used for gain control. This terminal needs to see a voltage of +5 to +9 volts, with the maximum gain being found at the +5 volts end of the range (this is opposite what is seen in other chips). The gain control pin is bypassed for RF signals.

The dc power supply is connected to pins 8 and 2 simultaneously. These pins are decoupled to ground for RF by capacitor C4. The ground for both signals and dc power are at pins 3 and 7. The V+ is isolated somewhat by a 100-ohm resistor (R3) in series with the dc power supply line. The V+ line is decoupled on either side of this resistor by electrolytic capacitors. C6 should be a 4.7- to 10-µF tantalum unit, while C7 is a 68-µF (or greater) tantalum or aluminum electrolytic capacitor.

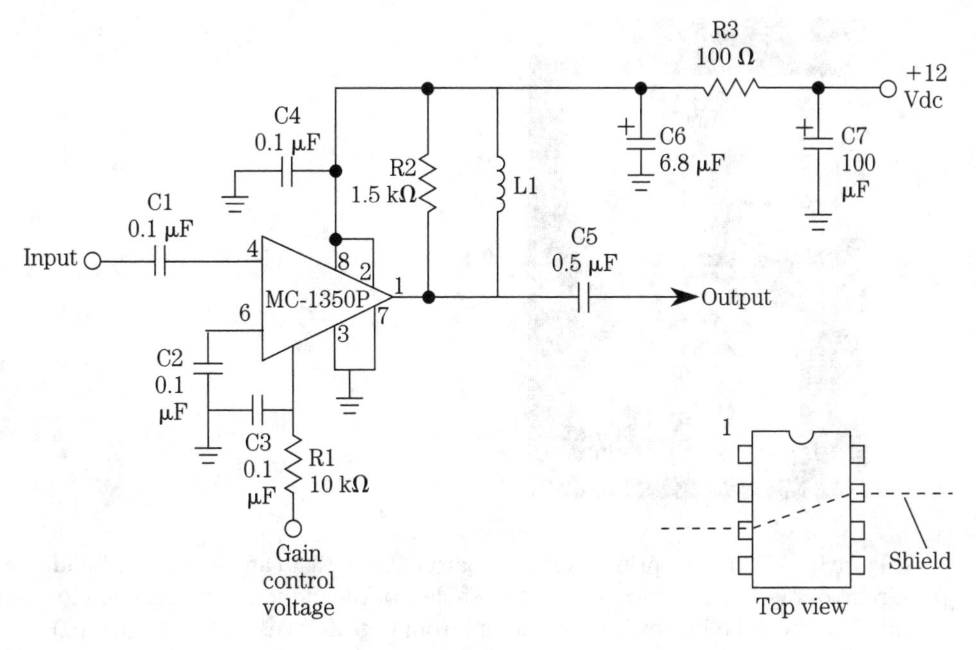

9-17 A. MC-1350P preamplifier.

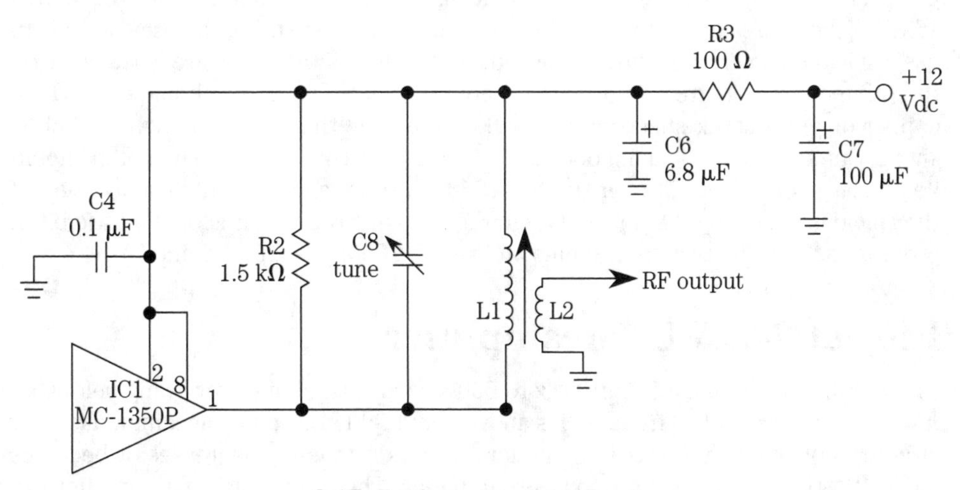

9-17 B. Alternate output circuit.

A partial circuit with an alternate output circuit is shown in Fig. 9-17B. This circuit is tuned, rather than broadband, so it might be used for IF amplification or RF amplification at specific frequencies. Capacitor C8 is connected in parallel with the inductance of L1, tuning L1 to a specific frequency. In order to keep the circuit from oscillating the resonant tank circuit is "de-Qed" by connecting a 2.2-kilohm resistor in parallel with the tank circuit. Although considered optional in Fig. 9-17A, it is not optional in this circuit if you want to prevent oscillation.

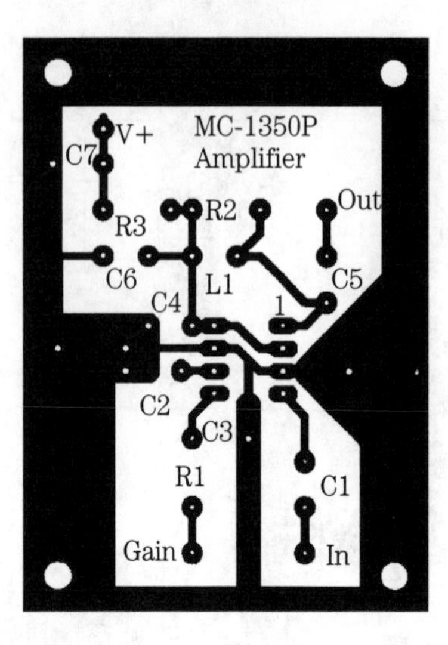

9-17 C. Printed circuit board pattern.

Figure 9-17C shows a printed circuit board pattern that can be used for building the circuit of Fig. 9-17A. The spacing of the holes for the inductor are designed to accommodate the Toko line of fixed inductors from Digi-Key (use size 7 or size 10).

The MC-1350P device has a disgusting tendency to oscillate at higher gains. One perfboard version of the Fig. 9-17A circuit that I built would not produce more than 16 dB without breaking into oscillation. One tactic of preventing the oscillation is to use a shield between the input and output of the MC-1350P. There are extra holes on the printed circuit pattern in order to anchor a shield. The shield should be made of copper or brass stock sheet metal, such as the type that can be bought at hobby shops. Cut a small notch along one edge of a piece of 1-inch stock. The notch should be just large enough to fit over the MC-1350P without shorting out. The location of the shield is shown by the dotted line in Fig. 9-17A. Note that it is bent a little bit in order to fit from the two ground pins on the MC-1350P (i.e., pins 2 and 7).

Project 9-9: VLF preamplifier

The VLF bands run from 5 or 10 kHz to 500 kHz, or just about everything below the AM broadcast band. The frequencies above about 300 kHz can be accommodated by circuitry not unlike 455 kHz IF amplifiers. But as frequency decreases, it becomes more of a problem to build a good preamplifier. This project is a preamplifier designed for use from 5 kHz to 100 kHz. This band contains a lot of Navy communications, as well as the most accurate time and frequency station operated by the National Institute of Science and Technology (NIST), WWVB. The operating frequency of WWVB is a very accurate 60 kHz, and is used as a frequency standard in many situations. The signal of WWVB is also used to update electronic clocks.

The circuit of Fig. 9-18A is similar to Fig. 9-17A, except that a large-value RF choke (L1) is used across the –IN and +IN terminals. Both of these chokes are 120-mH size 10 Toko units. If oscillation occurs, select a different value (82 mH or 100 mH) for one of these chokes. The problem is that the chokes have a capacitance be-

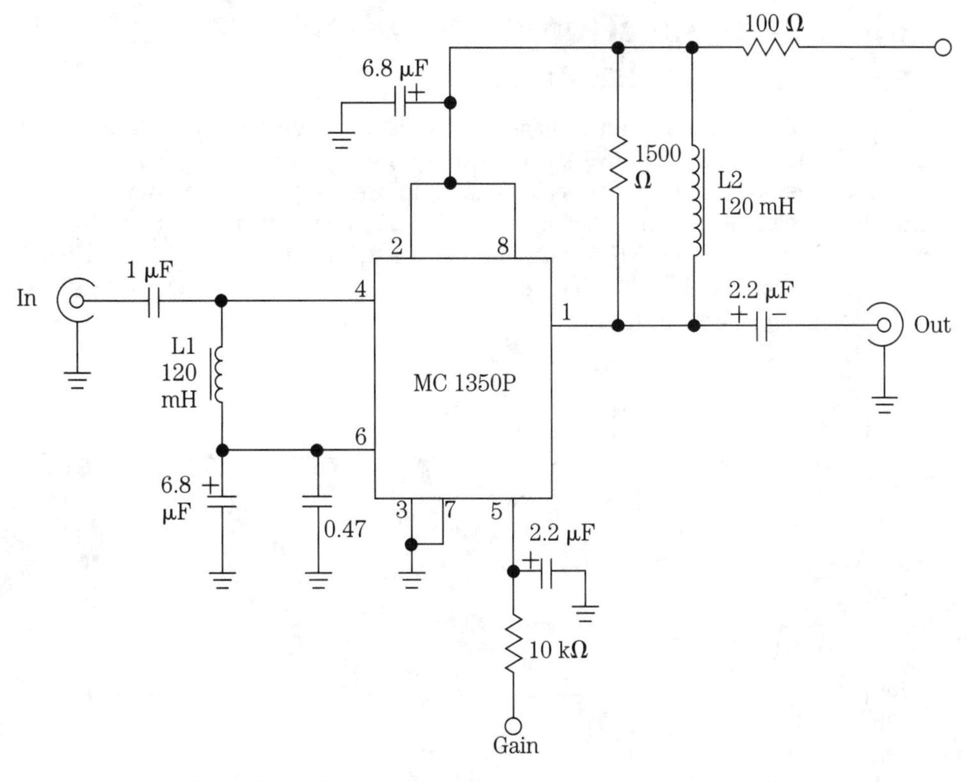

9-18 A. Circuit for an MC-1350P preamplifier (alternate version).

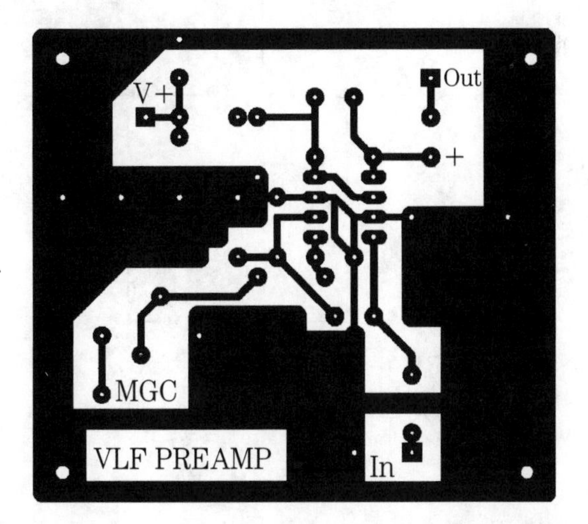

9-18 B. Printed circuit pattern.

tween windings, and that capacitance can resonate the choke to a frequency (this is the "self-resonance" factor). By using different values of choke in input and output circuits, we move their self-resonant frequencies away from each other, reducing the chance of oscillation.

Project 9-10: Electronic RF switch for IF and RF applications

In a number of applications, an RF signal must be switched between two different loads. Unfortunately, mechanical switches and relays are not very good at dealing with RF/IF signals. They tend to leak signal, and are a bit unpredictable. An electronic diode switching scheme is shown in Fig. 9-19. The central feature of this circuit is a trifilar broadband RF transformer with a 1:1:1 ratio. You can either wind your own using a T-37-2 or T-37-6 core (larger cores can also be used), using 12 trifilar turns of #30 or smaller wire.

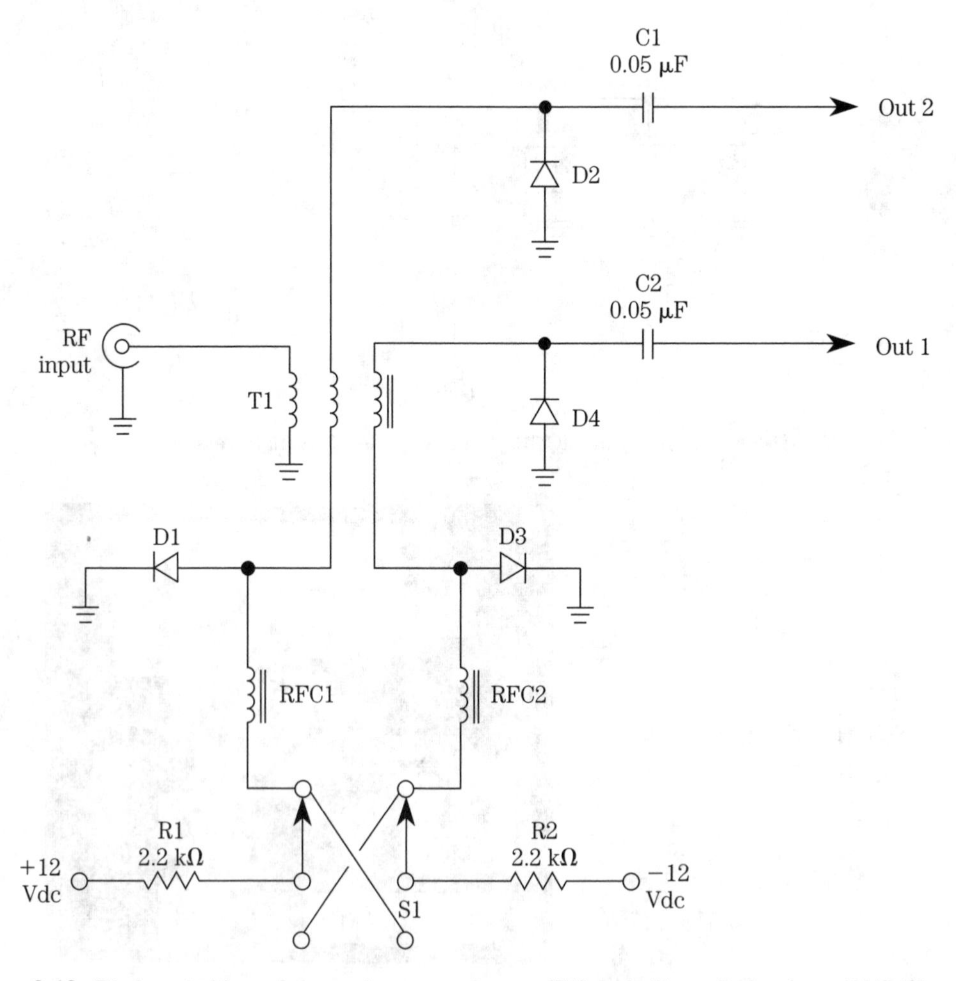

9-19 Diode switching scheme using a transformer (T1) and silicon diodes (e.g., 1N4148).

Signal is input to the circuit through one winding of the transformer. The output signal (OUT 1 or OUT 2) depends on which of the other two coils are activated by switching.

The switching occurs because of the action of diodes. Ordinary silicon diodes, such as 1N914 and 1N4148, can be used. However, Schottky PIN diodes are preferred. The NTE-553 and NTE-555 diodes can be used for this purpose. When forward-biased, the diode will conduct signal, and when reverse-biased it will block signal. When the switch (S1) is in the position shown, a positive voltage is applied to diodes D1 and D2, forward-biasing D1 and reverse-biasing D2. At the same time, a negative voltage reverse-biases D3 and forward-biases D4, preventing the signal in this path from reaching the output. Under these conditions, OUT 2 is activated and OUT 1 is inhibited. Setting the switch to the alternate position will reverse the situation.

This circuit can be used to select crystal or mechanical filters, attenuators, or different RF circuits. Wherever HF and lower signals need to be directed to either of two sources, this circuit can be used.

Project 9-11: IF mechanical filter switching

Shortwave receivers often have insufficient IF filtering when first "out of the box." The better receivers have optional filters that can be installed, but many other users have to rely on added filters. A good candidate is the Collins mechanical filter. These filters are legendary for their shape factor (steep skirts), so are considered superior for rejecting nearby interfering signals. Only the highest grade communications receivers have mechanical filters to set the IF bandpass.

One of the trade-offs in selecting filters is between reasonable fidelity and the ability to reject very close signals. If the filter has too wide a passband, then the fidelity will be fine, but nearby stations will cream the receiver reception. On the other hand, if the filter is too narrow, the adjacent channel stuff will be rejected but the fidelity will be messed up terribly. The solution is to have both filters available, and select whichever one suits the situation at any given time. In the circuit of Fig. 9-20, a pair of 455-kHz Collins mechanical filters are used: a 5.5-kHz bandwidth filter is used for normal AM reception, while a 2.7-kHz filter can be used for single-sideband (SSB) or when you are receiving AM under heavy interference and some reduction in fidelity is permissible.

The circuit of Fig. 9-20 is based on the circuit of Fig. 9-19A, with appropriate impedance matching added. The input and output impedances of the filters are on the order of 2000 ohms, so the impedance reflected across the transformers will be 2000 ohms also. In order to match these impedances to 50 ohms, standard for RF circuits, a pair of 36:1 ratio broadband RF transformers (T1 and T4) are added at the input and output circuits.

Figure 9-21 shows a pair of printed circuit board patterns that can be used with this circuit. The circuit in Fig. 9-21A retains both input and output impedance transformers, so can be used for 50-ohm circuits. The smaller version is more compact, but doesn't use the output impedance transformer. In the case where Fig. 9-21B is used, the output impedance is 2000 ohms . . . which is sufficient for use with most detector circuits other than simple diode envelope detectors.

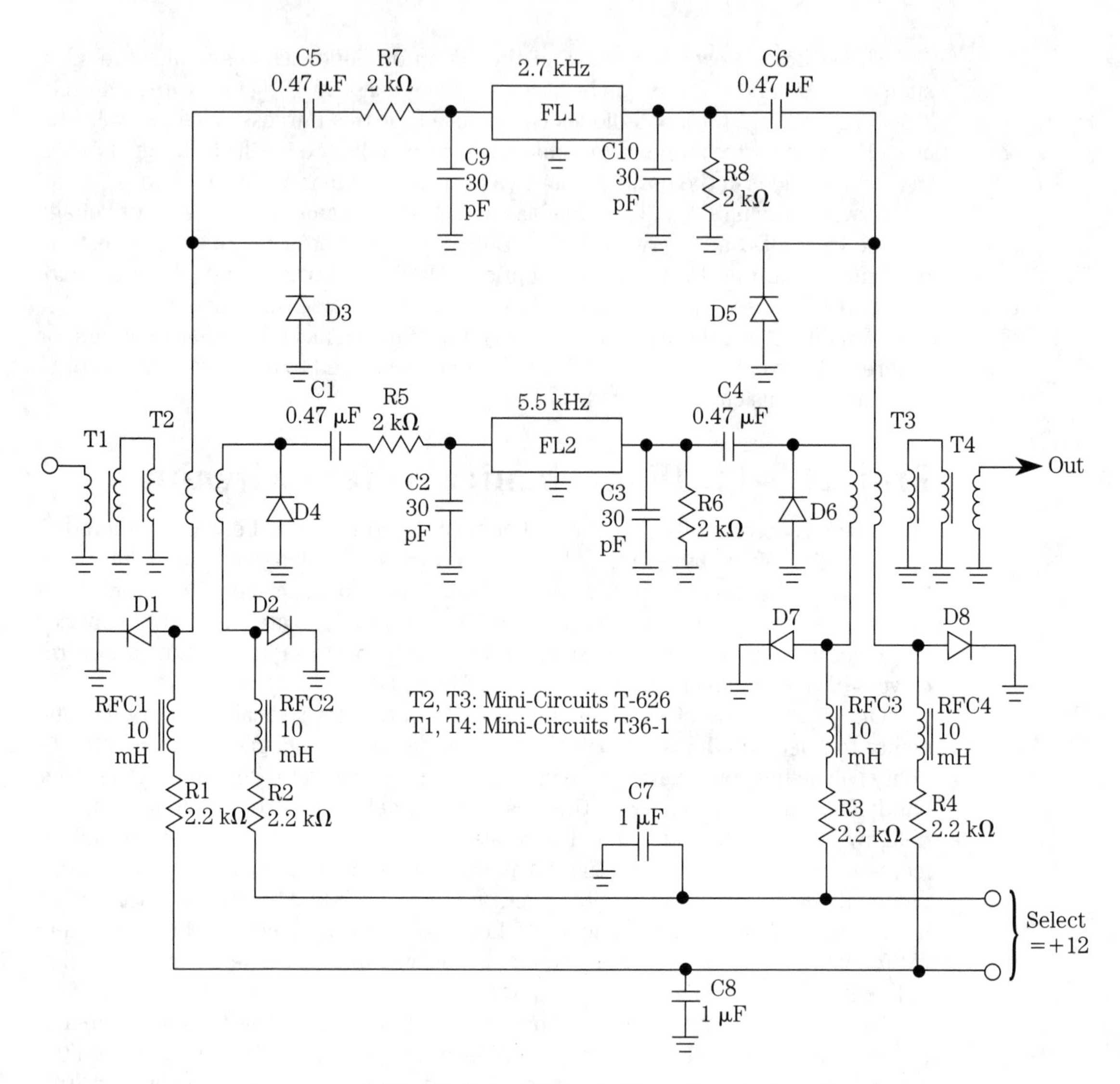

9-20 Bandpass filter switching scheme.

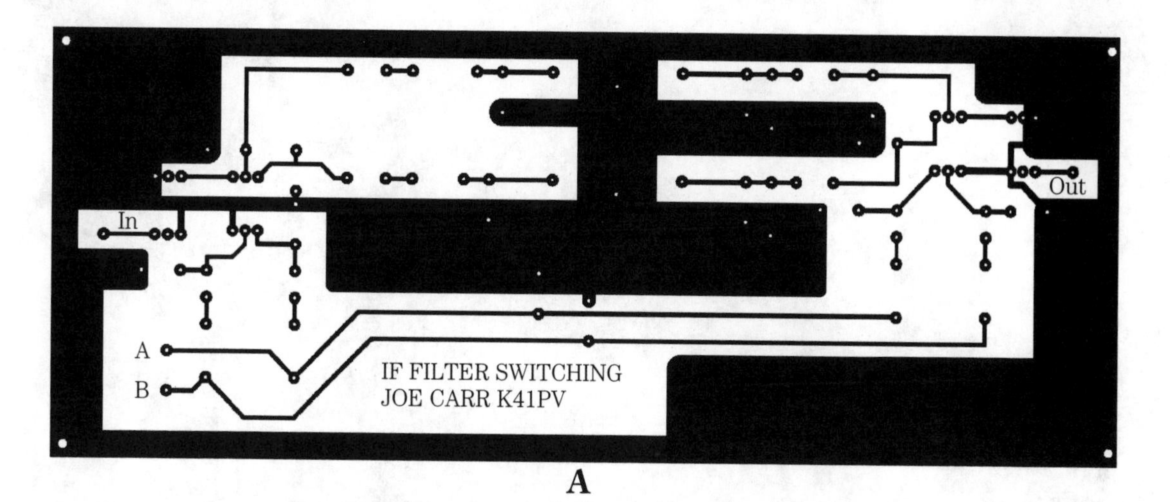

A

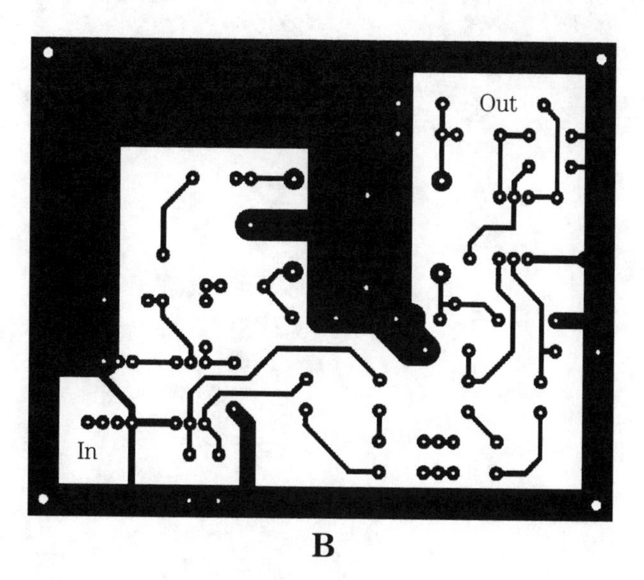

B

9-21 Filter printed circuit (Fig. 9-20). A. Style 1. B. Style 2.

10
RF mixer and frequency converter circuits

A *mixer* is a nonlinear circuit or device that permits frequency conversion (or "translation") by the process of heterodyning. Mixers are used in the front end of the most common form of radio (the superheterodyne); in certain electronic instruments; and in certain measurement schemes (receiver dynamic range, oscillator phase noise, etc).

The block diagram for a basic mixer system is shown in Fig. 10-1; this diagram is generic in form, but also represents the front end of superheterodyne radio receivers. The mixer has three ports: F_1 receives a low-level signal, and corresponds to the RF input from the aerial in radio receivers; F_2 is a high-level signal, and corresponds to the local oscillator (LO) in superhet radios; and F_3 is the resultant mixer product (corresponding to the intermediate frequency or "IF" in superhet radios). These frequencies are related by:

$$F_3 = mF_1 \pm nF_2 \qquad \textbf{[10-1]}$$

where

F_1, F_2, and F_3 are as described above
m and n are counting numbers (zero plus integers 0, 1, 2, 3, ...)

In any given circuit, m and n can be zero—or any integer—but in practical circuits it is common to consider only the first, second, and third-order products. For sake of simplicity, let's consider a first-order circuit ($m = n = 1$). Such a mixer would output four frequencies: F_1, F_2, $F_3 = F_1 + F_2$, and $F_3 = F_1 - F_2$. In terms of a radio receiver, these frequencies represent the RF input signal, the local oscillator signal, the sum IF and the difference IF. In radios, it is common practice to select either the sum or difference IF by filtering, and reject all others.

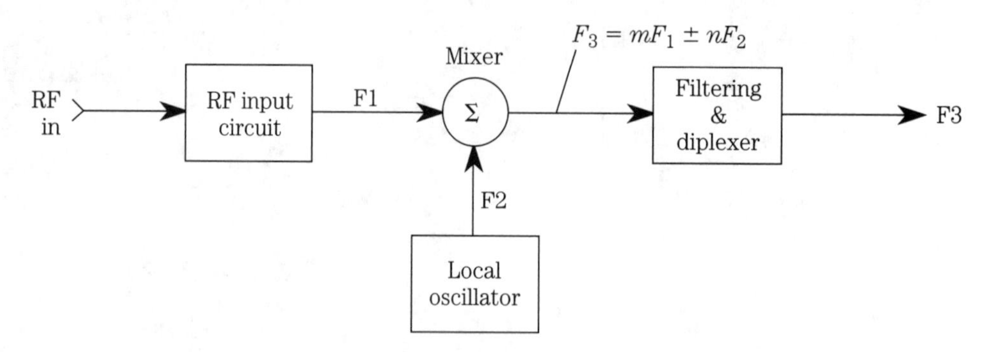

10-1 Mixer circuit block diagram.

A *diplexer* and *filter* stage is shown in Fig. 10-1. It is used to absorb unwanted mixer products and pass desired frequencies. The filter selects which frequencies are favored and which are rejected. Frequency-selective circuits are discussed below.

A *post-amplifier* stage is sometimes used following the diplexer because the insertion loss of most passive DBMs is considerable (5 to 12 dB). The purpose of this amplifier is to make up for the loss of signal level in the mixing process. Some active DBMs, incidentally, have a *conversion gain* figure instead of a loss. For example, the popular *Signetics* NE-602 device offers 20 dB of conversion gain.[1]

Diplexer circuits

The *RF mixer* is like most RF circuits in that it should be terminated in its characteristic impedance. Otherwise, a standing wave ratio (SWR) problem will result, causing signal loss and other problems. In addition, certain passive DBMs (of which, more below) will not work well if improperly terminated. A number of different diplexer circuits are known, but two of the most popular are shown in Figs. 10-2 and 10-3.

A *diplexer* has two jobs: it absorbs undesired mixer output signals, so they are not reflected back into the mixer, and it transmits desired signals to the output. In Fig. 10-2, these goals are met with two separate LC networks; a high-pass filter and a low-pass filter. The assumption in this circuit is that the difference IF is desired, so a high-pass filter with a cut-off above the difference IF is used to shunt the sum IF (plus and LO and RF signals that survived the DBM process) to a dummy load (R1). The dummy load shown in Fig. 10-2 is set to 50 ohms because that is the most common system impedance for RF circuits (in practice, a 51-ohm resistor might be used). The dummy load resistor can be a ¼-watt unit in most low-level cases—but regardless of power level, it must be a noninductive type (e.g., carbon composition or metal film).

The inductor (L_1) and capacitor (C_1) values in the high-pass filter are designed to have a 50-ohm reactance at the IF frequency. These values can be calculated from:

---

[1]"NE-602 Primer," Joseph J. Carr, *Elektor Electronics*, January 1992.

$$L_{\text{henrys}} = \frac{50\ \Omega}{2\pi F_{3\text{ Hz}}} \qquad [10\text{-}2]$$

$$C_{\text{farads}} = \frac{1}{2\pi F_{3\text{ Hz}}\,(50\ \Omega)} \qquad [10\text{-}3]$$

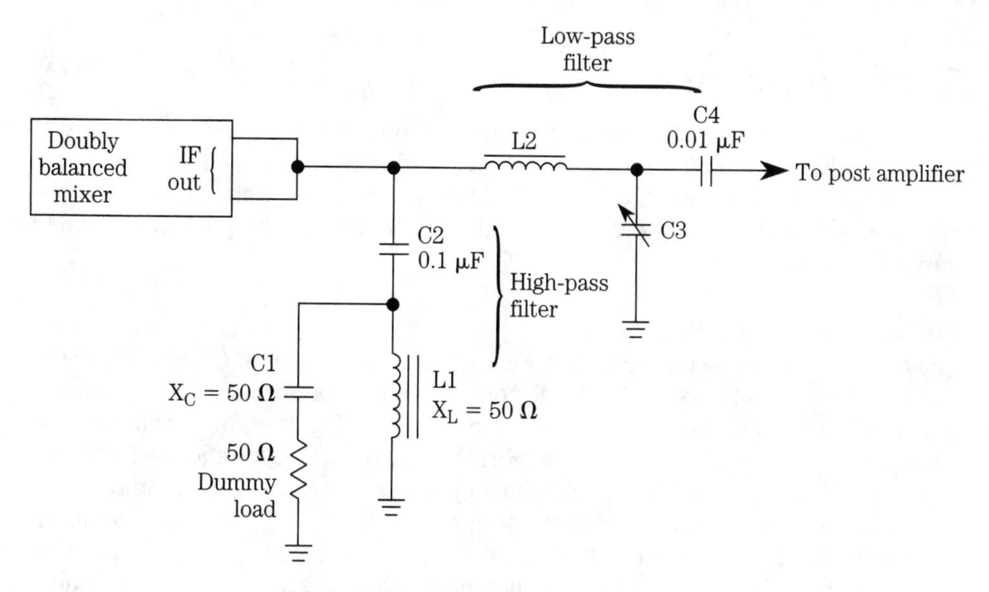

10-2 Diplexer circuit at output of mixer.

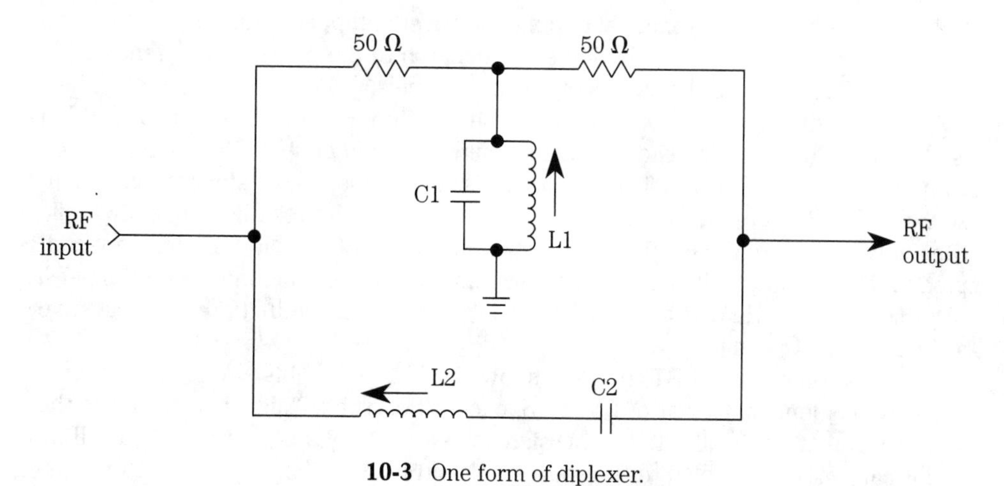

10-3 One form of diplexer.

The low-pass filter transmits the desired difference IF frequency to the output, rejecting everything else. Like the high-pass filter, the L and C elements of this filter are designed to have reactances of 50 ohms at the difference IF frequency.

Another popular diplexer circuit is shown in Fig. 10-3. This circuit consists of a parallel-resonant 50-ohm tank circuit (C1/L1), and a series-resonant 50-ohm tank circuit (C2/L2). The series-resonant circuit passes its resonant frequency while rejecting all others because its impedance is low at resonance and high at other frequencies. Alternatively, the parallel-resonant tank circuit offers a high impedance to its resonant frequency, and a low impedance to all other frequencies. Because C1/L1 are shunted across the signal line, it will short out all but the resonant frequency.

Types of mixer

There are a number of types of mixer circuit, but only a few generic classes: single-ended, singly balanced (or simply "balanced") and doubly balanced. Most low-cost superheterodyne radio receivers use single-ended mixers, although a few of the more costly "communications receiver" models use singly or doubly-balanced mixers for improved performance.

Single-ended mixers

The most commonly used form of mixer circuit is the *single-ended* mixer. These circuits are simple and low-cost, but offer only moderate performance. The simplest form of single-ended mixer is the diode mixer circuit of Fig. 10-4. In this circuit, a pn diode is connected such that its anode is driven with the RF signal and the LO signal, while the cathode is tuned to the IF frequency. Transformer T1 is shown in Fig. 10-4 as a broadband transformer, but in many cases, the transformer is tuned to the RF frequency. The LO signal must have a high level, on the order of +7.5 dBm, so that the diode is switched in and out of conduction by the LO signal. The output tuning is set to the IF frequency. In the version shown the diode is connected to the top of the LC tank circuit. In many practical circuits, however, a tapped inductor (see inset to Fig. 10-4) is used to improve the impedance match to the diode (which tends to be low). This form of mixer has a loss of several decibels.

A single-ended active mixer based on the junction field-effect transistor (JFET) is shown in Fig. 10-5. Instead of a conversion loss, this circuit offers a conversion gain of up to 10 dB. The JFET mixer circuit is very similar to the standard common-source JFET RF amplifier, except that the output is tuned to the IF frequency, while the input is tuned to the RF frequency. In addition, the LO signal is capacitively-coupled to the gate of the JFET. The LO signal amplitude must be on the order of +7.5 dBm so that the JFET can be switched in and out of conduction on successive swings of the LO signal.

A variation on the JFET theme is shown in Fig. 10-6. This circuit (shown only partially) is identical to that of Fig. 10-5, except that the LO signal is applied to the source terminal of the JFET. The LO signal must be 5 volts p-p, or about +18 dBm, and is capacitively-coupled to the source of the transistor. Part of the emitter resistance (R2) remains unbypassed in order to provide a good impedance match to the 50-ohm output of the signal source used as the LO.

Out final example of a single-ended active mixer is the MOSFET circuit shown in Fig. 10-7. The active device is a 40673 dual-gate RF MOSFET. In this circuit, the RF signal is applied to gate 1, while the LO signal is applied to gate 2. The LO signal

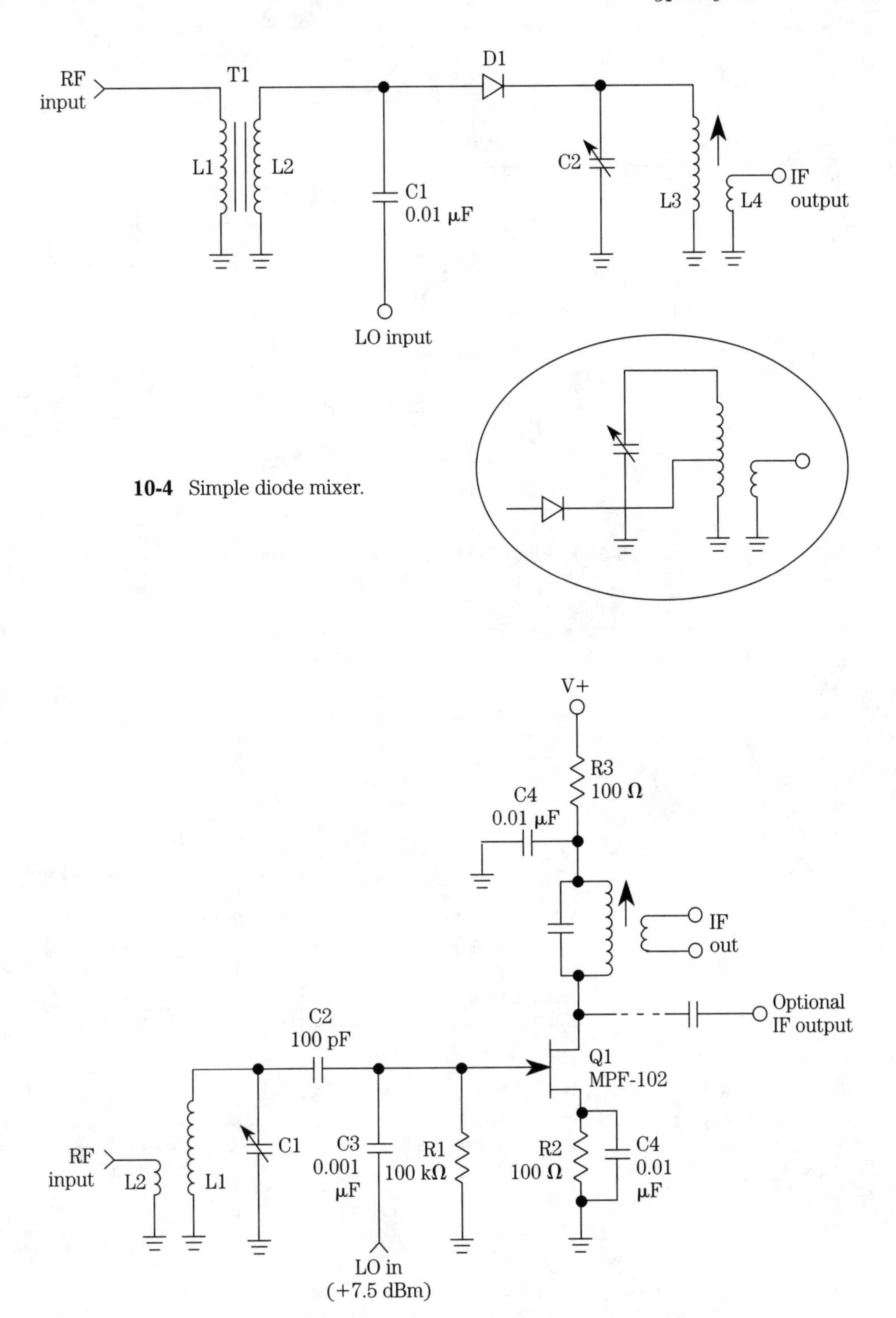

10-4 Simple diode mixer.

10-5 JFET mixer.

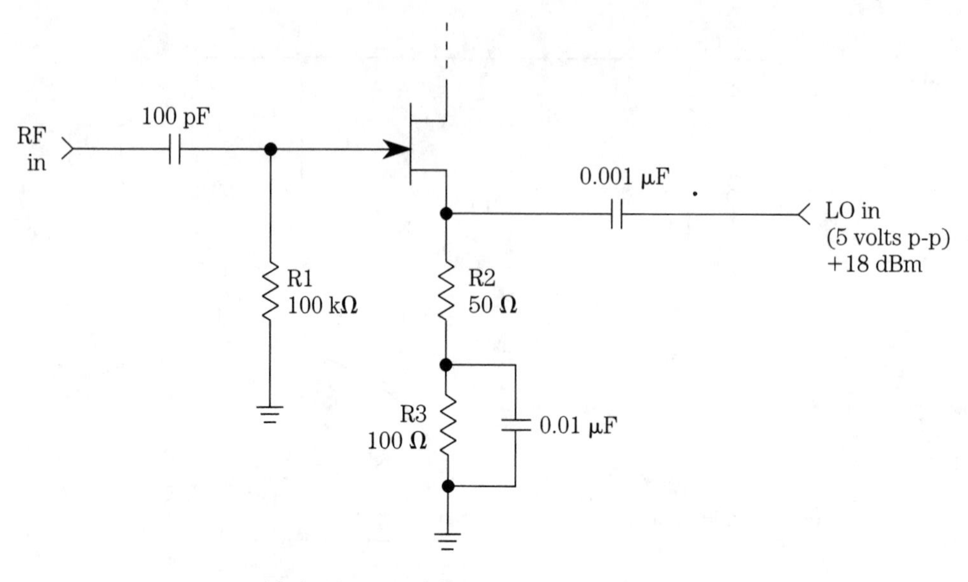

10-6 Alternate LO feed to the JFET mixer.

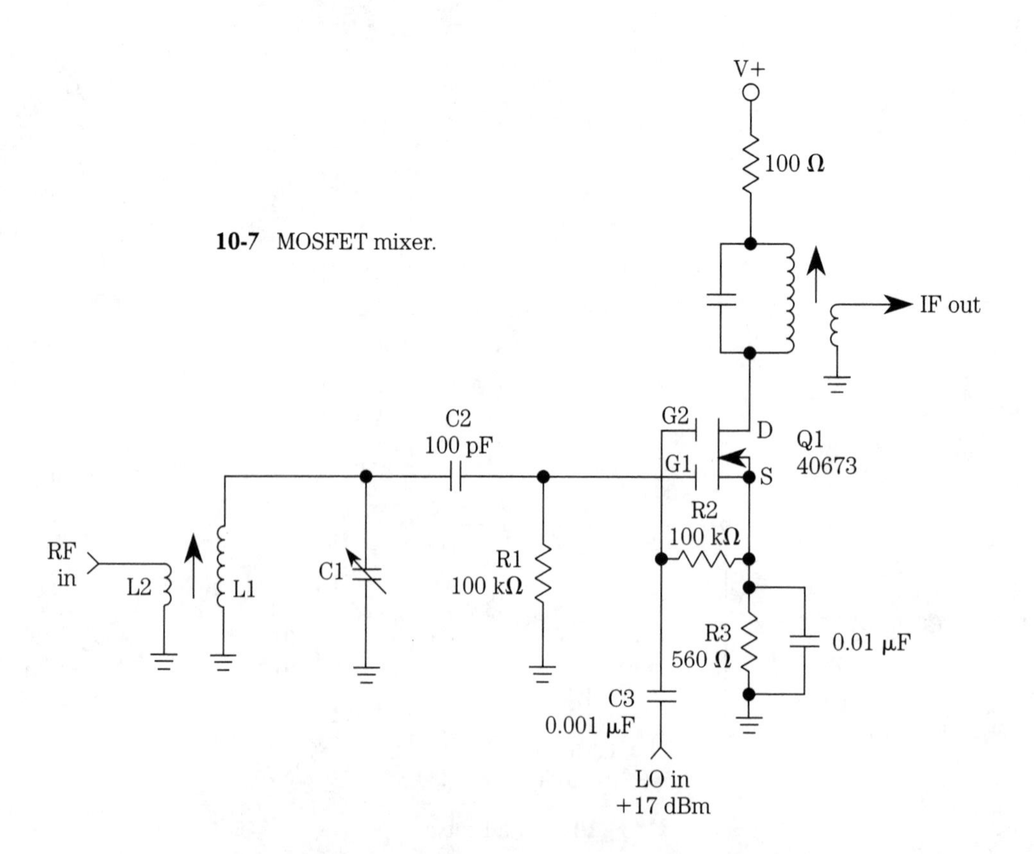

10-7 MOSFET mixer.

must have a level of +17 dBm in order to drive the MOSFET in and out of conduction. As in the previous FET cases, gate 1 is tuned to the RF frequency, while the drain (output) is tuned to the IF frequency.

Doubly-balanced mixers

The basic diode doubly-balanced mixer is shown in Fig. 10-8. This mixture consists of two diodes that are cross-connected to conduct on opposite halves of the applied signal cycle. A trifilar wound transformer, connected in the 4:1 manner, is used to interface the diodes with the RF and LO signals. The IF output of the circuit is at the junction of the two diodes. Signals from this point are normally fed to some kind of filter or IF tuned circuit. Like the passive diode mixture shown earlier, this circuit suffers conversion signal loss.

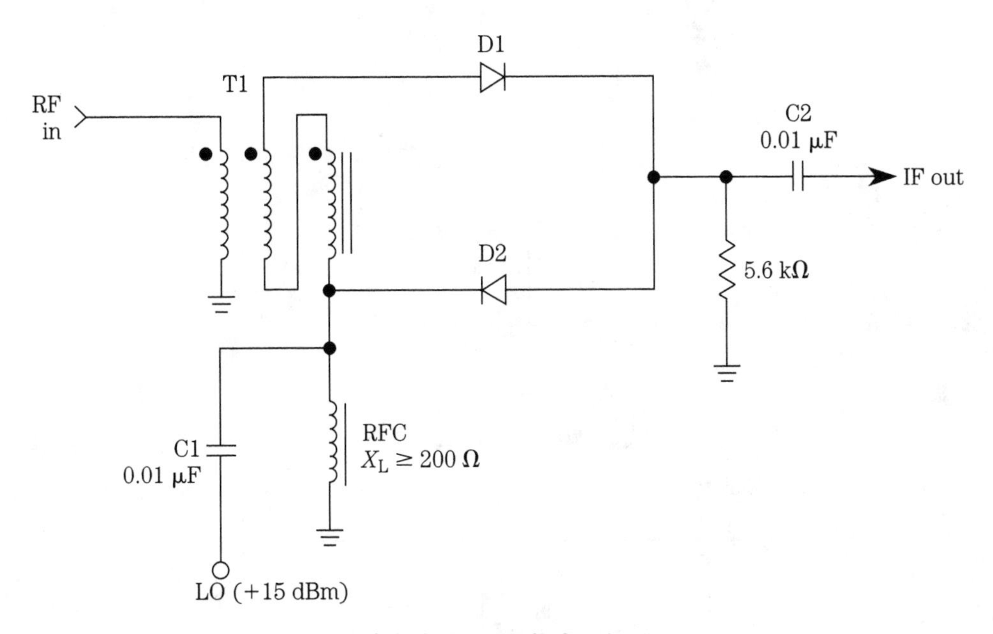

10-8 Full-wave diode mixer.

Active balanced mixers

An active balanced mixer is shown in Fig. 10-9A. This circuit is based on a pair of JFET devices that are arranged in a differential pair. In best practice, the two JFETs are closely matched and share a common thermal environment. These requirements are best met by using a dual JFET device, i.e., an integrated circuit type of device that contains a pair of independent JFETs. The gates of the two JFETs are driven out-of-phase with each other by signals from the trifilar wound input transformer (T1). The drain outputs of the JFETs are connected to a second trifilar wound transformer (T2).

All of the active mixer circuits discussed above have their equivalent in npn and pnp bipolar transistor versions. In those circuits, the RF will be fed to the base ter-

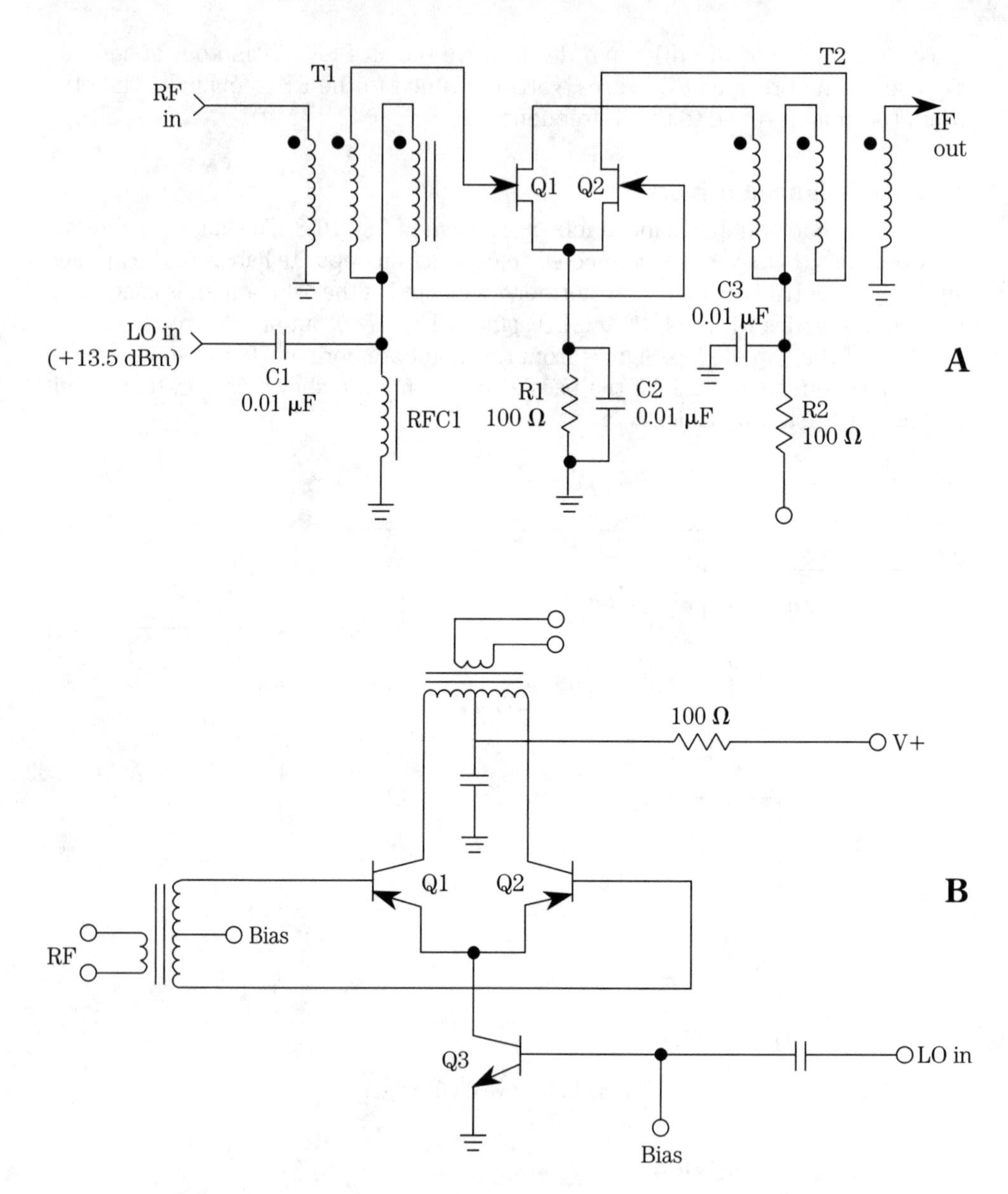

10-9 Full-wave balanced mixer. A. JFET version. B. Bipolar version.

minal, while the LO can be fed to either the base or emitter terminals. Of course, appropriate dc bias must be applied to the device. A differential mixer based on npn transistors is shown in Fig. 10-9B. Although this circuit is partial, it is representative of a wide range of actual devices. Perhaps the most commonly found is based on the CA-3028A integrated circuit. The CA-3028A contains all of the transistors needed, along with a bias network. Under most circumstances, only a pair of bias resistors to serve the Q1/Q3 common base supply is needed.

Double-balanced mixers (DBM)

One advantage of the double-balanced mixer (DBM) over other forms of mixers is that the DBM suppresses F1 and F2 components of the output signal, passing only the sum and difference signals. In a radio receiver using a DBM, the IF filtering and amplifier would only have to contend with sum and difference IF frequencies, and not bother with the LO and RF signals. This effect is seen in DBM specifications as the port-to-port isolation figure, which can reach 30 to 60 dB, depending on DBM model.

JFET and MOSFET doubly-balanced mixer circuits

Junction field-effect transistors (JFETs) and metal oxide semiconductor transistors (MOSFETs) can be arranged in a ring circuit that provides good doubly-balanced mixer operation. Figure 10-10 shows a circuit that is based on JFET devices. Although discrete JFETs (such as MPF-102 or its equivalent) can be used in this circuit with success, performance is generally better if the devices are matched, or are part of a single IC device (e.g., the U350 IC). With due attention to layout and input/output balanced, the circuit is capable of better than 30 dB port-to-port isolation over an octave (2:1) frequency change.

The inputs and output of this circuit are based on broadband RF transformers. These transformers are bifilar and trifilar wound on toroidal cores. The input circuit consists of two bifilar wound impedance transformers (T1 and T2). The LO and output circuits are trifilar-wound RF transformers. Part of the output circuit includes a pair of low-pass filters that also serve to transform the 1.5- to 2-kilohm impedance of the JFET devices to 50 ohms. As a result of the needed impedance transformation, the filters must be designed with different R_{in} and R_{out} characteristics, and that complicates the use of loop-up tables (which would be permitted if the input and output resistances were equal).

Figure 10-11 shows an integrated circuit DBM based on MOSFET transistors. This device was first introduced as the Siliconix Si8901, but they no longer make it. Today, the same device is made by Calogic Corporation [237 Whitney Place, Fremont, CA, 94539, USA; Phones 510-656-2900 (voice) and 510-651-3026 (fax)] under part number SD8901. The SD8901 comes in a seven-pin metal can package. The specifications data sheet for the SD8901 claims that it provides as much as 10 dBm improvement in the third-order intercept point over the U350 JFET design, or the diode ring DBM.

A basic circuit for the SD8901 is shown in Fig. 10-12. The input and output terminals are connected to center-tapped, 4:1 impedance ratio RF transformers. Although these transformers can be homemade (using toroidal cores), the Mini-Circuits Type T4-1 transformers were used successfully in an amateur radio construction project. As was true in other DBM circuits, a diplexer is used at the output of the SD8901 circuit.

The local oscillator inputs are driven in push-pull by fast rise time square waves. This requirement can be met by generating a pair of complementary square waves from the same source. In the chapter cited above, the circuit used a pair of high-speed TTL J-K flip-flops connected with their clock inputs in parallel, driven from a

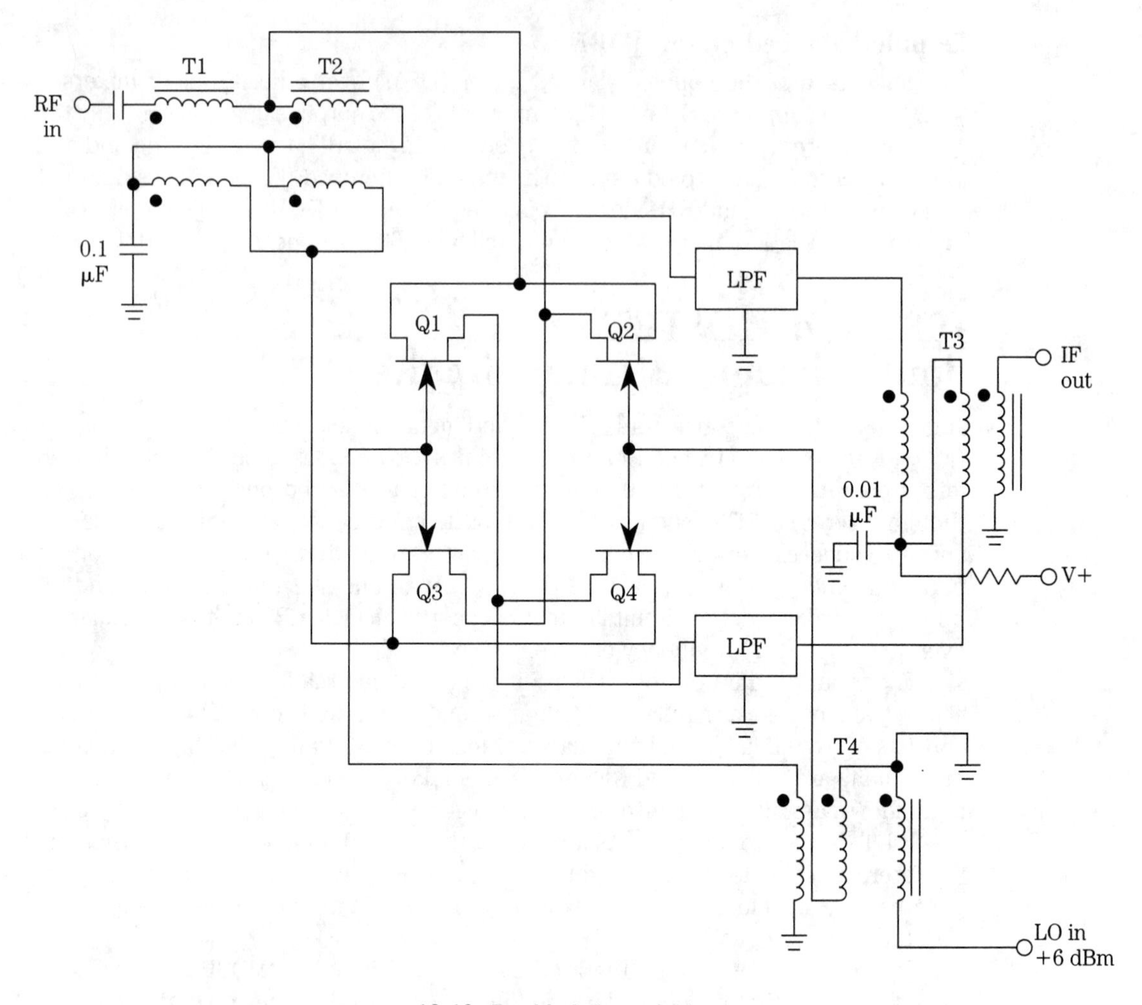

10-10 Double-balanced mixer.

variable frequency oscillator that operated at twice the required LO frequency. The complementarity requirement was met by using the Q-output of one J-K flip-flop and the NOT-Q output of its parallel twin.

The SD8901 device is capable of very good performance, especially in the dynamic range that is achievable. In the design by Makhinson cited previously, a +35 dBm third-order intercept point was achieved, along with a +16 dBm 1-dB output compression point, and a 1-dB output blocking desensitization of +15 dBm. Insertion loss was measured at 7 dB.

One problem with the SD8901 device is its general unavailability to amateur and hobbyist builders. Although low in cost, Calogic has a minimum order quantity of 100—and that makes it a little too expensive for most hobbyists to consider. A compromise that also works well is the MOS electronic switch IC devices, available on the

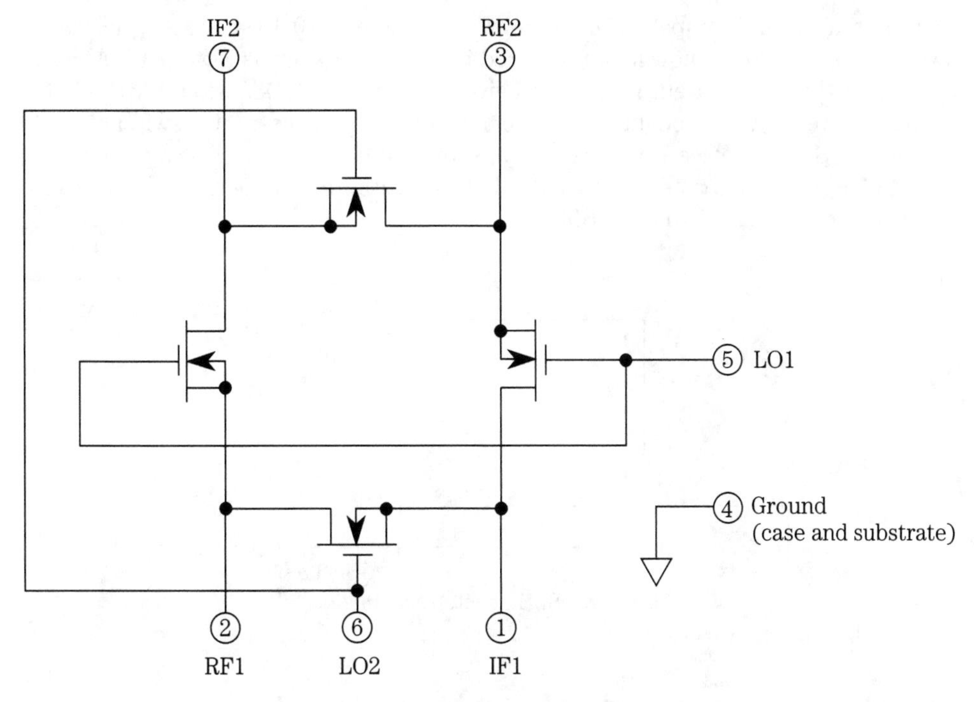

10-11 Si-8901/SD-8901 IC DBM pin-outs.

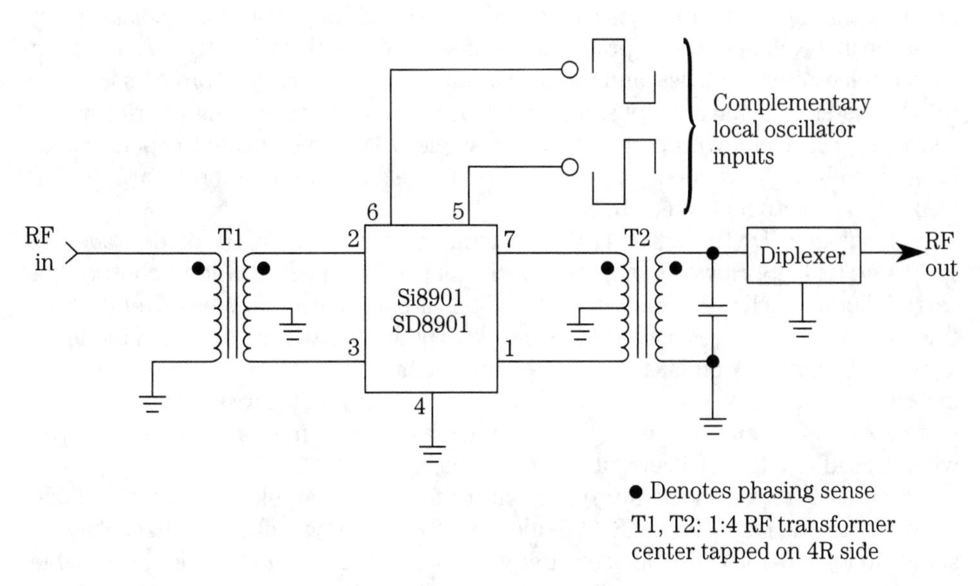

10-12 Si-8901/SD-8901 circuit.

market from several companies, including Calogic. Figure 10-13 shows a typical MOS switch. It is easy to see how it can be wired into a circuit such as Fig. 10-12. At least one top-of-the-line amateur radio transceiver uses a quad MOS switch for the DBM in the receiver. It would be interesting to see how well low-cost MOS switches such as the CMOS 4066 device would work. I've seen that chip work well as a doubly-balanced phase-sensitive detector in medical blood pressure amplifiers—and those circuits are closely related to the DBM.

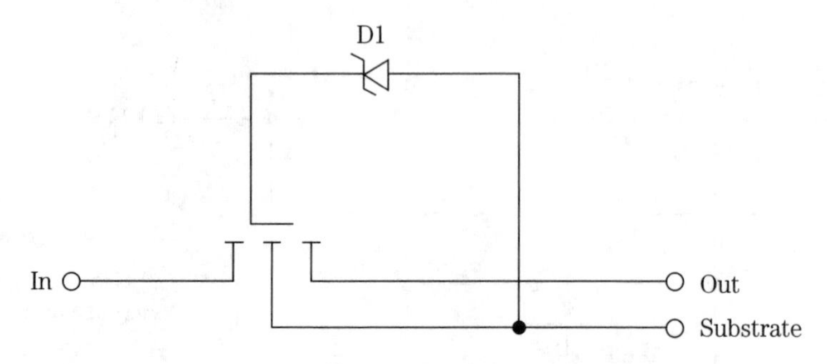

10-13 MOSFET switch schematic.

Doubly-balanced diode mixer circuits

One of the most easily realized doubly-balanced circuits, whether homebrew or commerical, is the circuit of Fig. 10-14. This circuit uses a diode ring mixer with balanced input, output and LO ports. It is capable of 30 to 60 dB of port-to-port isolation, yet is reasonably well-behaved in practical circuits. DBMs, such as Fig. 10-14, have been used by electronic hobbyists and radio amateurs in a wide variety of projects from direct-conversion receivers, to single sideband transmitters, to high-performance shortwave receivers. With proper design, a single DBM can be made to operate over an extremely wide frequency range; several models claim operation from 1 to 500 MHz, with IF outputs from dc to 500 MHz.

The diodes (D1 through D4) can be ordinary silicon VHF/UHF diodes, such as 1N914 or 1N4148. However, superior performance is expected when Schottky hot carrier diodes, such as 1N5820 through 1N5822 are used instead. Whichever diode is selected, all four devices should be matched. The best matching of silicon diodes is achieved by comparison on a curve tracer, but failing that, there should at least be a matched forward/reverse resistance reading. Schottky hot carrier diodes can be matched by ensuring that the selected diodes have the same forward voltage drop when biased to a forward current of 5 to 10 milliamperes.

Figure 10-15 shows the internal circuitry for a very popular commercial diode DBM device, the Mini-Circuits SRA-1 and SBL-1 series; a typical SRA/SBL package is shown in Fig. 10-16. These devices offer good performance, and are widely available to hobbyist and radio amateur builders. Some parts houses sell them at retail, as does Mini-Circuits (P.O. Box 166, Brooklyn, NY, 11235, USA: Phone 714-934-4500). I don't know the amount of their minimum order, but I've had Mini-Circuits in the

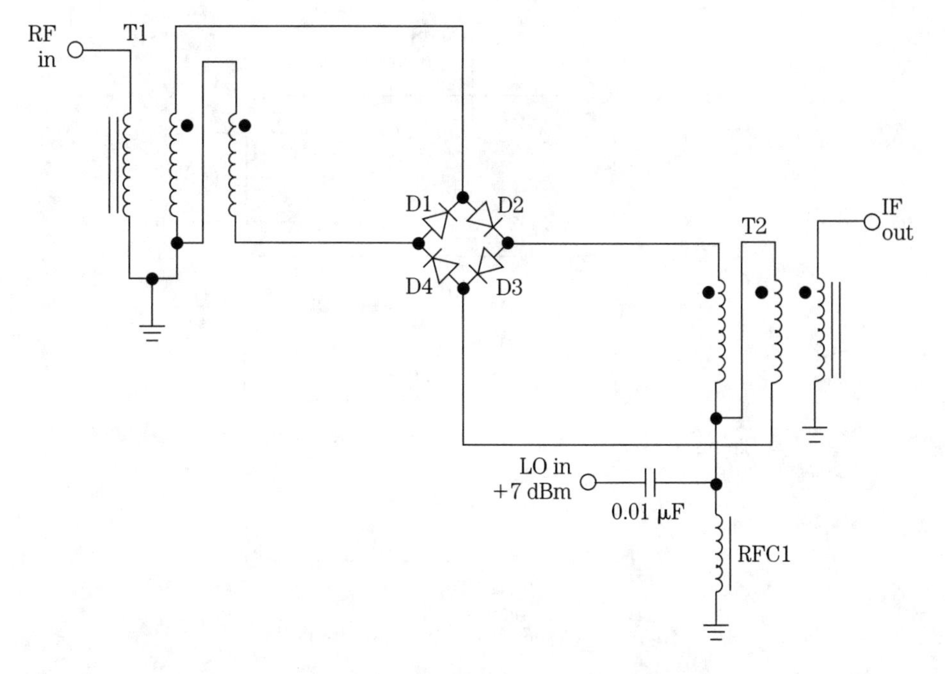

10-14 Diode-balanced mixer.

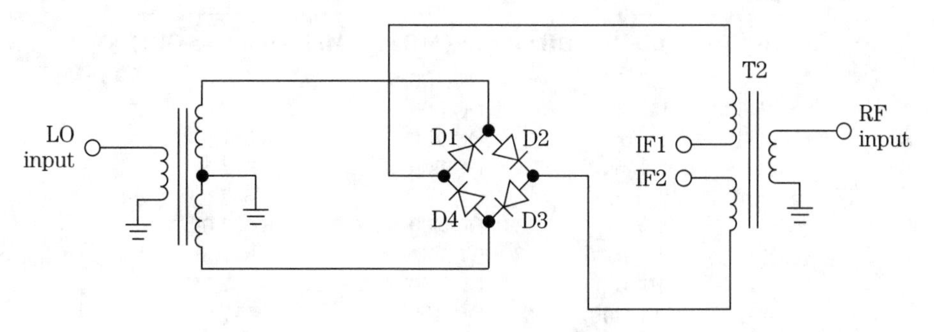

10-15 Double-balanced diode mixer.

USA respond to $25 orders on several occasions . . . which is certainly more reasonable than some other companies.

The packages for the SRA/SBL devices are similar, being on the order of 20-mm long with 5-mm pin spacing. The principal difference between the packages for SRA and SBL devices is in the height. In these packages, pin 1 is denoted by a blue bead insulator around the pin. Other pins are connected to the case or have a green (or other color) bead insulator. Also, the "MCL" logo on the top can be used to locate pin 1: the "M" of the logo is directly over pin 1. Table 10-1 shows the characteristics of several DBMs in the SRA and SBL series, while Table 10-2 shows the pin assignments for the same devices.

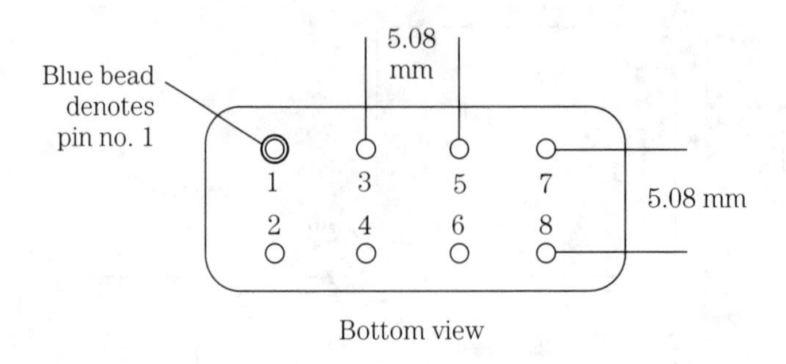

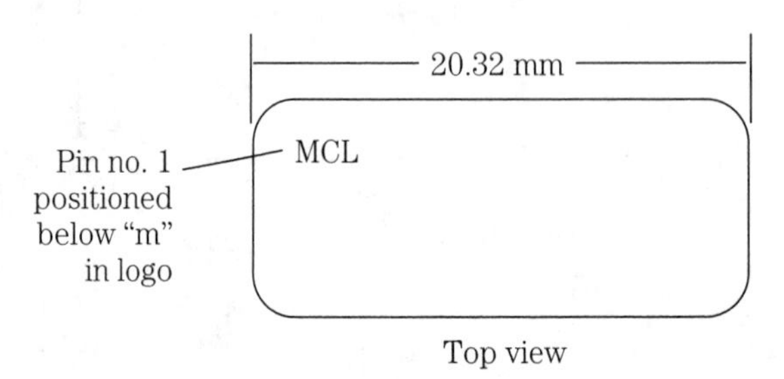

10-16 Package of the Mini-Circuits DBM.

Table 10-1

Type No.	LO/RF (MHz)	IF (MHz)	Mid-band loss (dB)
SRA-1	0.5–500	DC-500	5.5–7.0
SRA-1TX	0.5–500	DC-500	5.5–7.0
SRA-1W	1–750	DC-750	5.5–7.5
SRA-1-1	0.1–500	DC-500	5.5–7.5
SRA-2	1–1000	0.5–500	5.5–7.5
SBL-1	1–500	DC-500	5.5–7.0
SBL-1X	10–1000	5-500	6.0–7.5
SBL-1Z	10–1000	DC-500	6.5–7.5
SBL-1-1	0.1–400	DC-400	5.5–7.0
SBL-3	0.025–200	DC-200	5.5–7.5

Table 10-2. SRA/SBL DBM pinouts.

Type	Pin/function				
number	LO	RF	IF	GND	Case GND
SRA-1	8	1	3, 4	2, 5, 6, 7	2
SRA-1TX	8	1	3, 4	2, 5, 6, 7	2
SRA-1-1	8	1	3, 4	2, 5, 6, 7	2
SRA-1W	8	1	3, 4	2, 5, 6, 7	2, 5, 6, 7
SRA-2	8	3, 4	1	2, 5, 6, 7	2, 5, 6, 7
SRA-3	8	1	3, 4	2, 5, 6, 7	2

Type number	**LO**	**RF**	**Pin/function** **IF**	**GND**	**Case GND**
SBL-1	8	1	3, 4	2, 5, 6, 7	—
SBL-1-1	8	1	3, 4	2, 5, 6, 7	—
SBL-3	8	1	3, 4	2, 5, 6, 7	—

In the standard series of devices, the RF input can accommodate signals up to +1 dBm (1.26 mW), while the LO input must see a +7 dBm (5 mW) signal level for proper operation. Given the 50-ohm input impedance of all ports of the SRA/SBL devices, the RF signal level must be kept below 700 mV p-p, while the LO should see 1400 mV p-p. It is essential that the LO level be maintained across the band of interest, or else mixing operation will suffer. Although the device will work down to +5 dBm, a great increase in spurious output and less port-to-port isolation is found. Spectrum analyzer plots of the output signal at low LO drive levels show considerable second and third-order distortion products.

Note the IF output of the SRA/SBL devices. Although some models in the series use a single IF output pin, most of these devices use two pins (3 and 4), and they must be connected together externally for the device to work.

As is true with most DBMs and all diode ring DBM circuits, the SRA/SBL devices are sensitive to the load impedance at the IF output. Good mixing, and freedom from the LO/RF feedthrough problem, occurs when the mixer looks into a low VSWR load. For this reason, a good diplexer circuit is required at the output. In experiments, I've found that unterminated SBL-1-1 mixers produce nearly linear mixing when not properly terminated . . . and that's not desired in a frequency converter.

Figure 10-17 shows a typical SRA/SBL circuit: RF drive (≤ +1 dBm) is applied to pin 1, and the +7 dBm LO signal is applied to pin 8. The IF signal is output through pins 3 and 4, which are strapped together. All other pins (2, 5, 6, and 7) are grounded.

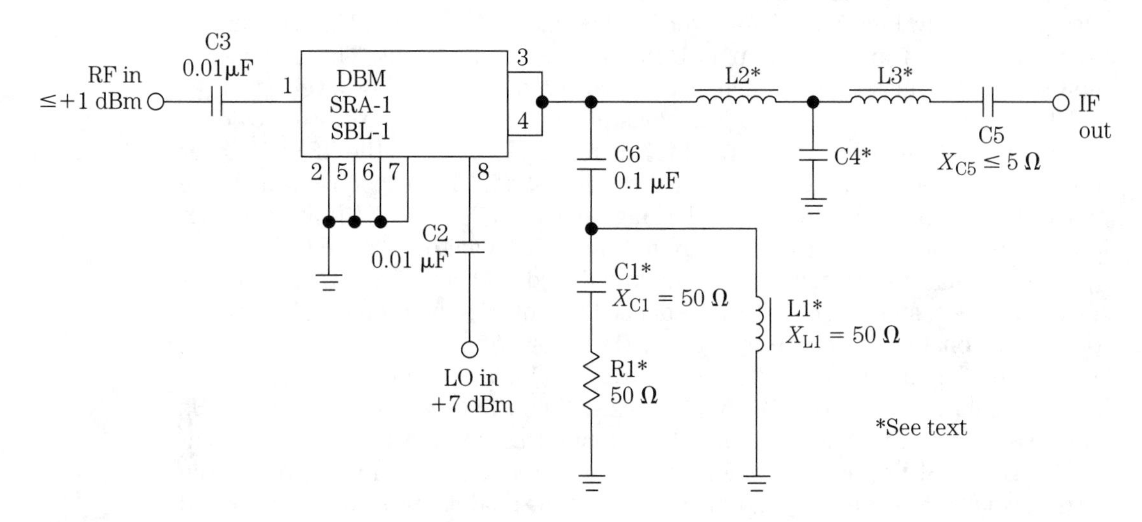

10-17 Circuit of Mini-Circuits DBM.

The diplexer circuit consists of a high-pass filter (C1/L1) that is terminated in a 50-ohm dummy load for the unwanted frequencies, and a low-pass filter (L2/L3/C4) for the desired frequencies. All capacitors and inductors are selected to have a reactance of 50 ohms at the IF frequency.

Sometimes, 1-dB resistor π-pad attenuators are used at the inputs and IF output of the DBM. In some cases, the input attenuators are needed to prevent overload of the DBM (overload causes spurious product frequencies to be generated, and may cause destruction of the device). In other cases, the circuit designer is attempting to "swamp out" the effects of source or load impedance variations. Although this method works, it is better to design the circuit to be insensitive to such fluctuations, rather than to use a swamping attenuator. The reason is that the resistive attenuator causes a signal loss and adds to the noise generated in the circuit (no resistor can be totally noise free). A good alternative is to use a stable amplifier with 50-ohm input and output impedances, and that is not itself sensitive to impedance variation, to isolate the DBM.

Mini-Circuits devices related to the SRA-1 and SBL-1 incorporate MAR-x series MMIC amplifiers internal to the DBM. One series of devices places the amplifier in the LO circuit, so that much lower levels of LO signal will provide proper mixing. Another series places the amplifier in the IF output port. This amplifier accomplishes two things: it makes up for the inherent loss of the mixer and it provides greater freedom from load variations that can affect the regular SRA/SBL devices.

Bipolar transconductance cell DBMs

Active mixers made from bipolar silicon transistors formed into Gilbert transconductance cell circuits are also readily available. Perhaps the two most common devices are the Signetics NE-602 device and the LM-1496 device (see chapter 11).

The LM-1496 device is shown in Figs. 10-18A through 18C. Figure 10-18A shows the internal circuitry, while Figs. 10-18B and 18C show the DIP and metal can packages, respectively. Pins 7 and 8 form the local oscillator (or "carrier" in communications terminology) input, while pins 1 and 4 form the RF input. These push-pull inputs are sometimes labeled "high level signal" (7 and 8) and "low level signal" (1 and 4) inputs. dc bias (pin 5) and gain adjustment (pins 2 and 3) are also provided.

Figure 10-19 shows the basic LM-1496 mixer circuit in which the RF and carrier inputs are connected in the single-ended configuration. The respective signals are applied to the input pins through dc-blocking capacitors C1 and C2; the alternate pin inputs in both cases are bypassed to ground through capacitors C3 and C4.

The output network consists of a 9:1 broadband RF transformer that combines the two outputs, and reduce their impedance to 50 ohms. The primary of the transformer is resonated to the IF frequency by capacitor C5.

Figure 10-20 shows a circuit that uses the LM-1496 device to generate double-sideband suppressed-carrier (DSBSC) signals. When followed by a 2.5- to 3-kHz bandpass filter that is offset from the IF frequency, this circuit will also generate single sideband (SSB) signals. In common practice, a crystal oscillator will generate the carrier signal (V_c), while the audio stages produce the modulating signal (V_m) from an audio oscillator or microphone input stage. I once saw a circuit very similar to this

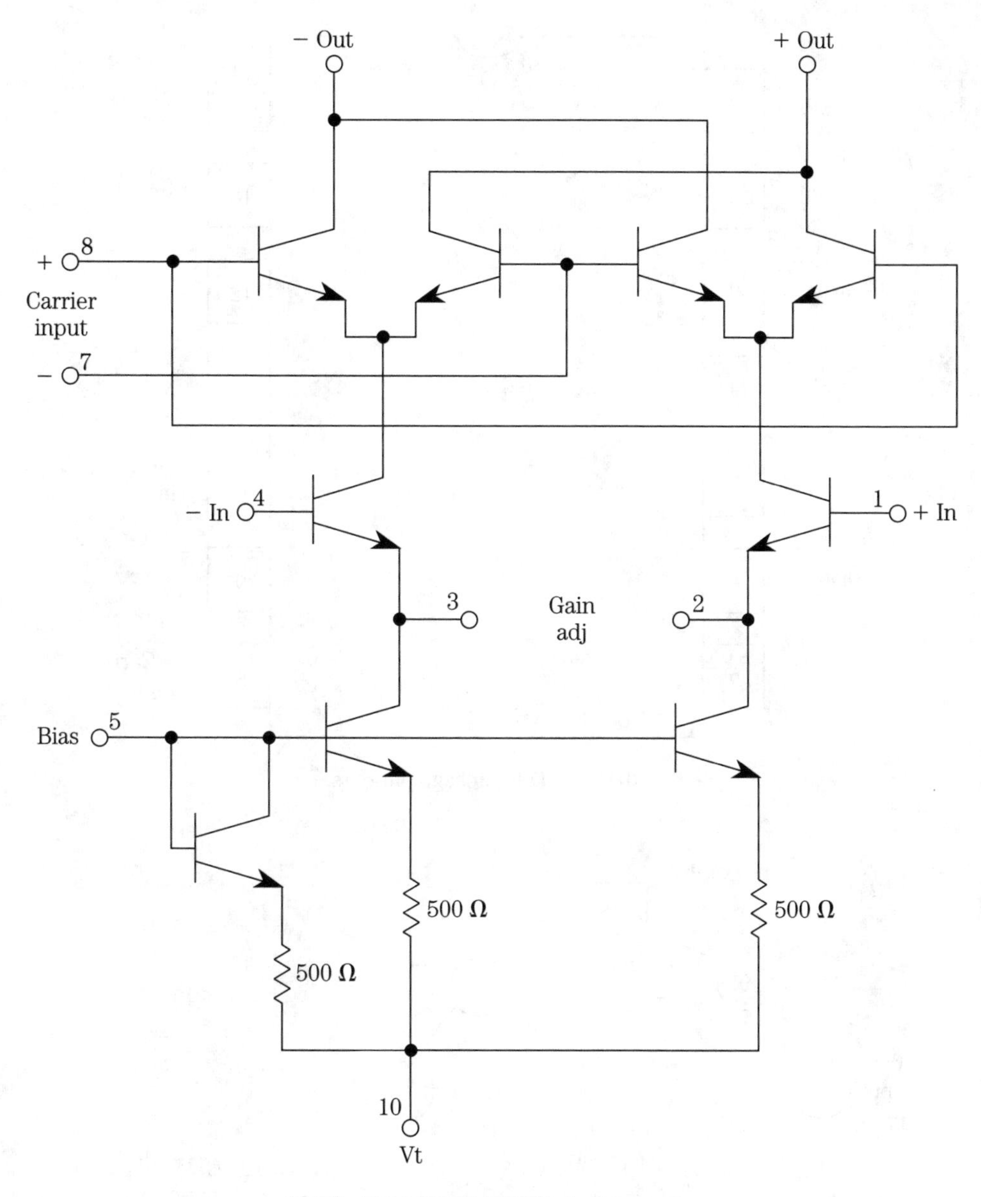

10-18 MC-1496G IC DBM. A. Schematic.

one in a signal generator/test set that was used to service both amateur radio and marine HF-SSB radio transceivers. The signal source was used to test the receiver sections of the transceivers. The carrier was set to 9 MHz, and both lower sideband (LSB) and upper sideband (USB) KVG crystal filters were used to select the desired sideband. A less expensive alternate scheme uses a single 9-MHz crystal filter, but uses two different crystals at frequencies either side of the crystal passband. One crystal generates the USB signal, while the other generates the LSB signal.

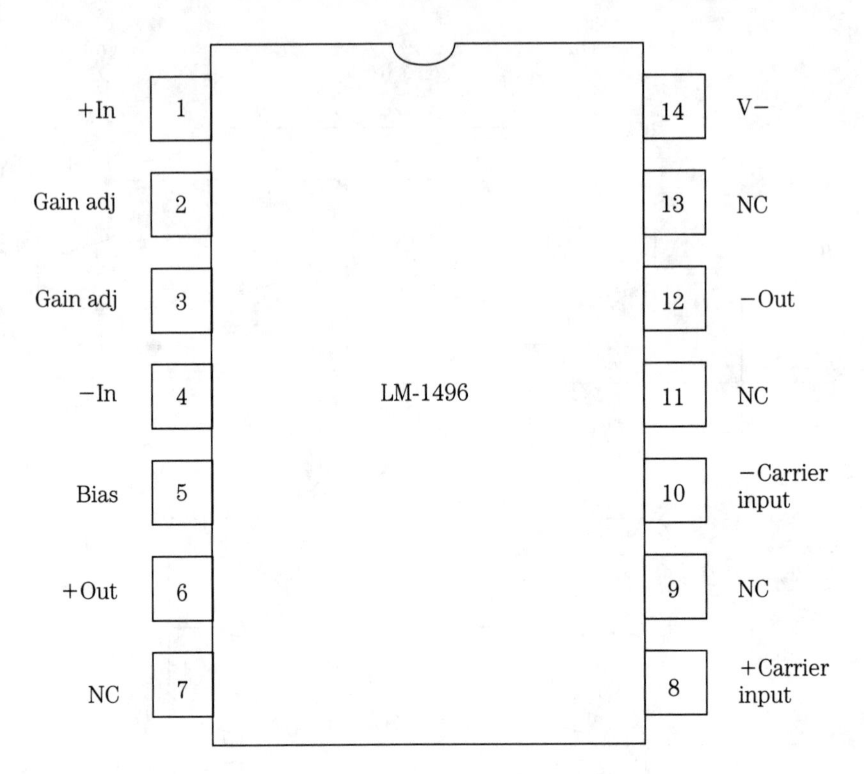

10-18 B. DIP package pin-outs.

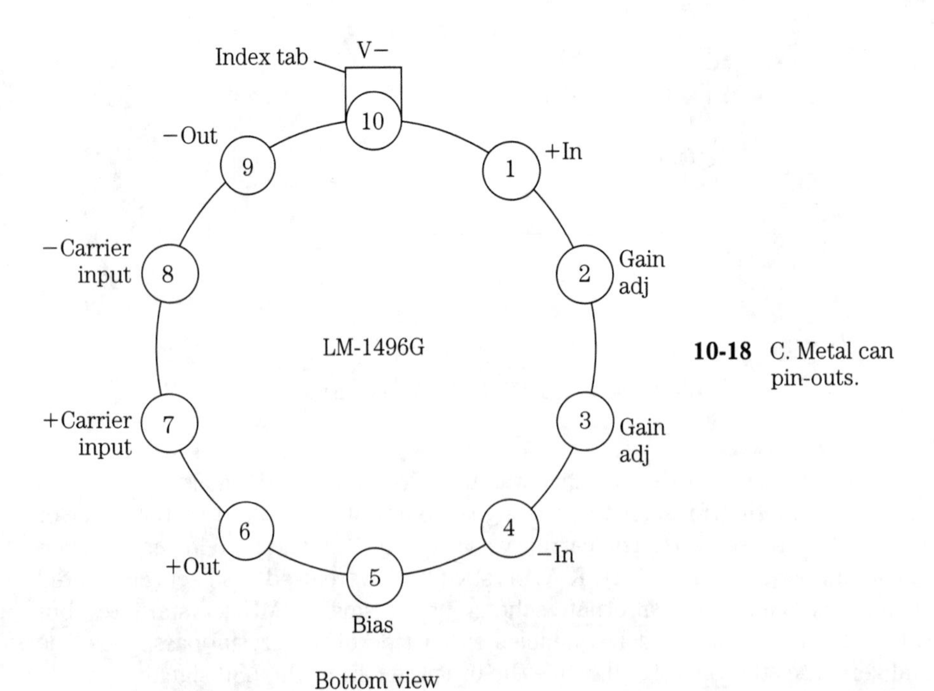

10-18 C. Metal can pin-outs.

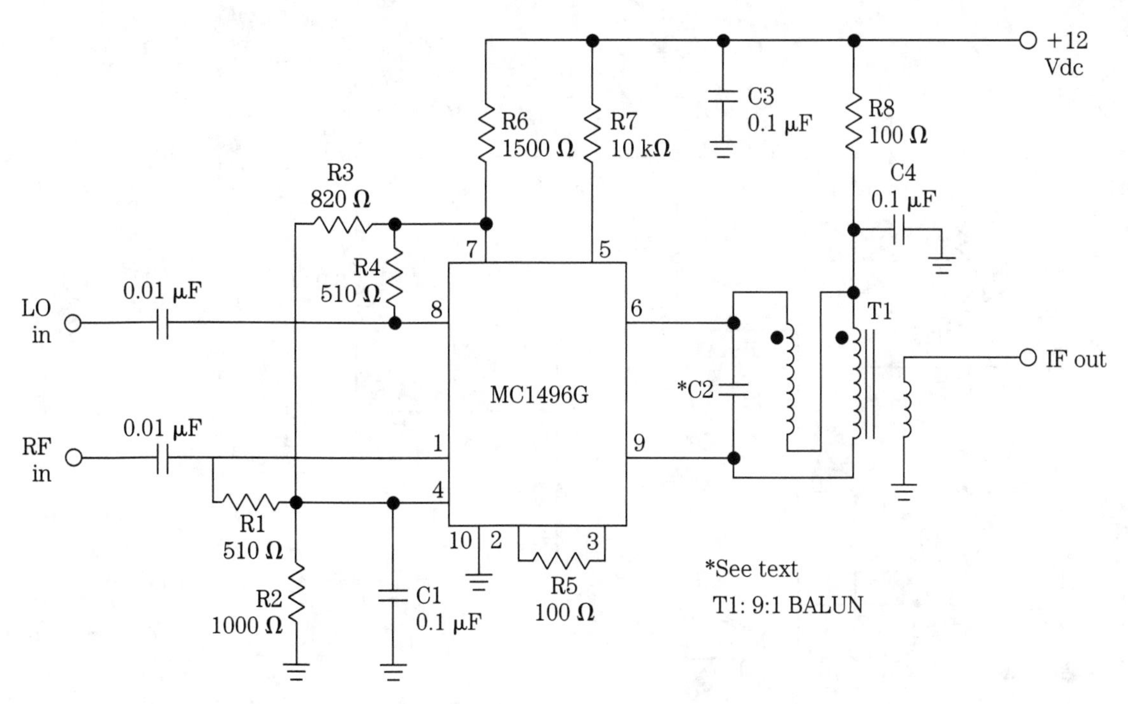

10-19 Circuit using the MC-1496G.

For single sideband to be useful, it has to be demodulated to recover the audio modulation. The circuit of Fig. 10-21 will do that job nicely. It uses an LM-1496 DBM as a *product detector*. This type of detector works on CW, SSB, and DSB signals (all three require a local oscillator injection signal), and produces the audio resultant from heterodyning the local carrier signal against the SSB IF signal in the receiver. All SSB receivers use some form of product detector at the end of the IF chain, and many use the LM-1496 device in a circuit similar to Fig. 10-21.

Pre-amplifiers and post-amplifiers

There is often justification for using amplifiers with DBM circuits. The inputs can be made more sensitive with preamplifiers. When an amplifier is used following the output of a passive DBM ("post-amplifier"), it makes up for the 5- to 8-dB loss typical of passive DBMs. In either case, the amplifier provides a certain amount of isolation of the input or output port of the DBM, which frees the DBM from the effects of source or load impedance fluctuations. In these cases, the amplifier is said to be acting as a *buffer amplifier*.

Figure 10-22 shows two popular amplifiers. They can be used for either preamplifier or post amplifier service because they each have 50-ohm input and output impedances. The circuit in Fig. 10-22A is based on the 2N5109 RF transistor, and provides close to 20 dB of gain throughout the HF portion of the spectrum. A small amount of stabilizing degenerative feedback is provided by leaving part of the emit-

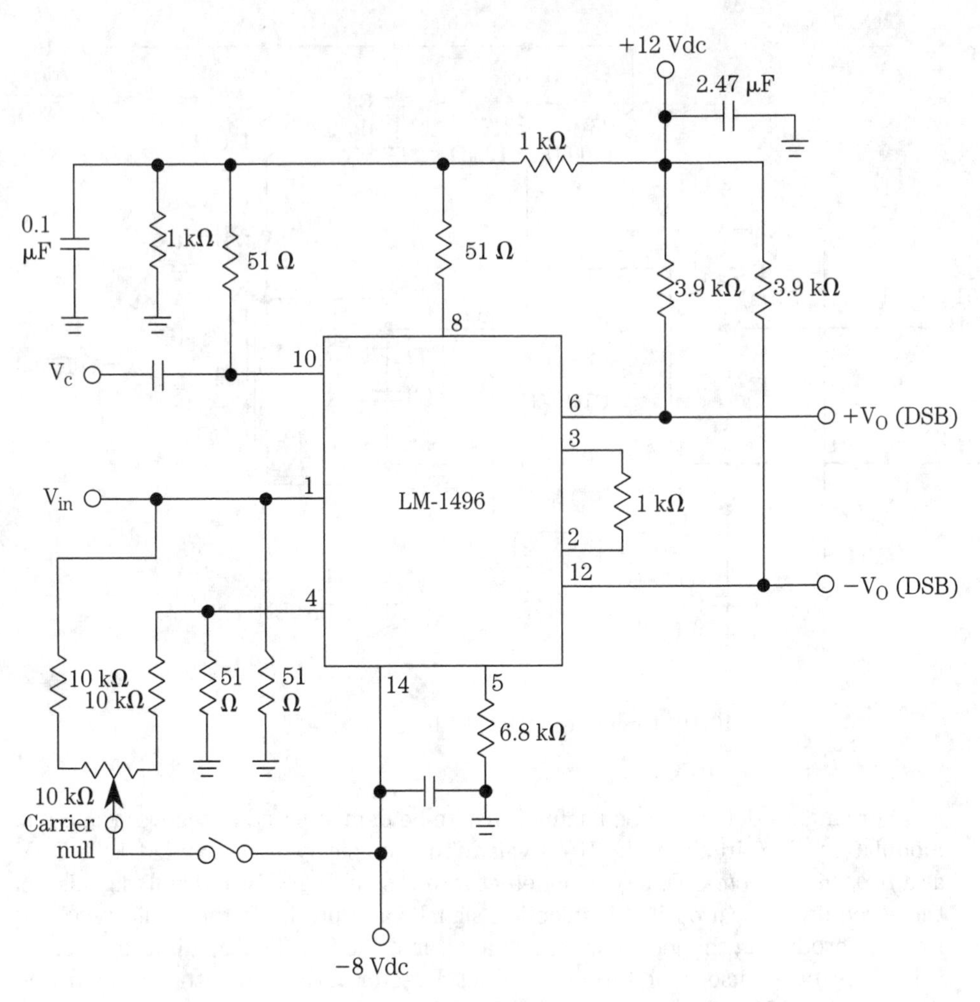

10-20 DSB generator using MC-1496G.

ter resistance (R3) unbypassed, while properly bypassing the remaining portion of the emitter resistance. Additional feedback occurs because of the 4:1 impedance transformer in the collector circuit of the transistor. This transformer can be home-brewed using an FT-44-43 or FT-50-43 toroidal ferrite core bifilar wound with seven to ten turns of #26 AWG enameled wire (or its equivalent in other countries).

The circuit in Fig. 10-22B is based on the Mini-Circuits MAR-x series of MMIC devices. These chips provide 13- to 26-dB gain, at good noise figures, for frequencies from near-dc to 1000 MHz (or more in some models, e.g., 1500 or 2000 MHz). The MAR-1 device shown in the circuit diagram is capable of 15 dB performance to 1000 MHz. The input and output capacitors can be disk ceramic types up to about 100 MHz, but above that frequency, "chip" capacitors should be used. Values of 0.01 μF should be used in the low HF region (<10 MHz), 0.001 μF can be used up to 100 MHz, and 100 pF above MHz. The RF choke (RFC1) should be 2.5 mH in the low HF region, 1 mH from about 10 MHz to 30 MHz, 100 μH from 30 MHz to 100 MHz and 10

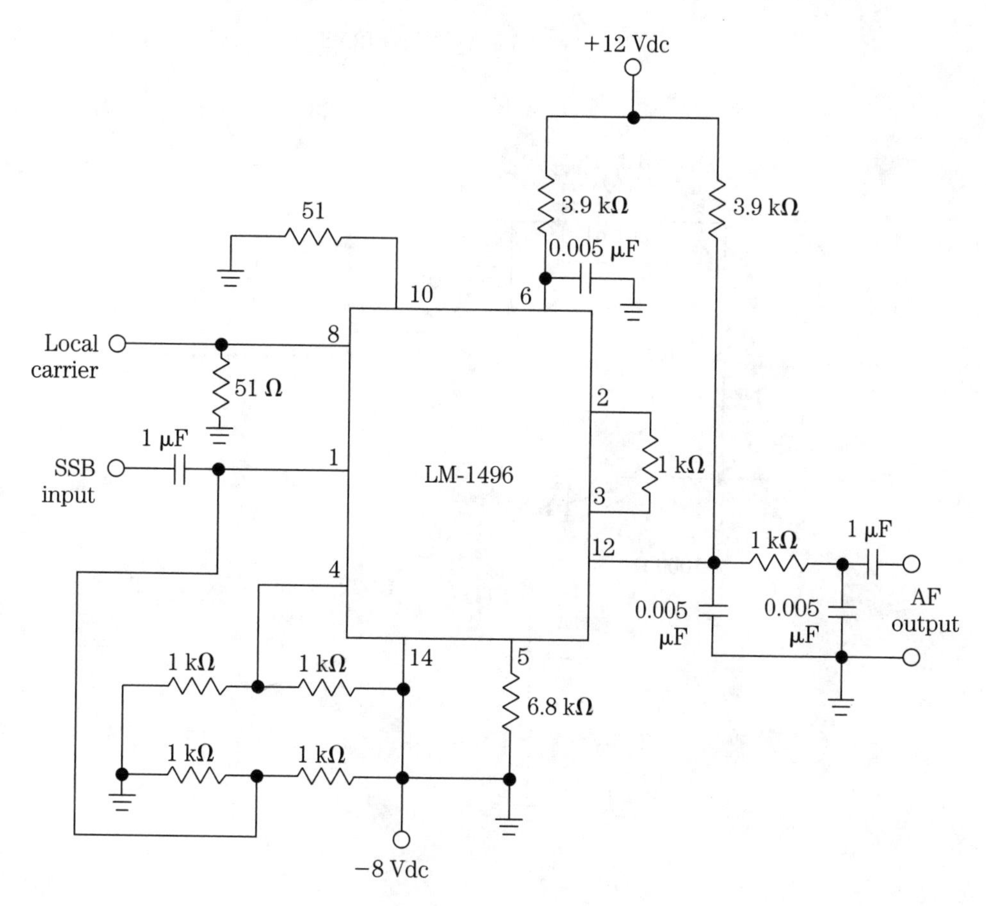

10-21 MC-1496G product detector.

μH above 100 MHz. These values are not crucial, and are given only as guidelines. While it might be a bit tricky to get a 1 mH RFC to operate well at 100 MHz, there is really no hard boundary in the frequency ranges given above.

Notes and references

1. Jacob Makhinson, "A High Dynamic Range MF/HF Receiver Front End," *QST*, February 1993, pp. 23–28.
2. In the U.K., contact Dale Electronics Ltd., Camberley, Surrey, 025 28 35094. In Netherlands, contact "Colmex" B.V., 8050 AA Hattem, Holland, phone (0) 5206-41214/41217.

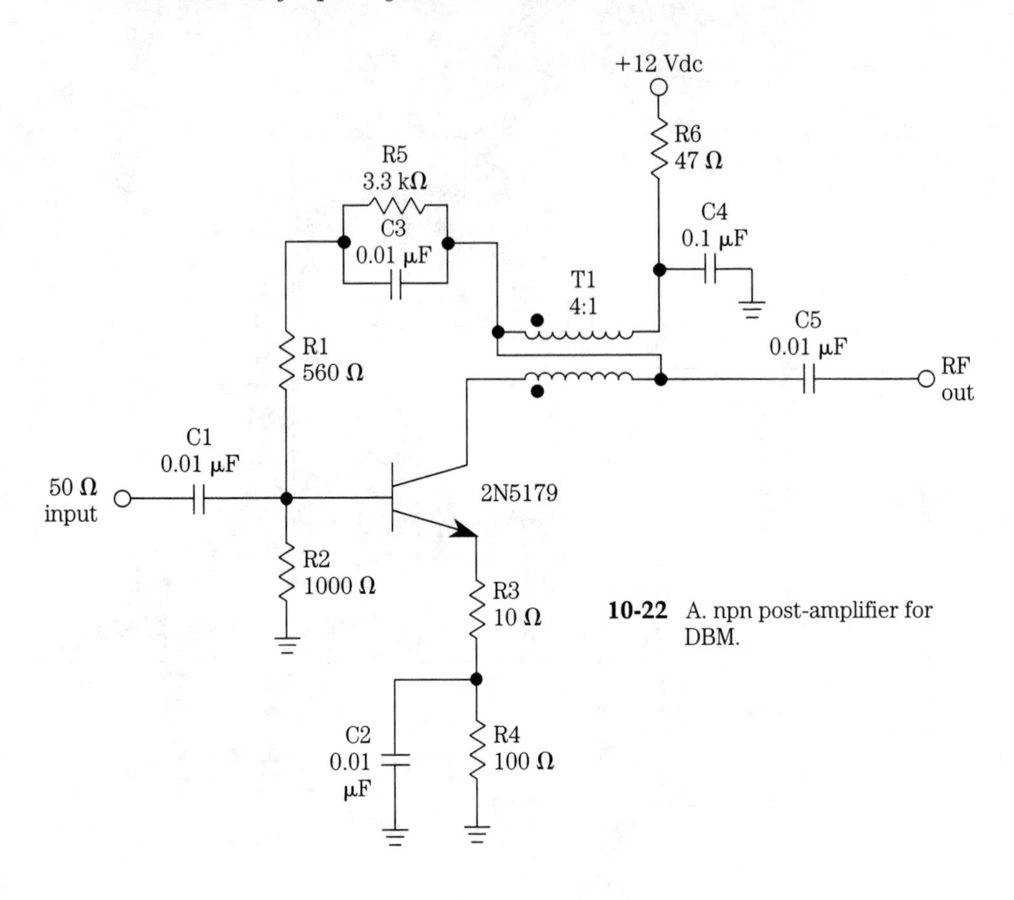

10-22 A. npn post-amplifier for DBM.

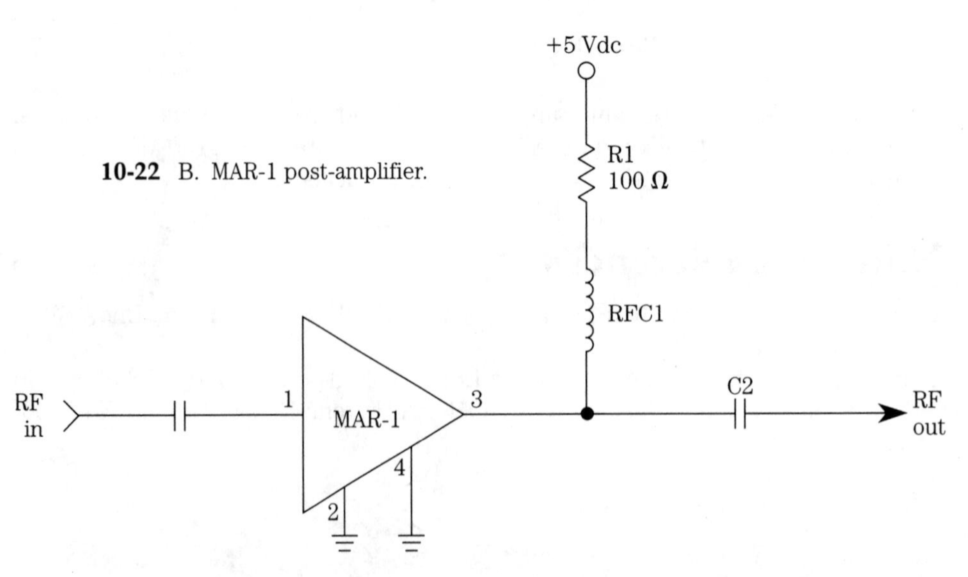

10-22 B. MAR-1 post-amplifier.

11
Using the NE-602 chip

The Signetics NE-602/SA-602 is a monolithic integrated circuit that contains a double balanced mixer (DBM), an oscillator, and an internal voltage regulator in a single eight-pin package (Fig. 11-1). The DBM section operates at frequencies up to 500 MHz, while the internal oscillator section works at frequencies up to 200 MHz. The primary uses of the NE-602/SA-602 are in HF and VHF receivers, frequency converters, and frequency translators. The device can also be used as a signal generator in many popular inductor-capacitor (LC) variable frequency oscillator (VFO), piezoelectric crystal (XTAL) or swept frequency configurations. In this chapter, we will explore the various configurations for the dc power supply, the RF input, the local oscillator, and the output circuits. We will also examine certain applications of the device.

The NE-602 version of the device operates over a temperature range of 0 to +70°C, while the SA-602 operates over the extended temperature range of –40 to +85°C. The most common form of the device is probably the NE-602N, which is an eight-pin mini-DIP package. Eight-lead SO surface-mount ("D-suffix") packages are also available. In this chapter, the NE-602N is featured—although the circuits also work with the other packages and configurations. By the time this book is published, it is expected that the improved NE-602AN and NE-602AD will be available.

Because the NE-602 contains both a mixer and a local oscillator, it can operate as a radio receiver frontend circuit. It features a good noise figure and reasonable third-order intermodulation performance. The noise figure is typically 5 dB at a frequency of 45 MHz. The NE-602 has a third-order intercept point on the order of –15 dBm referenced to a matched input—although it is recommended that a maximum signal level of –25 dBm (≈ 3.16 mW) be observed. This signal level corresponds to about 12.6 mV into a 50-ohm load, or 68 mV into the 1500-ohm input impedance of the NE-602. The NE-602 is capable of providing 0.2 μV sensitivity in receiver circuits without external RF amplification. One criticism of the NE-602 is that it appears to sacrifice some dynamic range for high sensitivity—a problem said to be solved in the "A" series (e.g., NE-602AN).

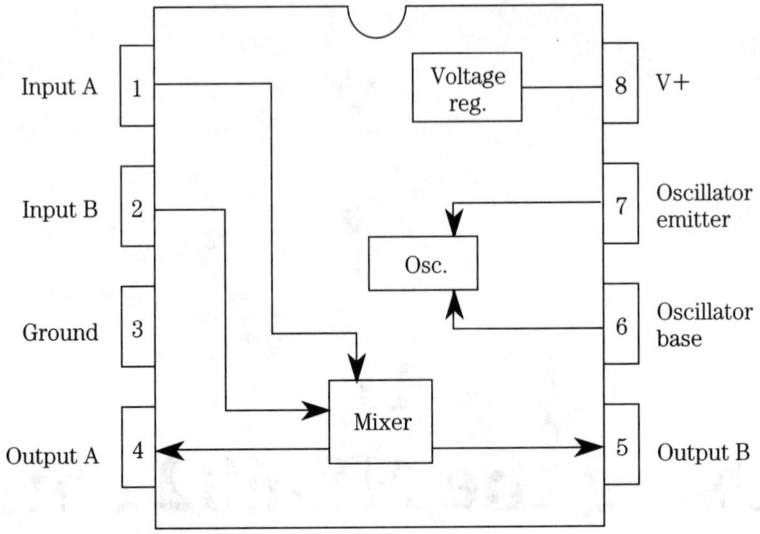

11-1 Block diagram of the NE-602 showing pin-outs.

Frequency conversion/translation

The process of frequency conversion is called *heterodyning*. When two signals of different frequencies (F1 and F2) are mixed in a nonlinear circuit, a collection of different frequencies will appear in the output of the circuit. These are characterized as F_1, F_2, and $nF_1 \pm mF_2$, where n and m are integers. In most practical situations, n and m are 1, so the total output spectrum will consist at least of F_1, F_2, $F_1 + F_2$, and $F_1 - F_2$. Of course, if the two input circuits contain harmonics, additional products are found in the output. In superheterodyne radio receivers, either the sum or difference frequency is selected as the *intermediate frequency* (IF). In order to make the frequency conversion possible, a circuit needs a local oscillator and a mixer circuit (both of which are provided by the NE-602).

The local oscillator (LO) consists of a VHF npn transistor with the base connected to pin 6 of the NE-602, and the emitter connected to pin 7; the collector of the oscillator transistor is not available on an external pin. It also has an internal buffer amplifier, which connects the oscillator transistor to the DBM circuit. Any standard oscillator circuit configuration can be used with the internal oscillator, provided that access to the collector terminal is not required. Thus, Colpitts, Clapp, Hartley, Butler, and other oscillator circuits can be used with the NE-602 device, while the Pierce and Miller oscillator circuits cannot.

The double-balanced mixer (DBM) circuit shown in Fig. 11-2 consists of a pair of cross-connected differential amplifiers (Q1/Q2 with Q5 as a current source; similarly Q3/Q4 with Q6 working as a current source). This configuration is called a *Gilbert transconductance cell*. The cross-coupled collectors form a push-pull output (pins 4 and 5) in which each output pin is connected to the V+ power supply terminal through 1500-ohm resistances. The input is also push-pull, and likewise is

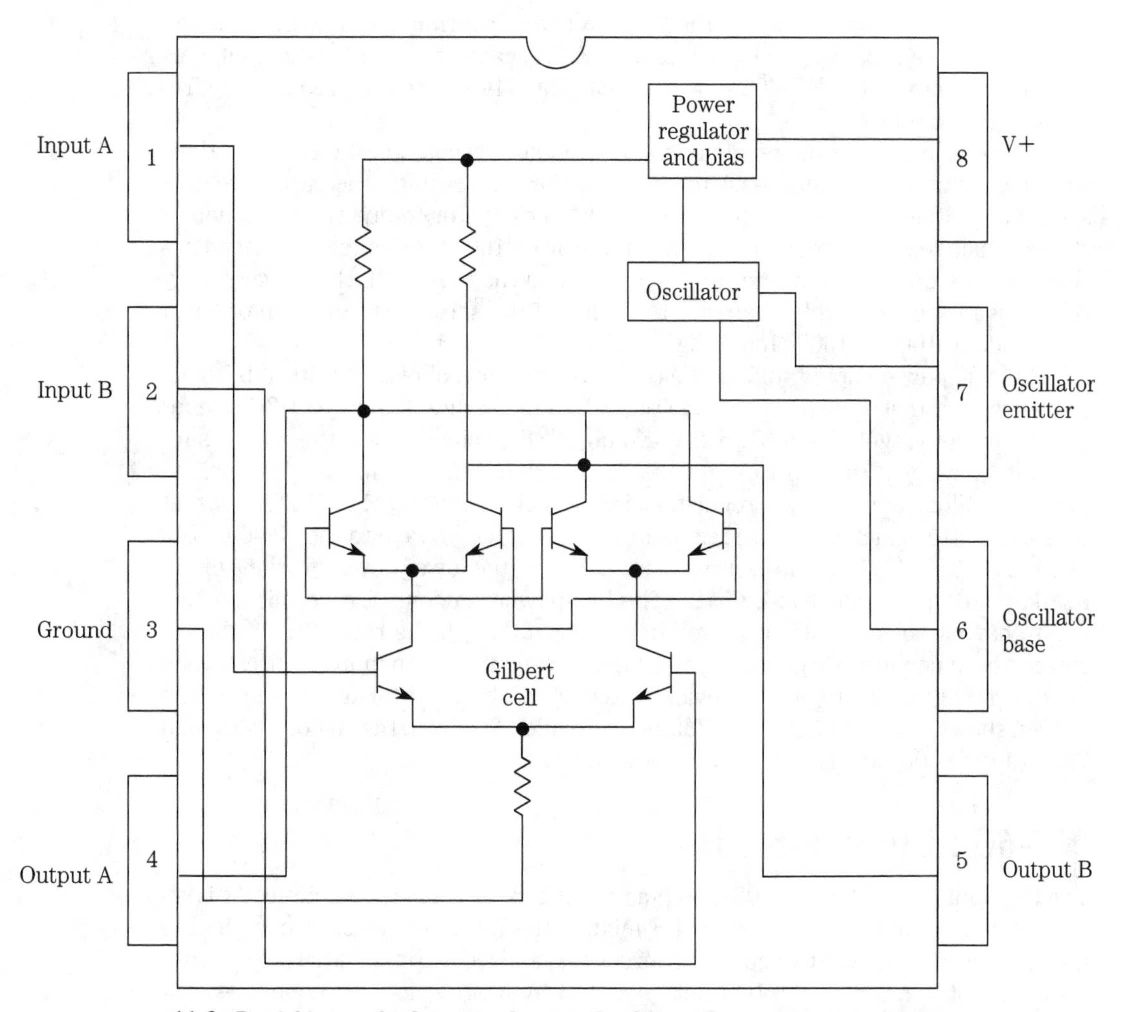

11-2 Partial internal schematic showing the Gilbert Transconductance Cell.

cross-coupled between the two halves of the cell. The local oscillator signal is injected into each cell half at the base of one of the transistors.

Because the mixer is "double-balanced," it has a key attribute that makes it ideal for use as a frequency converter or receiver front end: suppression of the LO and RF input signals in the outputs. In the NE-602 chip, the output signals are $F_1 + F_2$, and $F_1 - F_2$; neither LO nor RF signals appear in the output in any great amplitude. Although some harmonic products appear, many are also suppressed because of the DBM action.

dc power supply connection on the NE-602

The V+ power supply terminal of the NE-602 is pin 8, and the ground connection is pin 3; both must be used for the dc power connections. The dc power supply range is +4.5 to +8 volts dc, with a current drain ranging from 2.4 to 2.8 mA.

It is highly recommended that the V+ power supply terminal (pin 8) be bypassed to ground with a capacitor of 0.01 μF to 0.1 μF. The capacitor should be mounted as close to the body of the NE-602 as is practical; short leads are required in radio frequency (RF) circuits.

Figure 11-3A shows the recommended power supply configuration for situations where the supply voltage is +4.5 to +8 volts. For best results, the supply voltage should be voltage-regulated. Otherwise, the local oscillator frequency might not be stable, which leads to problems. A series resistor (≈ 100 to 180 ohms) is placed between the V+ power supply and the V+ terminal on the NE-602. If the power supply voltage is raised to +9 volts, increase the value of the series resistance an order of magnitude to 1000 to 1500 ohms (Fig. 11-3B).

If the dc power supply voltage is either unstable, or is above +9 volts, it is highly recommended that a means of voltage regulation be provided. In Fig. 11-3C a zener diode is used to regulate the NE-602 V+ voltage to 6.8 volts dc, even though the supply voltage ranges from +9 to +18 volts (a situation found in automotive applications). An alternative voltage regulator circuit is shown in Fig. 11-3D. This circuit uses a three-terminal IC voltage regulator to provide V+ voltage to the NE-602. Because the NE-602 is a very low-current drain device, the lower power versions of the regulators (e.g., 78Lxx) can be used. The low-power versions also permit the NE-602 to have its own regulated power supply, even though the rest of the radio receiver uses a common dc power supply. Input voltages of +9 to more than +28 volts dc, depending on the regulator device selected, can be used for this purpose. The version shown in Fig. 11-3D uses a 78L09 to provide +9 volts to the NE-602, although 78L05 and 78L06 can also be used to good effect.

NE-602 input circuits

The RF input port of the NE-602 uses pins 1 and 2 to form a balanced input. As is often the case in differential amplifier RF mixers, the RF input signals are applied to the base terminals of the two current sources (Q5 and Q6 in Fig. 11-2). The input impedance of the NE-602 is 1500 ohms, shunted by 3 pF at lower frequencies—although in the VHF region, the impedance drops to about 1000 ohms.

Several RF input configurations are shown in Fig. 11-4; both single-ended (unbalanced) and differential (balanced) input circuits can be used with the NE-602. In Fig. 11-4A, a capacitor-coupled, untuned, unbalanced input scheme is shown. The signal is applied to pin 1 (although pin 2 could have been used instead) through a capacitor, C1, which has a low impedance at the operating frequency. The signal level should be less than –25 dBm, or about 68 mV rms (180 mV peak-to-peak). Whichever input is used, the alternate input is unused and should be bypassed to ground through a low-value capacitor (0.001 μF to 0.1 μF, depending on the frequency).

A wideband transformer-coupled RF input circuit is shown in Fig. 11-4B. In this configuration, a wideband RF transformer is connected such that the secondary is applied across pins 1 and 2 of the NE-602, with the primary of the transformer connected to the signal source or antenna. The turns ratio of the transformer can be used to transform the source impedance to 1500 ohms (the NE-602 input imped-

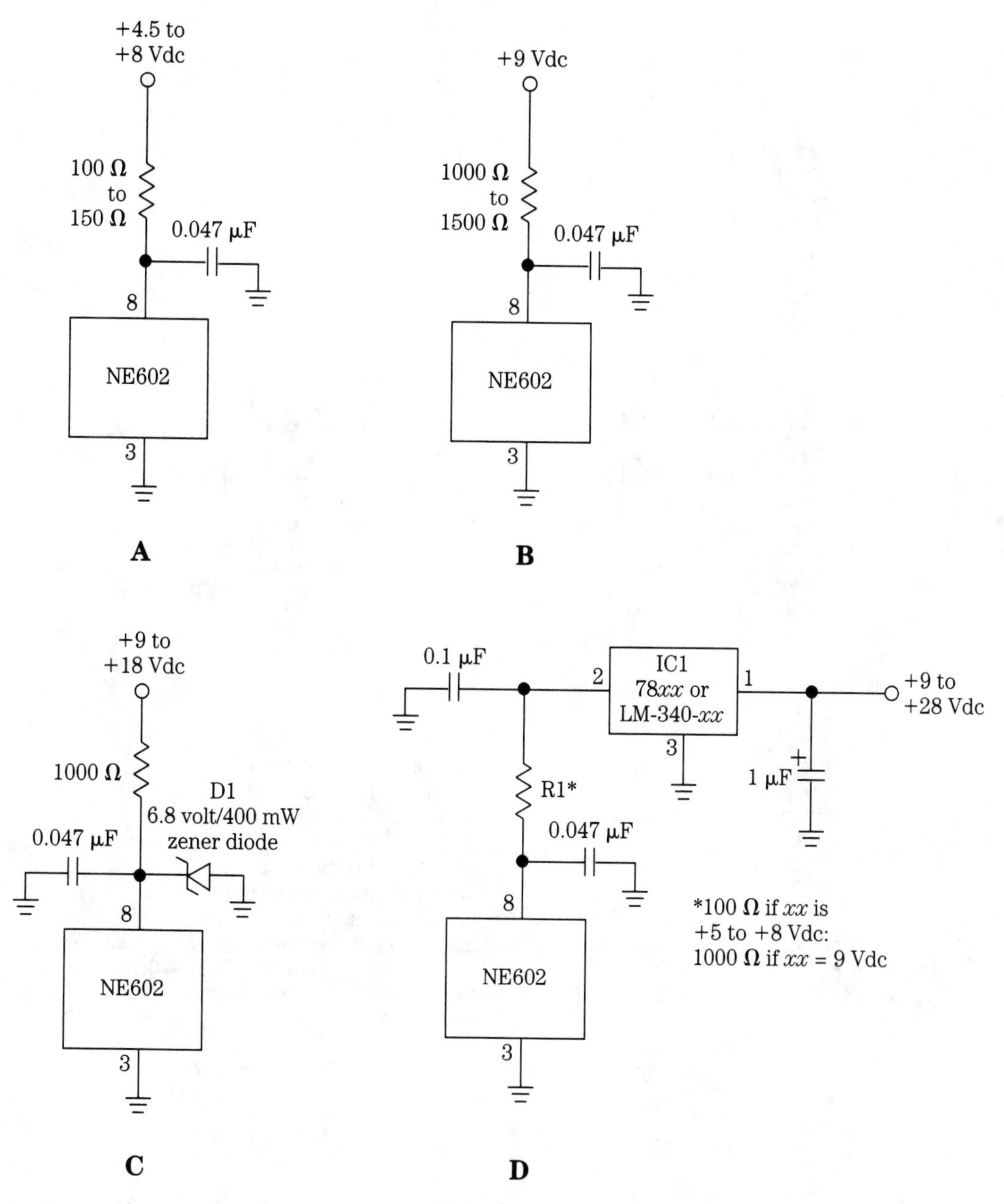

11-3 dc power supply configurations for the NE-602. A. For supplies $+4.5 \leq V \leq +8$ volts dc. B. For +9 volt supplies. C. Zener diode regulator for +9 to +18 volt dc supplies. D. Three-terminal IC voltage regulator for supplies from +8 to +28 volts.

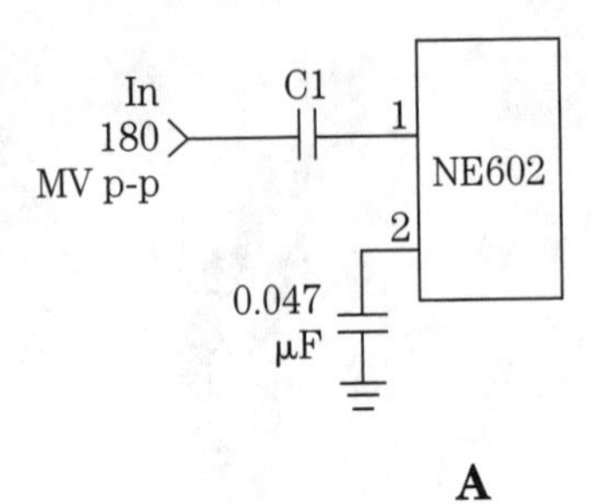

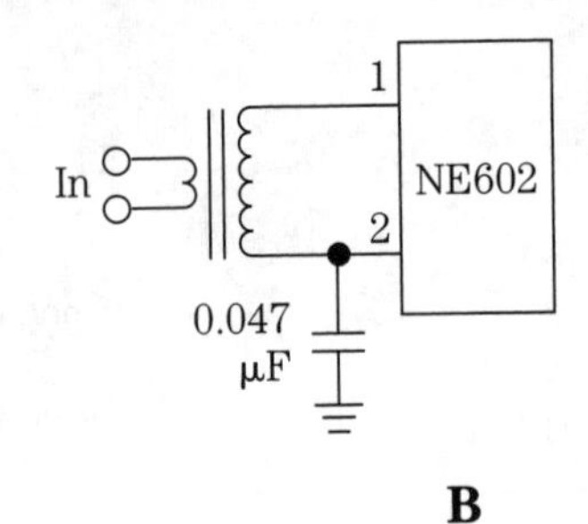

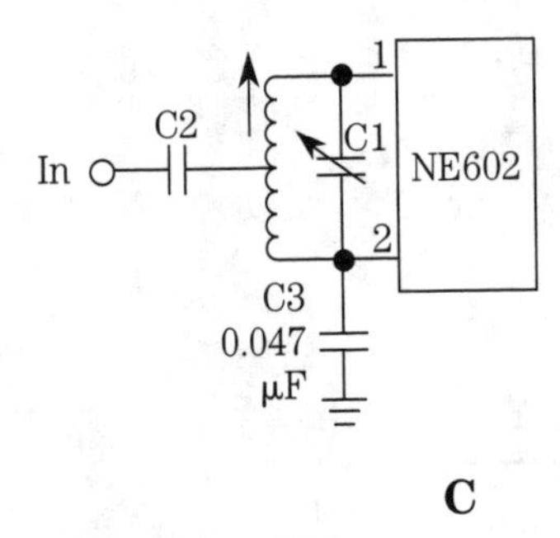

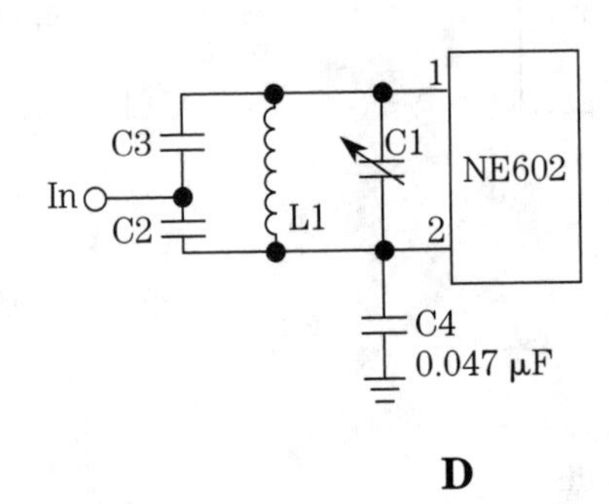

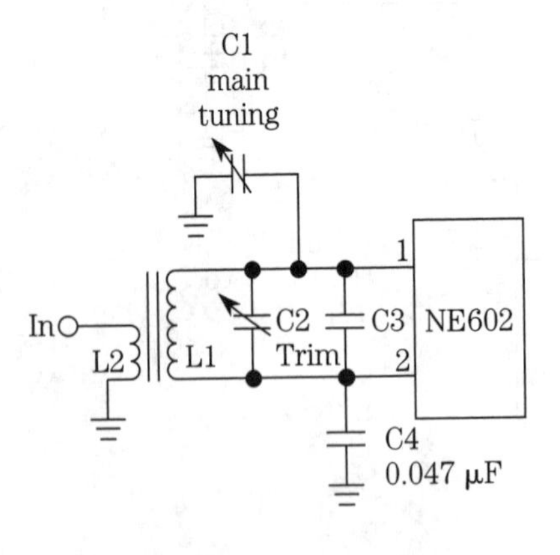

11-4 NE-602 input circuit configurations. A. Direct, untuned input ($V_{in} \leq 180$ mV p-p). B. Broadbanded RF transformer couples signal and transform antenna impedance to 1500 ohms. C. Tuned input uses a tap on the inductor for impedance matching. D. Tuned input uses a tapped capacitor voltage divider for impedance matching. E. Tuned transformer input that uses a grounded frame variable capacitor.

ance). Either conventional or toroid core transformers can be used for T1. As in the previous circuit, one input is bypassed to ground through a low-reactance capacitor.

Tuned RF input circuits are shown in Figs. 11-4C, 4D, 4E, and 5. Each of these circuits performs two functions: it selects the desired RF frequency while rejecting others, and it matches the 1.5-kΩ input impedance of the NE-602 to the source or antenna system impedance (e.g., 50 ohms). The circuit shown in Fig. 11-4D uses an inductor (L1) and capacitor (C1) tuned to the input frequency, as do the other circuits, but the impedance-matching function is done by tapping the inductor; a dc-blocking capacitor is used between the antenna connection and the coil. A third capacitor, C3, is used to bypass one of the inputs (pin 2) to ground.

Another version of the circuit is shown in Fig. 11-4D. It is similar in concept to the previous circuit, but uses a tapped capacitor voltage divider (C2/C3) for the impedance-matching function. Resonance with the inductor is established by the combination of C1, the main tuning capacitor, in parallel with the series combination of C2 and C3:

$$C_{\text{tune}} = C_1 + \frac{C_2 C_3}{C_2 + C_3} \quad \textbf{[11-1]}$$

The previous two circuits are designed for use when the source or antenna system impedance is less than the 1.5-kΩ input impedance of the NE-602. The circuit shown in Fig. 11-4E can be used in all three situations: input impedance lower than, higher than, or equal to the NE-602 input impedance—depending on the ratio of the number of turns in the primary winding (L2) to number of turns in the secondary winding (L1). The situation shown schematically in Fig. 11-4E is for the case where the source impedance is less than the input impedance of the NE-602.

The secondary of the RF transformer (L1) resonates with a capacitance made up of C1 (main tuning), C2 (trimmer tuning or bandspread), and a fixed capacitor C3. An advantage of this circuit is that the frame of the main tuning capacitor is grounded. This feature is an advantage because most tuning capacitors are designed for grounded frame operation, so construction is easier. In addition, most of the variable-frequency oscillator circuits (discussed in chapter 13) used with the NE-602 also use a grounded-frame capacitor. The input circuit of Fig. 11-4E can therefore use a single dual-section capacitor for single knob tuning of both the RF input and the local oscillator.

Figure 11-5 shows a tuned-input circuit that relies, at least in part, on a voltage-variable capacitance (varactor) diode for the tuning function. The total tuning capacitance that resonates L2 (the transformer secondary) is the parallel combination of C1 (trimmer), C2 (a fixed capacitor), and the junction capacitance of varactor diode D1. The value of capacitor C3 is normally set to be large compared with the diode capacitance so that it will have little effect on the total capacitance of the series combination C_3/C_{D1}. In other cases, however, the capacitance of C3 is set close to the capacitance of the diode so it becomes part of the resonant circuit capacitance.

A varactor diode is tuned by varying the reverse bias voltage applied to the diode. Tuning voltage V_t is set by a resistor voltage divider consisting of R1, R2, and R3. The main tuning potentiometer (R1) can be a single-turn model, but for best resolution of the tuning control, use a ten- or fifteen-turn potentiometer. The fine tuning potentiometer can be a panel-mounted model for use as a bandspread control, or a trimmer model for use as a fine adjustment of the tuning circuit (a function also shared by trimmer capacitor C1).

The voltage used for the tuning circuit (V_A) must be well-regulated, or the tuning will shift with variations of the voltage. Some designers use a separate three-terminal IC regulator for V_A, but that is not strictly necessary. A more common situation is to use a single low-power 9-volt three-terminal IC voltage regulator for both the

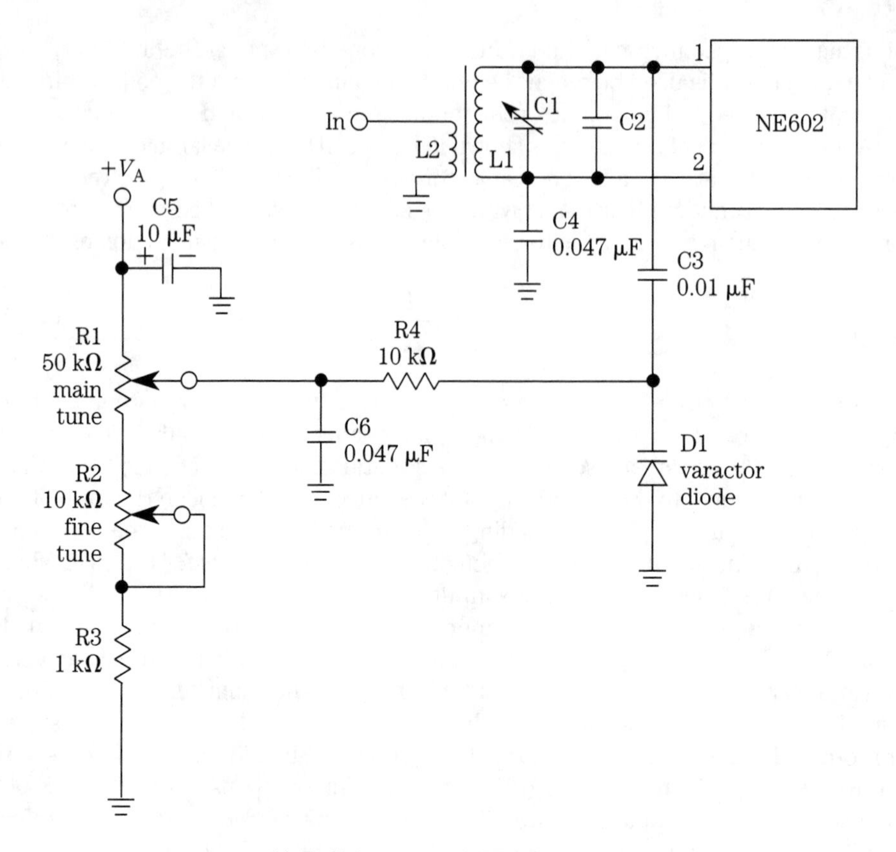

11-5 Voltage-tuned RF input circuit.

NE-602 and the tuning network. However, it will only work when the diode needs no more than +9 volts dc for correct tuning of the desired frequency range. Unfortunately, many varactor diodes require a voltage range of about +1 volt to +37 volts to cover the entire range of available capacitance.

In due course, when oscillator circuits are discussed, you will also see a version of the Fig. 11-5 circuit that is tuned by a sawtooth waveform (for swept frequency operation) or a digital-to-analog converter (for computer-controlled frequency selection).

NE-602 output circuits

The NE-602 output circuit consists of the cross-coupled collectors of the two halves of the Gilbert transconductance cell (Fig. 11-2), with output available on pins 4 and 5. In general, it doesn't matter which of these pins is used for the output; in single-ended output configurations only one terminal is used, and the alternate output terminal is ignored. Each output terminal is connected internally to the NE-602 to V+ through separate 1.5 kΩ resistors.

Figure 11-6A shows the wideband, high-impedance (1.5 kΩ) output configuration. Either pin 4 or 5 (or both) can be used. A capacitor is used to provide dc block-

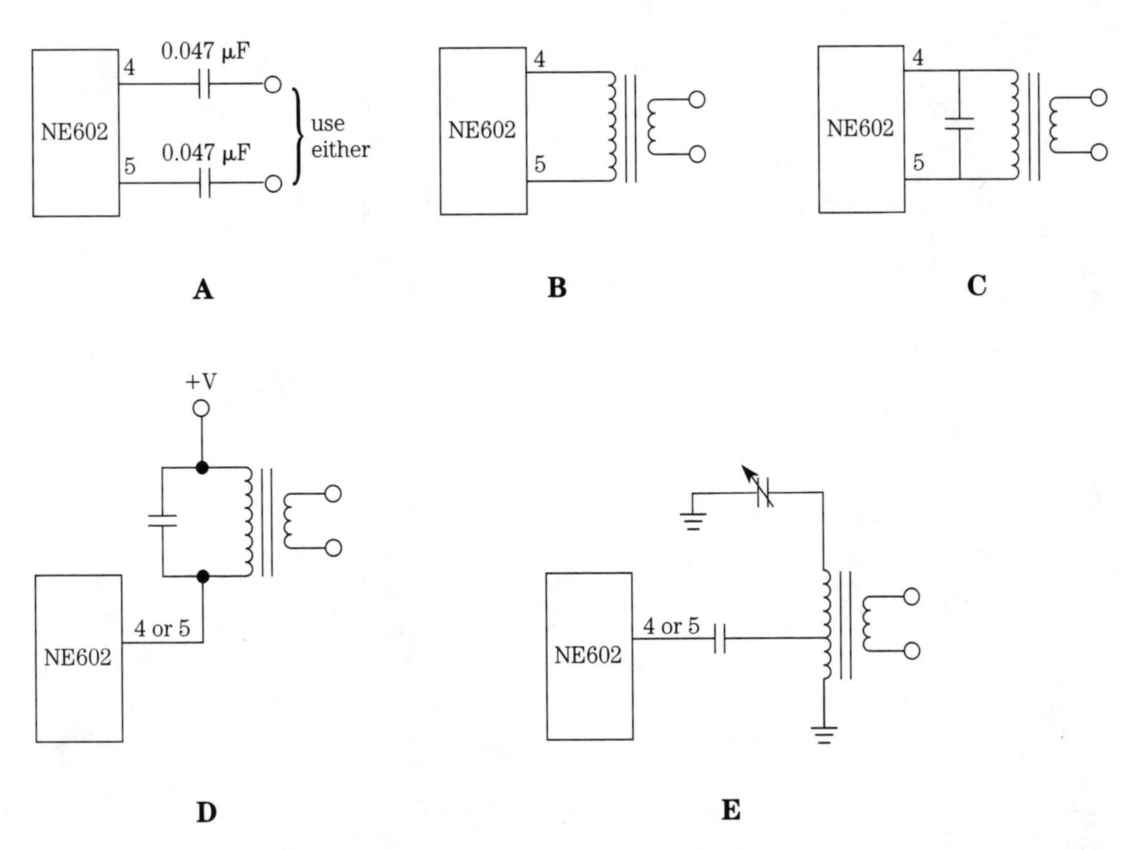

11-6 Output circuit configurations. A. Direct capacitor coupled output (untuned). B. Broadband transformer coupled output. C. Tuned transformer output. D. Tuned transformer to V+. E. Grounded tuned transformer output.

ing. This capacitor should have a low reactance at the frequency of operation, so values between 0.001 μF and 0.1 μF are generally selected.

Transformer output coupling is shown in Fig. 11-6B. In this circuit, the primary of a transformer is connected between pins 4 and 5 of the NE-602. For frequency converter or translator applications, the transformer could be a broadband RF transformer, wound on either a conventional slug-tuned form or a toroid form. For direct-conversion autodyne receivers, the transformer would be an audio transformer. The standard 1:1 transformers used for audio coupling can be used. These transformers are sometimes marked as to impedance ratio rather than turns ratio (e.g., 600 Ω : 600 Ω, or 1.5 kΩ : 1.5 kΩ).

Frequency converters and translators are the same thing, except that the "converter" terminology generally refers to a stage in a superhet receiver, while "translator" is more generic. For these circuits, the broadband transformer will work, but it is probably better to use a tuned RF/IF transformer for the output of the NE-602. The resonant circuit will reject all but the desired frequency product; e.g., the sum or difference "IF" frequency. Figure 11-6C shows a common form of resonant output

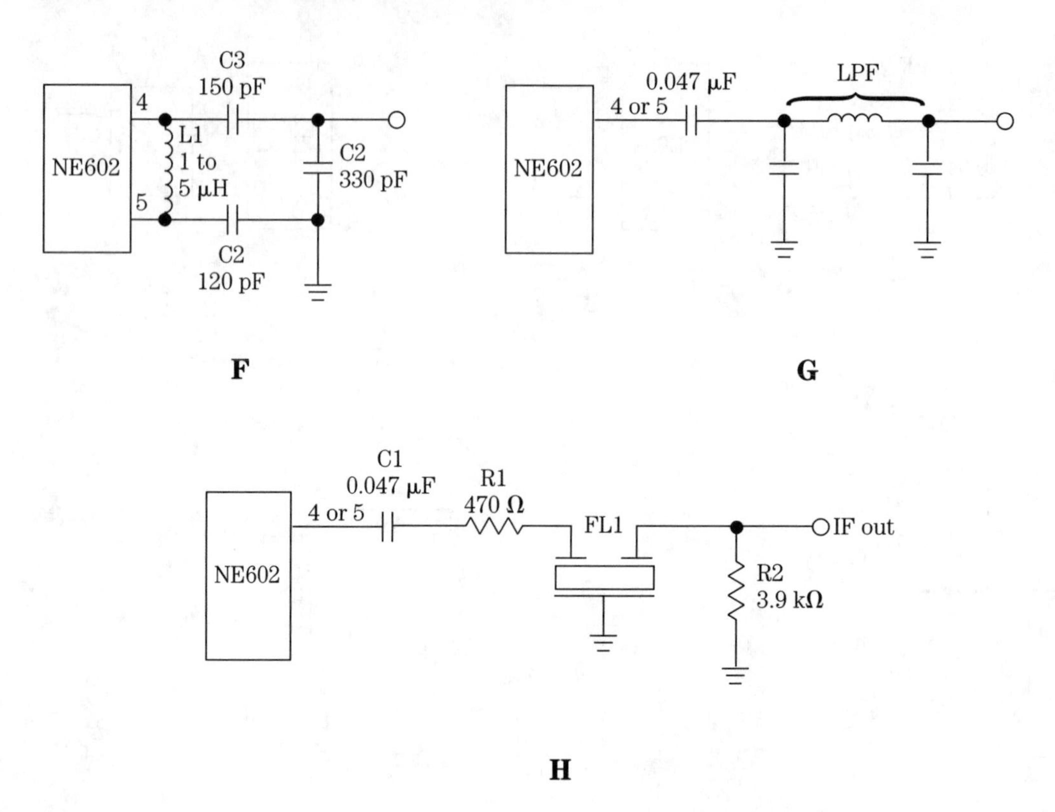

11-6 Continued. F. Tapped capacitor tuned output (VHF circuits). G. Low-pass filter output. H. Filter output.

circuit for the NE-602. The tuned primary of the transformer is connected across pins 4 and 5 of the NE-602, while a secondary winding (which can be tuned or untuned) is used to couple signal to the following stages.

A single-ended RF tuned transformer output network for the NE-602 is shown in Fig. 11-6D. In this coupling scheme, the output terminal of the IC is coupled to the V+ dc power supply rail through a tuned transformer. Perhaps a better solution to the single-ended problem is the circuit of Fig. 11-6E. In this circuit, the transformer primary is tapped for a low impedance, and the tap is connected to the NE-602 output terminal through a dc-blocking capacitor. These transformers are readily available in either 455 kHz or 10.7 MHz versions, and may also be made relatively easy.

Still another single-ended tuned output circuit is shown in Fig. 11-6F. In this circuit, one of the outputs is grounded for RF frequencies through a capacitor. Tuning is a function of the inductance of L1 and the combined series capacitance of C1, C2, and C3. By tapping the capacitance of the resonant circuit, at the junction of C2–C3, it is possible to match a lower impedance (e.g., 50 Ω) to the 1.5-kΩ output impedance of the NE-602.

The single-ended output network of Fig. 11-6G uses a low-pass filter as the frequency selective element. This type of circuit can be used for applications such as a heterodyne signal generator in which the local oscillator frequency of the NE-602 is heterodyned with the signal from another source applied to the RF input pins of the

IC. The difference frequency is selected at the output when the low-pass filter is designed such that its cut-off frequency is between the sum and difference frequencies.

In Fig. 11-6H, an IF filter is used to select the desired output frequency. These filters are available in a variety of frequencies and configurations, including the Collins mechanical filters that were once used extensively in high-grade communications receivers (260 kHz, 455 kHz, and 500 kHz center frequencies). Current high-grade communications receivers typically used crystal IF filters centered on 8.83 MHz, 9 MHz, 10.7 MHz, or 455 kHz (with bandwidths of 100 Hz to 30 kHz). Even broadcast radio receivers can be found using IF filters. Such filters are made of piezoceramic material, and are usually centered on either 260 or 262.5 kHz (AM auto radios), 455 or 460 kHz (other AM radios) or 10.7 MHz (FM radios). The lower-frequency versions are typically made with 4, 6, or 12 kHz bandwidths, while the 10.7 MHz versions have bandwidths of 150 to 300 kHz (200 kHz being most common).

In the circuit of Fig. 11-6H, it is assumed that the low-cost ceramic AM or FM filters are used (for other types, compatible resistances or capacitances are needed to make the filter work properly). The input side of the filter (FL1) in Fig. 11-6H is connected to the NE-602 through a 470-ohm resistor and an optional dc-blocking capacitor (C1). The output of the filter is terminated in a 3.9-kΩ resistor. The difference IF frequency resulting from the conversion process appears at this point.

One of the delights of the NE-602 chip is that it contains an internal oscillator circuit that is already coupled to the internal double-balanced mixer. The base and emitter connections to the oscillator transistor inside the NE-602 are available through pins 6 and 7, respectively. The internal oscillator can be operated at frequencies up to 200 MHz. The internal mixer works at frequencies up to 500 MHz. If higher oscillator frequencies are needed, you can use an external local oscillator. An external signal can be coupled to the NE-602 through pin 6, but must be limited to less than about –13.8 dBm, or 250 mV across 1500 ohms.

In the next section, we will take a look at some of the practical local oscillator (LO) circuits that can be successfully used with the NE-602, including one that allows digital or computer control of the frequency. Oscillator circuits are discussed in greater detail in chapter 13.

NE-602 local oscillator circuits

There are two general methods for controlling the frequency of the LO in an oscillator circuit: inductor-capacitor (LC) resonant tank circuits or piezoelectric crystal resonators. We will consider both forms, but first the crystal oscillators.

Figure 11-7A shows the basic Colpitts crystal oscillator. It will operate with fundamental mode crystals on frequencies up to about 20 MHz. The feedback network consists of a capacitor voltage divider (C_1/C_2). The values of these capacitors are crucial, and should be approximately:

$$C_1 = \frac{100}{\sqrt{F_{MHz}}}\ \text{pF} \qquad \textbf{[11-2]}$$

$$C_2 = \frac{1000}{F_{MHz}}\ \text{pF} \qquad \textbf{[11-3]}$$

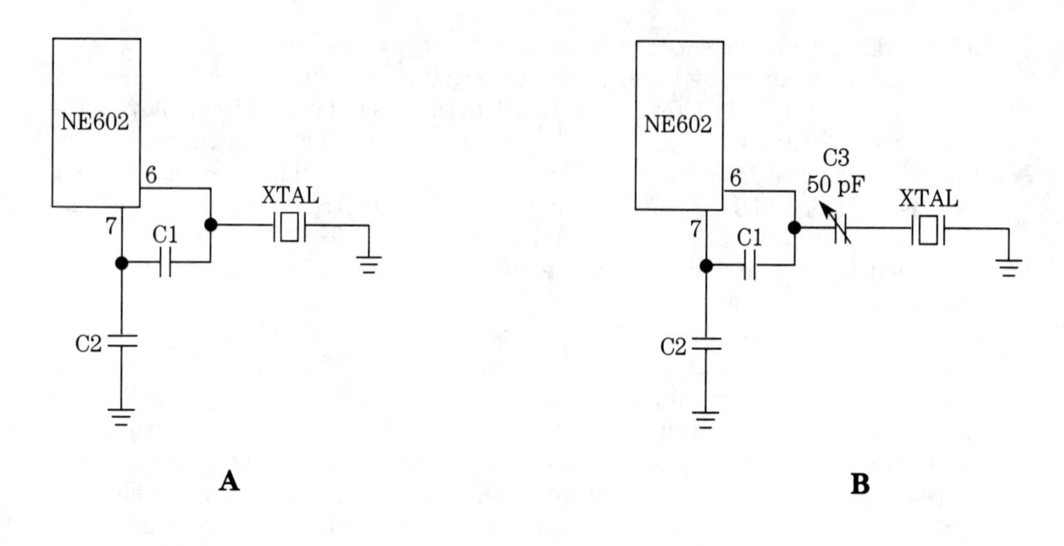

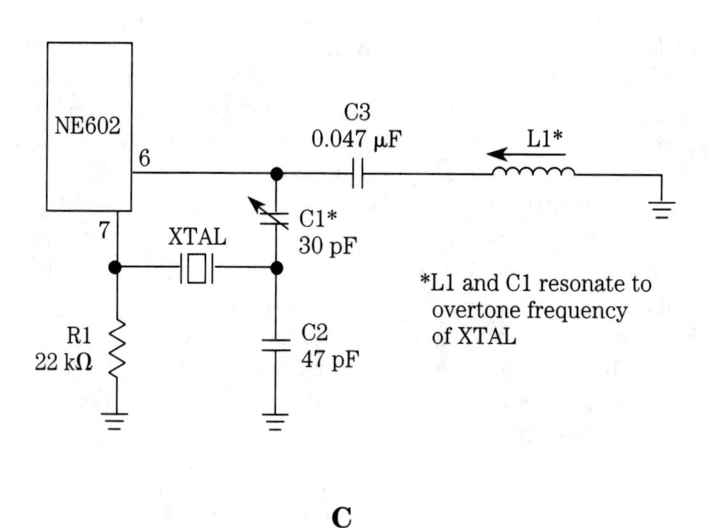

11-7 Local oscillator circuits for the NE-602. A. Simple Colpitts crystal oscillator. B. Colpitts crystal oscillator with adjustable frequency control. C. Butler overtone oscillator for low-band VHF.

The values predicted by these equations are approximate, but work well under circumstances where external stray capacitance does not dominate the total. However, the practical truth is that capacitors come in standard values—and these may not be exactly the values required by Eqs. 11-2 and 11-3.

When the capacitor values are correct, the oscillation will be consistent. If you pull the crystal out, and then reinsert it, the oscillation will restart immediately. Alternatively, if the power is turned off and then back on again, the oscillator will always restart. If the capacitor values are incorrect, then the oscillator will either fail

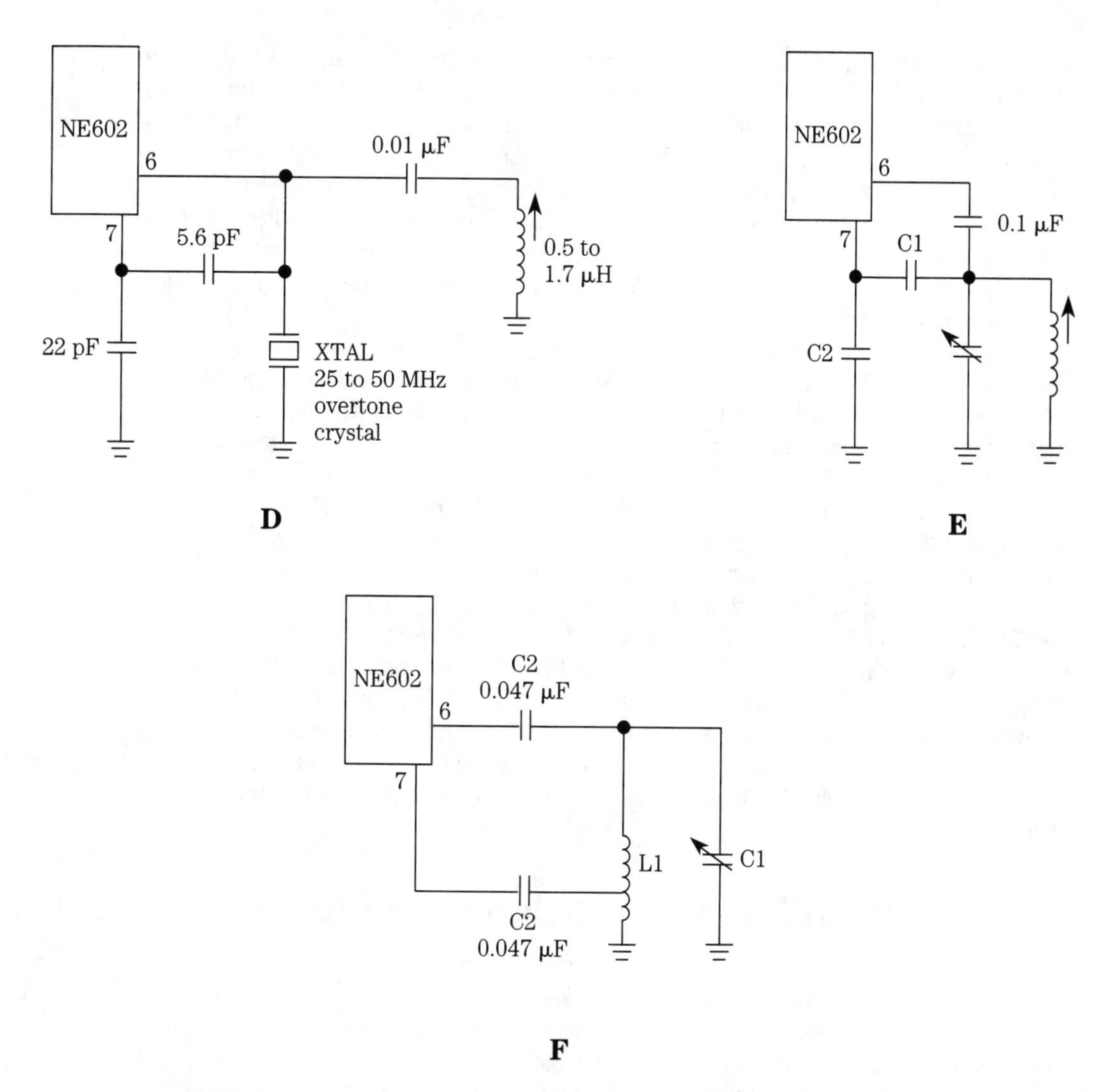

11-7 D. Additional overtone oscillator. E. Colpitts VFO. F. Hartley VFO.

to run at all, or will operate intermittently. Generally, an increase in the capacitances will suffice to make operation consistent.

A problem with the circuit of Fig. 11-7A is that the crystal frequency is not controllable. The actual operating frequency of any crystal depends, in part, on the circuit capacitance seen by the crystal. The calibrated frequency is typically valid when the load capacitance is 20 or 32 pF, but this can be specified to the crystal manufacturer at the time of ordering. In Fig. 11-7B, a variable capacitor is placed in series with the crystal in order to set the frequency. This trimmer capacitor can be adjusted to set the oscillation frequency to the desired frequency.

The two previous crystal oscillators operate in the fundamental mode of crystal oscillation. The resonant frequency in the fundamental mode is set by the dimensions of the slab of quartz used for the crystal; the thinner the slab, the higher the frequency. Fundamental mode crystals work reliably up to about 20 MHz, but above 20 MHz, the slabs become too thin for safe operation. Above about 20 MHz, the thinness of the slabs of fundamental mode crystal causes them to fracture easily. An alternative is to use *overtone mode* crystals. The overtone frequency of a crystal is not necessarily an exact harmonic of the fundamental mode, but is close to it. The overtones tend to be close to odd integer multiples of the fundamental (3rd, 5th, and 7th). Overtone crystals are marked with the appropriate overtone frequency, rather than the fundamental.

Figures 11-7C and 7E are overtone mode crystal oscillator circuits. The circuit shown in Fig. 11-7C is the *Butler oscillator*. The overtone crystal is connected between the oscillator emitter of the NE-602 (pin 7) and a capacitive voltage divider that is connected between the oscillator base (pin 6) and ground. There is also an inductor in the circuit (L1), and this inductor must resonate with C1 to the overtone frequency of crystal Y1. Figure 11-7C can use either 3rd or 5th overtone crystals up to about 80 MHz. The circuit in Fig. 11-7D is a third overtone crystal oscillator that works from about 25 to 50 MHz, and is simpler than Fig. 11-7C.

A pair of *variable-frequency oscillator* (VFO) circuits are shown in Figs. 11-7E and 7F. The circuit in Fig. 11-7E is the Colpitts oscillator version, while Fig. 11-7F is the Hartley oscillator version. In both oscillators, the resonating element is an inductor-capacitor (LC) tuned resonant circuit. In Fig. 11-7E, however, the feedback network is a tapped capacitor voltage divider, while in Fig. 11-7F, it is a tap on the resonating inductor. In both cases, a dc-blocking capacitor to pin 6 is needed in order to prevent the oscillator from being dc-grounded through the resistance of the inductor.

Voltage-tuned NE-602 oscillator circuits

Figure 11-8 shows a pair of VFO circuits in which the capacitor element of the tuned circuit is a *voltage-variable capacitance diode*, or *varactor* (D1 in Figs. 11-8A and 8B). These diodes exhibit a junction capacitance that is a function of the reverse-bias potential applied across the diode. Thus, the oscillating frequency of these circuits is a function of tuning voltage V_t. The version shown in Fig. 11-8A is the parallel-resonant Colpitts oscillator, while that in Fig. 11-8B is the series-tuned Clapp oscillator.

Conclusion

The NE-602 is a well-behaved RF chip that will function in a variety of applications from receivers, to converters, to oscillators, to signal generators.

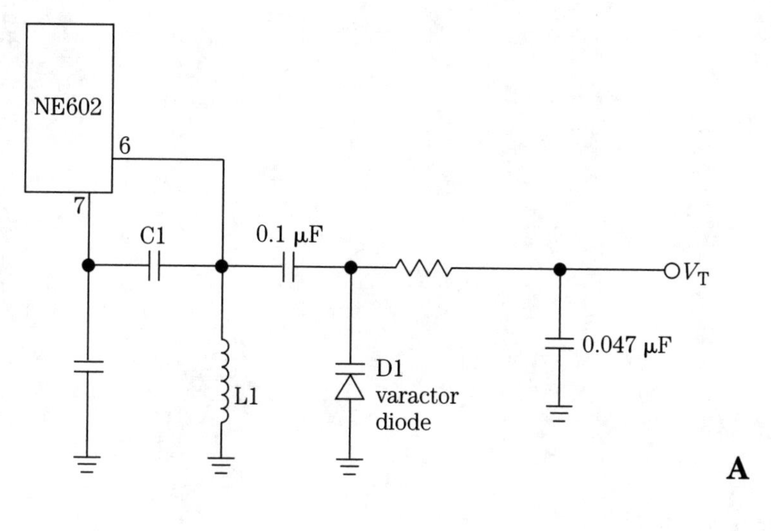

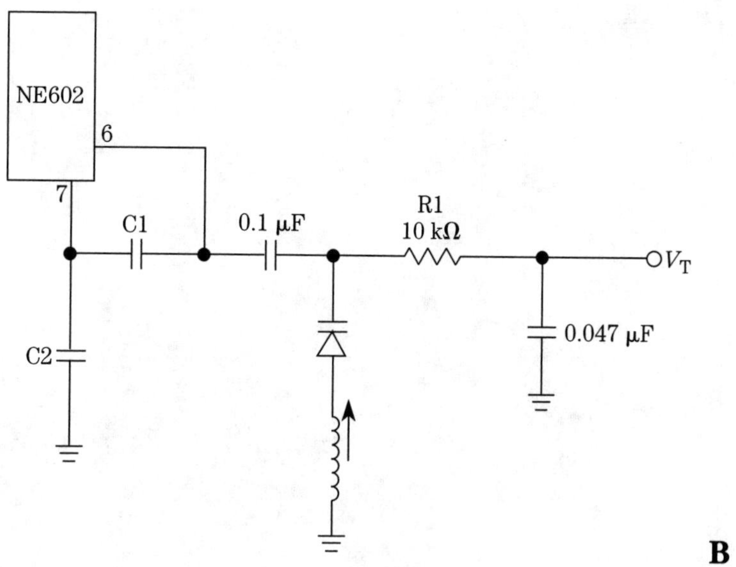

11-8 Voltage-tuned (varactor) VFO circuits. A. Colpitts. B. Clapp.

12
Direct-conversion radio receivers

Theory and projects

The direct-conversion, or *synchrodyne*, receiver was invented in the late 1920s, but only with the advent of modern semiconductor technology has it come into its own as a real possibility for good-performance receivers. Although most designs are intended for novices, and lack certain features of high grade superheterodyne receivers, the modern direct-conversion receiver (DCR) is capable of very decent performance. A case can be made for the assertion that the modern DCR is capable of performing as good as many middle-grade ham and SWL communications receivers. Although that assertion might seem very bold, experience bears it out. While no one, least of all me, would represent the DCR as capable of the best possible performance, modern DCR designs are no longer in the hobbyist curiosity category. In this chapter, you will find the basic theory of operation and some of the actual designs tried on the workbench. You are also advised to read chapter 10 on mixer circuits.

Basic theory of operation

The DCR is similar to the superheterodyne in underlying concept: the receiver radio frequency (RF) signal is translated in frequency by nonlinear mixing with a *local oscillator* (LO) signal ("heterodyning"). Figure 12-1 shows the basic block diagram for the front end of both types of receiver. The mixer is a nonlinear element that combines the two signals, F_{RF} and F_{LO}. The output of the mixer contains a number of different frequencies that obey the relationship:

$$F_o = mF_{RF} \pm nF_{LO} \qquad \textbf{[12-1]}$$

where

F_o is the output frequency
F_{RF} is the frequency of the received radio signal

F_{LO} is the frequency produced by the local oscillator (All frequencies in same units)
m and n are integers (0, 1, 2, 3, ...)

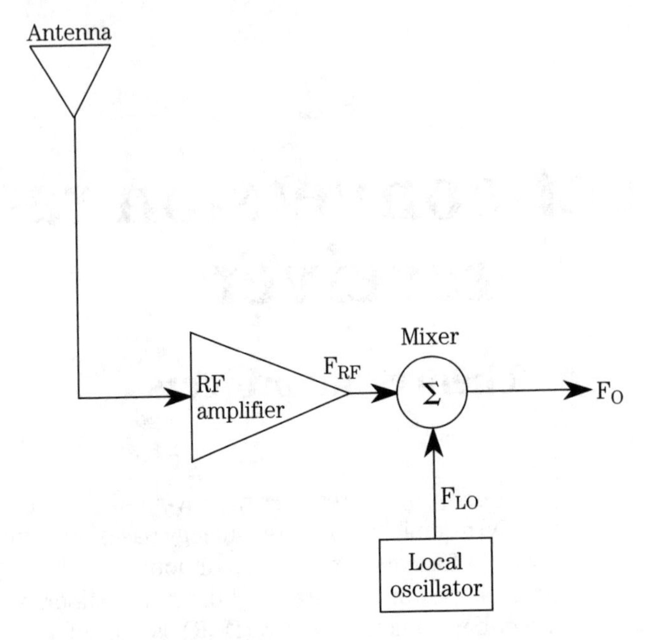

12-1 Basic mixer-local oscillator circuit.

All frequencies other than F_{RF} and F_{LO} are *product frequencies*. In general, we are only interested in the cases where m and n are either 0 or 1, so the output frequency spectrum of interest is limited to F_{RF} and F_{LO} plus the product frequencies $F_{RF} + F_{LO}$, and $F_{RF} - F_{LO}$. The latter two are called *sum* and *difference intermediate frequencies* (IF). Other products are certainly present, but for purposes of this discussion they are regarded as negligible. They are not regarded as negligible in serious receiver design, however.

In a superheterodyne radio receiver, a tuned bandpass filter will select either the sum IF or the difference IF, while rejecting the other IF—the LO and RF signals. Most of the gain (which helps determine sensitivity) and the selectivity of the receiver are accomplished at the IF frequency. In older receivers, it was almost universally true that the difference IF frequency was selected (455 kHz and 460 kHz being very common), but in modern communications receivers, either or both might be selected. For example, it is common to use a 9-MHz IF amplifier on high frequency (HF) band shortwave receivers. On bands below 9 MHz, the sum IF is selected—while on bands above 9 MHz, the difference IF is selected. A popular combination on amateur radio receivers uses a 9-MHz IF combined with a 5- to 5.5-MHz variable frequency oscillator. To receive the 75/80-meter band (3.5 to 4.0 MHz), the sum IF is used. The same combination of LO and IF frequencies will also receive the 20-meter (14.0 to 14.4 MHz) band if the difference IF (i.e., 14.0 – 5 = 9 MHz) is used.

In a DCR, on the other hand, only the difference IF frequency is used (see Fig. 12-2). Because the DCR LO operates at the same frequency as the RF carrier, or on a nearby frequency in the case of CW and SSB reception, the difference frequency represents the audio modulation of the radio signal. Amplitude modulated (AM) signals are accommodated by zero-beating the LO to the radio signal making

$$F_{LO} = F_{RF(carrier)}$$

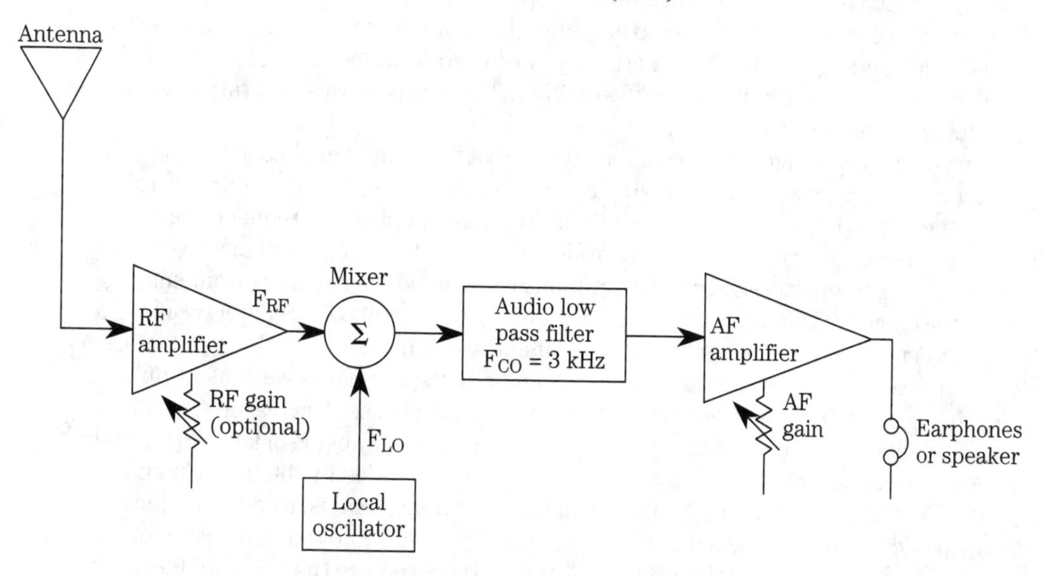

12-2 Partial block diagram for a direct conversion receiver.

Thus, only the recovered upper and lower sidebands will pass through the system, and they are at audio frequency.

For CW signals (Morse code on-off telegraphy), and single sideband signals, it is necessary to offset the LO frequency slightly to recover the signal. For the CW case, one must select a comfortable tone (which is an individual preference). In my own case, I am most comfortable using an 800-Hz (0.8 kHz) note when copying CW, so I offset the LO from the RF by 800 Hz. For example, when copying a CW signal at, say, 3650 kHz, the LO will be tuned to either 3649.2 kHz or 3650.8 kHz. In either case, the beat note heard in the output is 800 Hz. Single sideband (SSB) reception requires an offset on the order of 1.8 to 2.8 kHz for proper reception.

As was true in the superheterodyne receiver, the majority of the gain and selectivity in the DCR is provided by the stages after the first mixer. While the superheterodyne uses the IF amplifier chain for this purpose, followed by second detection and audio amplification, the DCR must use only the audio amplifier chain. Thus, it becomes necessary to provide some very high-gain audio amplifiers and audio bandpass filtering in the DCR design.

One implication of DCR operation is the lack of single signal operation. Both CW and SSB signals will appear on both sides of the zero-beat point ($F_{RF} = F_{LO}$ exactly). While this feature can be a problem, it has at least one charming attribute on SSB reception: the DCR will receive LSB signals on one side of zero-beat and USB signals on the other side of zero-beat. There have been attempts to provide single-signal reception of SSB signals on DCRs[1] by using audio and VFO phasing circuits (in the manner of the phasing method of SSB generation). That approach greatly increases the complexity of the receiver, which may make other design approaches more reasonable than DCRs.[2]

The most basic implementation of the DCR (Fig. 12-2) requires only a mixer stage, a local oscillator and an audio amplifier. If the mixer has a high enough output signal level, and high impedance earphones are used to detect the audio, some designs can make do without the audio amplifier. These are, however, a rarity and the one version that I tried did not work very well.

In some DCR designs, there will be an optional RF input signal conditioning consisting of either a low-pass filter, high-pass filter, or bandpass filter (as appropriate) to select the desired signal or reject undesired signals. Without some frequency selection at the front end, the mixer is wide-open with respect to frequency, and may be unable to prevent some unwanted signal (or spurious combinations of signals) from entering the receiver circuits. Some designs include more than one style of filter. For example, a popular combination uses a single-staged tuned resonant circuit at the input of the mixer to select the RF signal to be received and a high-pass filter—with a cut-off frequency F_{CO} of 2200 kHz—to exclude AM broadcast band signals. The reason for such an arrangement is that the AM signal may be quite intense, usually being of local origin, and is therefore capable of overriding the minor selectivity provided by the tuned circuit.

The RF amplifier used in the front end is also optional, and is used to provide extra gain and possibly some selectivity. The gain is needed to overcome losses or inherent insensitivity in the mixer design. Not all mixers require the RF amplifier, so it is frequently deleted in published designs. In general, RF amplifiers are used only in DCRs operating above 14 MHz. Below 14 MHz, signals tend to be relatively strong, and man-made noise tends to be much stronger than inherent mixer noise.[3]

Problems associated with DCR designs

Over the years, a number of articles have been published in the popular technical press on DCR radio receiver designs, some of which are cited in this book. In addition, the major amateur radio communications handbooks typically discuss DCR designs. Several problems are cited as being common in the DCR receiver designs. While these problems are very real, careful design and construction can render them harmless. The list of observed problems includes hum, microphonics, poor dynamic range, low output power levels (which makes for uncomfortable listening) and unwanted detection of AM broadcast band signals ("AM breakthrough").[4]

Hum

Hum is due to the alternating current (ac) power lines, and can either be radiated into the DCR through the antenna circuit, radiated into the wiring of the set, caused by ground loops, or it might be communicated to the DCR circuits as ripple in the dc power supply. In the first two instances, the hum will have a frequency of 50 Hz (Europe) or 60 Hz (North America) (which are the ac power line frequencies), while in the latter two cases, it will be 100 Hz (Europe) or 120 Hz (North America) if full-wave-rectified dc power supplies are used. All hum problems are aggravated by the high-gain audio amplifiers used in DCRs, and for this reason, bandpass filtering that attenuates signals below 300 Hz is highly recommended. (300 Hz is considered the low end –3 dB point for speech waveforms; the upper –3 dB point is usually specified as 3000 Hz.)

Hum signals received through the antenna circuits are best handled by use of high-pass RF filtering that does not admit energy in the 50/60 Hz region. The front end of many DCRs is wide open with respect to frequency, so improper choices can leave the receiver sensitive to hum. Some mixer circuits, such as the double-balanced mixer (DBM) diode designs, are inherently insensitive to hum because of the nature of the inductors (and their cores) used in the associated inductors. If the hum is radiated into the circuit from the environment, then shielding the circuit against electrical fields will usually solve the problem.

Finally, we have the case where hum is received via the dc power supply and is ripple related (probably the most common cause). *Ripple* is the residual rectified ac (or "pulsating dc") riding on the output voltage of a dc power supply after all filtering and voltage regulation is done. The worst hum occurs when the dc power supply ripple modulates the LO signal. In that case, hum becomes worse as frequency increases.[5] A combination of good voltage regulation (which tends to limit ripple as if it were a very large filter capacitor) and proper grounding techniques will solve the problem.

Figure 12-3 shows one grounding sin that must be avoided at all costs. In the properly grounded receiver all of the grounds will be connected to point "B" (single-point or star-grounded). In actual practice, however, the nature of printed circuit or perforated wire board point-to-point wiring is such that the "star" ground concept is not achievable in its fullest sense, so there may be one or two additional points of grounding. Care must be taken to prevent ground loops in such cases.

The great sin in Fig. 12-3 is grounding the dc power supply to the antenna input ground point, while the amplifier is grounded elsewhere. Even with very low resistance ground tracks, a considerable signal is created when the dc current required by the DCR flows from "A" to "B" and then into the receiver. For example, when the receiver draws 50 mA, and the track has a dc resistance of 0.05 ohms (neither number is unreasonable), a voltage drop of 2.5 mV will exist. While this voltage seems small, it is not so small when followed by the 80- to 120-dB gain typically found in a radio receiver. Under these conditions, even the smallest ripple waveform riding on the dc power supply voltage can cause massive hum in the receiver's output!

A solution to the hum problem is shown in Fig. 12-4. Single-point grounding is employed in order to reduce ground loops. In addition, a toroidal decoupling choke is placed in the dc power supply leads.[6] This choke consists of two bifilar windings of 20 turns each over a toroidal core that has a permeability (μ) between 600 and 1500. The wire used to make the bifilar windings must be sufficiently large to accommodate the current requirements of the DCR circuitry.

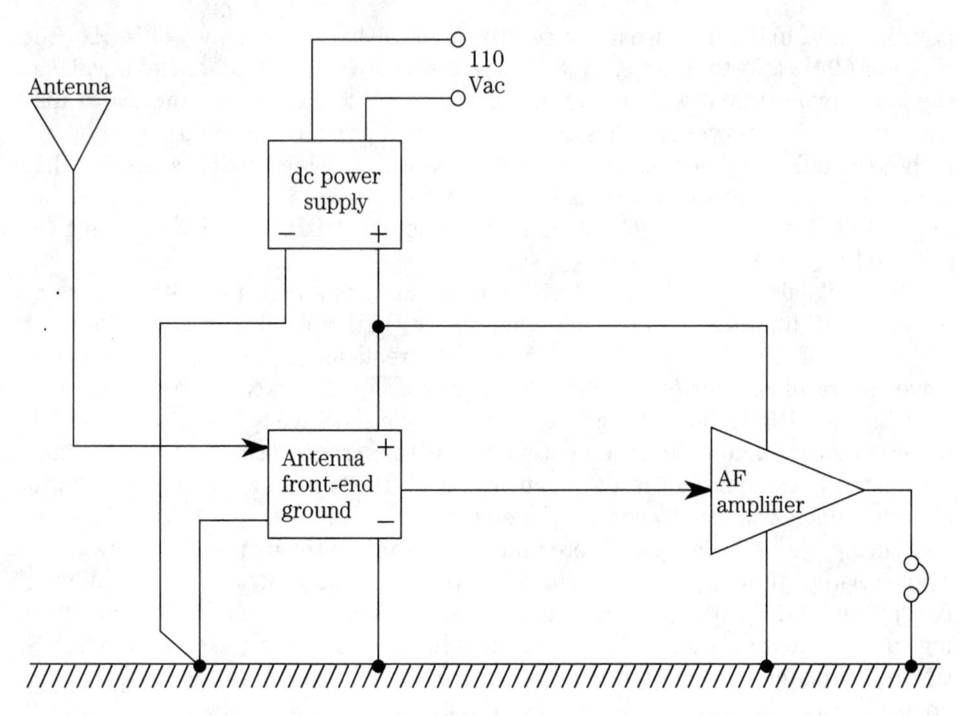

12-3 Improper wiring of stages in direct-conversion receiver.

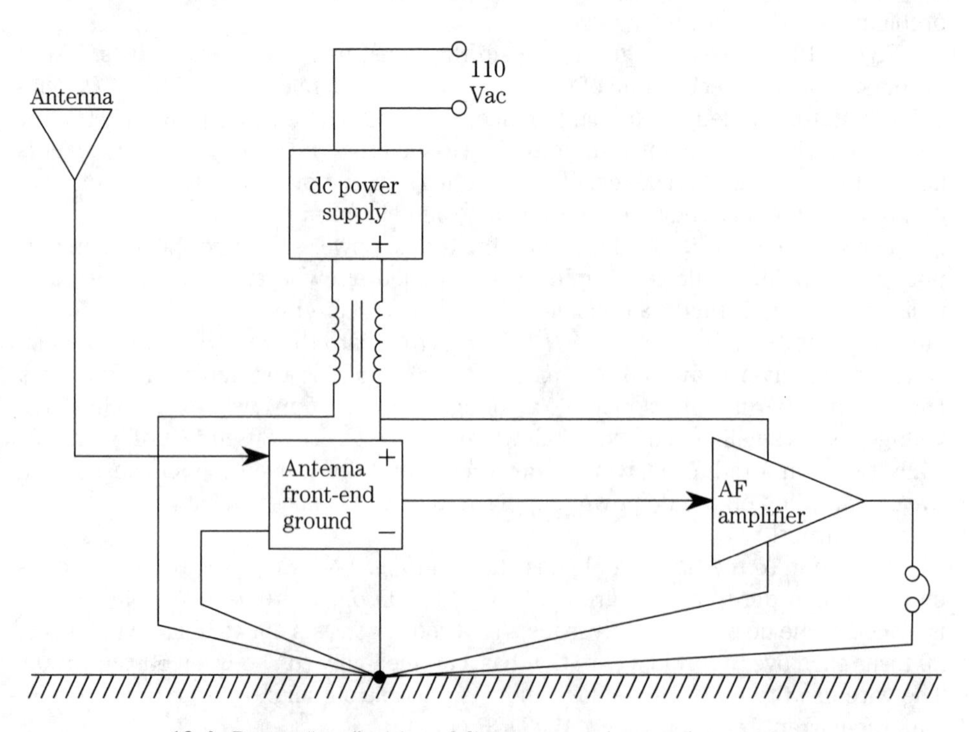

12-4 Proper "star" wiring of direct-conversion receiver stages.

Of course, using batteries for the dc power supply will solve the problem of ripple-induced hum. Even so, good grounding technique is a fundamental requirement for building a good receiver.

Microphonics

Microphonics are a highly sensitive, damped, echo-like ringing sound in the output when the receiver is vibrated or jarred, and continues until the vibration dies out. It is caused by modulation of the signal by the vibration. Microphonics are generated by the 80- to 120-dB gain associated with a typical DCR. Any modulation of the signal therefore produces change in the output signal. Two principal sources of microphonics are the movement of the frequency setting components (mostly the inductor) in the LO, and changes in coupling between circuit elements (including wires). Ordinarily, the latter would not occur—but in DCRs, there is usually a tremendous amount of gain, so such possibilities exist.

Dynamic range problems

Dynamic range is basically the difference between the maximum and minimum signals that the receiver can accommodate, and is usually expressed in decibels (dB). A description of dynamic range found in Hayward and DeMaw describes it as:[7]

$$\text{Dynamic range} = \frac{2(P_i - MDS)}{3} \qquad \textbf{[12-2]}$$

where

P_i is the signal level associated with the third-order intercept point of the receiver

MDS is the minimum discernible signal

Dynamic range is important in all bands, but there are cases where it is most important. For example, when receiving any frequency when the receiver is in close proximity to an AM broadcast station, a ham operator, or other transmitter site—regardless of the frequency being received. This problem becomes especially acute when the front-end of the receiver is wideband, and therefore provides little or no discrimination against the unwanted frequencies. However, also recognize that front end tuning does not necessarily completely eliminate unwanted out of band signals that are very strong, especially if only one tuned circuit is provided.

Although normally associated with very large signals, another case where dynamic range improves signals is on very crowded bands where there are a large number of weak, moderate and strong signals mixed together. One source asserts that the dynamic range of signals received on a 40-meter (7 to 7.3 MHz) half-wavelength dipole is greater than the dynamic range of an audio compact disk.[8] Given the mixture of powerful international broadcasters and ham operators (both kilowatt and "QRP" stations), that claim is probably close to the mark. Intermodulation products and other spurious signals on these bands are often stronger than legitimate signals. While none of the myriad signals present would tax the receiver's capabilities by itself, the net result of all of them being simultaneously present is poor performance. Being able to discriminate weak signals from strong signals, especially when using a wideband front end, is a benefit of having a high dynamic range receiver. When the dynamic range of the receiver is insufficient for the task at hand, the output will be distorted and possibly very weak if desensitization occurs.

Several factors contribute to the dynamic range of the DCR, but the primary focus should be on the characteristics of the mixer circuit. Campbell provides three criteria for achieving a good dynamic range in the DCR:[9]

- Proper termination of the mixer over a wide band.
- Avoidance of resistors in the signal path.
- Restriction of the system bandpass prior to the audio preamplifier (which also helps hum rejection).

Any RF device, such as a mixer, should be terminated in its output impedance for maximum signal transfer, minimum reflections, and proper operation. Indeed, many circuits will not perform as advertised unless they are properly terminated. Resistors in the signal path are to be avoided, especially prior to the high gain audio section, because resistors generate noise of their own. Even a "perfect" resistor will generate thermal noise (4KTBR), and practical resistors add some additional types of noise to the mix.

Restriction of the system bandpass prior to audio amplification reduces the unwanted noise components of the signal prior to processing in the high gain of the audio stages. The signal-to-noise ratio (SNR) is thereby improved. This same trick is used by some authorities to justify using an "antenna-to-mixer" approach on expensive shortwave superheterodyne receivers. In those designs, the main bandpass filtering is at the output of the mixer, and the mixer input is driven directly from the bandpass filters in the front end of the receiver (no RF amplifier). This philosophy was used in the Squires-Sanders HF vacuum tube receivers made in the United States some years ago. While it was a novel approach at the time, it has become one of several possible standard approaches today.

Many designers of DCRs use a filter between the output of the mixer and the input of the preamplifiers. If the filter is matched to the impedance of the mixer and the input of the preamplifier, and if it contains no resistors in series with the signal path, it largely meets Campbell's criteria stated above. The *diplexer* filter (more on this later) largely meets these criteria.

AM breakthrough

Stations in the AM broadcast band (550 to 1620 kHz), and in many regions of the world VLF broadcast band also, tend to be both relatively high-powered and very local to the receiver (thus very strong). As a result, many users of radio receivers for other bands are afflicted with the phenomenon of AM breakthrough. This term refers to any of several phenomena, but the end result is AM interference to reception. Sometimes, the problem is due to simple overload of the front end of the receiver, causing desensitization. In other cases, there will be intermodulation problems—and in still other cases, there will be severe distortion of the desired signals. In a great many cases, the AM BCB signal will be demodulated and interfere with the audio from the desired station. AM breakthrough is often accompanied by howling and screeching sounds from the receiver, while at other times, the effects are quite subtle. A frequently seen form of this problem is a strong background "din" caused by demodulation of the AM BCB signal in nonlinear elements in the circuit, particularly pn junctions.[10] The effects are usually more severe in simple DCR designs because the front end may be wideband.

Several approaches are used to overcome AM breakthrough. First, the DCR should be well-shielded so that there is no "antenna effect" from circuit wiring. Even so, coaxial cable should be used from the DCR printed circuit board to the antenna connector. In addition to shielding, which by itself does not affect signals riding in on the antenna, there should be at least some filtering of the input signal in an LC network. The network can be either a bandpass filter that passes an entire segment of the spectrum but restricts the rest, or tuned to a single frequency.

Another approach, which can be used in conjunction with either or both of the above approaches, is to place a high-pass filter that rejects the AM broadcast band in series with the signal path. Figure 12-5 shows two such filters that can be made with easily available components. Both of these filters are designed to be installed right at the antenna connector, either inside or outside of the receiver cabinet.

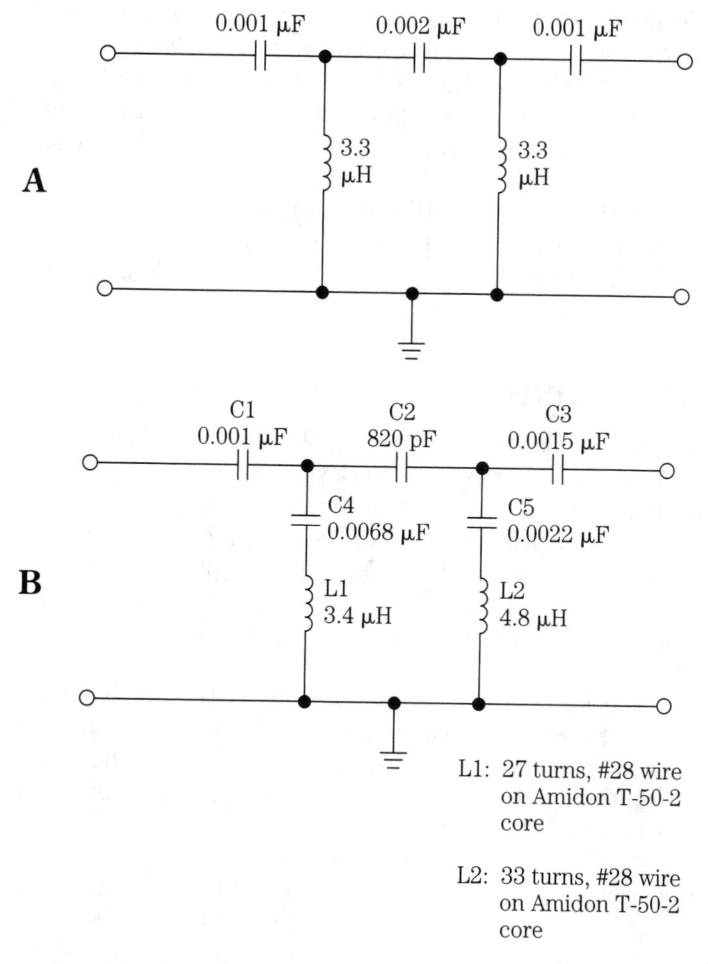

12-5 Two filters for eliminating AM broadcast band interference.

The AM rejection filter of Fig. 12-5A can be built with ordinary disk ceramic, silver mica capacitors or Panasonic V-series mylar capacitors. Good performance is achievable with 5% tolerance units, although it is also a good practice to match them using a digital capacitance meter or bridge. If silver-mica units are selected, it is possible to select 1000 pF (0.001 μF) capacitors, using two in parallel for the 0.002 μF capacitors (C1 and C3). The inductors of this circuit are each 3.3 μH, and can be either shielded inductors with regular cores, or unshielded inductors with toroidal cores. One combination that I've used is the Amidon Associates (P.O. Box 956, Torrance, CA 90508, USA) Type T-50-2 (RED) toroidal core wound with 29 turns of #26 enameled wire.

The filter in Fig. 12-5B is a little more complex, but offers as much as 40 dB of AM suppression in the HF bands. It is a high-pass filter with a cut-off near 2200 kHz. Note that the coils have different inductance values, and are each in series with a capacitor. These coils are wound on Amidon T-37-2 toroidal forms.

Both of these filter circuits should be built inside of a well-shielded enclosure. Most well-made aluminum sheet metal or die-cast project boxes will do nicely. However, be wary of the sort of sheet metal box that has no overlapping edges at the mating surfaces of the two halves of the box. A well-made sheet metal box will have at least 5 mm of overlapping flange built onto either the top or bottom half of the box. These boxes are a bit more "RF tight" when joined together. Even so, additional screws may prove necessary.

When laying out the filter on a perfboard or printed circuit board, be sure to use good construction practices. That is, lay out the components in a line from one end to the other, without excessive space between them and without doubling back in such a way that the output is close to the input.

Low-audio output

One of the frequent complaints made about DCRs is that the audio level is too low for comfortable listening. The mixer output level is very low, so a DCR typically requires a large amount of gain in the audio amplifier chain to produce even minimal levels of power to earphones or loudspeakers. Additional gain, and a reasonable power amplifier can be provided—but only at the risk of exacerbating all of those problems that come along with high gain in the first place. Attention to good layout, design and grounding practices can greatly ease some of these problems without undue burden.

The use of proper filtering of the signal prior to the audio preamplifier will greatly enhance the enjoyment of listening to a DCR, and reduce some of the problems associated with low power levels. The goal is to structure the bandpass of the receiver to that which is needed to pass the information content of the modulation, but not the noise outside of that passband.

Figure 12-6 shows the block diagram for a simple but reasonable DCR, based on the principles developed in this chapter. Shortly, we will take a closer, more detailed look at some specific circuits to fill the blocks—but for now the block diagram description will suffice. Note that this description is functional, rather than stage-by-stage, so some of the blocks in most designs will be combined into a single stage.

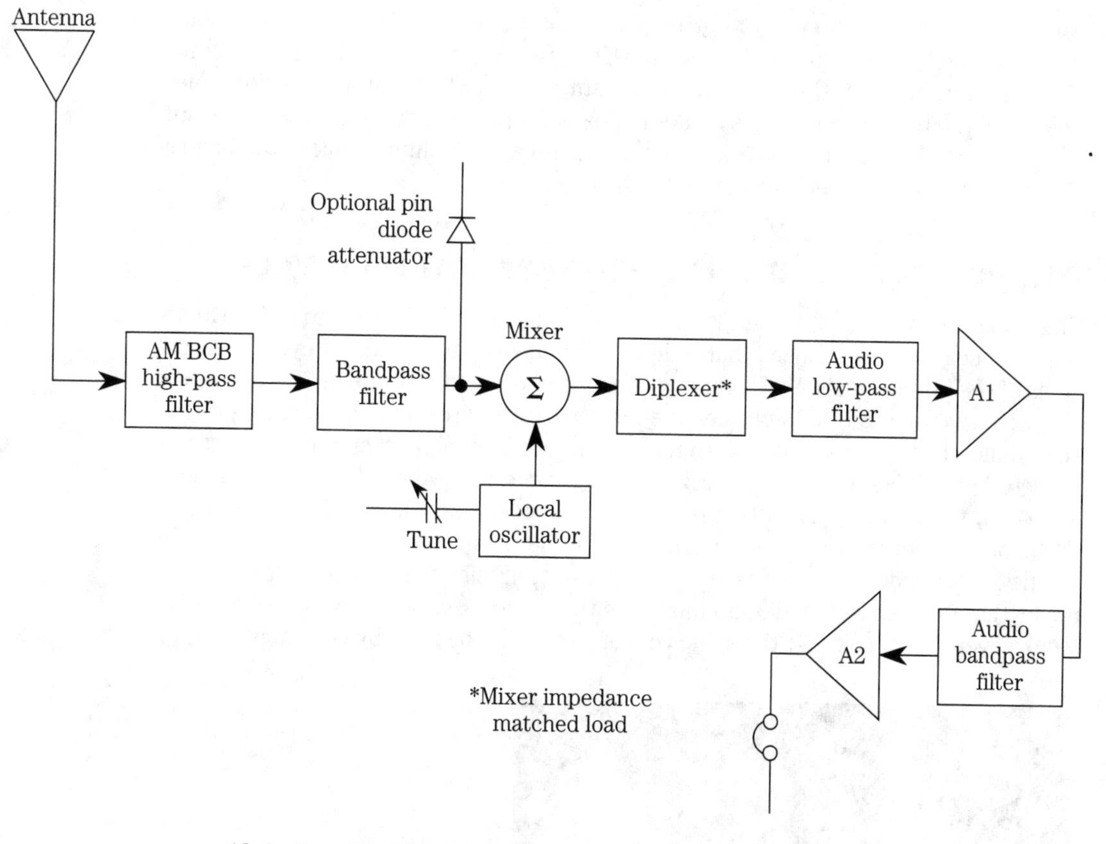

12-6 Complete block diagram for a direct-conversion receiver.

The front end of the DCR of Fig. 12-6 consists of an AM suppression high-pass filter, followed by either a tuned filter or bandpass filter at the frequency of choice. An optional pin diode attenuator circuit provides a degree of manual gain control for RF signals. The local oscillator is a variable frequency oscillator (VFO) that tunes the entire band of interest. It must be relatively well-designed and be free of drift and noise problems. The LO/VFO signal is mixed with the RF path signals from the chain of filters at the mixer circuit to produce a down-converted signal that contains the modulation that we are attempting to recover.

The function following the mixer provides bandpass filtering to limit the effects of out-of-band signals, noise and other artifacts, while also providing the mixer with appropriate impedance termination at all frequencies (in practice, the mixer will be well-terminated above and below the audio "base band" of interest—typically 300 to 3000 Hz—but is not well-terminated in the base band). Amplifiers A1 and A2 provide the bulk of the necessary audio frequency gain, while a filter between them keeps the bandwidth correct for good noise suppression.

In practice, amplifiers A1 and A2, and the audio bandpass filter, will be all part of one stage—although such is not a requirement. Hayward and others advocate the use of a post-filter at the output of the high gain stages. This filter should have a bandwidth matched to the system bandwidth, although one source recommends "slightly wider" without being too specific. The purpose of the post-filter is to reduce the noise generated in the high-gain audio amplifiers (which can be considerable). Following the high-gain stages is an audio power amplifier that boosts the signal sufficiently for either earphones or a loudspeaker. An audio volume control will be provided at this point, or perhaps earlier in the chain.

Mixer circuits in direct-conversion receivers

The principal element in any direct conversion receiver (DCR) is the mixer. The mixer is a nonlinear circuit element that exhibits changes of impedance of cyclical excursions of the input signals. When mixing is linear, one signal will ride on the other (see Fig. 12-7A) as an algebraic sum (i.e., the two waveforms are additive), but the product (i.e., multiplicative) frequencies are not generated. In nonlinear mixing, the classical amplitude modulated waveform (Fig. 12-7B) is produced when the two frequencies are widely separated. A mixer that produces product frequencies can be used either in DCRs or in superheterodyne receivers. In superhet terminology, it is common to call the frequency translation mixer that produces the IF signal a *first detector*, and the mixer that recovers the audio modulation either a *product detector* or a *second detector*, even though exactly the same sort of circuit can be used by either case.

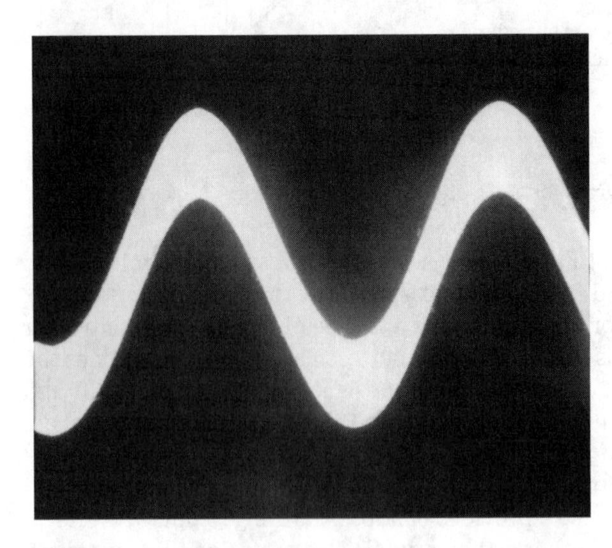

12-7 A. Linearly mixed signals.

A number of mixer circuits are used in radio receivers, and most are candidates for use in direct-conversion receivers. As you will see in chapter 10, however, not all mixers are created equal, so some are better suited to DCR applications than others. In nearly all cases, the output circuit of the mixer will be a low-pass filter that passes audio frequencies, but not RF frequencies.

Two issues dominate mixer selection for DCR service: *sensitivity* and *dynamic range*. The former determines how small a signal can be detected, while the latter determines the ratio between the minimum detectable signal and the maximum al-

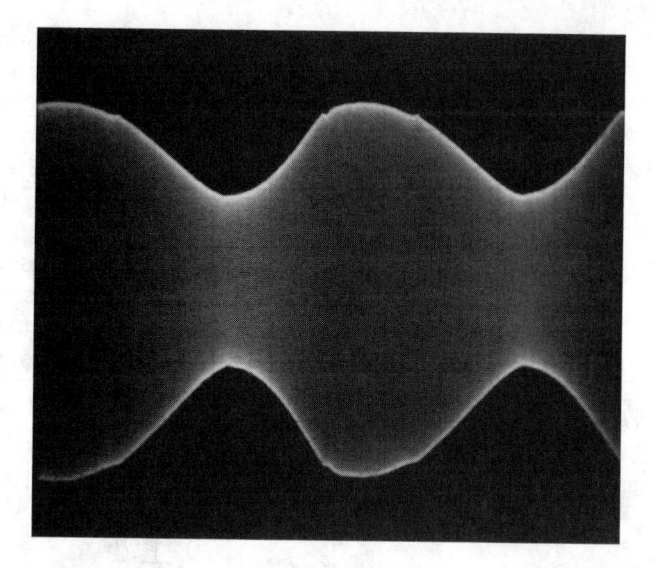

12-7 B. Nonlinearly mixed signals result in the AM modulation waveform when two frequencies are wide apart.

lowable signal. Some passive mixers produce so much loss, so much noise, and require so much signal strength to operate in the first place that they are simply not suited to DCR design unless adequate preamplification is provided. Such detectors can sometimes be put to good use in superheterodyne receivers because they are preceded by the gain of the front end and the gain of the IF amplifier chains, which can be considerable.

There are two issues to resolve when selecting a mixer element for a DCR. First, there is always the possibility of radiation of the local oscillator (LO) signal back through the antenna. In order to prevent this problem, it is necessary to keep the mixer unilateralized, i.e., signal flowing in only one direction through the circuit. Some mixers are inherently good in this respect, while others are a bit problematical. In cases where LO radiation might occur, it is recommended that an RF amplifier be used ahead of the mixer, regardless of whether or not it is needed for purposes of improving sensitivity or selectivity.

The second problem that must be resolved is transmission of the RF and LO signals to the output of the mixer. You normally only want the first order sum and difference frequencies. Many forms of simple mixer circuit are particularly bad in this respect, while others are considerably better. Theoretically, any mixer can be used for the front end of the DCR, however, the simple half-wave rectifier diode envelope detectors are not at all recommended. Representative mixer circuits can be found in chapter 10.

Considerations for DCR good designs

It probably does not surprise many readers that principles of good design result in superior DCR performance. Some of these principles were discussed by Campbell[11] and others.[12] Even relatively simple DCR designs, including those based on the Signetics NE-602 integrated circuit double-balanced modulator[13] and the popular LM-

386 audio amplifier, have proven to be very sensitive and free of hum and microphonics, even though that combination is not without critics. Dillon's design, which was tested in the laboratories of the American Radio Relay League (ARRL), proved remarkably free of the problems often associated with simple DCR designs.[14]

One method for terminating the mixer is to place a resistor-capacitor (RC) network across the "IF OUT" terminals of the mixer and ground (see Fig. 12-8). The SBL-1 mixer is designed for 50-ohm input and output impedances, so the device is terminated in its characteristic impedance at RF frequencies by the 51-ohm resistor (R1). Because capacitor C1 has a value that produces a high reactance at audio frequencies (AF), and a low reactance at RF, the mixer is terminated for any residual LO and RF signal (which are absorbed by R1), but AF is transmitted to the low-pass filter.

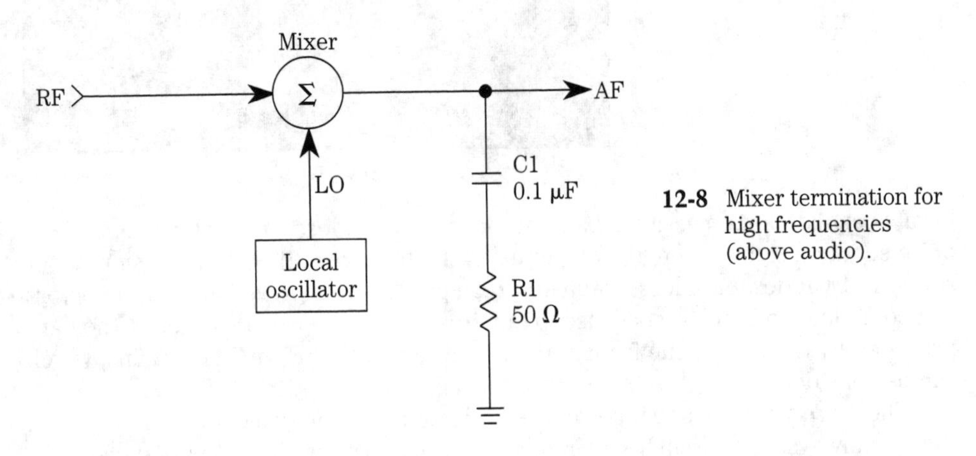

12-8 Mixer termination for high frequencies (above audio).

Some practical design approaches

The literature on DCRs has several popular approaches, and each has its own place. Some of the simpler designs are based on the combination of a Signetics NE-602 device and an LM-386 IC audio section. Others are based on different IC devices such as the Signetics TDA7000 or some other product. These chips were designed for the cellular telephone and "cordless" telephone markets as receiver front ends. Still others are based on commercial or homebrew double-balanced modulators. In this section, we will examine several of these approaches.

The NE-602 type of DCR is relatively easy to build, and provides reasonable performance for only a little effort. The NE-602 chip is relatively easy to obtain, and for the most part is well-behaved in circuits (i.e., it does what it is supposed to do). It has about 20 dB of conversion gain, so it can help overcome some circuit losses, and reduces slightly the amount of gain required of the audio amplifier that follows. The NE-602 can provide very good sensitivity (on the order of 0.3 μV), is relatively easy to obtain, but lacks something in dynamic range. Although the specifications of the device allow it to accept signals up to –15 dBm, at least one source recommended a maximum signal level of –25 dBm.[15] At higher input signal levels, the NE-602 tends to fall apart.[16] The newer NE-612 is basically the same chip, but has improved dynamic range. While I've not personally tried the newer variety, it is reputed to be a greatly improved device compared to the NE-602.

Figure 12-9 shows the basic circuit of the simplest form of NE-602 DCR. The input and LO circuits can be constructed in any of several configurations, although that shown here is probably the most common.[17] The output signal is taken from either pin 4 or pin 5, and is fed to an audio amplifier. This circuit configuration will work, but it is not recommended. There will be a fairly large noise and image signal component, and no filtering is provided.

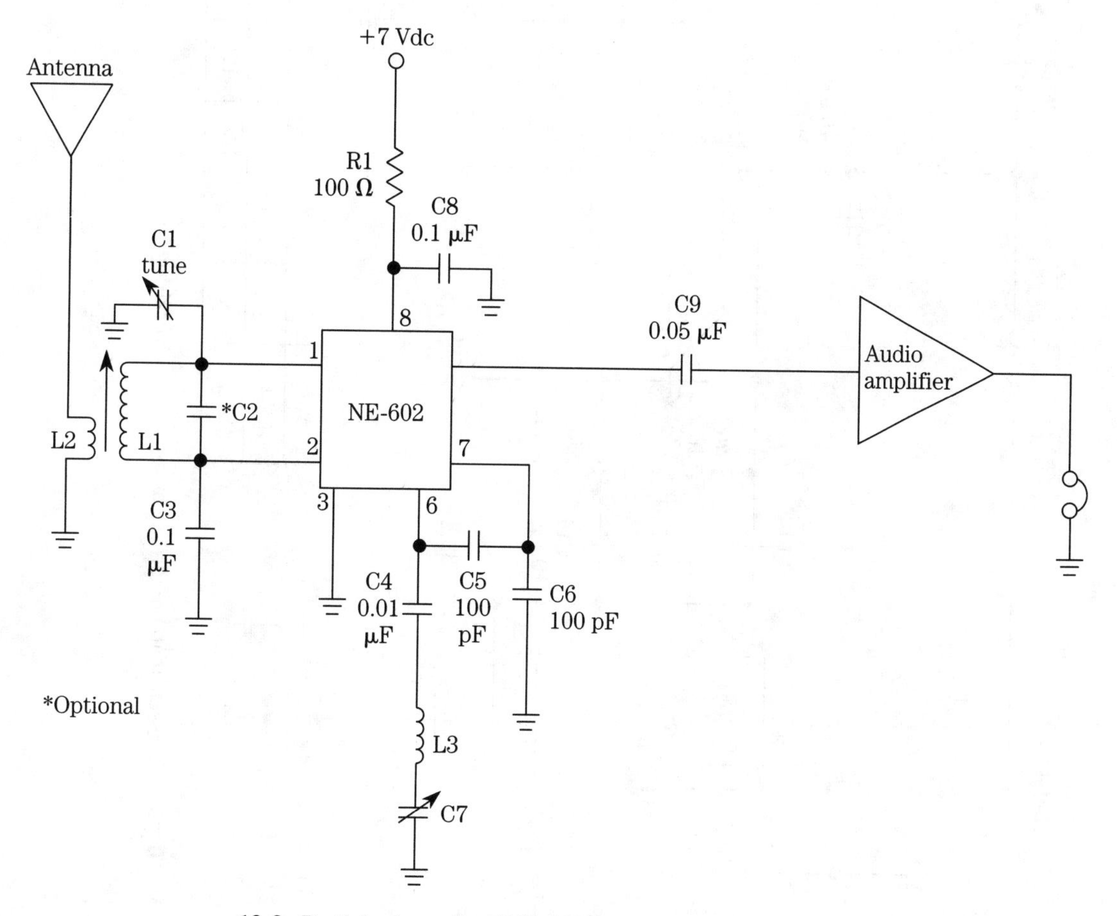

12-9 Partial schematic of NE-602 direct-conversion receiver.

The Dillon design shown in Fig. 12-10 uses the push-pull outputs of the NE-602 (i.e., both pins 4 and 5), and is superior to the single-ended variety. According to Dillon, the balanced output approach improves the performance, especially in regard to AM BCB breakthrough rejection. Also helping the breakthrough problem is the use of a 0.047-μF capacitor placed across the output terminals of the NE-602.[18]

Daulton takes exception to the use of the NE-602 as the DCR front end, and prefers instead to use the Signetics TDA-7000 chip. While functionally similar to the NE-602, the TDA-7000 is more complex and is said to deliver superior performance with respect to dynamic range and signal overload characteristics. Figure 12-11

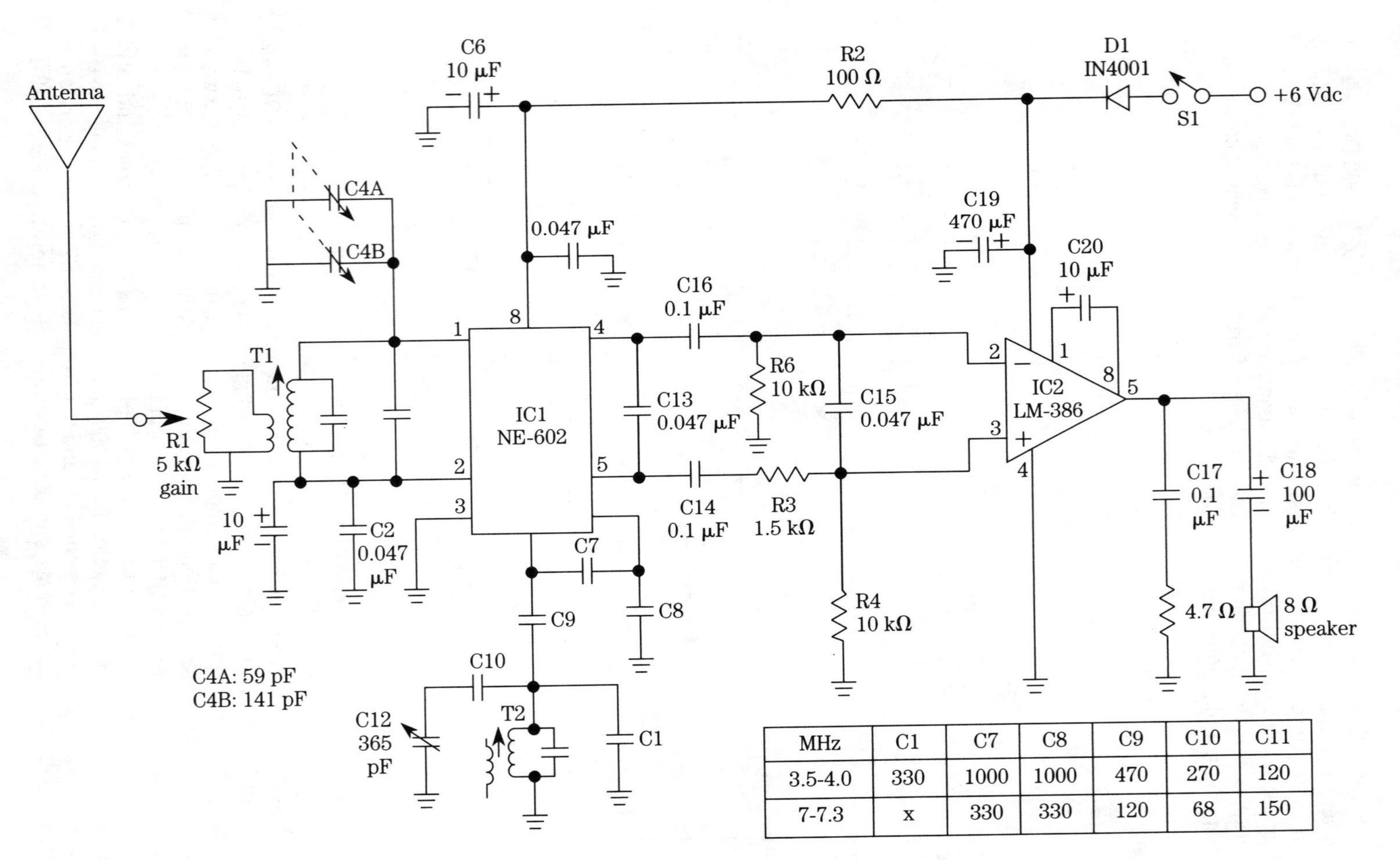

MHz	C1	C7	C8	C9	C10	C11
3.5-4.0	330	1000	1000	470	270	120
7-7.3	x	330	330	120	68	150

12-10 Complete schematic of direct-conversion receiver using NE-602.

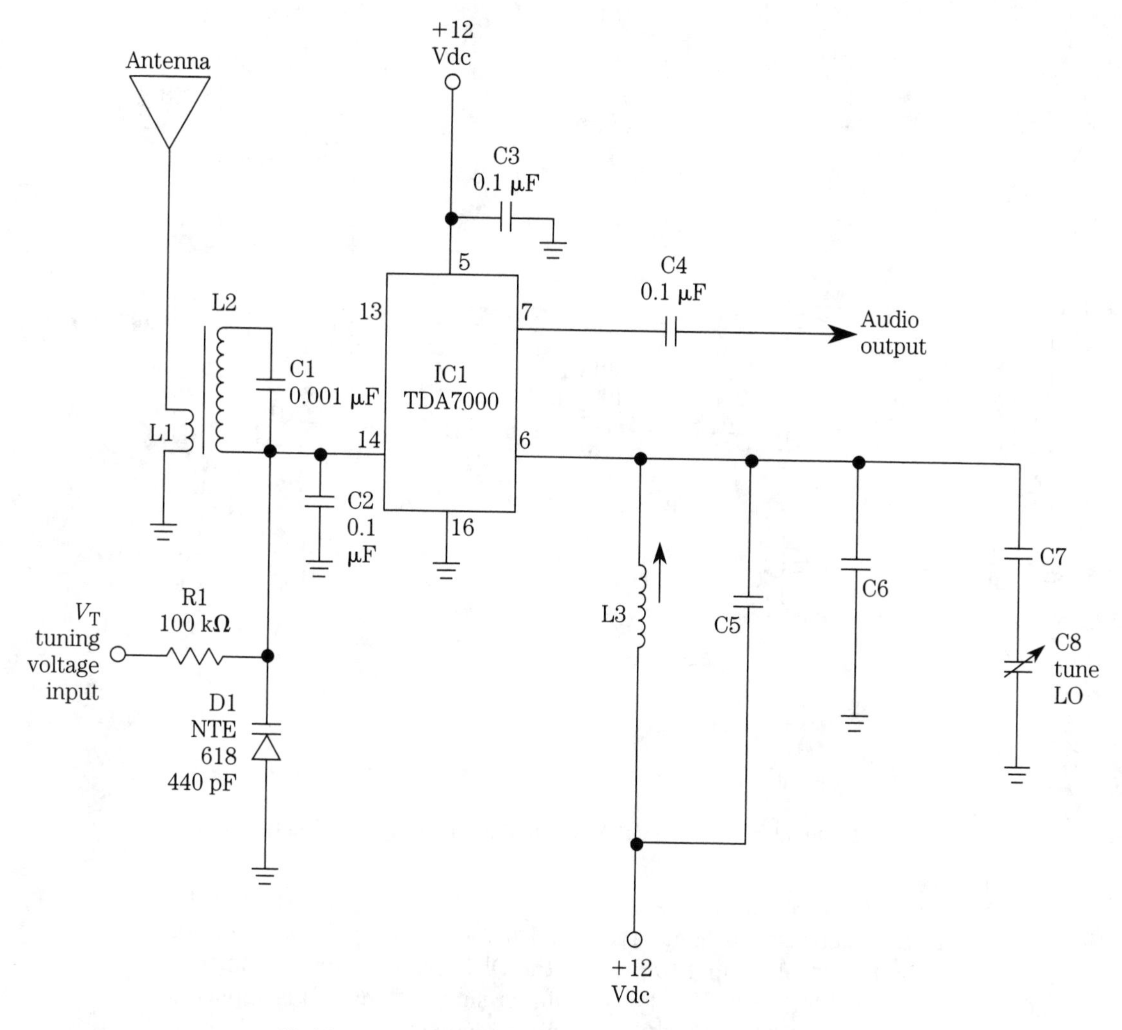

12-11 Direct-conversion receiver using TDA-7000 chip.

shows a DCR front end circuit that is based on the TDA-7000 after Daulton's design. This circuit uses the same balanced front end as other designs and, like the typical NE-602 design, uses the internal oscillator for the variable frequency oscillator (VFO). The circuit following this front end should be of the sort typically found in the NE-602 designs. This particular variant uses the internal operational amplifiers of the TDA-7000 to provide active bandpass filtering.

A simplified variant on the Lewallen design[19] is shown in Fig. 12-12 (the original design included a QRP transmitter as well as the DCR). The front end of this variant consists of an RF transformer with a tuned secondary winding (L1A). This secondary is tuned to resonance by capacitor C1, and is tapped at the 50-ohm point in order to match the input impedance of the double-balanced mixer (DBM).

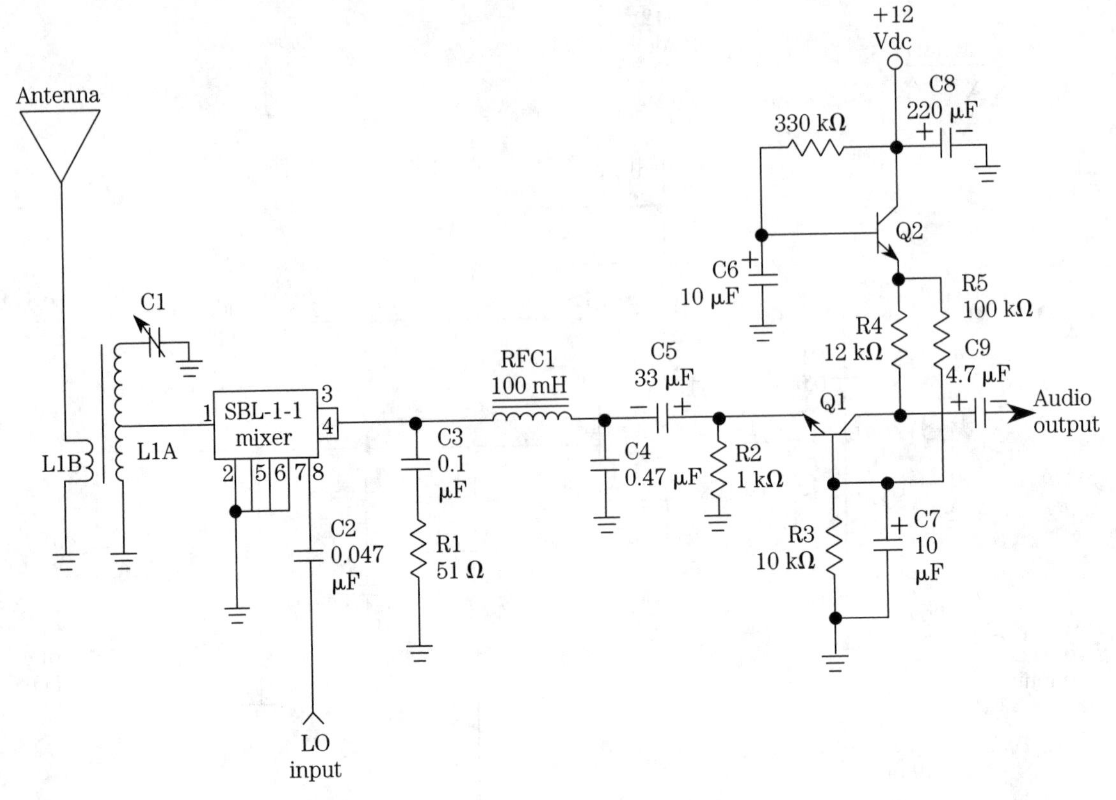

12-12 Direct-conversion mixer and first audio using the SBL-1-1 DBM.

The particular DBM selected here is a Mini-Circuits SBL-1-1, although in the original article Lewallen used a homebrew DBM made from diodes and toroidal transformers. The RF signal is input to pin 1 of the DBM (which can accommodate signal levels up to +1 dBm), while the local oscillator signal ("VFO IN") is applied to pin 8 through capacitor C2. The VFO/LO signal must be on the order of +7 dBm.

The design of Fig. 12-12 uses two methods for matching the output of the mixer circuit. First, there is an RC network (R1/C3) that matches high frequencies to 51 ohms (the capacitor limits operation to high frequencies). The second method used here is to use a grounded-base input amplifier (Q1) to the audio chain. Such an amplifier applies input signal across the emitter-base path, and takes output signal from the collector-base path (the base being grounded for audio ac signals through capacitor C5). This preamplifier is equipped with an active decoupler circuit consisting of transistor Q2 and its associated circuitry. The input side of the grounded base audio amplifier consists of an LC low-pass filter (C4/RFC1) that passes audio frequencies, but not the residual VFO and RF signals from the DBM. In the original design, Lewallen followed the grounded-base amplifier with a direct-coupled operational amplifier active low-pass filter.

The Campbell design[20] extends the concepts from Lewallen. Figure 12-13 shows the block diagram for a portion of Campbell's direct conversion receiver. The front

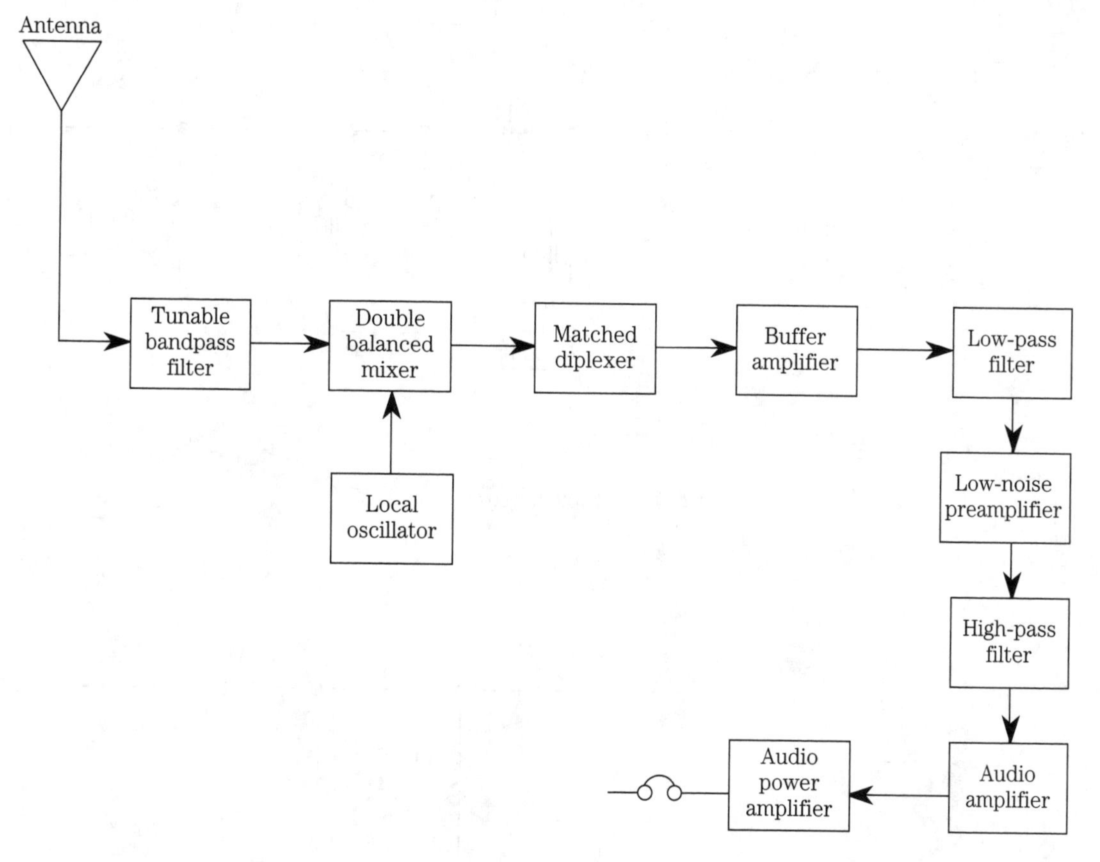

12-13 Block diagram for a complex direct-conversion receiver.

end consists of a double-balanced mixer followed by a matched diplexer filter that provides a 50-ohm input impedance from dc to 300 Hz, and from 3000 Hz to some upper high frequency beyond the range of interest. The diplexer also passes the standard communications audio bandwidth (300 to 3000 Hz) to the matched grounded-base amplifier. Finally, there is an LC audio bandpass filter prior to sending the signal on to the high-gain audio amplifier stages.

Figure 12-14A shows the passive diplexer used by Campbell. It consists of several inductor, resistor, and capacitor elements that form both low-pass and high-pass filter sections. The values of the inductors (L1, L2, and L3) are selected with their dc resistance in mind, so it is important to use the originally specified components, or their exact equivalents in replicating the project. Campbell used *Toko* Type 10RB inductors: L1 is Toko 181LY-392J, L2 is Toko 181LY-273J and L3 is Toko 181LY-273J. These coils are available from Digi-Key (P.O. Box 677, Thief River Falls, MN 56701-0677, USA; Voice No. 1-800-344-4539; fax no. 218-681-3380).

Campbell's article supplied me with another example of the "digital myth," i.e., the concept that the digital implementation of a function is always superior over the analog version. He points out that the dynamic range of an LC filter is set by the inductors. The low-end is the thermal noise currents created by the circuit resistance

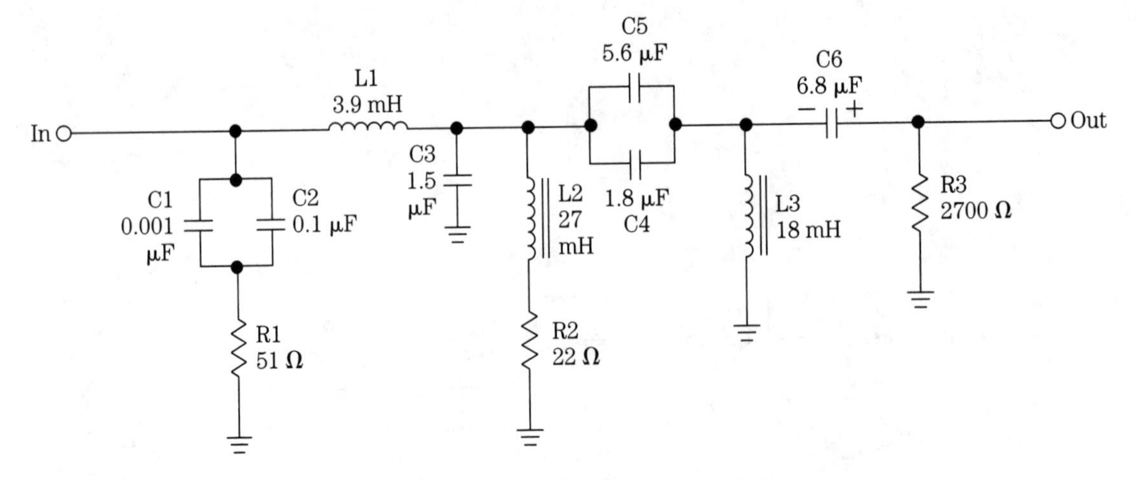

12-14 A. Diplexer circuit for DBM.

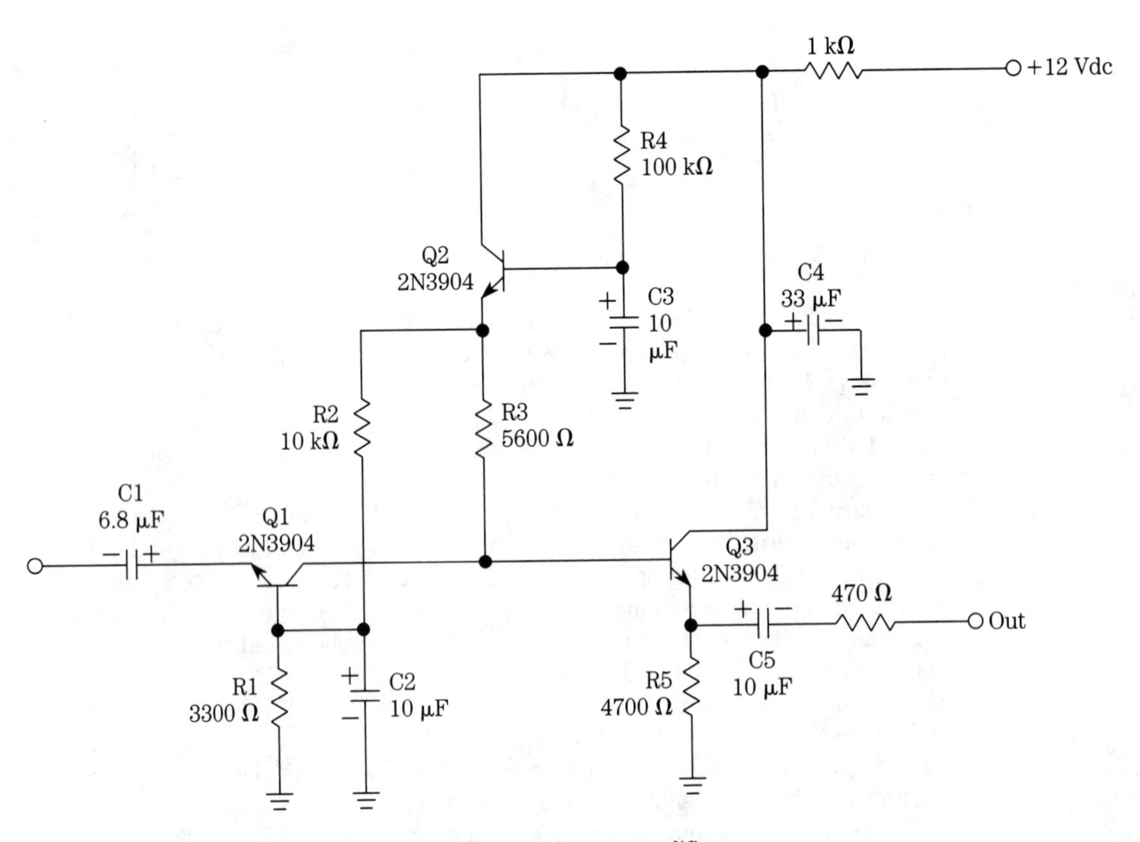

12-14 B. Audio postamplifier.

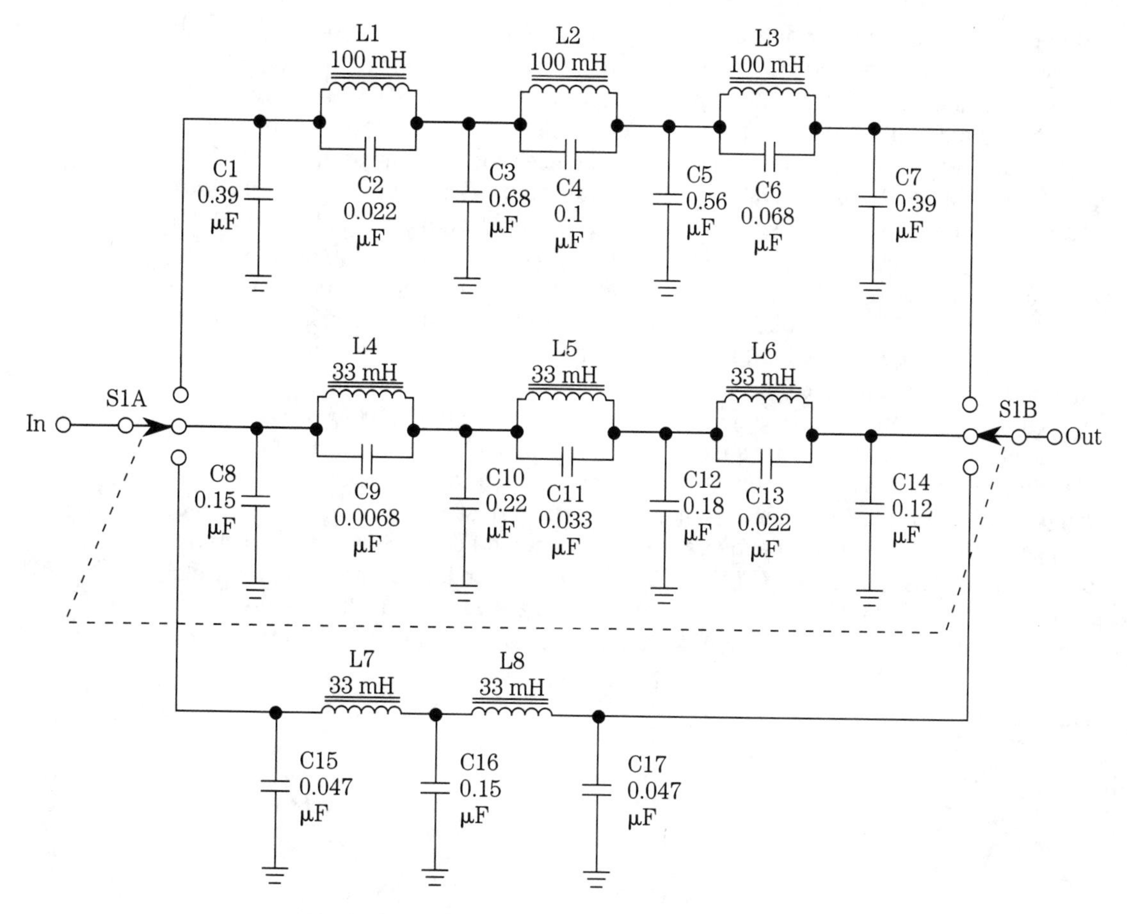

12-14 C. Three audio bandpass filters.

(4KTBR), while the upper end is set by the saturation current of the inductor cores. For the parts selected by Campbell, he claims this range to be 180 dB. By contrast, an expensive 24-bit audio A/D converter provides only 144 dB of dynamic range.

The matched 50-ohm audio preamplifier shown in Fig. 12-14B is an improved version of the Lewallen circuit. According to Campbell, this circuit provides about 40 dB of gain, and offers a noise figure of about 5 dB. The range of input signals it will accommodate ranges from about 10 nV to 10 mV, without undue distortion. These specifications make the amplifier a good match to the DBM. Like the Lewallen circuit, the Campbell circuit uses a grounded-base input amplifier (Q1), and an active decoupler (Q2). But Campbell also adds an emitter follower/buffer amplifier (Q3).

A set of three passive audio filters, which can be switched into or out of the circuit, is shown in Fig. 12-14C. These filters are designed for termination in an impedance of 500 ohms. Three different bandpasses are offered: 1 kHz, 3 kHz, and 4 kHz. The 4-kHz filter is a fifth-order Butterworth design, while the 3-kHz filter is a seventh-order Elliptical design after Niewiadomski.[21] The 1000-Hz design is scaled from the

3000-Hz design. Campbell claims that these filters offered a shape factor of 2.1:1, with an essentially flat passband "... with rounded corners, no ripple, and no ringing."[22]

Campbell implied the use of switching, as shown in Fig. 12-14C, but did not actually show the circuitry. As shown here, the switching involves use of a pair of ganged SP3P rotary switches. PIN diode switches can be used for this purpose (chapter 17).

A complex DCR was designed by Breed, and reported in the amateur radio literature as a direct conversion single sideband receiver.[23] The single sideband (SSB) mode is properly called *single-sideband suppressed-carrier amplitude modulation*, for it is a variant of AM that reduces the RF carrier and one of the two AM sidebands to negligible levels. This mode is used in HF transmissions because it reduces the bandwidth required by half, and removes the carrier that produces heterodyne squeals on the shortwave bands.

There are two methods of generating SSB. The most common today uses a double-balanced modulator to combine a fixed carrier and the audio signal to produce a double-sideband suppressed-carrier output signal; the unneeded sideband is then removed by filtering. The older and more complex variant uses a phasing method of SSB generation. Breed uses the inverse process to demodulate SSB signals in a clever, but complex, receiver design (Fig. 12-15). This circuit splits the incoming RF signal into two components and then feeds them both to separate mixers. These mixers are driven 90 degrees out of phase by a VFO that produces –45° and +45° outputs. The re-

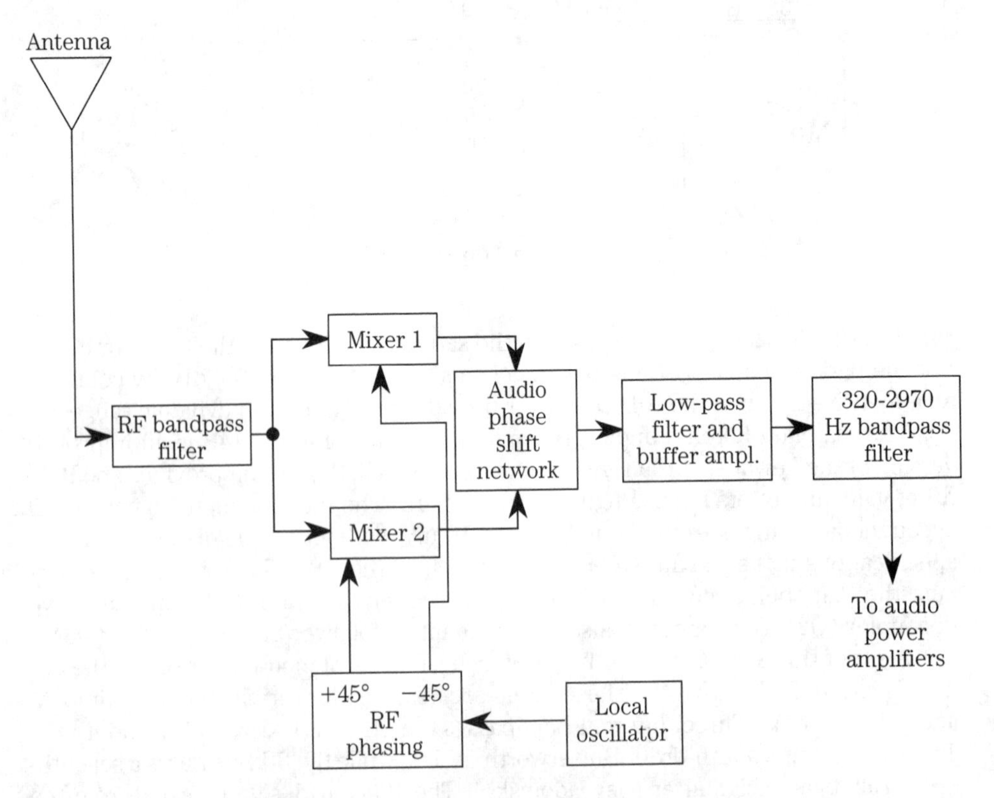

12-15 Phased CW/SSB direct-conversion receiver.

spective outputs of the mixers are amplified and then fed to bilateral 90° audio phase shift networks where they are recombined. The output of the phase shift network is filtered in a low-pass filter and bandpass filter, to provide the recovered modulation.

Audio circuits

The audio chain in the direct-conversion receiver tends to be very high gain in order to compensate for the low output levels usually found on the mixer circuits. The principal job of the audio amplifier is to increase the signal level by an amount that will create comfortable listening level, while also tailoring the bandpass characteristics of the overall receiver to limit noise and other artifacts. Although any number of discrete and integrated circuits (IC) are suitable, most designers today tend to use the IC versions. (Campbell uses a discrete circuit with IC preamplifiers. Most other designers today use all-IC circuits.) Figure 12-16 shows a simple LM-386 design, while the published literature shows many other designs as well.[24]

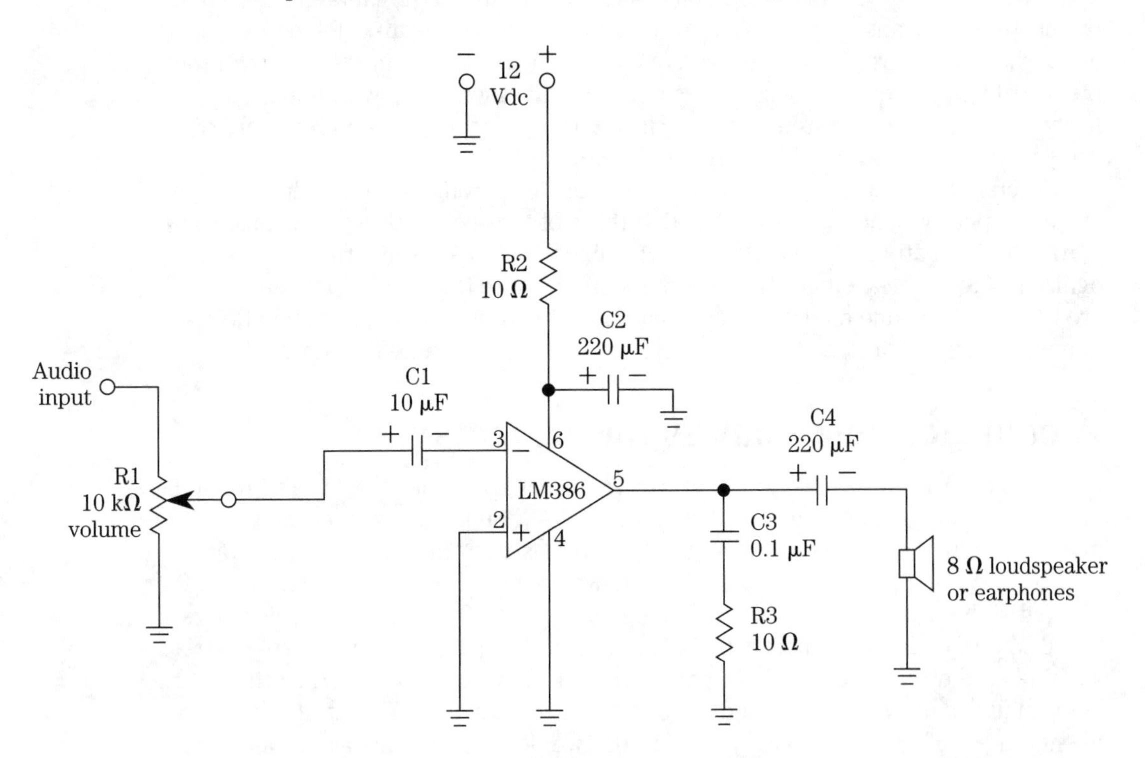

12-16 LM-386 audio power amplifier stage.

The LM-386 design of Fig. 12-16 is the single-ended configuration for the LM-386 low-power audio stage. This IC device contains both preamplifiers and power amplifiers for a nominal output power of 250 mW. (Variants of the LM-386 include the LM-386-3, which produces 500 mW, and the LM-386-4, which produces 700 mW.) The LM-386 series of audio power ICs are easy to use, but because of the high gain needed, it will oscillate if layout is not correct, or if grounding is not proper. There are two basic circuit configurations for the LM-386. The differential version was shown in Fig. 12-10 (Dillon's design), while Fig. 12-16 is the more common single-ended design. The gain of the circuit can be either 46 dB (X200) when capacitor C2 is used, or 26 dB (X20) when C2 is deleted (leave pins 1 and 8 open-circuited).

Local oscillator circuits for direct-conversion receivers

The local oscillator (LO) for a continuously tunable receiver of any description is basically a variable frequency oscillator (VFO). Although higher-grade receivers today typically use frequency synthesis techniques for generating the LO signal, the standard inductor-capacitor (LC) controlled VFO still has appeal for less complex receivers. The VFO used for the LO in receivers is pretty much the same as the VFO in transmitters, so transmitter VFOs are frequently used. There are some cases, however, where a receiver LO has at least one specification that is more rigid than the transmitter equivalent: many receivers have a requirement for low FM phase noise. In the main, however, amateur radio applications of direct conversion receivers typically use the transmitter VFO for the receiver as well.

Several VFO designs are used for receiver LOs: Armstrong, Hartley, Colpitts/Clapp and an amplitude-limiting design. The first three of these circuits are recognized according to the nature of their respective feedback networks, while the other is recognized by the special connection of a transformer. Note that the Colpitts and Clapp are basically the same circuit, except that the Colpitts uses a parallel-tuned LC frequency setting network and a Clapp oscillator uses a series-tuned LC network.

A common test chassis for projects

Part of the decision to build and test several direct-conversion receivers involved having a common chassis for all three designs. Although not very elegant, being made of scrap aluminum chassis and bottom plates from my "junk box," it was at least low-cost and effective. Figure 12-17 shows the receiver test bed front panel. It is fitted with a Jackson Brothers calibrated dial with a 10:1 fast/slow vernier drive. The 6.35 mm (0.25 inch) shaft coupling on the vernier drive is used to turn either a variable air-dielectric capacitor or potentiometer, depending on whether the DCR being built is a mechanically tuned or voltage-tuned version. For most of the experiments, the voltage-tuned variety was used. Two additional controls are also provided, and both are potentiometers. The pot to the right of the tuning knob is a volume control, while that to the left is the RF tuning control (for voltage-tuned front end circuits).

Three circuits are provided for the test bed: two dc power supplies and an LM-386 audio power amplifier. One dc power supply (Fig. 12-18) is used to provide +12 volt dc regulated power to the circuits of the DCRs used on the test bed. It uses a 7812 three-terminal IC voltage regulator, and works from a +15 volt dc, or higher source. In

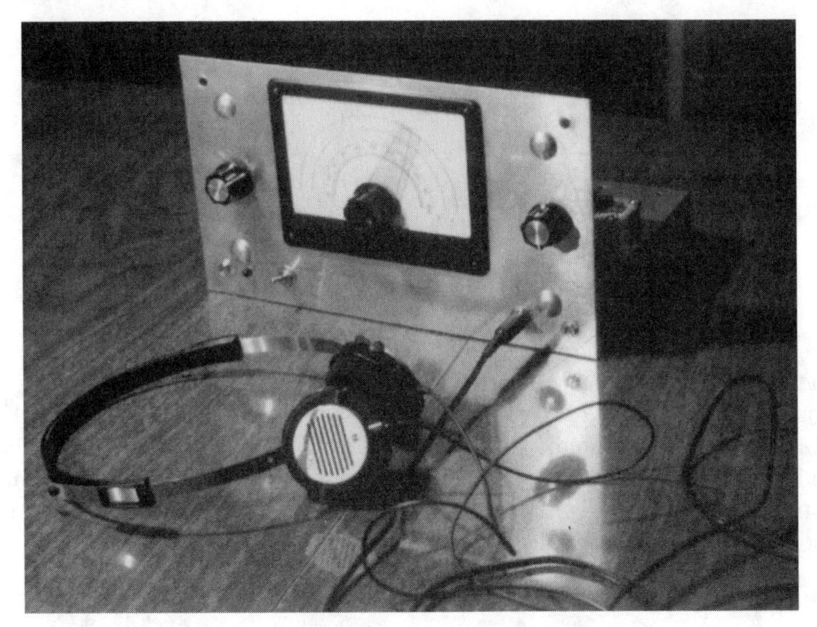

12-17 Photo of direct-conversion receiver.

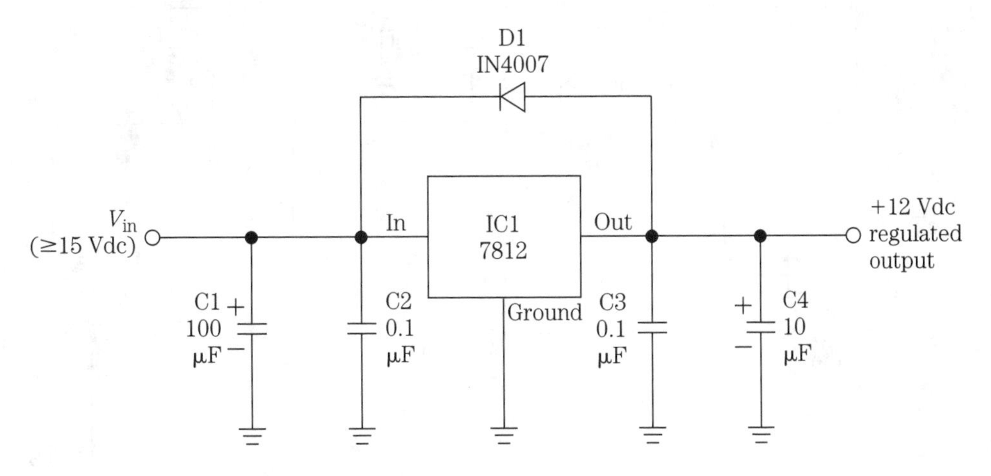

12-18 dc power-supply circuit.

the case of this project, raw power was provided by a series of three 6-volt lantern dry-cell batteries connected in series. The second dc power supply (Fig. 12-19) consists of a 78L12 low-power, three-terminal voltage regulator and a potentiometer. The potentiometer is for main tuning, and is ganged to the main dial of the chassis. In normal use, this potentiometer is used to tune the local oscillator (LO) of the DCR.

The audio amplifier is the LM-386 low-power "audio system on a chip" device. The LM-386 uses a minimum of external components, and includes both the audio preamplifiers and the power amplifiers to produce between 250 mW and 700 mW of

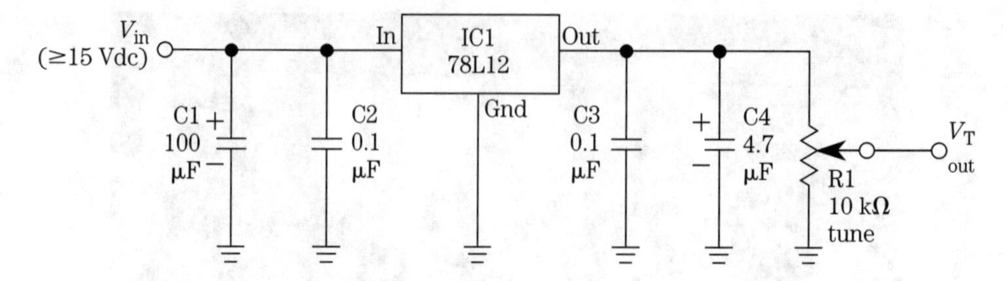

12-19 Voltage tuning power supply.

audio power, depending on the particular device specified. The circuit for the audio section is shown in Fig. 12-20, while its location on the test bed chassis is shown in Fig. 12-21. Note that the audio section has its own +12 volt regulator. This is an optional feature, but does serve to keep load variations in the audio amplifier from coupling to the rest of the circuitry. The audio section is the small circuit board on the left side of the chassis—right by the audio volume control.

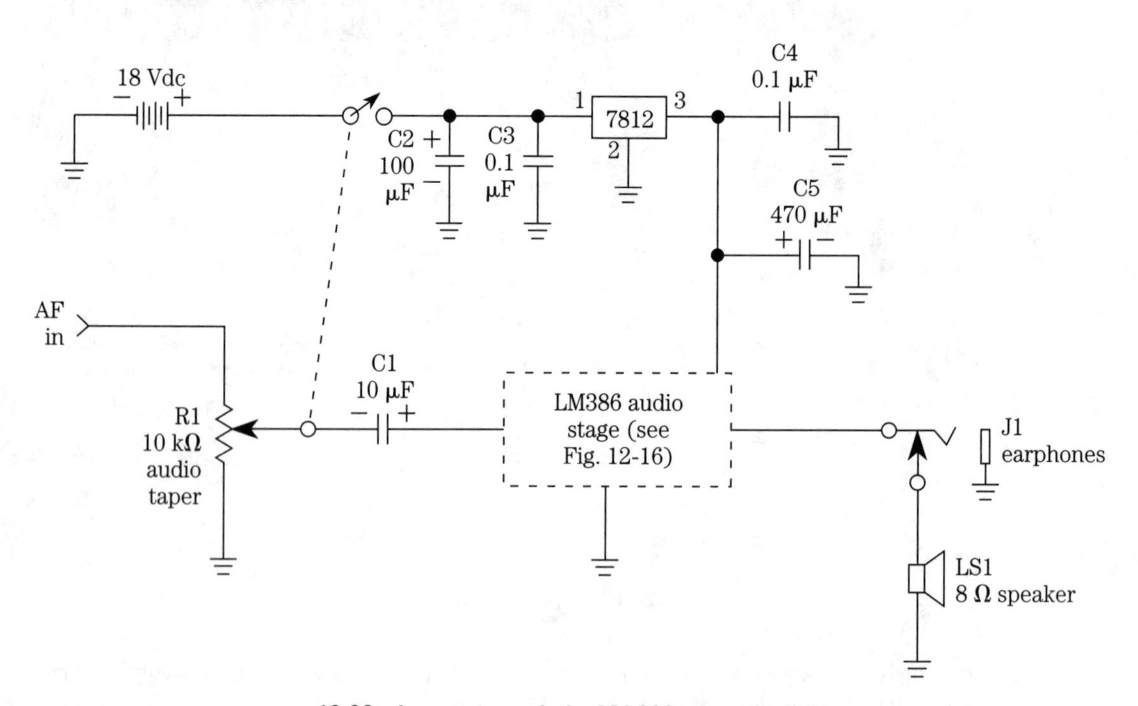

12-20 dc power supply for LM-386 power amplifier.

Figure 12-21 shows the rear view of the DCR test bed chassis. As mentioned, the audio amplifier section is shown on the left. The gray metal box contains either the variable air-dielectric capacitor, or the potentiometer and power supply (Fig. 12-19), either of which may be used to tune the local oscillator of the DCRs being tested. The +12 volt dc power supply is located on the small printed circuit board to the rear of

12-21 Rear view of DCR test bed chassis.

the tuning box. Space is provided to the right of the tuning box for the circuitry of the DCR being tested. This board is changed from one project to the other.

Wiring board construction used two different methods. The audio amplifier, the tuning dc power supply and +12 Vdc regulated power supply were built on Radio Shack "universal" printed circuit boards. The DCRs, on the other hand, were wired using Vector perfboard with a hole grid on 0.100 inch centers, or an equivalent product.

Three different antennas were used for testing the DCRs in this article: a 5BTV Cushcraft 0.5 wavelength ham band vertical, an outdoor 30-m (100-foot) random length end-fed wire, and a 6-m (20-foot) wire strung across the ceiling of my basement workshop. Interestingly enough, on the HF bands there was not a large difference between the two outdoor antennas' performance and only slightly more difference between the outdoor antennas and the indoor antenna. On the VLF band, however, the random length wire was clearly superior to the other two antennas.

Notes

1. Gary A. Breed, "A New Breed of Receiver," *QST*, Jan. 1988, pp. 16–23.
2. *The ARRL Handbook for the Radio Amateur—Sixty-Fifth Edition*, American Radio Relay League (Newington, CT, USA, 1988).
3. *Ibid* (*ARRL Handbook* pp. 12–7 and 12–8).
4. Rick Campbell, KK7B, "High Performance Direct-Conversion Receivers," *QST*, August 1992, pp. 19–28.

5. *Ibid* (*ARRL Handbook*).
6. After W. Hayward in *ARRL Handbook* (op-cit).
7. Wes Hayward and Doug DeMaw, *Solid-State Design for the Radio Amateur*, American Radio Relay League (Newington, CT, USA, 1977).
8. *Ibid* (Campbell).
9. *Ibid* (Campbell).
10. Roy W. Lewallen, W7EL, "An Optimized QRP Transceiver," *QST*, August 1980, pp. 14–19.
11. Rick Campbell, KK7B, "High Performance Direct-Conversion Receivers," *QST*, August 1992, pp. 19–28.
12. Roy W. Lewallen, W7EL, "An optimized QRP Transceiver," *QST*, August 1980, pp. 14–19; Paul G. Daulton, K5WMS, "The Explorer: HF Receiver for 40 and 80 Meters," *73 Amateur Radio Today*, August 1992, pp. 30–34; John Dillon, WA3RNC, "The Neophyte Receiver," *QST*, February 1988, pp. 14–18.
13. Michael A. Covington, "Single-Chip Frequency Converter," *Radio-Electronics*, April 1990, pp. 49–52; Joseph J. Carr, "NE-602 Primer," *Elektor Electronics USA*, Jan. 1992, pp. 20–25.
14. *Ibid* (Dillon).
15. *Ibid* (Covington).
16. *Ibid* (Daulton).
17. *Ibid* (Carr).
18. Telephone conversation between the author and John Dillon, 27 August 1992. See also Dillon's article (*op-cit.*).
19. *Ibid* (Lawallen).
20. *Ibid* (Campbell).
21. S. Niewiadomski, "Passive Audio Filter Design," *Ham Radio*, Sept 1985, pp. 17–30; cited in Campbell (*op-cit*).
22. This claim seems excessive. I've ordered the correct parts and will be investigating these filter designs within the next few months.
23. Gary A. Breed, "A New Breed of Receiver," *QST*, Jan. 1988, pp. 16–23.
24. Ray Marston, "Audio Amplifier ICs," *Radio-Electronics*, April 1990, pp. 53–57. See also the *National Semiconductor Linear Data Books*, or the data books of other major IC manufacturers for applications information and device data.

13
RF oscillator and signal generator circuits

In this chapter, we will examine oscillator circuits that generate radio frequency signals. They can be used as workbench signal generators, local oscillators in receivers, or to control the operating frequency of transmitters.

Variable frequency oscillator

Variable frequency oscillators (VFOs) are radio frequency signal generators that can be continuously tuned by using an inductor connected to either an air-variable capacitor, mica "trimmer" capacitor, or a voltage-tuned variable capacitance diode (*varactor*). The VFO differs from the crystal oscillators in that the frequency can be varied in the VFO—while in the crystal oscillator, it is either fixed or variable over only a tiny region.

VFO circuits can be used as signal generators in test equipment, to control radio transmitters, and as the local oscillator in superheterodyne or direct-conversion receiver projects. They can also be used in other applications where a continuously variable source of RF energy is needed. In this chapter, you will find some practical circuits that are based on easily obtained components—as well as some general guidelines for using and modifying the circuits for your own use.

RF oscillator basics

Both VFOs and crystal oscillators are part of a class of circuits called *feedback oscillators*. Figure 13-1 shows the basic configuration of this type of circuit; it consists of an amplifier, with open-loop gain A_{vol}, and a feedback network (which is usually

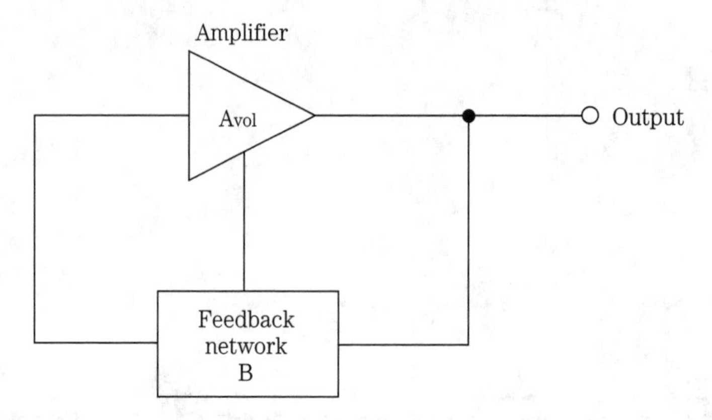

13-1 Feedback oscillator block diagram.

frequency selective) with a "gain" of β. If two conditions are met, the circuit will oscillate. These conditions are called *Barkhausen's criteria*:

- The loop gain is unity or greater
- The feedback signal arriving back at the input is in-phase with the input signal (phase-shifted 360 degrees)

For most practical circuits, with the 180 degrees provided by an inverting amplifier, there must be an additional 180 degrees of phase shift provided by the feedback network in order to meet the latter criterion. The nature of an inductor-capacitor (LC) tuned circuit is that it can provide this 180-degree phase shift at only one frequency, so the circuit will oscillate cleanly on that frequency.

We will consider two general categories of feedback RF oscillator here: Hartley oscillators and Colpitts oscillators (also, a subset of the Colpitts oscillator is the Clapp oscillator). The basic configurations are shown in Fig. 13-2. The circuit in Fig. 13-2A is the *Hartley oscillator*. It is identified by the fact that the resonant feedback network contains a tapped inductor (which effectively forms an inductive voltage di-

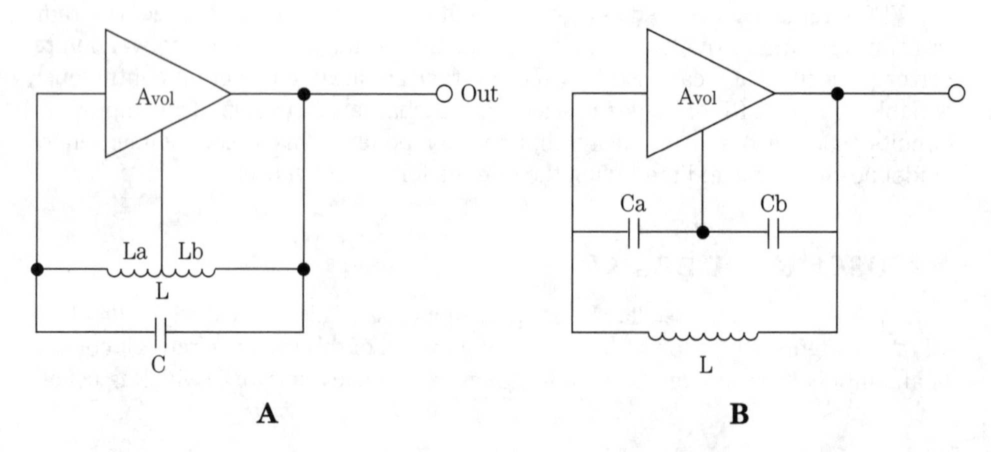

13-2 A. Hartley oscillator. B. Colpitts oscillator.

vider). The *Colpitts* and *Clapp circuits* (Fig. 13-2B) are identified by the fact that the feedback network contains a tapped-capacitance voltage divider. In both cases, the feedback voltage divider is part of the resonant LC tuning network.

It is a good idea to use only regulated dc power supplies with VFO (or any oscillator) circuits. In fact, most experts agree that a single regulator serving the oscillator is best because it is not affected by load changes in other circuits on the same side of the regulator. This is because most oscillators shift frequency a slight amount when the power supply voltages change. If you are a ham radio operator familiar with CW (i.e., Morse code) transmission, you will recognize such variation as the transmission defect called "chirp." Although not all the oscillator circuits in this chapter show the regulator, it is a good idea to use one anyway.

Some basic circuit configurations

The basic circuits discussed above can be configured in several different ways. In this chapter, we will use one of three active devices as the amplifier portion: junction field-effect transistor (JFET), metal oxide field-effect transistor (MOSFET), and the Signetics NE-602 RF balanced mixer integrated circuit. The JFET and MOSFET parts were selected to be easily available to as many readers as possible. The basic JFET is the MPF-102 device, while the MOSFET is the 40673 device. If you buy parts from a distributor who carries a radio-TV service replacement line, such as ECG or NTE, you will find these parts easily available in those lines as well. The MPF-102 is replaced by the ECG-312 and the NTE-312, while the 40673 is replaced by the ECG-222 and NTE-222 devices.

Figure 13-3 shows two input configurations, one each for MPF-102 and 40673. In Fig. 13-3A is the circuit for the JFET device. It consists of a gate resistor of 100 kilohms to ground, and a diode (1N914, 1N4148, or equivalent). In many cases, you will need to use a capacitor in the gate circuit (C1), especially if there is a dc source or ground directly in the circuit. In order to prevent loading of the tuned circuit, it is customary to make the coupling capacitor small compared to the tuning capacitor; typically values from 2 to 10 pF are use for HF and MW VFO circuits.

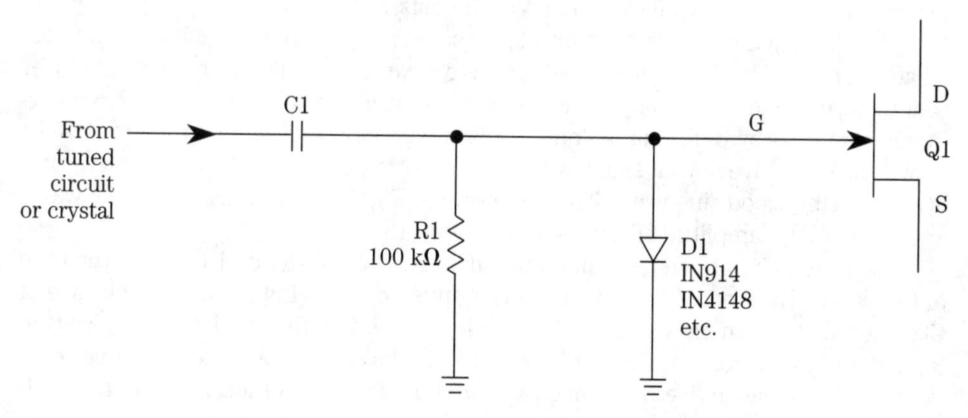

13-3 Diode amplitude limiting circuit. A. JFET version.

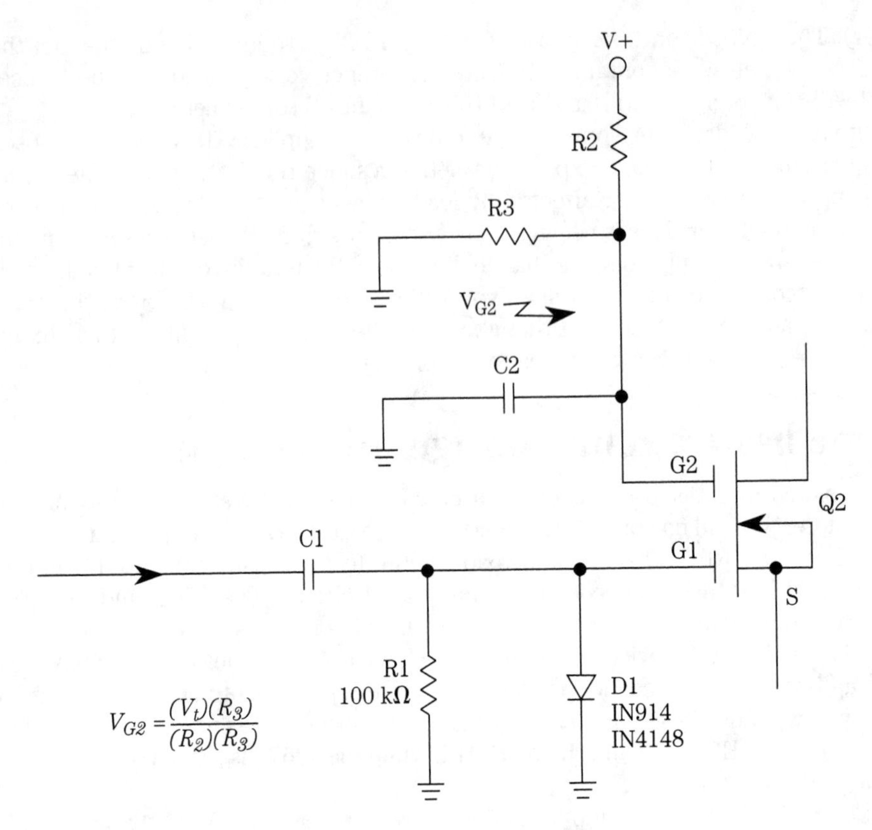

13-3 B. MOSFET version.

The same circuit design can be used on the MOSFET, but there will also need to be a dc-bias circuit for the second gate, G2. The bias network shown in Fig. 13-3B (R2/R3) is set to bias G2 to about ⅛ the V+ voltage. I've used equal value resistors (10 kΩ and 10 kΩ) for R2/R3. A bypass/decoupling capacitor (C2) is used to set G2 to a low impedance for RF, while keeping it at the bias voltage for dc.

The diode in the input circuit perplexes some people when they first see it in oscillator circuits. The function of the diode is to clean up the signal, and make it closer to a low-harmonic sine wave (all non-sine wave, or distorted sine waves, have harmonic content—by definition a "pure" sine wave has no harmonics). Figure 13-4A shows the sine wave output signal with the diode connected. Note that the waveform is a reasonably good sine wave. The waveform in Fig. 13-4B is a distorted sine wave, and has a higher amplitude than the case of Fig. 13-4A.

The tuned circuit can be either a parallel-tuned circuit (Fig. 13-5A), in the case of Colpitts or Hartley oscillators, or a series-tuned circuit (Fig. 13-5B) in the case of Clapp oscillators. In the case of the Hartley oscillator, the inductor (L1) will be tapped. In any LC resonant circuit, resonance is that point were the inductive reactance (X_L) and capacitive reactance (X_C) are equal to each other. Because these elements cancel out, the impedance of such a circuit is resistive.

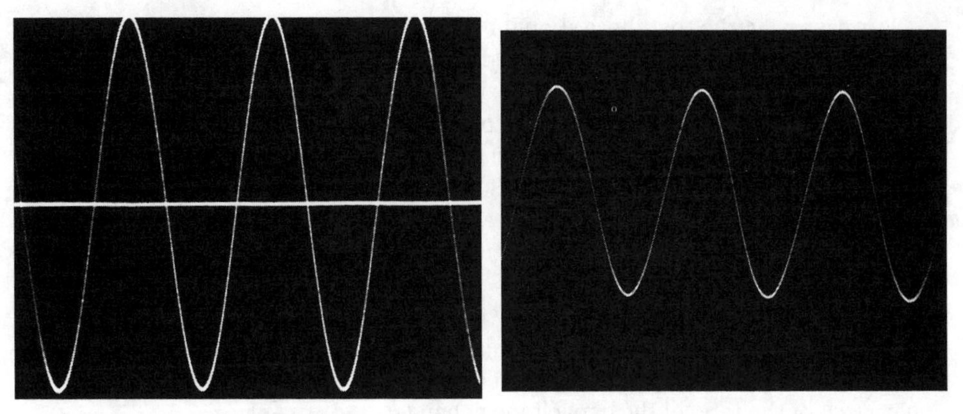

13-4 Oscillator output. A. Without diode. B. With diode.

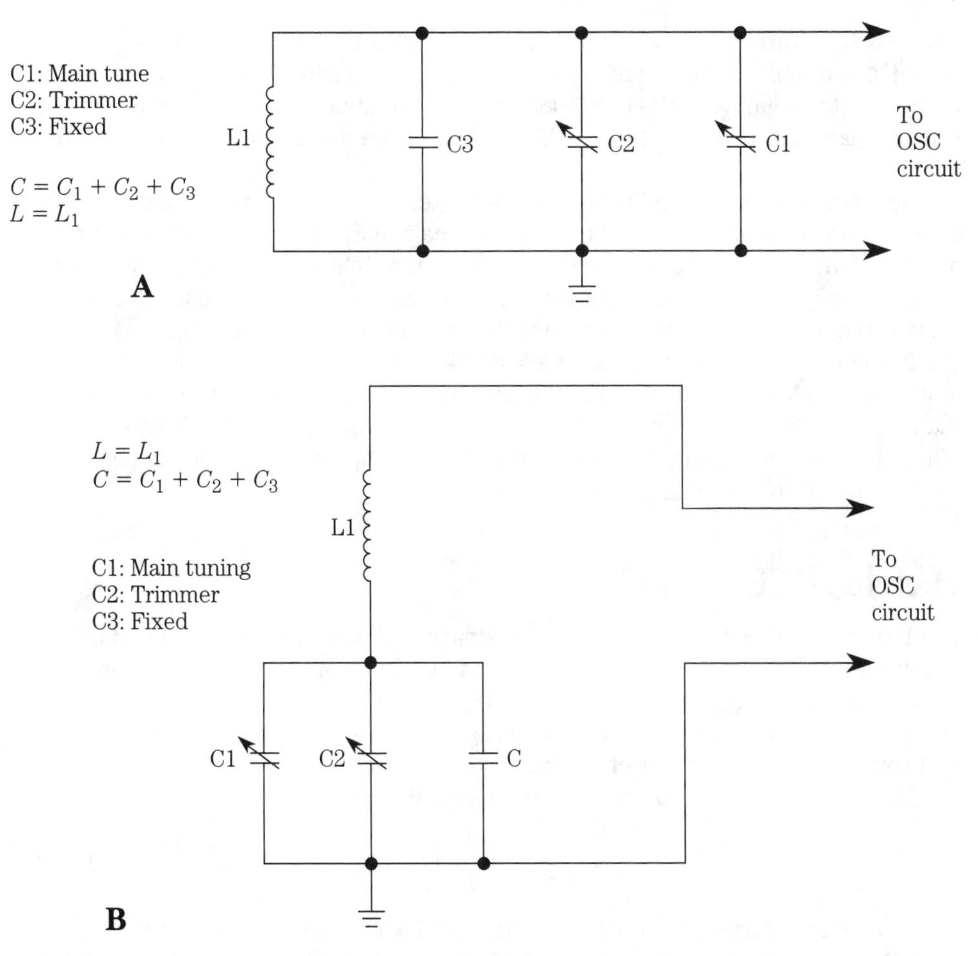

13-5 LC oscillator-tuning circuit. A. Parallel-tuned. B. Series-tuned.

It is generally true that there should be a high C/L ratio, so it is common practice to select a relatively small inductance and match it with a higher capacitance. Many of the oscillators in this chapter are designed for the middle of the 1- to 10-MHz band, and use inductances on the order of 3.3 to 7 μH. These inductances are relatively easy to obtain when using either "solenoid" (cylindrical) or "toroidal" coil forms. See the Amidon Associates (P.O. Box 956, Torrance, CA 90508) catalog for information on winding both sorts of coils. Also, Barker & Williamson (10 Canal Street, Bristol, PA, 19007; phone 215-788-5581) makes air core coil stock that is suitable for most HF oscillators. Both fixed value and variable inductors can be used for these circuits.

The total capacitance value can be made up from more than one capacitor, and this is generally the best practice. The change of frequency is proportional to the square root of the ratio of the capacitance change:

$$\frac{F_{max}}{F_{min}} = \sqrt{\frac{C_{max}}{C_{min}}} \qquad [13\text{-}1]$$

That is why a variable capacitor for the AM broadcast band consists of a 25-pF to 365-pF air-variable capacitor shunted by a 25-pF (or so) trimmer capacitor (which is usually set to around 15 pF). The 3.08:1 ratio of maximum to minimum capacitance is more than sufficient to cover the 3.02:1 ratio of the maximum and minimum frequencies of the AM broadcast band.

Selecting values of L and C is a somewhat tedious and iterative affair. It is best to sit down with a calculator and make a few trials. Part of the problem is that both fixed and variable capacitors come in standard values that may or may not be exactly what's needed. Juggle the inductance (which is easy to wind to a custom value) to provide the desired frequency change with the available variable capacitors.

It is common practice to make the total of the fixed capacitors—plus the maximum values of the variable (main tuning and trimmer) capacitors—somewhat larger than the total required to resonate at the lowest frequency in the desired range. When the trimmer is set to a value less than maximum, the total capacitance will be close to the desired value.

Hartley JFET VFO

The Hartley oscillator is identified by a feedback path that includes a tapped inductor; the inductor is also part of the resonant tuning circuit of the oscillator. Figure 13-6 shows a simple Hartley oscillator based on the MPF-102 junction field effect transistor (JFET). The output signal is taken through a small-value capacitor (to limit loading) connected to the emitter of the transistor.

The frequency of oscillation is set by the combined effect of L_1, C_1, and C_2:

$$F_{Hz} = \frac{1}{2\,\pi\,\sqrt{(L_1)\,(C_1 + C_2)}} \qquad [13\text{-}2]$$

To resonate at 5 MHz with a 5-μH inductor, a total capacitance of about 200 pF is required. Because of stray capacitances, and errors in the values of actual capacitors,

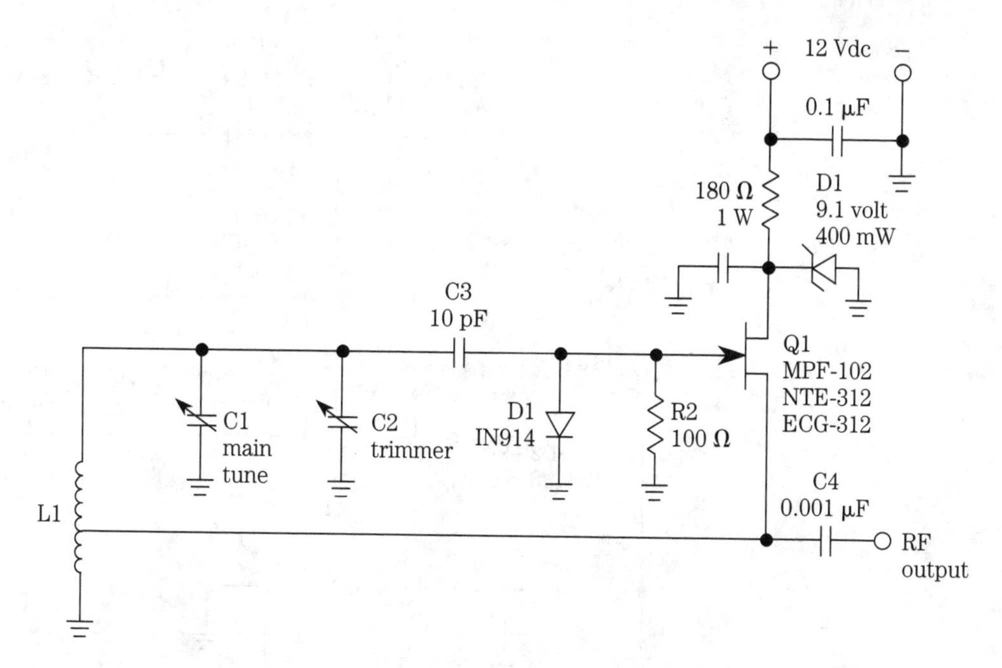

13-6 JFET Hartley oscillator.

it is common practice to use more total capacitance than needed, and use variable capacitors to trim. For example, we could use a 140-pF variable capacitor for the main tuning, and an 80-pF trimmer to set the maximum to the required value.

Figure 13-7 shows a 5-MHz Hartley VFO circuit based on the MPF-102 JFET transistor. It is very similar to the previous circuit in basic concept, but there are some differences. The most significant difference is the use of a variable capacitance diode (*varactor*) instead of the main tuning capacitor. The diode shown here is a 14-pF to 440-pF varactor used to tune the AM broadcast band in radio receivers. Another significant difference is that the output signal is taken from a secondary winding on the tuning inductor. This winding consists of fewer turns than the lower portion of the main tuning inductor. You should not load down this circuit by connecting a varying load resistance across the output winding.

The circuit of Fig. 13-7 can be made to sweep the entire frequency range by applying a sawtooth waveform that rises from 0 volts to +12 volts. This type of circuit makes it relatively easy to build a sweep generator circuit, or a swept-tuned radio receiver. In general, the sweep rate should be around 40 Hz if the detector has a narrow band filter. Other sweep frequencies are usable as well, but case must be taken not to "ring" any following resonant circuits or filters because of a too-fast sweep rate.

The tuning properties of a varactor LC circuit are nonlinear because of the nature of varactor diodes. It is recommended that a graph of tuning voltage versus frequency be made, and that only the linear portion of the curve be used for sweep purposes.

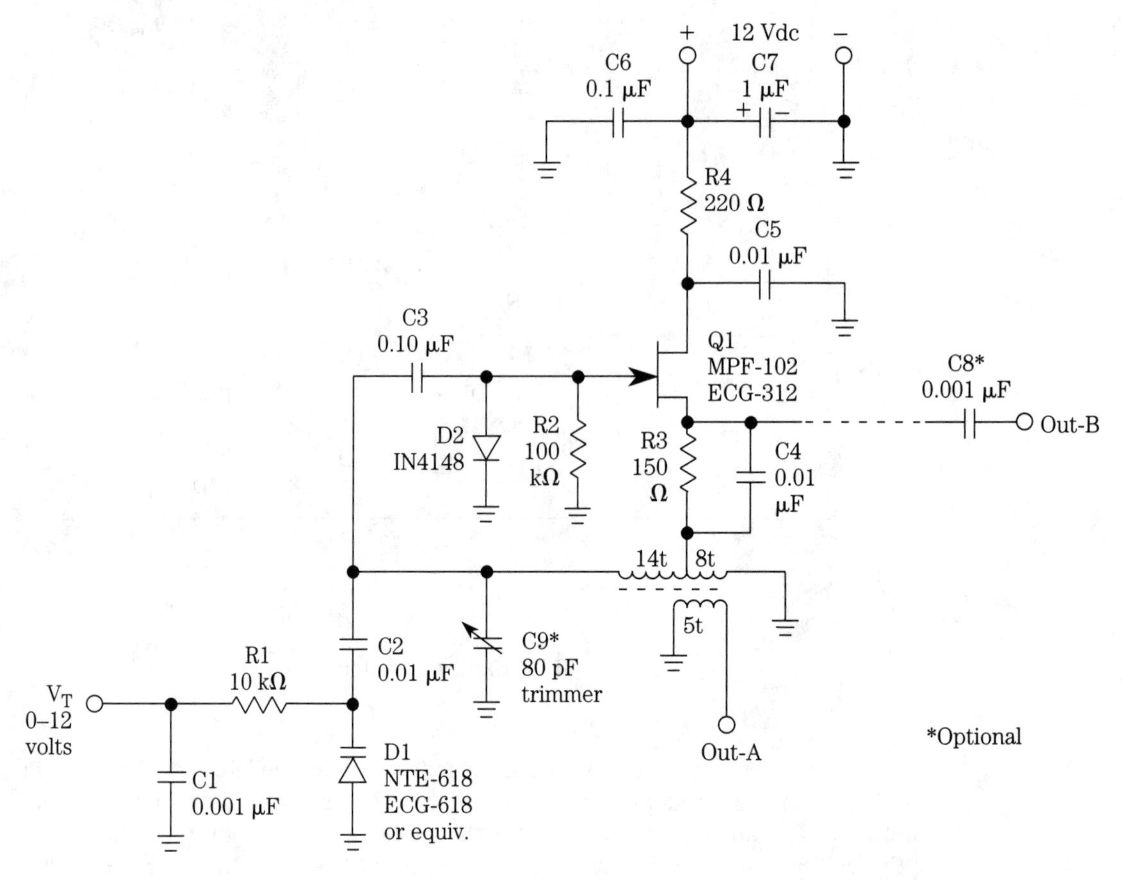

13-7 Voltage-tuned JFET Hartley oscillator.

The circuit in Fig. 13-8A is capable of producing up to several volts (that's volts) of signal in the 1 to 10 MHz range. The actual tuning range depends on the particular components used. The heart of the oscillator is an MPF-102 JFET (Q1). Two tuning schemes are provided. For a limited range, the main tuning is provided by a 365 pF AM broadcast band tuning capacitor, and the tuning range is 5000 to 5500 kHz. In that case, the varactor diode (D2) is disconnected and not used. The alternate scheme deletes C1, and uses either a 365 pF variable or the varactor circuit (shown) at point "A." This version has a rather wider tuning range for the same capacitance change. In the previous circuit, the tuning range was reduced by the capacitor divider action of C4 and C5.

The circuit of Fig. 13-8A can be modified by adding or subtracting capacitors from the tuning network. As shown, C1, C2, C3, C4, and C5 are all part of the tuning circuit. The 126 pF fixed capacitor (needed for 5 to 5.5 MHz operation) is made up from parallel 82 pF and 47 pF capacitors. A version of the circuit using the NTE-618 or ECG-618, 440 pF diode was built without the mica trimmer capacitor (C2). It produced a voltage (V_t) versus frequency characteristic shown in Fig. 13-8B (this curve is for a circuit with C2 removed). The varactor is comfortable over a range of 0 to +12

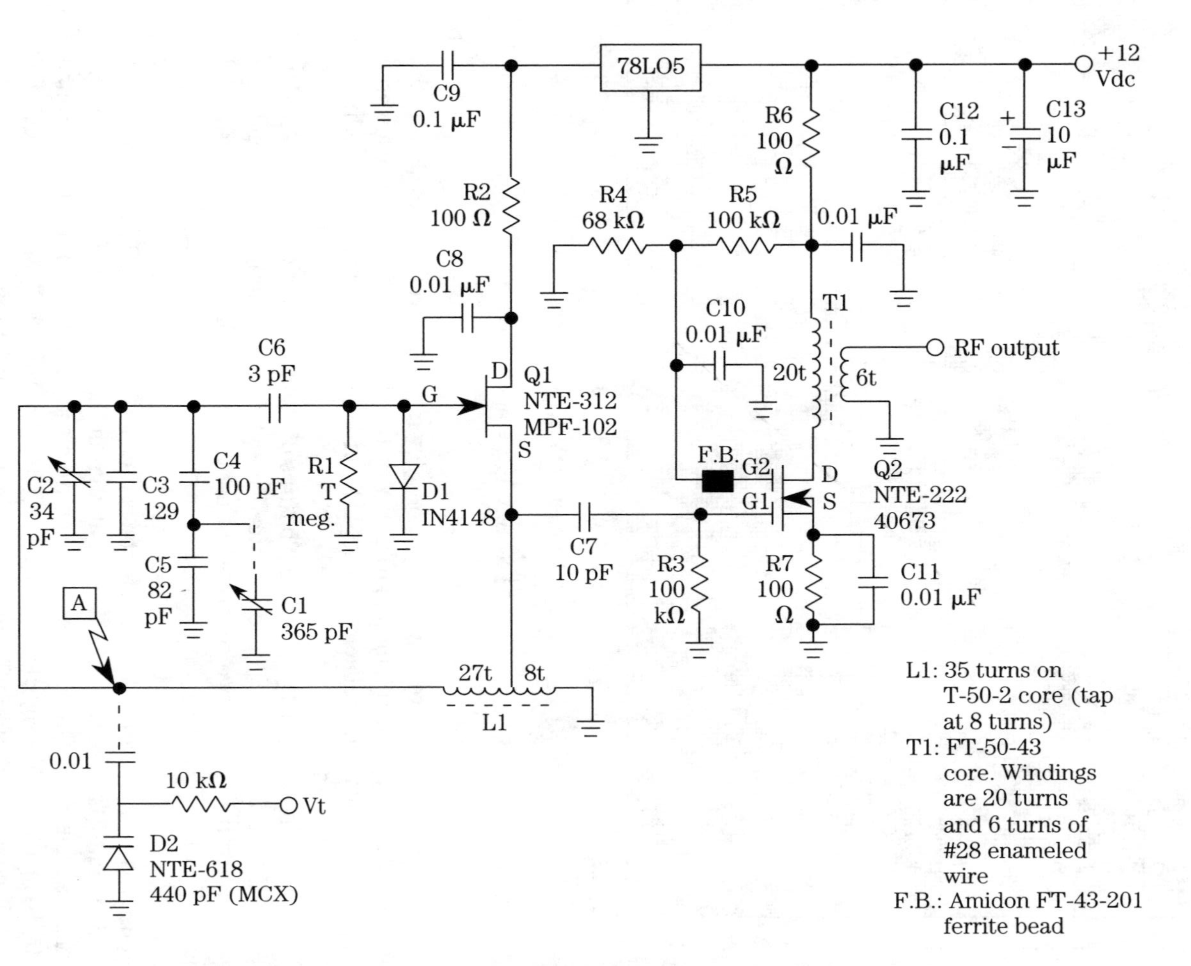

13-8 A. Voltage-tuned JFET Hartley oscillator with output buffer amplifier.

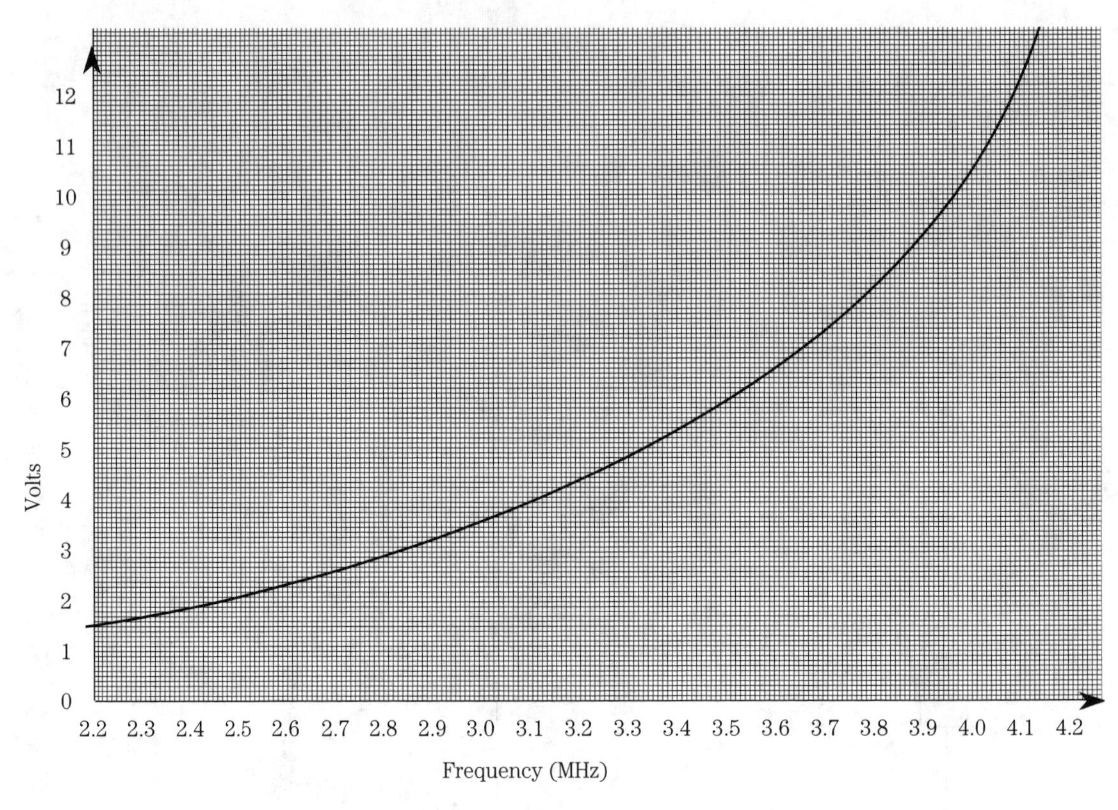

13-8 B. Voltage tuning curve.

volts (although my dc power supply only goes down to +1.26 volts, which is why the offset is at 2.2 MHz). The tuning range is 2.2 MHz to 4.1 MHz, but this can be lowered by increasing any of the capacitors (e.g., C3), or by either eliminating or reducing the capacitors (C1 to C5).

The main inductor (L1) consists of 35 turns of #28 enameled wire on an Amidon T-50-2 (RED) toroidal core. The tap is placed at eight turns from the ground end. The tap is formed by winding two separate but contiguous windings: one of 8 turns and one of 27 turns. These two windings are connected together electrically at the point where they come together. The two wire ends of the two coils are soldered together and used as a single wire to connect to the source ("S") of the JFET oscillator.

The drain ("D") of the JFET is kept at a ground potential for RF signals by the bypass capacitor, C8. Output signal is taken from the source ("S") terminal through a 10 pF capacitor. This signal is fed to gate 1 ("G1") of the 40673 dual-gate MOSFET that is used as an output buffer amplifier stage. This circuit produces a 44-dB gain, and is responsible for making the signal voltage so large. The output transformer (T1) consists of an Amidon Associates FT-50-43 core wound with 20 turns of #28 enameled wire for the primary, and 6 turns of the same wire for the secondary.

The output signal of Fig. 13-8A is huge as RF oscillators go, so it might have to be attenuated in some cases. This job can be done by connecting an attenuator resistor pad in series with the output signal, or by reducing the dc bias voltage on the MOSFET gate 2. This latter job can be done by reducing the value of R4 until the desired output signal is achieved (do not reduce it below about 1 kilohm).

Clapp VFO circuit

The Colpitts and Clapp oscillators are very similar to each other in that both depend on a capacitor voltage divider (C4 and C5 in Fig. 13-9) for feedback. The difference in the two oscillator circuits is that a Colpitts oscillator uses a parallel-resonant LC circuit, while a Clapp oscillator uses a series-resonant LC circuit. The circuit in Fig. 13-9 is classified as a Clapp oscillator because of the series-tuned LC network. This circuit can be used from 0.5 to 7 MHz, and even higher if C4 and C5 are reduced to about 100 pF.

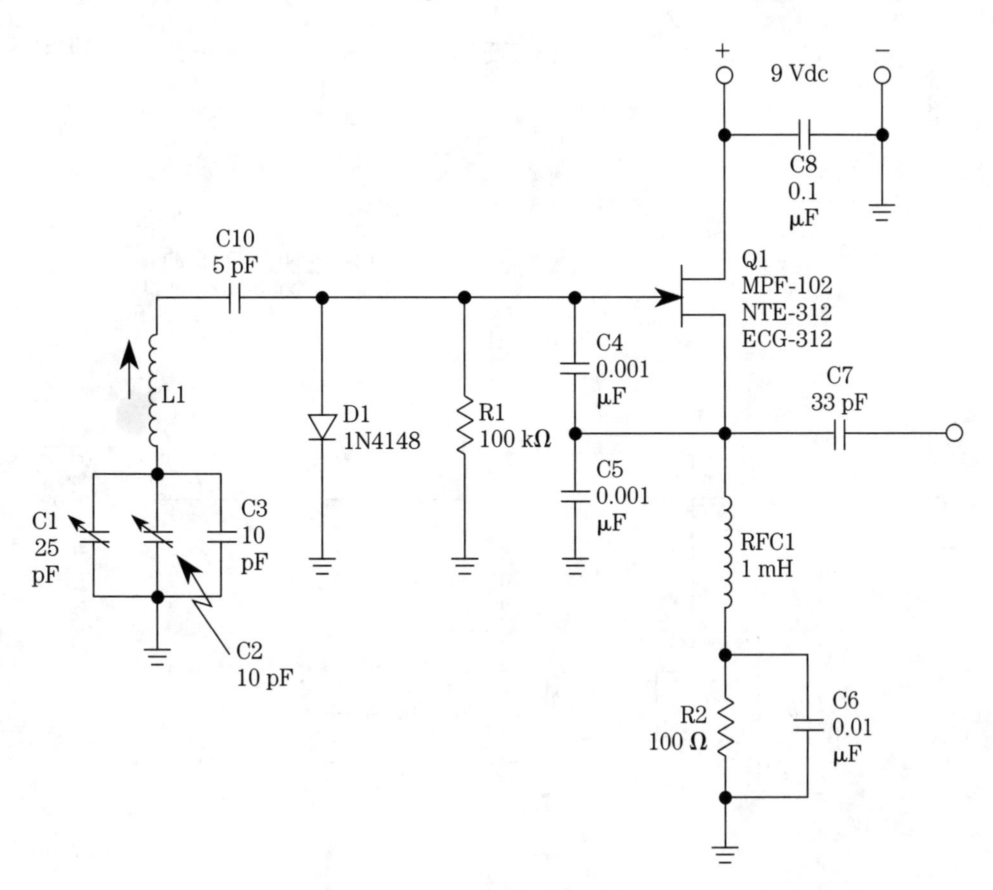

13-9 JFET series-tuned Clapp oscillator.

The output circuit from this circuit is taken from the source ("S") of the JFET oscillator transistor. The output circuit includes an RF choke (RFC1) that builds the output amplitude. Bias voltage for the JFET is provided by R2, and the source-drain current flowing through it.

NE-602 VFO circuits

The Signetics NE-602AN device is a double-balanced modulator (DBM) and an oscillator-integrated circuit. Normally, it is used as the RF front end of radio receivers, but if the DBM is unbalanced by placing a 10-kilohm resistor from the RF input (pin 1) to ground, it will function as an oscillator that produces about 500 mV of output signal.

Figure 13-10 shows an NE-602AN Colpitts oscillator circuit. Three capacitors (C1, C2, and C3) are used in this circuit, rather than two, because of a need for dc blocking. These capacitors should be equal to each other, and have a value on the order of 2400 pF/F_{MHz}. The inductor should have an approximate value of 7 μH/F_{MHz}. The tuning capacitor, C4, should have a value that will resonate with the selected inductor:

$$C_4 = \frac{1}{4\,\pi^2 f^2 L_1} \qquad [13\text{-}3]$$

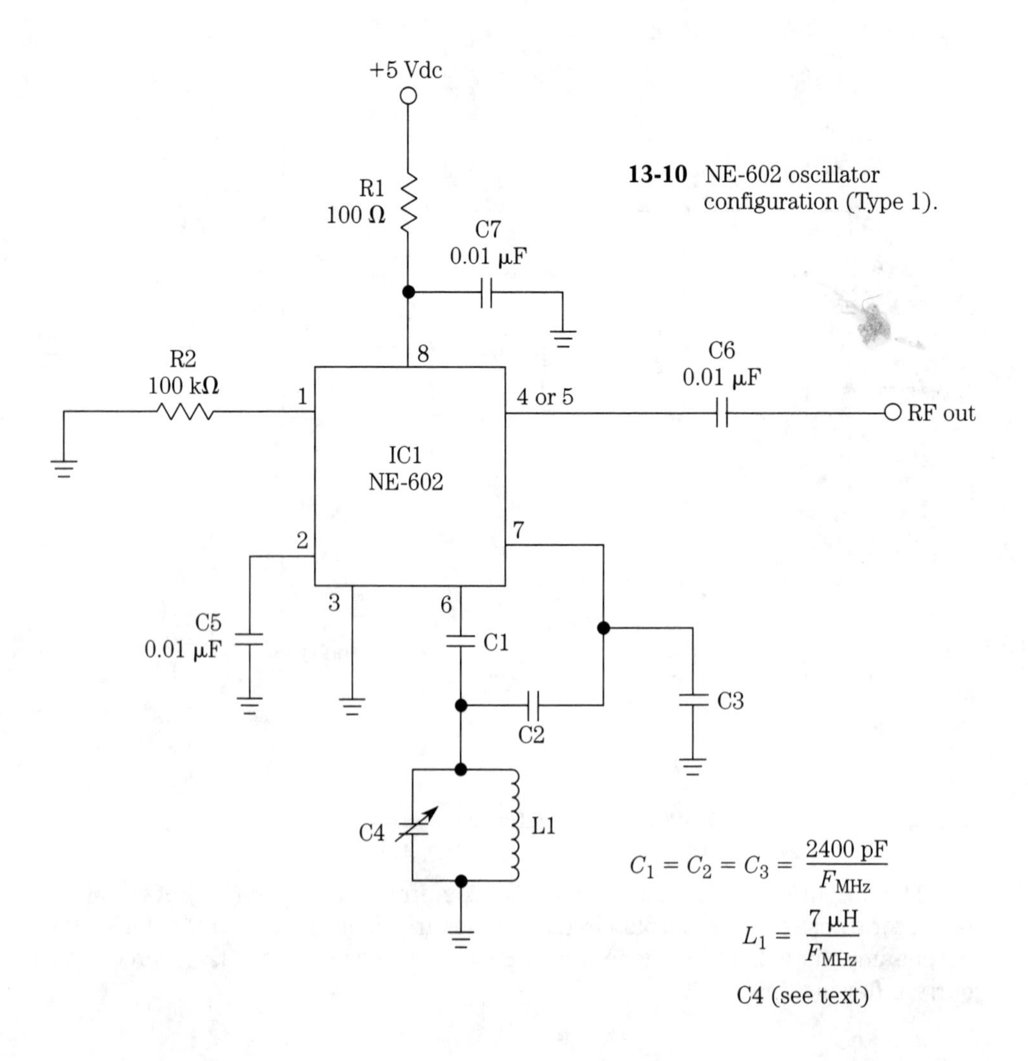

13-10 NE-602 oscillator configuration (Type 1).

For example, a 5000 kHz (5 MHz) oscillator should have network capacitors of 2400 pF/5 MHz, or 480 pF (use 470 pF standard value capacitors). The inductor should be 1.4 μH (17 turns on an Amidon T-50-2 (RED) core). To resonate with the 1.4 μH inductor requires 723 pF. But 470 pF/2, or 236 pF, are already in the circuit because of the series network C2/C3. Thus, a variable capacitor (C4) of 723 pF – 236 pF, or 487 pF is used.

An NE-603AN Hartley oscillator is shown in Fig. 13-11. This circuit is identified by the tapped coil in the LC network. The value of the inductor is about 10 μH/F_{MHz}, and is tapped from ¼ to ⅓ of the way from the ground end. The capacitor needs to resonate at the desired frequency. For our 5 MHz example, an inductor of 2 μH is used, which means 20 turns of wire on a T-50-2 (RED) core.

A voltage-tuned Clapp NE-602AN oscillator is shown in Fig. 13-12. This circuit uses a varactor diode to set the operating frequency. With the 100 pF capacitors shown, this circuit has oscillated from about 6 MHz to about 15 MHz, using an NTE-614 (33 pF) diode.

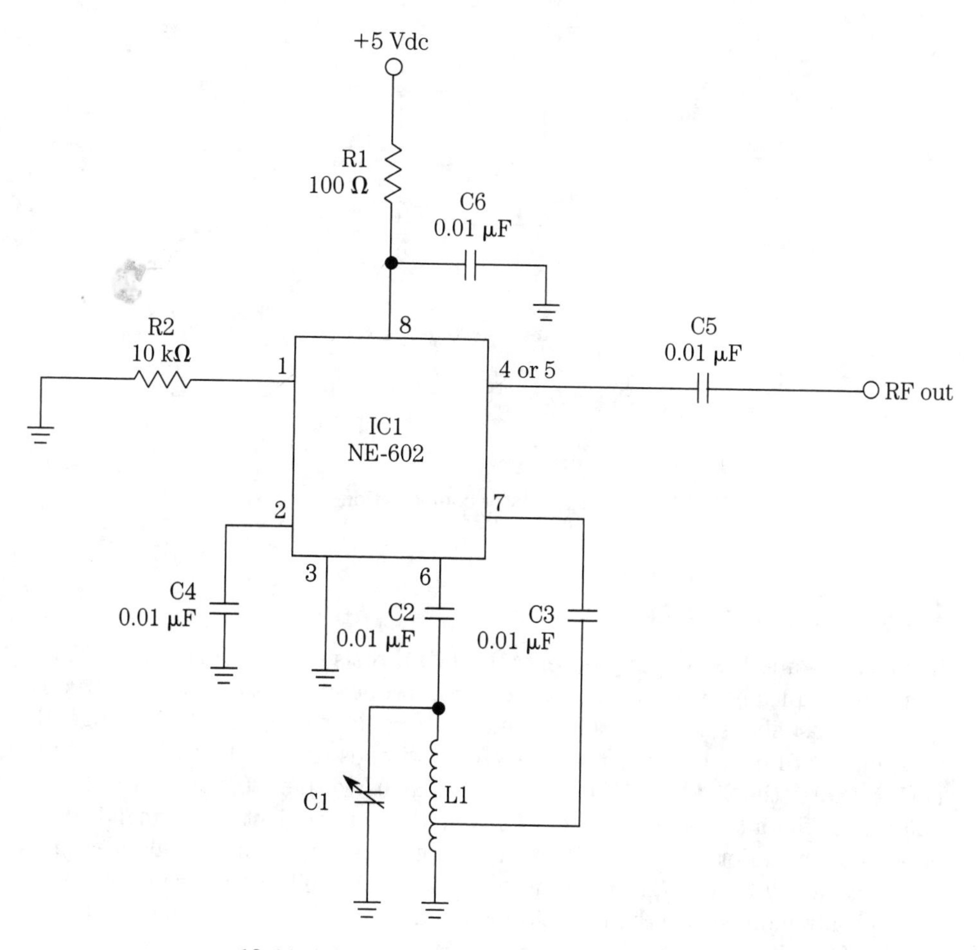

13-11 NE-602 oscillator configuration (Type 2).

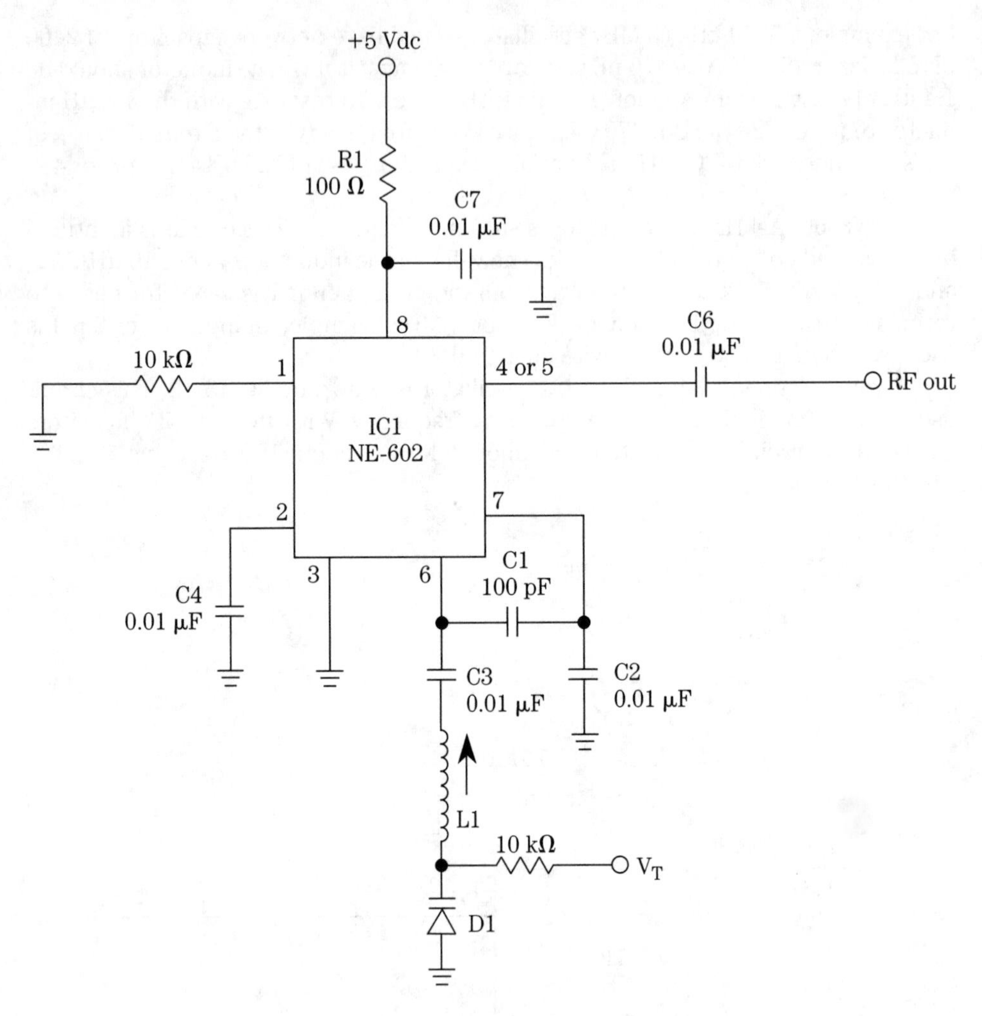

13-12 NE-602 oscillator configuration (Type 3).

Crystal oscillator

In the audio and low-RF range (i.e., <100 kHz), resistor and capacitor (RC) elements are typically used to set the operating frequency of oscillators. But as frequency rises above 20 kHz, or so, into the radio frequency (RF) range the components of choice for frequency setting switches to inductors and capacitors (LC) circuits (the 20 kHz to 100 kHz region may use either RC or LC). But LC circuits are difficult to make with precision, and they are subject to thermal drift and other problems. For operation where "rock solid" operation on a single frequency is needed, a *crystal oscillator* is required. Oscillators built with crystal resonators are typically more stable than LC circuits, and can easily be built to operate on a specific frequency.

Piezoelectric crystals

Crystal resonators are based on the phenomenon called *piezoelectricity*, i.e., the generation of an electrical potential from mechanical deformation of the crystal surface. If a slab of the right kind of crystal at rest (Fig. 13-13A) is deformed in a certain direction, a positive potential will appear across its faces (Fig. 13-13B). And when the same crystal is deformed in the opposite direction, the polarity of the voltage across its faces reverses (Fig. 13-13C). Thus, when the crystal is wiggled back and forth an ac voltage appears across the faces (Fig. 13-13D).

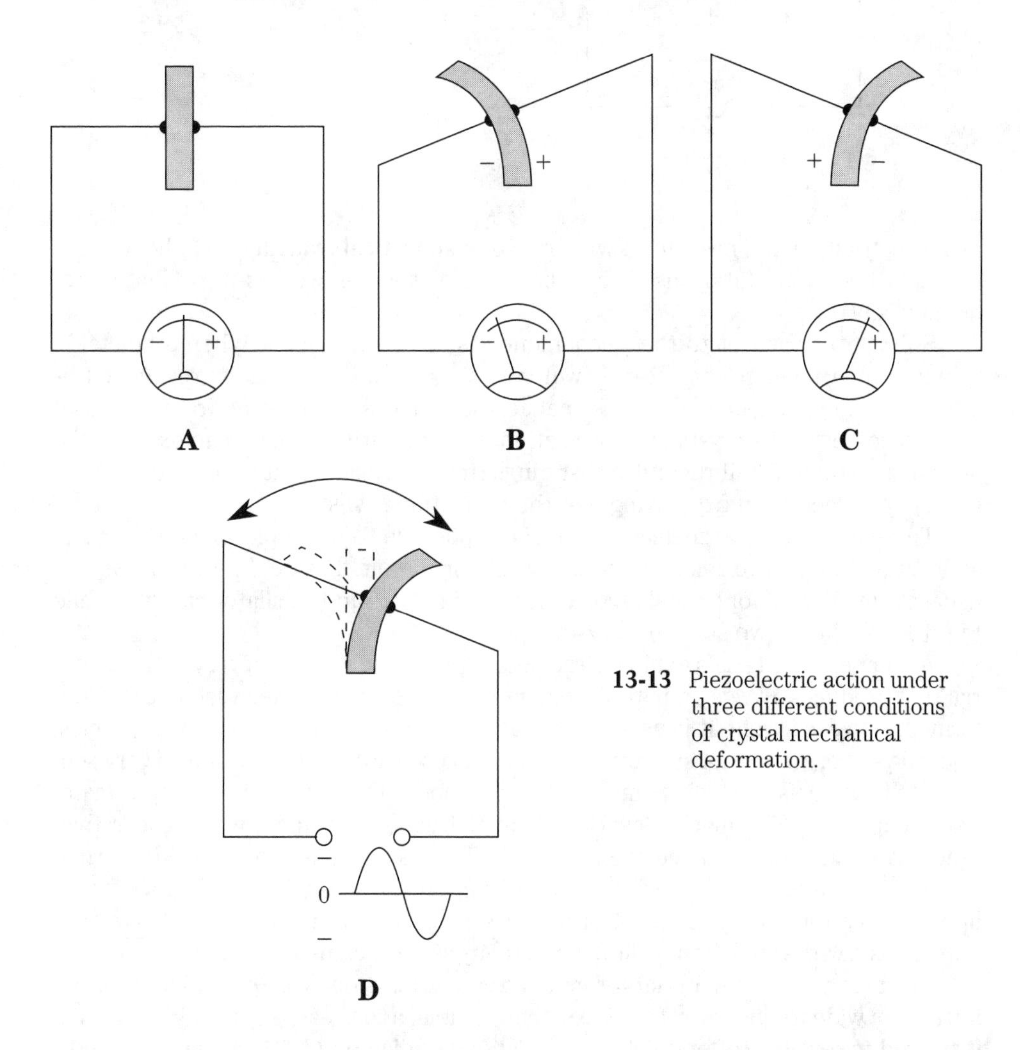

13-13 Piezoelectric action under three different conditions of crystal mechanical deformation.

The inverse action also occurs: when an ac voltage is applied to the faces of the crystal, it will deform in a direction determined by the polarity of the voltage (Fig. 13-14). Something special happens when the frequency produced by the oscillator matches the natural mechanical resonance of the crystal—the process become very efficient, and little energy is required to keep the motion going. This aspect of

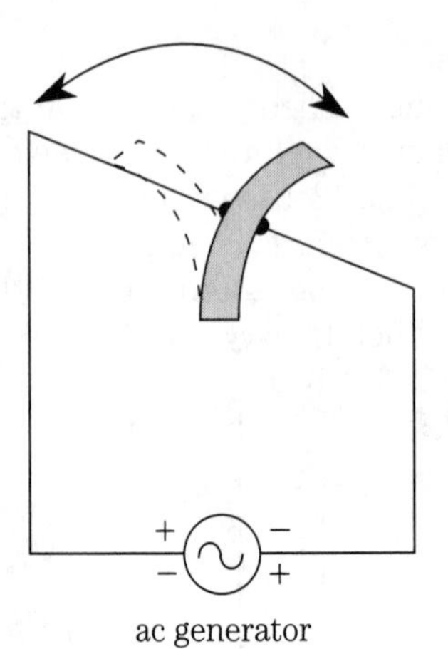

13-14 Crystal response to ac signal.

piezoelectricity is the basis for a wide range of acoustical transducers, phono pick-ups and the crystal filters used in radio receiver sets, as well as in oscillator frequency controls.

Still another aspect to the phenomenon is sometimes seen. When a crystal is "pinged" by a momentary pulse, it will vibrate back and forth at its resonant frequency, producing a sine wave ac signal across its faces at the same frequency. Because of losses in the crystal, the oscillation dies out fairly quickly in an exponential decaying manner. But if the pulse that pings the crystal is repeated often enough to prevent the oscillation from dying out, the oscillation is sustained.

These aspects of piezoelectricity make it possible to use a piezoelectric crystal as a frequency control element in an oscillator circuit. Figure 13-15A shows the usual circuit symbol for crystal resonators used in filters and oscillator circuits, while Fig. 13-15B shows two standard crystal holders.

A number of materials exhibit piezoelectricity. Rochelle salt is a very active material that produces a large amplitude voltage-per-unit of strain when mechanically deformed. However, while it is used in crystal phonograph pick-ups and microphones, Rochelle salt crystal is not suitable for RF crystal oscillators. The material is very sensitive to heat, moisture, aging, and mechanical shock. The next highest output material is tourmaline. This material works well at all frequencies, and it works better than other materials over the range of 3 to 90 MHz. There's only one problem with tourmaline: it cost an arm and a leg, as you will discover quickly enough if you buy a tourmaline necklace for your "significant other." It seems that tourmaline crystals are very popular as a variegated (red, yellow, and green) semiprecious gemstone.

The best practical material for crystals is quartz. It behaves much like tourmaline over a wide frequency range, is relatively stable, and is easily available. Although it is used in jewelry (often mislabeled "Cape May diamond," "Herkimer diamond," "Arkansas diamond" in the colorless varieties, "topaz"—which it's not—in the yellow variety ("citrine" is the correct name, "topaz" is a bit of a trade fraud), and "smoky quartz" in the variety that looks like smoked glass), it is low in cost because it is plentiful. The only costly quartz crystal is the purple or violet variety, which is called amethyst.

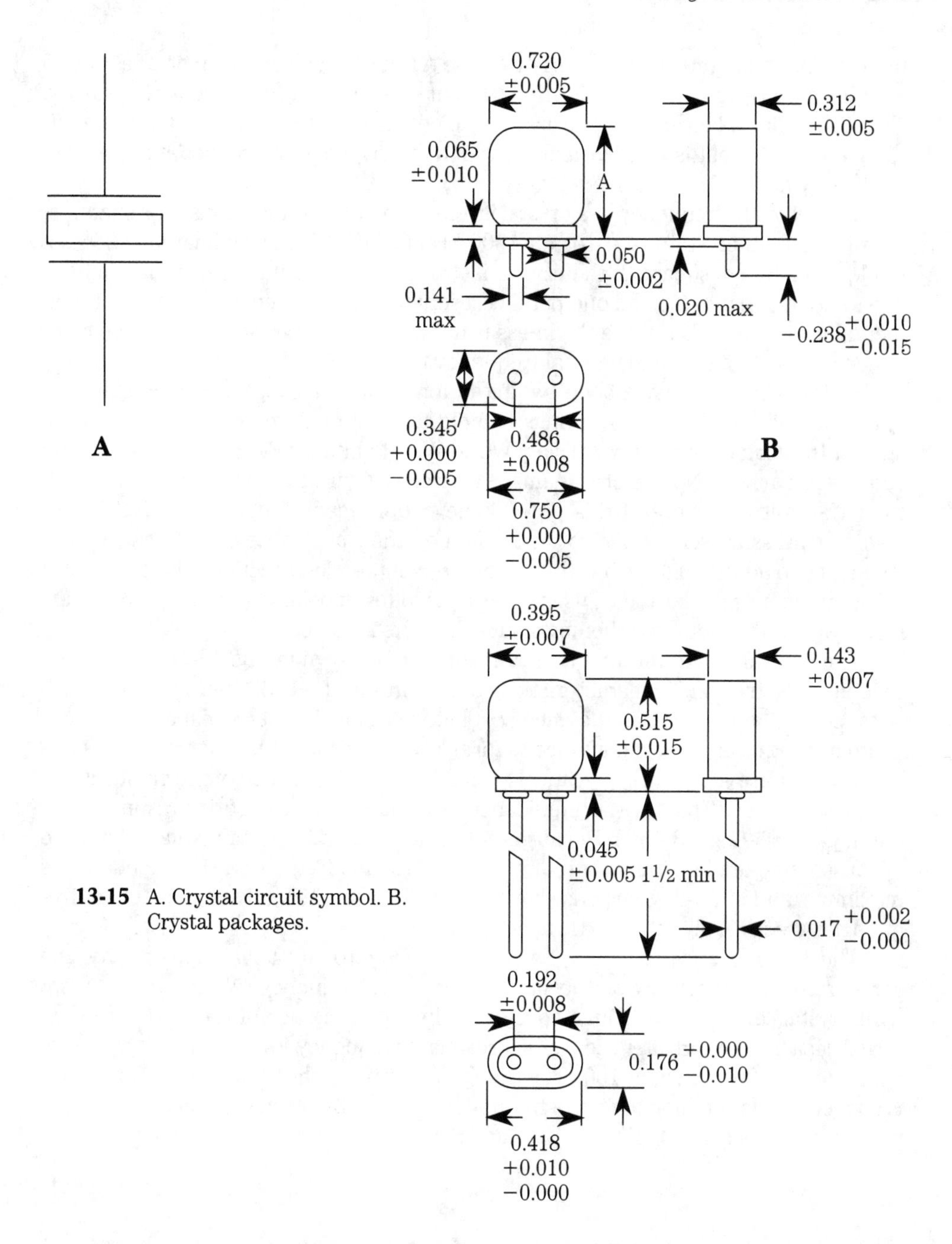

13-15 A. Crystal circuit symbol. B. Crystal packages.

The quartz crystal is hexagonal in shape, and pointed at both ends (if perfect, natural crystals are often broken or cut off on at least one end). A series of three axes are created with the Z-axis reference being from one tip point to the other; this axis is also called the *optic axis*. Crystal slabs for use as resonators are taken with different "cuts" through the crystal body. The X and Y cuts are through the X and Y axis, respectively. These are not favored for radio work, however, because they have undesirable temperature characteristics. The AT cut is made at an angle of about 35 degrees from the Z axis. There is also a BT cut (not shown) that is sometimes used. The AT-cut has a better temperature coefficient (ΔF/F ppm) by an order of magnitude, but the BT cut is usually thicker (which means it is more robust at higher frequencies where AT-cut "rocks" are very thin).

The resonant frequency of a crystal is a function of its dimensions. For example, a typical quartz crystal resonator for 1000 kHz (1 MHz) is approximately 0.286 cm thick, and 2.54 cm square. If the crystal is ground to uniform thickness, it will have one series resonant (F_s) and one parallel resonant (F_p) frequency. These are fundamental frequencies. But if the thickness is not uniform, there may be spurious resonances other than the fundamental frequency.

Historically, there have been two basic forms of mounting for the crystal. The older method used a pair of springs to hold a brass or silvered copper electrode against the surface of the crystal slab. World War II vintage "FT-243" crystal mounts (once popular with Novice-class hams, who were required to use crystal control on their transmitters) were of this type. Some people made "rubber crystals" by installing a pressure screw to vary the tension on the slab. These devices caused the frequency to adjust slightly, but this is not a recommended practice. The other form of mount, more popular today, uses silver electrodes deposited onto the crystal surface. Wire connections can then be soldered to the surface.

The equivalent circuit for a crystal resonator is shown in Fig. 13-16A, while the reactance vs. frequency characteristic is shown in Fig. 13-16B. There is a series resistance (R_s), and a series inductance (L_s) in the circuit. The series capacitance (C_s) combines with the series inductance to form a series-resonant frequency. At this frequency, because $-X_{CS}$ and $+X_{LS}$ cancel each other, the impedance of the crystal is the series resistance. That is, the impedance is minimum at the series-resonant frequency, F_s (see Fig. 13-16B). Because there is also a parallel capacitance (C_p) there will also be a parallel-resonant frequency (F_p). At this frequency, the impedance is maximum, and a 180-degree phase shift occurs. The parallel and series-resonant frequencies are typically 1 to 15 kHz apart.

The design of any particular oscillator is selected to take advantage of either the series-resonant frequency or the parallel-resonant frequency. When parallel-resonant crystals are used, you must specify the load capacitance of the crystal (an external capacitance can alter the parallel resonant frequency a small amount). Typical values are 20, 30, 50, 75, or 100 pF, although for most applications the 30 (or 32) pF is specified. It is common to form a frequency adjuster by placing a small trimmer capacitor in series (Fig. 13-17A) or parallel (Fig. 13-17B) with the crystal.

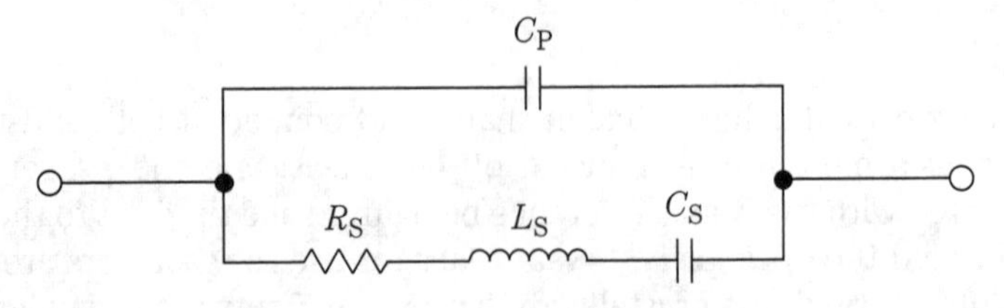

13-16 A. Equivalent crystal circuit.

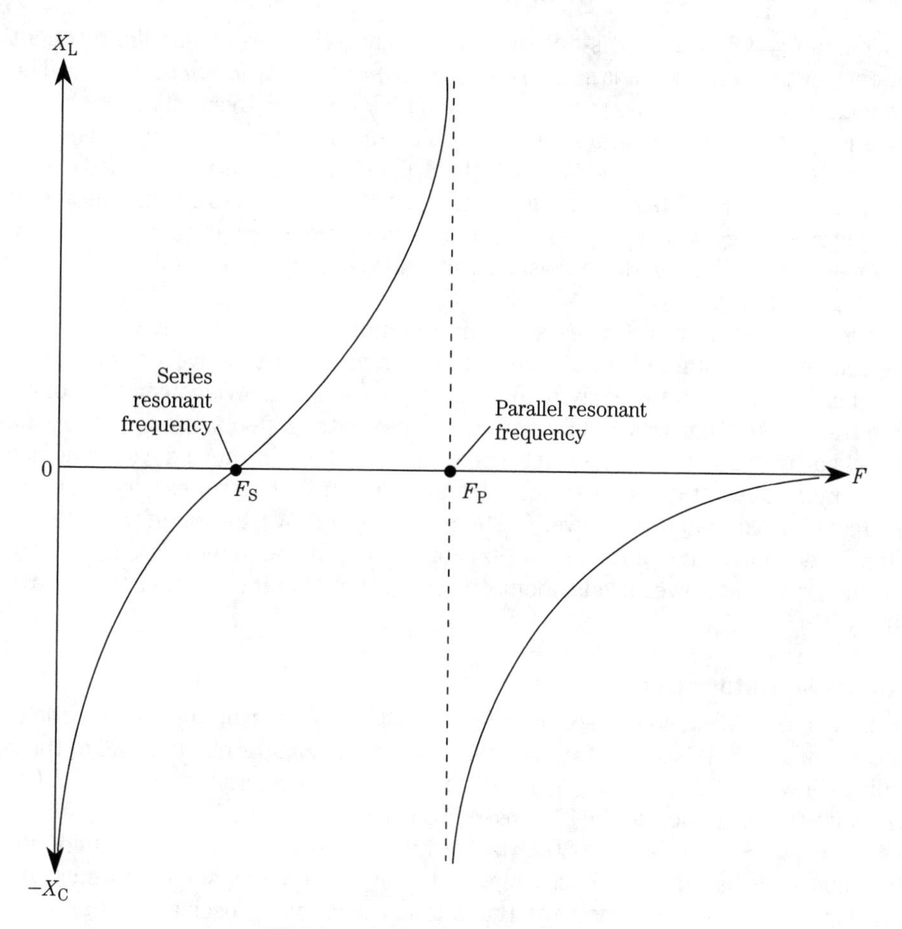

13-16 B. Impedance vs. frequency plot.

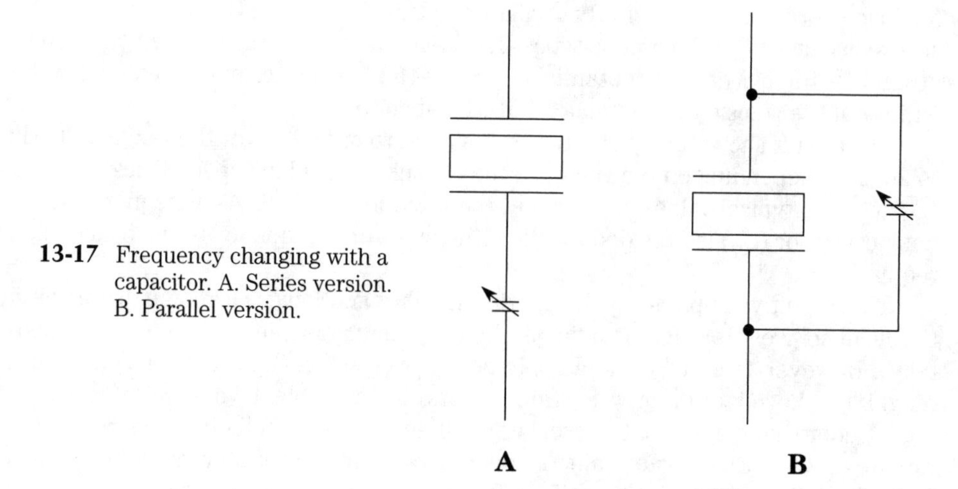

13-17 Frequency changing with a capacitor. A. Series version. B. Parallel version.

When a crystal oscillator is operated at the natural series or parallel-resonant frequency of the oscillator, it is said to be a *fundamental frequency oscillator*. The fundamental mode is used up to frequencies of 20 MHz or so. In some cases, the oscillator is operated on or near a harmonic of the fundamental frequency. These are called *overtone oscillators*, and typically the 3rd, 5th, or 7th overtone is used over a range of 20 to 90 MHz. When ordering overtone crystals, be sure to specify the actual operating frequency, not the apparent fundamental frequency. Dividing the actual frequency of, say, a 5th overtone crystal by five does not yield the parallel mode fundamental frequency.

Crystals typically want to see a certain minimum drive power in order to operate reliably (i.e., start when the circuit is turned on). But drive power can be overdone. I recall from my early ham radio days a fellow in Wisconsin who operated a transmitter that was a 700-watt crystal oscillator. Even keeping the crystal in an oil bath wouldn't stop the fractures! It was common in those days for novice hams to build 30-watt crystal oscillator transmitters from 6L6 tubes. Crystals typically have a maximum drive power of 200 microwatts (μW), although those under 1000 kHz may have maximum dissipations of 100 μW. It is common practice to operate the crystal at power levels about one-half the maximum in order to improve stability.

Crystal oscillator circuits

Now let's take a look at some oscillator circuits that use either fundamental mode or overtone mode crystals as the frequency-controlling resonator element. All of these circuits will work with common, "garden variety" silicon transistors, JFETs or MOSFETs, with the exception of the TTL-based IC oscillator circuit.

Figure 13-18 shows a 1- to 20-MHz crystal oscillator that is easy to build, and very simple. It is based on an Si npn bipolar transistor and a crystal operated in the parallel mode. The feedback network that allows the circuit to oscillate is the capacitive voltage divider network, consisting of C1 and C2, which is effectively shunted across the crystal (Y1). Because this circuit uses a capacitive voltage divider for the feedback network, it is a Colpitts oscillator circuit. Capacitors C1 and C2 should be silver-mica or NPO disk ceramic types. The collector of the transistor is bypassed to ground for RF, but has a dc potential of +5 to +15 volts dc. Output is taken from the emitter of the transistor through a 0.001 μF capacitor.

The circuit shown in Fig. 13-19A is designed to operate over the range 500 kHz to 20 MHz, depending on the values of the capacitors used in the feedback network (C_a and C_b); typical values are given in the table in Fig. 13-19A. A frequency-trimming capacitor (C_T) is provided to adjust the operating frequency to the exact value required.

The circuit will operate with superior stability and lower harmonic distortion if the feedback resistor (R1) is adjusted higher (the value can be found experimentally). However, this tactic should only be followed when the oscillator is free-running. If it is keyed, or otherwise turned on and off, a problem will result if R1 is too high. Under that condition, the oscillator will not rise to its full output amplitude as rapidly as when the resistor value is lower. Dropping the resistor value lower than 2200 ohms, however, might have the effect of over-driving the crystal.

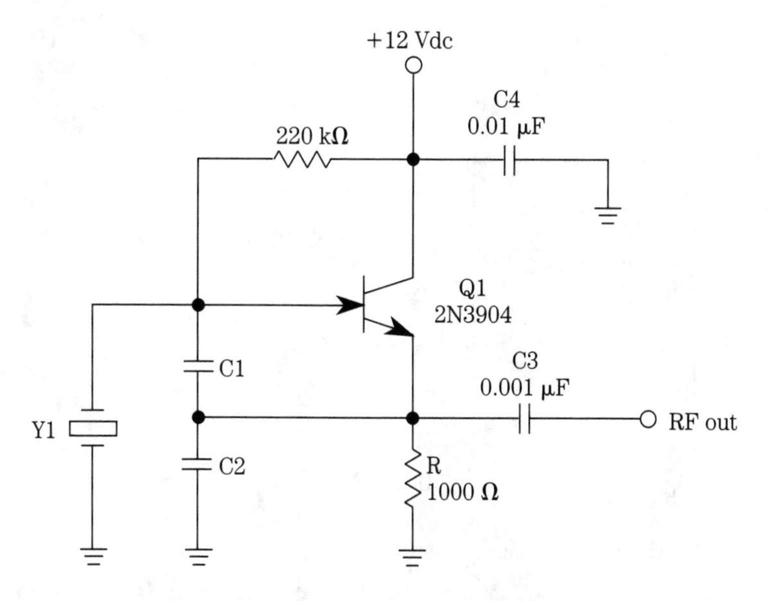

13-18 Simple crystal oscillator for 1-20 MHz.

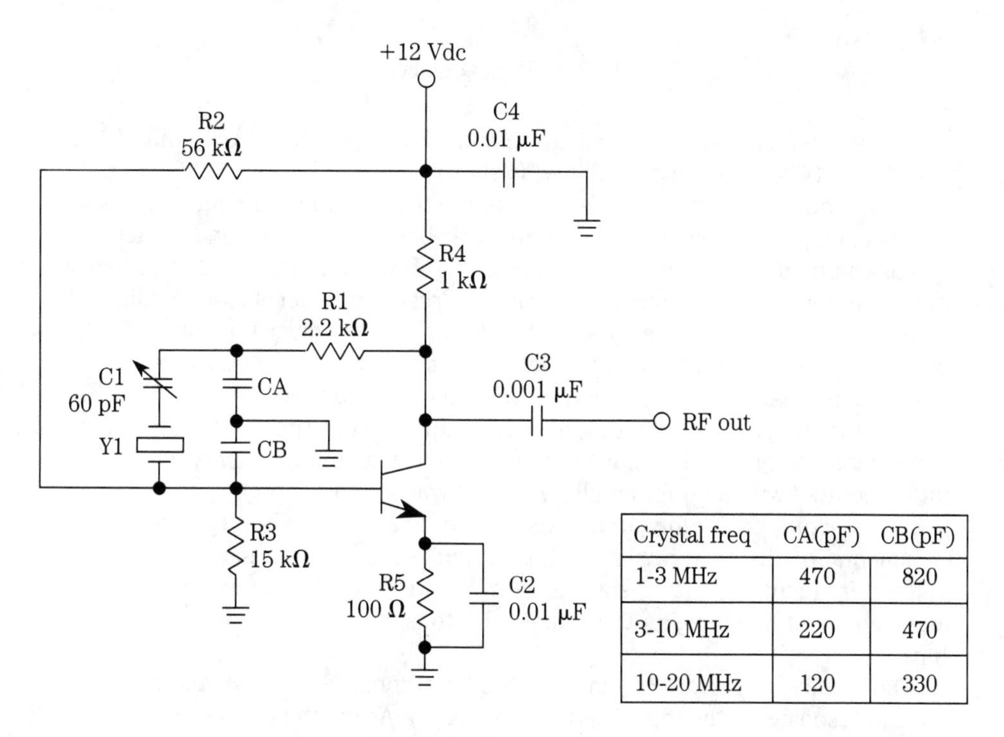

Crystal freq	CA(pF)	CB(pF)
1-3 MHz	470	820
3-10 MHz	220	470
10-20 MHz	120	330

13-19 A. Pierce oscillator.

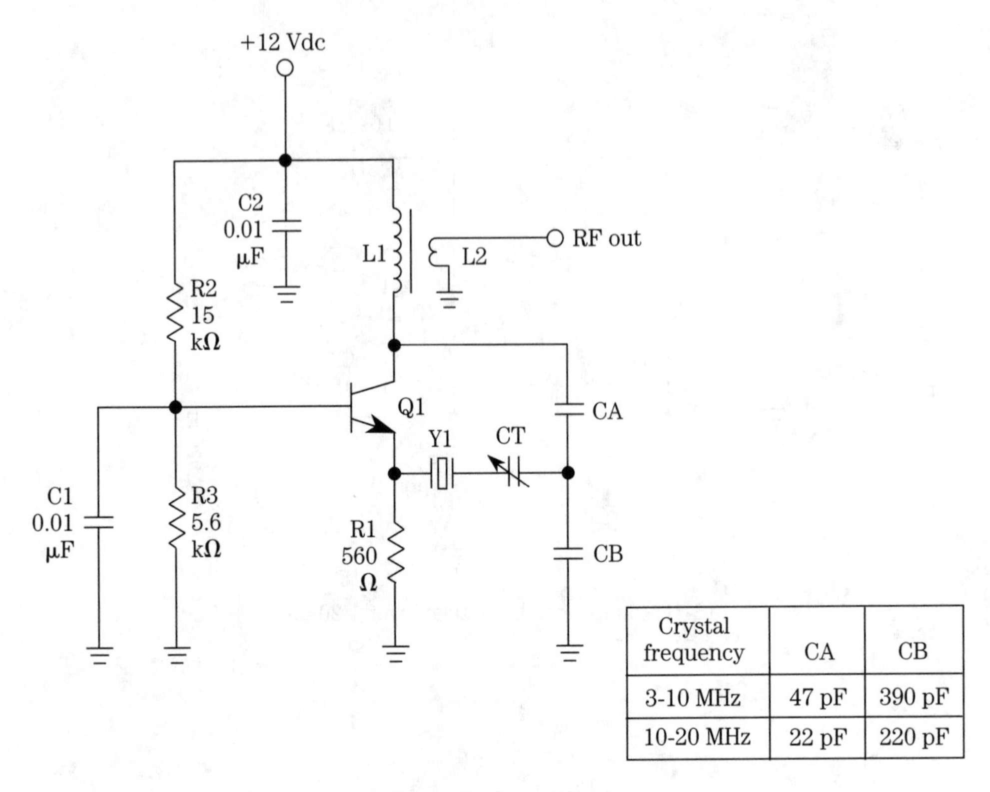

Crystal frequency	CA	CB
3-10 MHz	47 pF	390 pF
10-20 MHz	22 pF	220 pF

13-19 B. Butler oscillator.

A fundamental frequency oscillator circuit that is capable of providing 10 parts per million (PPM) frequency stability, is shown in Fig. 13-19B. In this circuit, the crystal is connected between the emitter of the transistor and the junction point on the capacitor voltage divider feedback network; both series-mode and parallel-mode crystals can be used. The ratio of the feedback network capacitors can be adjusted empirically for best (most stable) operation. Crystal drive level can be adjusted by making R3 any value between 100 and 1000 ohms. The lower the value of R3, the lower the crystal dissipation and the better the stability. Inductor L1 is resonated to the crystal frequency by C_a. The circuit will fail to start if this coil is misaligned. It is common practice to find a setting near resonance where the crystal oscillator will start reliably every time it is powered up, and has maximum stability when the circuit is operated with a buffer amplifier (about which, more later).

An *overtone oscillator* circuit is shown in Fig. 13-20. Although similar to the fundamental oscillator in Fig. 13-19B, it produces an output frequency at the third overtone frequency of the crystal. As with the previous circuit, crystal drive power can be adjusted by changing the value of R1 to some value between 100 and 1000 ohms.

The variable inductor (L2) in Fig. 13-20 is resonated with C3, and must be selected to resonate on the third overtone frequency. As in the previous case, set L2 to a point where the oscillator starts reliably and is stable. This coil will pull the fre-

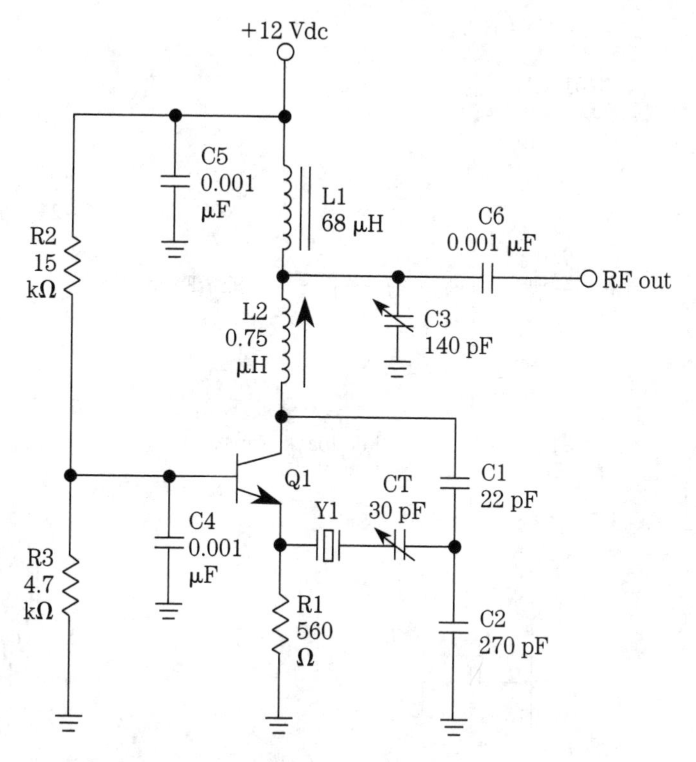

13-20 Butler oscillator.

quency somewhat, so don't adjust the frequency trimmer capacitor (C_T) the final time until after the correct adjustment point for L2 is found. After that, don't change the setting of L2. Again, a buffer amplifier is highly recommended.

A Pierce oscillator circuit is shown in Fig. 13-21. *Pierce oscillators* are identified by having the crystal connected between the output and input of the active device. Because a junction field effect transistor (JFET) is used in Fig. 13-21, the crystal is connected between the drain and gate; in a circuit using a bipolar transistor the crystal is connected between the collector and base. The capacitor (C2) in series with the crystal is used in a dc-blocking function. In some low-voltage transistor circuits, this capacitor can be eliminated—although most authorities would agree that it is highly desirable. Capacitor C1 is used to adjust the feedback level. For most crystals, a 100 pF unit is usually sufficient.

A pair of *Miller oscillator* circuits are shown in Fig. 13-22. The version in Fig. 13-22A uses a capacitor output circuit, while the version in Fig. 13-22B uses a link-coupled output from L1. Again, a JFET is used as the active device, even though a properly biased bipolar npn or pnp device could also be used. The Miller oscillator circuit is identified by having the crystal in a parallel-mode connection, with a parallel-resonant output tuned-tank circuit, and no capacitive voltage divider feedback network. This circuit is quite popular, but it escapes me just why that's so. It seems that the Miller oscillator is subject to frequency and output amplitude instabilities,

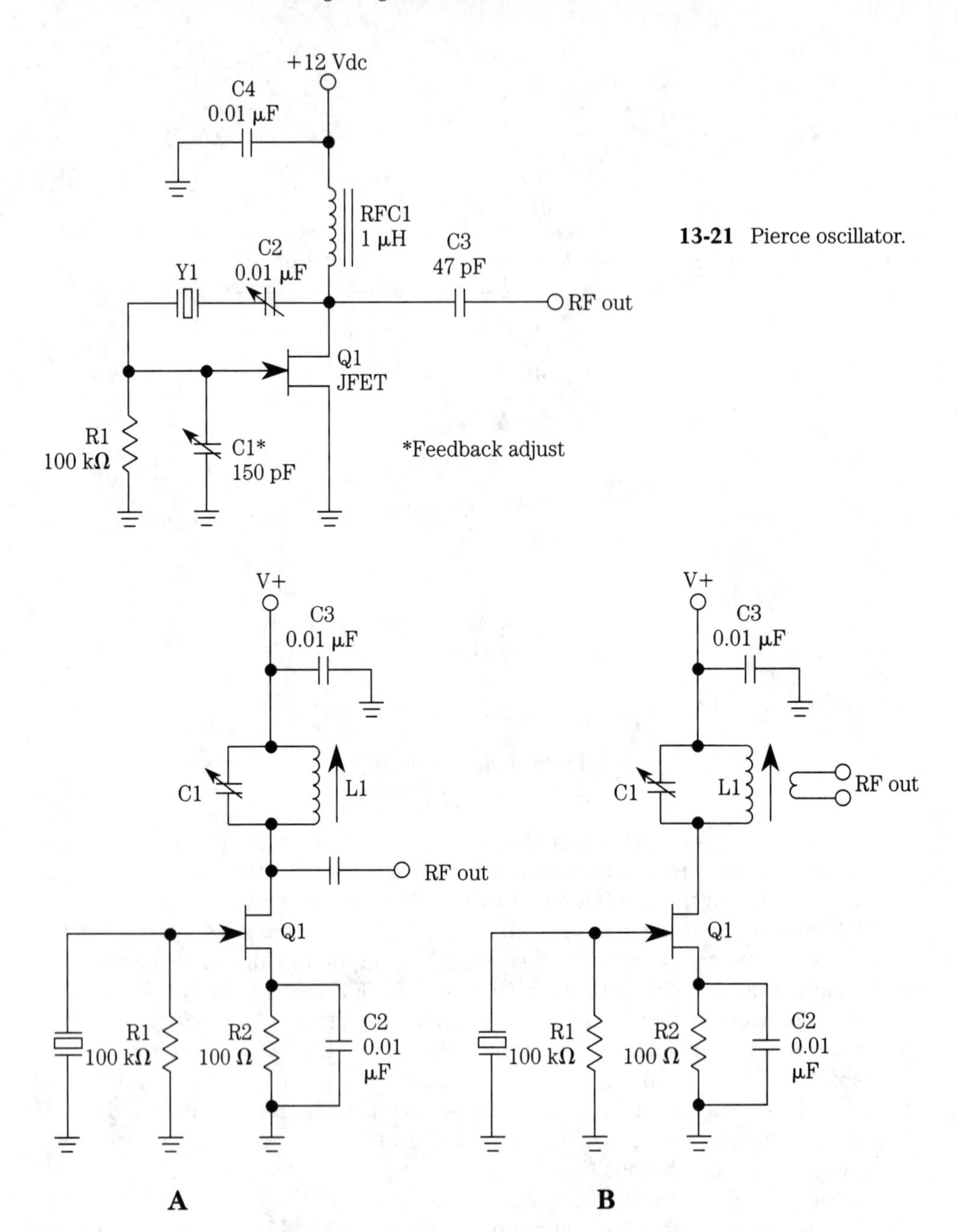

13-21 Pierce oscillator.

13-22 Two forms of Miller oscillator circuit.

and suffers badly from load impedance variations. The setting of the output tuned circuit (L1/C1 in both Figs. 13-22A and 22B) is critical to proper operation.

A *transformer-coupled output crystal oscillator* circuit is shown in Fig. 13-23. This circuit is based on a common 2N3904 (or equivalent) npn transistor. The output transformer (T1) is wound on an FT-37-43 toroidal core. The primary consists of

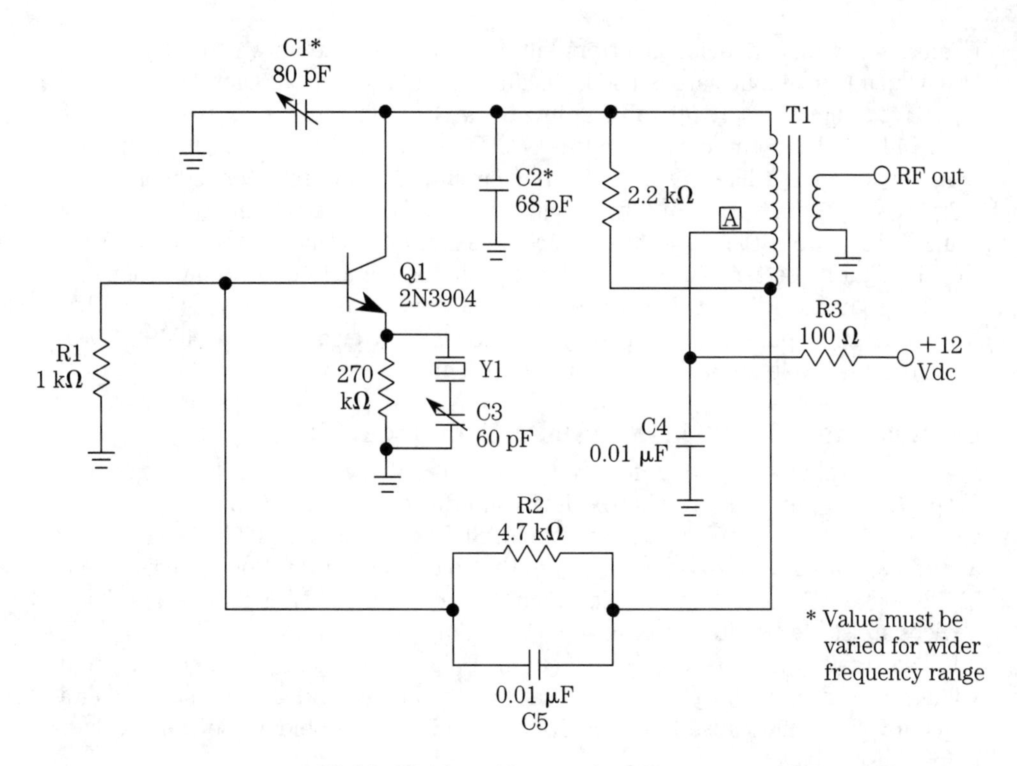

13-23 Highly stable crystal oscillator.

about 20 turns of #26 enameled wire, while the secondary is 2 turns of #26 wound over the primary. The primary is taped such that one end is 7 turns, and the other end is 13 turns from the common point ("A"). The V+ voltage is applied to point "A," at the junction of the two halves of the primary winding.

A *TTL-compatible crystal oscillator*, the type used as a clock in digital circuits and computers, is shown in Fig. 13-24. This circuit can be built with any set of TTL

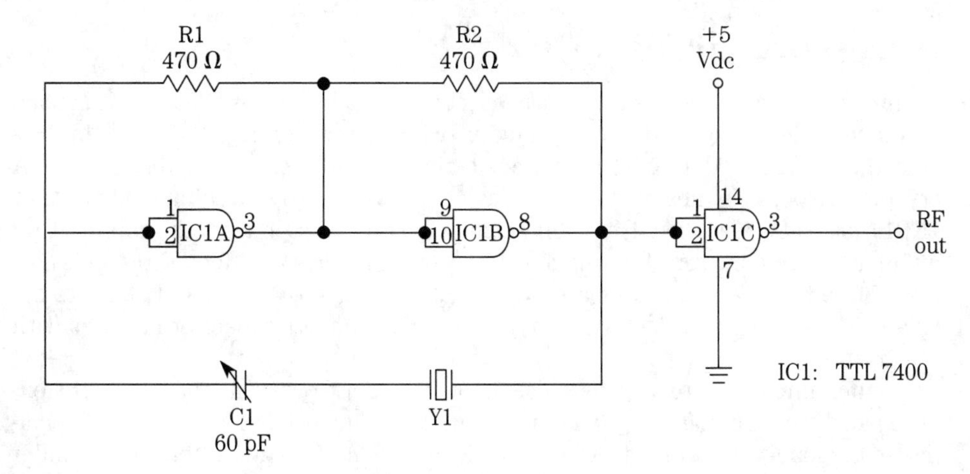

13-24 TTL crystal oscillator.

(transistor-transistor logic) inverters, although in the case shown the Type 7400 NAND quad two-input gate is used. Because the two inputs of each gate are connected together, they are operated as inverters. Because four NAND gates are inside each 7400 package, there will be a free NAND gate for use elsewhere in the circuit. Some crystals don't like to work with this circuit, and I've experienced some difficulty making them work properly at higher frequencies (over about 10 MHz). If you want a very stable TTL clock frequency, or experience reliable starting problems, then it might be better to use one of the other fundamental mode circuits, then convert the signal to TTL with a voltage comparator that has TTL output (i.e., an LM-311 with a 2700 pull-up resistor to +5 volt dc supply), or a 4050B or 4049B CMOS operated at +5 volts, or a TTL Schmitt trigger chip.

Oven-controlled 1000 kHz crystal calibrator circuit

Figure 13-25 shows the circuit for a very stable frequency standard that can be used for calibration purposes. The active devices in this circuit are a pair of *metal oxide semiconductor field-effect transistors* (MOSFETs); Q1 is the oscillator while Q2 is a buffer amplifier used to isolate Q1 from the cold, cruel world. Both transistors are 3N128 MOSFETs, or a service replacement type such as the NTE-220 device. These devices are static sensitive, so use "ESD" handling procedures.

The stability is enhanced in this circuit by using a separate integrated circuit voltage regulator (IC1) to keep the voltage applied to Q1 and Q2 very stable. Variations in voltage can cause frequency pulling and other problems, so the regulator helps substantially.

The real key to stability in this circuit, however, is the crystal oven. These devices are available from some radio parts stores, but are increasingly difficult to obtain. I've seen them sold at hamfests new in the last year, but otherwise they tend to be expensive and hard to come by . . . but they're worth it. The typical oven keeps the crystal at a constant temperature of 75 or 80 degrees Celsius. When specifying the crystal, specify a 20 pF load capacitance and inform the vendor that you need it calibrated at the oven temperature (75/80 C).

Accessory circuits

In this section, we will take a look at a circuit that is not an oscillator, but is used with crystal oscillators—the multichannel switch circuit of Fig. 13-26. In this circuit, there is more than one crystal available—but only one is used at a time. A crystal is selected by heavily forward-biasing its associated switching diode (either 1N914 or 1N4148 are used). When selector switch S1 applies +12 volts dc to a diode, through a current-limiting and isolation resistor, the affected diode is forward-biased. The crystal associated with that diode is grounded, so it starts to oscillate. The exact oscillation frequency is set by a trimmer capacitor in series with the crystal.

Buffer amplifiers are used with oscillators in order to prevent changes in the external load from pulling the oscillating frequency. Many oscillator circuits are sensitive to the load-pulling effect. A *buffer amplifier* is nothing more than an amplifier stage that prevents changes in the ultimate load from being felt by the oscillator cir-

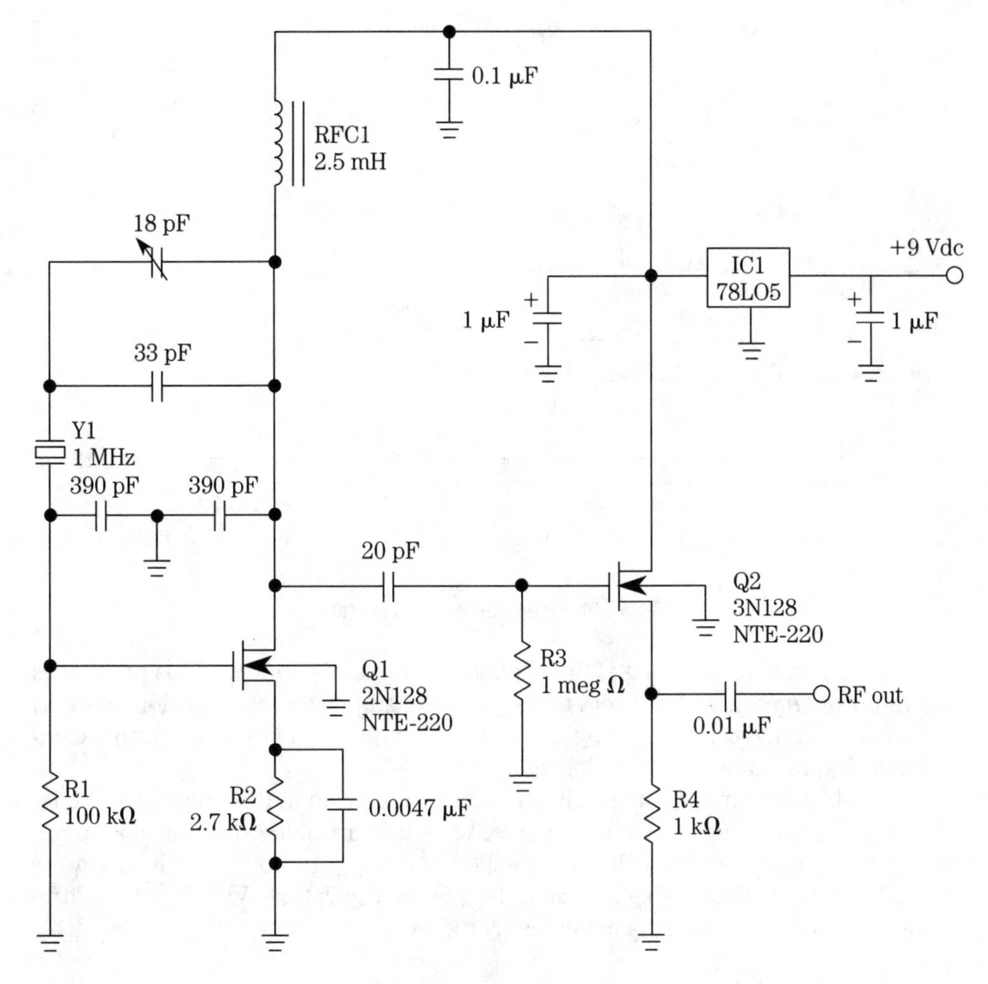

13-25 MOSFET Pierce oscillator.

cuit. While many of the oscillators in this chapter will operate well without a buffer amplifier, it is always good practice to use one.

Crystal oscillators provide a superior means of providing stable, accurate RF frequencies. They are relatively easy to build, were mostly well-behaved, and should work well.

Building stable RF oscillators

Radio frequency oscillator stability is always important in radio circuits, and in both CW and (especially) single-sideband circuits it is crucial. Frequency stability is one of the principal specifications that defines the quality of radio receivers and transmitters, as well as signal generators and other RF devices. A radio or signal generator that changes frequency without any help from the operator is said to *drift*. Frequency stability refers generally to freedom from frequency changes over a rela-

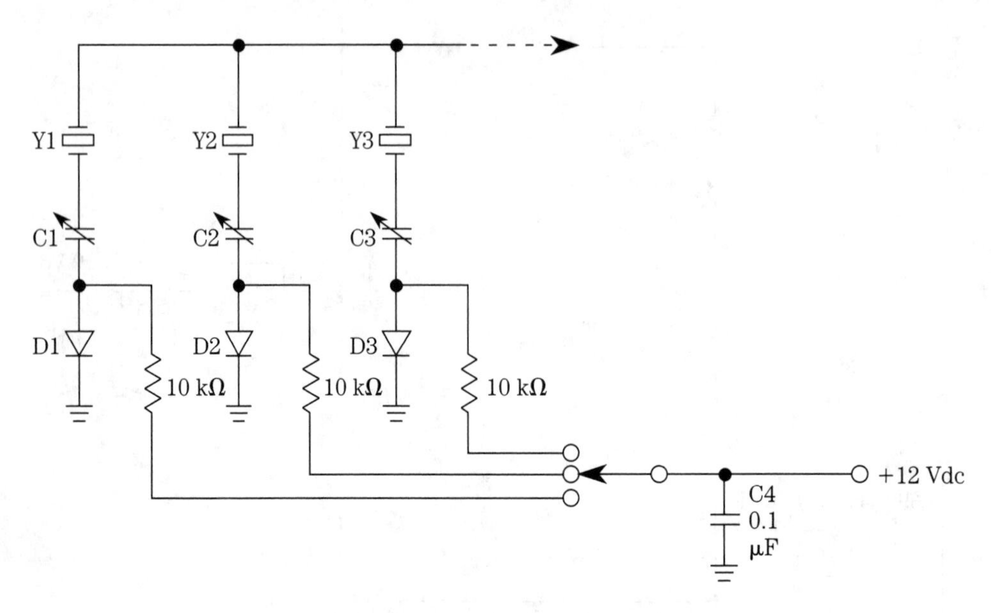

13-26 Voltage selection of crystals.

tively short period of time (e.g., few seconds to dozens of minutes). This problem is different from *aging*, which refers to frequency change over relatively long periods of time (i.e., hertz/year) caused by aging of the components (some electronic components tend to change value with long use).

Several factors involved in oscillator stability are given in this chapter as guidelines. If these guidelines are followed, it will result in a stable oscillator more often than not. For the most part, the comments below apply to both crystal oscillators and LC tuned oscillators (e.g., variable frequency oscillators, VFOs), although in some cases one or the other is indicated by the text.

Temperature

Excessive temperatures and temperature variation have a tremendous effect on oscillator stability. Avoid locating the oscillator circuit near any source of heat within the unit. In other words, keep it away from power transistors or IC devices, rectifiers, lamps, or other sources of heat. There was one kit-built tube-type ham radio SSB transceiver (not a Heathkit, by the way) that drifted terribly because the VFO circuit was located only two inches or so from the vacuum tubes of the IF amplifier chain. The heat was tremendous. Using a bit of insulation to cover the VFO housing noticeably reduced the heat-induced drift of that rig!

I've seen more than a few radios come into a shop where I once worked that had ¼-inch sheets of styrofoam glued over all surfaces of the VFO housing (Fig. 13-27). In those days, the styrofoam job was nearly always a mess because only salvaged coffee cups and ¾-inch builder's styrofoam was easily available. These could be cut down to size with a hobby knife, razor knife, or other sharp tool. Today, however, art supplies stores sell a kind of poster board that makes the job real easy. This new

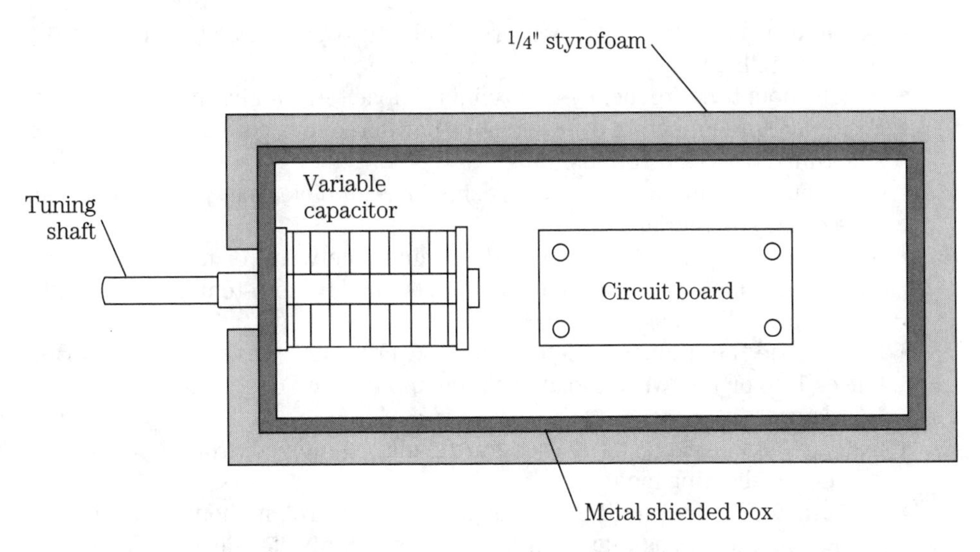

13-27 Heat-shielded VFO.

poster board is glued to a backing of styrofoam to give it substance enough for self-support. It is very easy to cut using hobby or razor knives (I used a scalpel in some experiments). Cut the pieces to size, then glue them to the metal surface of the shielded cabinet using contact cement, or some other form of cement that will cause metal and paper to adhere together.

Operate the oscillator at as low a power level as can be tolerated to prevent self-heating of the active device and associated components. It is generally agreed that a power level on the order or 10 mW is sufficient. If a higher power is needed, a buffer amplifier can be used (always a good idea, anyway).

The oscillator should be constructed to prevent variation in temperature. In some cases, this might mean using styrofoam or cork sheeting to line the shielded box that houses the oscillator (Fig. 13-27), while in others, it might mean ensuring that the location of the box within the cabinet of the equipment served by the oscillator is at a constant temperature. At one time, it was common practice to use a constant-temperature oven for crystal oscillators. The oven kept the piezoelectric resonator at a constant temperature of 75°C or 80°C. I've only seen one (now very obsolete) piece of equipment that housed an LC VFO resonating circuit in an oven environment. In some equipment, the internal temperature will build up to a certain level and then remain stable as long as there are no air currents circulating.

Some general guidelines are:

- Avoid excessive temperatures in the oscillator. Also avoid wide variation in temperature.
- Use only as much feedback as needed to ensure quick-starting of the oscillator.
- Use an output buffer amplifier to isolate the VFO circuit from changes in the external load.

- Use an IC voltage regulator that serves only the oscillator device (but not the buffer amplifier).
- Rigidly mount the frequency-determining capacitors and inductors.
- Prefer air-core inductors over ferrite or powdered iron core inductors; prefer slug-tuned coils over toroids.
- Trimmer and main tuning capacitors should be air dielectric types, rather than mica or other materials.
- Small capacitors in the frequency-determining network, or used as coupling from the frequency-determining network, should be zero-temperature coefficient types (NPO disk ceramics preferred).
- Lightly load the frequency-determining LC network by using a small capacitance (1-10 pF) between the tank circuit and the gate or base of the oscillator transistor.
- If an air-variable capacitor is used for the main tuning control, then it should be a double-bearing model.
- If a varactor diode is used for the main tuning control, it should use a tuning voltage supply that is regulated by a varactor-controller device such as the MVS-460-2/ZTK33B.[1]

These guidelines are neither exhaustive nor absolute, but following them as closely as possible will result in superior VFO stability.

Other criteria

In general, VFOs should not be operated at frequencies above about 12 MHz. For higher frequencies, it is better to use a lower frequency VFO and heterodyne it against a crystal oscillator to produce the higher frequency. For example, one common combination uses a 5- to 5.5-MHz main VFO for all HF bands in SSB transceivers. To make a 20-meter VFO, use the 5- to 5.5-MHz VFO against a 9 MHz crystal for 14 to 14.5 MHz.

Use only as much feedback in the oscillator as is needed to ensure that the oscillator starts quickly when turned on (or keyed in the case of a CW transmitter), and does not "pull" in frequency when the load impedance changes. In some cases, a small-value capacitor is used between the LC resonant circuit and the gate or base of the active device to prevent drift by lightly loading the tuned circuit. The most common means of doing this job is to use a 3 to 12 pF NPO disk ceramic capacitor. However, for best stability, use an air-dielectric trimmer capacitor (2-12 pF) and adjust it for the minimum value that ensures good starting and freedom from frequency changes under varying load conditions.

A buffer amplifier, even if it is a unity gain emitter-follower, is also highly recommended. It will permit building up the oscillator signal power (if that is needed) without loading the oscillator, and isolates the oscillator from variations in the output load conditions.

---

[1]Maplin Professional Supplies, P.O. Box 777, Rayleigh, Essex, SS6 8LU, UK.

Power supply voltage variations have a tendency to frequency-modulate the oscillator signal. Because dynamic circuit conditions often result in a momentary transient drop in the supply voltage, and because line voltage variations can cause both transient drops and peaks, it is a good idea to use a voltage-regulated dc power supply on the oscillator.

It is a very good idea to use a voltage regulator to serve the oscillator alone, even if another voltage regulator is used to regulate the voltage applied to other circuits (see Fig. 13-28). Although this double-regulation approach may have been a cost burden in "long ago" times, it is now reasonable. For most low-powered oscillators, a simple low-power "L-series" (e.g., 78L06) three-terminal integrated circuit voltage regulator is sufficient (see U2 in Fig. 13-28). The L-series devices provide up to 100 mA of current at the specified voltage, and this is enough for most oscillator circuits.

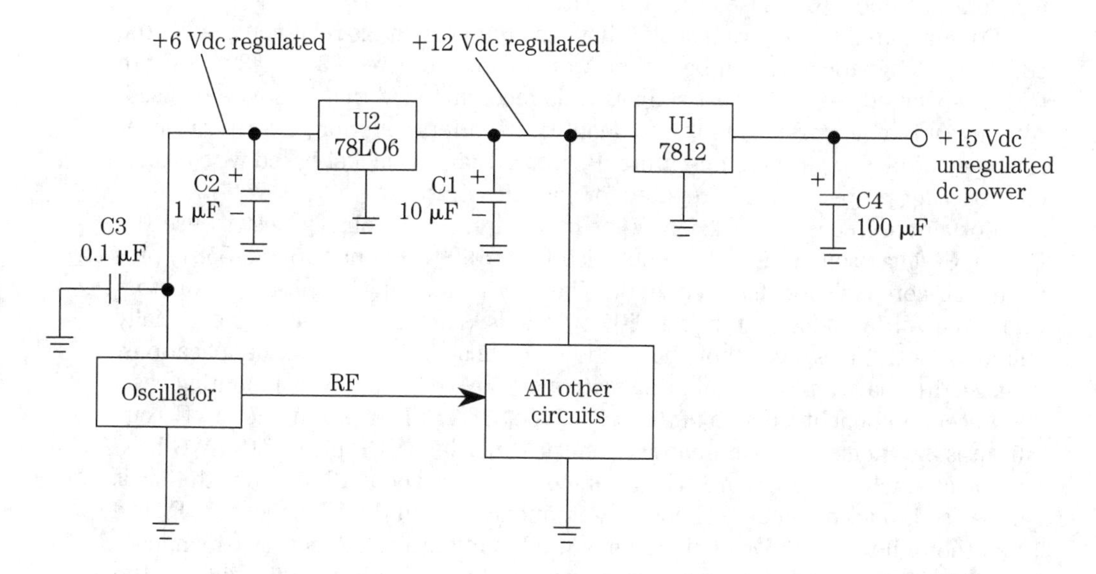

13-28 Voltage regulation scheme for oscillators.

Capacitors on both the input and output sides of the voltage regulator (C1 and C2 in Fig. 13-28) add further protection from noise and transients. The values of these capacitors are selected according to the amount of current drawn. The idea is to have a local supply of stored current to temporarily handle sudden demand changes, allowing time for a transient to pass, or the regulator to "catch-up" with the changed situation.

The frequency-setting components of the oscillator can also affect the stability performance. The inductor should be rigidly mounted so as to prevent vibration. While this requirement means different things to different styles of coil, it is nonetheless important.

Air-core coils are generally superior to those with either ferrite or powdered iron cores because the magnetic properties of those types of core are affected by temperature variation. Of those coils that do use cores, slug-tuned coils are said to be best because they can be operated with only a small amount of the tuning core actually inside of the windings of the coil, thereby reducing the vulnerability to temperature effects. Still, toroidal cores have a certain endearing charm, and can be used wherever the ambient temperature is relatively constant. The Type SF material is said to be the best in this regard, and it is easily available as Amidon Associates (P.O. Box 956, Torrance, CA, 90508) Type 6 material. For example, you could wind a T-50-6 core and expect relatively good frequency stability.

One source recommends tightly winding the coil wire onto the toroidal core, then annealing the assembly. This means placing it in boiling water for several minutes, then removing it and allowing it to cool in ambient room air while it sits on an insulated pad. I haven't personally tried it, but the source reported a remarkable freedom from inductor-caused thermal drift.

For most applications, especially where the temperature is relatively stable, the coil with a magnetic core can be wound from enameled wire (#20 to #32 AWG are usually specified)—but for best stability, it is recommended that Litz wire be used. Although it is a bit hard to get in small quantities, it offers superior performance over relatively wide changes in temperature. Be aware that this nickel-based wire is difficult to solder properly, so be prepared for some frustration.

For air-core coils, use #22 or larger bare wire. It is probably best to use the Barker & Williamson (10 Canal Street, Bristol, PA, 19007; Phone 215-788-5581) prewound air-core coils for this service. B&W makes a wide range of air-core inductors under the Airdux, Miniductor and Pi-Dux brands. The Pi-Dux coils are especially suited to use in VFOs, even though intended for transmitter pi-network applications, because they have a plastic mounting plate incorporated that makes mounting easy.

Recently I bought a length of the B&W Type 816A Pi-Dux product for a VFO circuit. It is 3$\frac{3}{16}$ inches long, and has sixteen turns per inch (16 tpi) of #18 AWG bare wire on a 1-inch diameter core. The total inductance is about 17 μH, so with taps it can be used to accommodate almost any frequency within the HF spectrum. Figure 13-29 shows how the Pi-Dux coil was mounted in my project. A pair of 1-inch insulated stand-offs provided adequate clearance for the coil, and held it rigidly to the chassis. The lucite mount shown in Fig. 13-29 is integral to the B&W Type 816A Pi-Dux coil. Other forms of mounting will also work, so long as it is rigid.

The trimmer capacitors used in the circuit should be air-dielectric types (such as the E.F. Johnson & Co. printed circuit mounted trimmers), rather than ceramic or mica dielectric trimmers.

The small fixed capacitors used in the oscillator should be either NPO disk ceramics (i.e., zero temperature coefficient), silver-mica or polystyrene types. Some people dislike the silver-mica types because they tend to be a bit quirky with respect to temperature coefficient. Even out of the same batch, they can have widely differing temperature coefficients on either side of zero.

Sometimes, you will find fixed capacitors with other than zero temperature coefficient in an oscillator frequency determining circuit. These are manufactured to make temperature compensated oscillators. The temperature coefficients of certain critical capacitors are selected to create a counterdrift that cancels out the natural drift of the circuit.

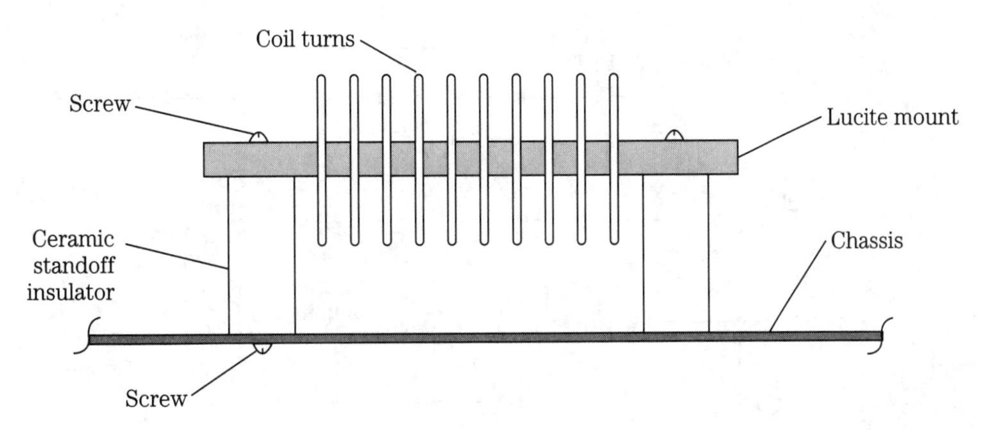

13-29 Air-core coil mounting.

The main tuning air-variable capacitor should be an old-fashioned double-bearing (i.e., bearing surface on each end-plate) type made with either brass or iron stator and rotor plates (not aluminum). The capacitor should be made ruggedly, and if possible use surplus military VFO capacitors. An excellent choice, where still available, is the tuning capacitors from the World War II AN/ARC-5 series of airborne transmitters and receivers.

Voltage-variable capacitance diodes ("varactors") are often used today as a replacement for the main tuning capacitor. If this is done, then it becomes crucial to temperature-control the environment of the oscillator. Temperature variations will result in changes in the diode pn-junction capacitance, and contributes much to thermal drift.

Another requirement for varactor-based oscillator circuits is a clean, noise free, separately regulated dc power supply for the tuning voltage. In most cases, you can use a low-powered three-terminal IC fixed-voltage regulator for this purpose, or select an LM-317 programmed for the specific voltage.

Figure 13-30 shows a sample VFO circuit with several stability-enhancing features. This circuit is a Hartley oscillator, as identified by the fact that the feedback to the JFET transistor (Q1) is supplied by a tap on the tuning inductance (L1). The position of this tap is usually between 10 percent and 35 percent of the total coil length. The exact position is sometimes a trade-off between stability and output power. Always opt for stability unless it is absolutely impossible to provide a power boost (as needed) with a buffer amplifier.

The capacitors in Fig. 13-30 are selected according to the criteria discussed above. The main tuning capacitor (C1) is a 100-pF air-dielectric variable with heavy duty double-bearing construction. The trimmer (C2) is used to set the exact frequency, especially when a dial is used and must be calibrated. Several of the fixed capacitors are marked with an asterisk as NPO disk ceramic, polystyrene, or silver-mica (in order of preference).

A buffer amplifier is provided, and it serves two basic functions: it boosts the low output power from the oscillator to a higher level, and it isolates the oscillator from changing load impedances.

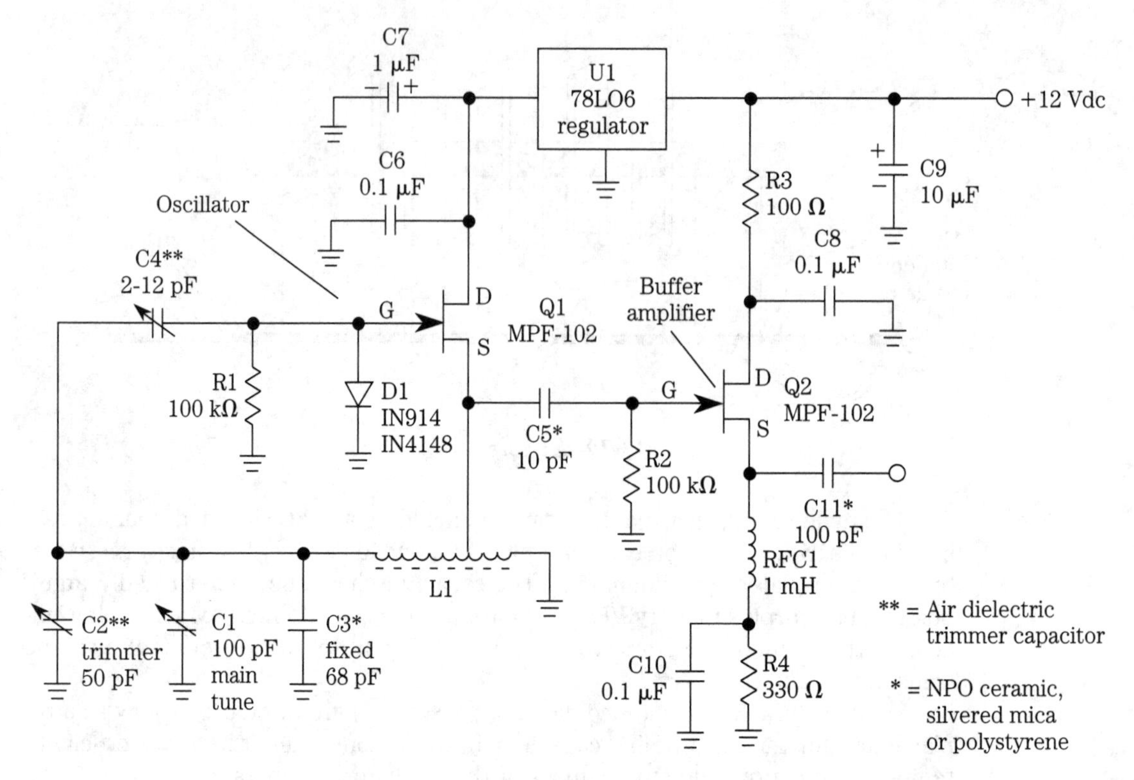

13-30 Hartley VFO circuit.

Note that the voltage regulator serves the oscillator, but not the buffer amplifier. These regulators come in both metal and plastic TO-92 packages, and generate very little heat. Nonetheless, it is still a good idea to mount the regulator (U1) away from the oscillator circuit.

The small trimmer capacitors (C2 and C4) are air-dielectric type, rather than mica or ceramic. The purpose of C4 is to provide dc-blocking to the transistor gate circuit. It is such a small value because we want to lightly load the LC tuned circuit. This trimmer is adjusted from a position of minimum capacitance (i.e., with the rotor plates completely unmeshed from the stator plates), and is then advanced to a higher capacitance as the oscillator is turned on and off. The correct position is one where the oscillator starts immediately without fail every time dc power is applied.

Oscillator circuits should be stable for best operation. If you follow these guidelines, your circuits will be highly successful. While these techniques do not exhaust the possibilities for stable oscillator construction, they are a good start. These represent a practical collection of weapons for your use.

Build a ham-band 5 to 5.5 MHz VFO

Figure 13-31 shows the basic circuit for the Colpitts oscillator. This circuit is based on the junction field-effect transistor (JFET), although with proper biasing a pnp or

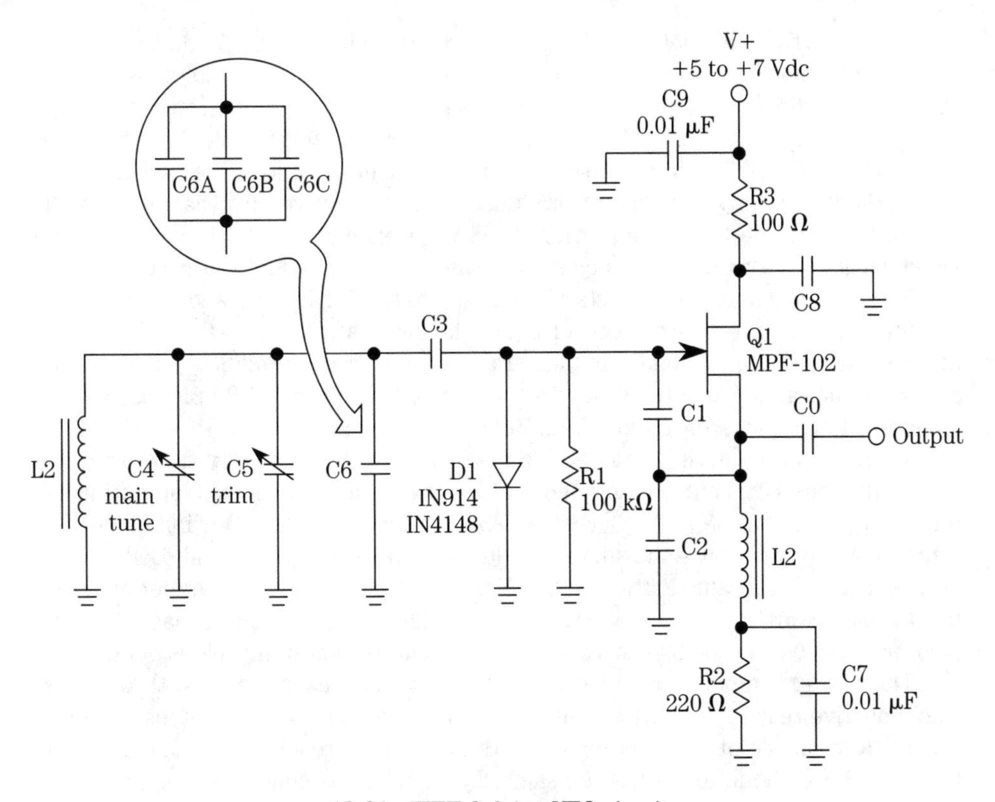

13-31 JFET Colpitts VFO circuit.

npn bipolar transistor could also be used. The device to use for Q1 is the MPF-102 or MPF-106 devices. These can be replaced by either NTE-312 or ECG-312 devices (these are widely available service shop replacement devices).

The drain terminal of Q1 is kept at RF ground potential by decoupling capacitor C8. For most HF work, the value of C8 can be 0.01 μF—although at frequencies lower than 3.5 MHz, you might want to use 0.1 μF; similarly, above 30 MHz use 0.001 μF. Resistor R2 serves to self-bias the JFET; source-drain current flowing in the JFET creates a voltage drop that places the source terminal of Q1 more positive than the gate terminal, which in effect is the same as applying a small negative bias directly to the gate. The bias resistor (R2) is bypassed by capacitor C7. The value of C7 is selected to have a capacitive reactance that is less than ⅒ the resistance of R2, or <22 ohms, at the lowest frequency of oscillation. For frequencies down to 1.8 MHz, the capacitor must be at least 0.004 μF (or larger where possible—for good measure, I usually make it 0.01 μF).

The source terminal of the JFET is kept above RF ground by the action of RF choke RFC1. The general rule for this part is to have a minimum inductive reactance of 4700 ohms at the lowest frequency of oscillation. For our ham bands (1.8 MHz and up), a minimum value of 415 μH will suffice—but again, most builders prefer to use a 1 mH or 2.5 mH device rather than the bare minimum.

There are three elements to the gate circuit: dc components, feedback network components, and frequency-setting components. The dc components consist of resistor R1 and diode D1. The resistor prevents gate circuit currents from building up on the coupling capacitor (C3), which could eventually form a blocking bias that shuts off the JFET. The diode is used to regulate the amplitude of the oscillator output. It rectifies strong signals and produces a small negative bias that reduces the gain of the JFET. In this manner, the amplitude remains relatively constant over a wider range of operating frequencies (some amplitude variation is normal).

The feedback network consists of capacitor voltage divider C1 and C2. These capacitors should have a reactance of about 45 ohms at the lowest oscillation frequency. The value of C1 and C2 in picofarads is found from $10^6/283f_{MHz}$. Thus, for an 8 MHz oscillator, use about 440 pF; for 5 to 5.5 MHz VFOs, use 700 pF (use 680 pF standard values); and for 3.5 to 4 MHz VFOs, use 1000 pF.

There are two theories for selecting the value of C3. One theory says to make it a value that has a capacitive reactance of about 260 ohms at the lowest operating frequency. For a 5 to 5.5-MHz VFO, then a 122-pF value (use 120 pF) is indicated. The other philosophy is to use the smallest value of capacitor that will still oscillate reliably, which means a value in the 1 to 20 pF range. This value is found experimentally, but I usually start at 5 pF and work upwards. In general, I've found that a 5 pF capacitor works well in the 3.5- to 10-MHz range, where most of my VFOs operate.

The tuning components consist of inductor L1 and capacitors C4 through C6. The inductive reactance of L1 should be on the order of 140 to 250 ohms (depending on convenience of coil winding), and the capacitive reactance of the combined C4 through C6 should be about the same. For a 5 MHz oscillator, the approximate value for L1 should be:

$$L_1 = \frac{X_{L\,(ohms)}}{2\,\pi f_{MHz}} = 4.46\ \mu H$$

The capacitor values must be selected very carefully. First of all, the total capacitance of $C_4 + C_5 + C_6$ should resonate with $L1$ at the desired operating frequencies. Capacitor C4 is the main tuning capacitor, and is typically a shaft-driven air-variable. It has minimum (C_{4min}) and maximum (C_{4max}) values. Capacitor C5 is a small-value trimmer, and is typically screw driver (or tuning wand) driven. It also has minimum and maximum values. A typical arrangement is to make the value of *C*5 at the midway point in its range (no more than 10-20 percent of the maximum value of *C*4, although more is certainly permissible). Capacitor C6 is optional, depending on the design. It is used to pad the total capacitance needed for resonance. It can also be used for setting such things as temperature compensation—but more of that later. In addition, C6 can be multiple capacitors (C6A, C6B, C6C, etc.) if an odd value is needed.

Also, keep in mind that a few picofarads must be allocated to stray capacitance. The exact amount is highly dependent on the method of construction used, but about 7 pF seems like a good starting point for most ham HF calculations.

The total value of capacitance needed is a function of what resonates with L1 at the minimum and maximum operating frequencies. Keep in mind that the ratio of maximum-to-minimum frequencies is the square root of the ratio of maximum-to-minimum capacitance.

For example, in the AM band an 18 to 365 pF capacitor is used, and there are strays plus trimmer (C5) capacitors of about 25 pF. This capacitance ratio is (365 + 25)/(18 + 25), or 9.07:1; the square root of 9.07 is 3.012—which is just about right for the AM BCB ratio of 1620 kHz/540 kHz. In the relationship above, you must "tap dance" a little in order to get the band you want, with little overlap. That is, you must adjust the values of the capacitance, and possibly the inductance, in order to get the correct range. This is done on paper by trial and error. For a standard ham SSB rig VFO, the operating frequency is 5 to 5.5 MHz, so the frequency ratio is 5.5/5, or 1.1. Thus, the capacitance ratio in the tuning circuit is 1.21:1, or 1.22:1—if you want a little overlap.

Let's do a little math. By the rule above, our 5-5.5 MHz VFO should have an inductance value of about 4.46 µH. By $C_{pF} = 10^6/39.5f^2L_{\mu H}$, the value of $C_4 + C_5 + C_6$ at 5.5 MHz is 187 pF. Thus, because a 1.22:1 ratio is desired, a maximum capacitance (when C4 is fully closed) should be (187 pF) (1.22) = 228 pF, or a change of 41 pF. A standard 50 pF variable capacitance, with a minimum value of 7 to 9 pF, should work well for the main tuning capacitor. With about 7 pF of stray capacitance, we still need to account for 171 pF. This can be made up with an 80 pF trimmer (40 pF midscale), and any combination of fixed capacitors that total 130 pF. Thus, a 30 pF and a 100 pF in parallel will do the job. Note that some sloppiness will result because capacitors are available in standard values . . . that's why the trimmer is wider range than necessary.

The output capacitor C_o is selected for a reactance of about 750 ohms at the lowest operating frequency. For a 5 to 5.5 MHz VFO, a 42 pF value is needed (use 39 pF or 47 pF standard values).

In the final "cut-n-try" on the bench, adjust the exact values of the trimmer capacitor (C5) and the inductance of L1 to find the exact match needed. For our first trial, though, use the following component values:

C_1, C_2	680 pF
C_3	5 pF or 120 pF (depending on philosophy)
C_4	50 pF variable
C_5	60 or 80 pF trimmer (set to about 40 pF)
C_{6A}	100 pF
C_{6B}	27, 30 or 33 pF as available
C_o	39 or 47 pF
L_1	4.46 µH

All the capacitors that can affect the operating frequency (C1, C2, C3, C4, C5, C6, and C_o) should be NPO disk ceramic, silver-mica, or polyethylene, for the best temperature drift performance. Be wary of silver-mica units, however, because they can individually be quite poor in the tempco department. I'd stick with NPO ceramic or polyethylene units . . . unless you can hand-select the silver-mica units.

In general, all capacitors that could have an effect on the operating frequency should be either NPO disk ceramics, silver-mica, or polyethylene dielectric types for minimum drift. The silver-mica capacitors are somewhat suspect because there is a wide variation in temperature coefficient (which is measured in parts-per-million-per-degree-centigrade—PPM/°C) from low to high in these capacitors. Some individuals are very good, others stink. That's good advice, normally, but where a temperature drift occurs, it could be less than optimum.

The principal drift element (besides the capacitors) is the tuning inductor. Most oscillators drift lower in frequency from turn-on to temperature stability (about an hour later). It is common practice to use a capacitor with a negative temperature coefficient (e.g., N750 or EIA P3K) in parallel with the main tuning capacitor. In the circuit of Fig. 13-31, for example, C3 or C4 can be the temperature coefficient capacitor. The procedure for finding the value of the temperature compensation capacitor is relatively straightforward. First, arrange a test set-up where the oscillator is not affected by stray air currents. I place the VFO inside a styrofoam "picnic cooler" (the sort intended for a six-pack is usually suitable). A digital frequency counter is allowed to warm up to stability (about 1.5 hours) before the test begins. The procedure is:

1. Turn the oscillator on, adjust it to the high end, and measure the oscillation frequency ten seconds later. Call this frequency F_o.
2. Wait 1.5 hours and then measure the frequency again (keep the oscillator running and the frequency counter turned on all this time); call this frequency F_1.
3. Now, turn the oscillator off (leave counter on), and let the oscillator cool back to room temperature (about one hour).
4. Replace capacitor C4 with a capacitor of the same value, but with a temperature coefficient of N750 (negative temperature coefficient of 750 PPM); call this capacitance $C4_{trial}$. Turn on the oscillator, readjust it to the original F_o. Wait 1.5 hours and measure the frequency; call this frequency F_2.
5. Replace $C4_{trial}$ with an N750 capacitor with a capacitance of:

$$C_{4new} = C_{4trial}\left(\frac{f_1}{f_1 - f_2}\right)$$

If $f_2 < f_1$, or:

$$C_{4new} = C_{4trial}\left(\frac{f_1}{f_1 + f_2}\right)$$

If $f_2 > f_1$. Install the new capacitor, and readjust the trimmer capacitor or the inductance of the coil.

Another means for temperature compensating a VFO is shown in Fig. 13-32. This circuit is similar to one that was used in the old Hallicrafters HT-32 AM/CW/SSB high-frequency transmitter a number of years ago. In this circuit, a portion of the total frequency-setting capacitance is made up from two small disk ceramic capacitors: C4 is an NPO (zero temperature coefficient) and C5 is an N1500 (negative 1500 PPM) unit. Capacitor C6 is a differential-variable air-dielectric capacitor. These devices have two variable capacitor sections on the same shaft, arranged so that one is increasing in capacitance (with shaft rotation) and the other decreases a like manner. The sum of C6A and C6B remains constant with shaft rotation. By adjusting C6 one can crank in a variable temperature coefficient from NPO to N1500. The total capacitance of C4/C5/C6A and C6B remains constant as C6 is varied (if $C_4 = C_5$).

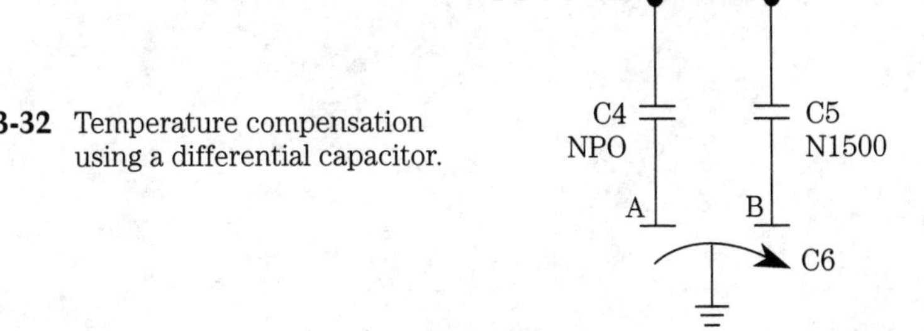

13-32 Temperature compensation using a differential capacitor.

The adjustment procedure is to set C6 to midrange, so that equal amounts of C6A and C6B are in the circuit.

1. Turn the counter on and let it warm up for 1.5 hours. Turn the oscillator on, adjust it to the high frequency end, and then note the frequency (f_o) after ten seconds.
2. Wait 1.5 hours and measure the frequency again (f_1). If f_1 is lower than f_o, then turn C6 in a direction to crank in a little more N1500 characteristic.
3. Turn the oscillator off, and let cool for one hour.
4. Turn oscillator on, and readjust C1 to f_o. Wait 1.5 hours, and measure the operating frequency (f_2). If less than 500 Hz difference, then do nothing (unless you want to attempt the best stability, on the order of 100 Hz).
5. If f_2 is considerably different from f_1, then adjust C6 to crank in more or less of the N1500 characteristic, as needed to cancel the drift.

If this procedure is followed, a reasonably good-performing VFO or XO will be obtained.

14
Radio frequency filters

Radio frequency filters are inductor-capacitor networks that pass one band of frequencies, while rejecting all other frequencies. RF filters are used in a wide variety of applications: keeping harmonics within the transmitter, preventing out-of-band signals from getting into the receiver, cleaning up the output of frequency mixers, etc., etc., etc.

There are four basic types of RF filters: low-pass filter (LPF), high-pass filter (HPF), band-pass filter (BPF) and rejection filter (called notch filters when the rejection band is narrow, and bandstop filters when the rejection band is wider).

The LPF and HPF frequency responses are shown in Figs. 14-1A and 14-1B, respectively. In the LPF (Fig. 14-1A), all signals from dc to some cut-off frequency (F_c) are passed—but above F_c, the response falls off to the point where there is little signal passing. The cut-off frequency is usually defined as the point where the frequency response falls off –3 dB from the in-band response. The HPF characteristic is shown in Fig. 14-1B, and is exactly the opposite of the LPF: it rejects all frequencies below its cut-off frequency, while passing all frequencies above F_c. Note that these curves are a bit idealized; real RF filters are not so smooth in either the passband or outside of it.

If you start "raw" and design your own filter, the task is daunting indeed. But if you use tables of values for "normalized" generic filters, the job becomes a lot easier. I recently tried my hand at building a number of RF filters for different purposes (only some are related to ham radio)—all of which illustrate the design principles involved. You can use these same methods to design filters for your own use.

The tables give the values for the normalized case where F_c = 1 MHz; the inductances are given in microhenrys (μH) and the capacitances in picofarads (pF).

Example: the low-pass case

In this example let's look at my 3000-kHz LPF (Fig. 14-2). It was used in a sweep-signal generator I designed for the AM broadcast band, and for common AM IF

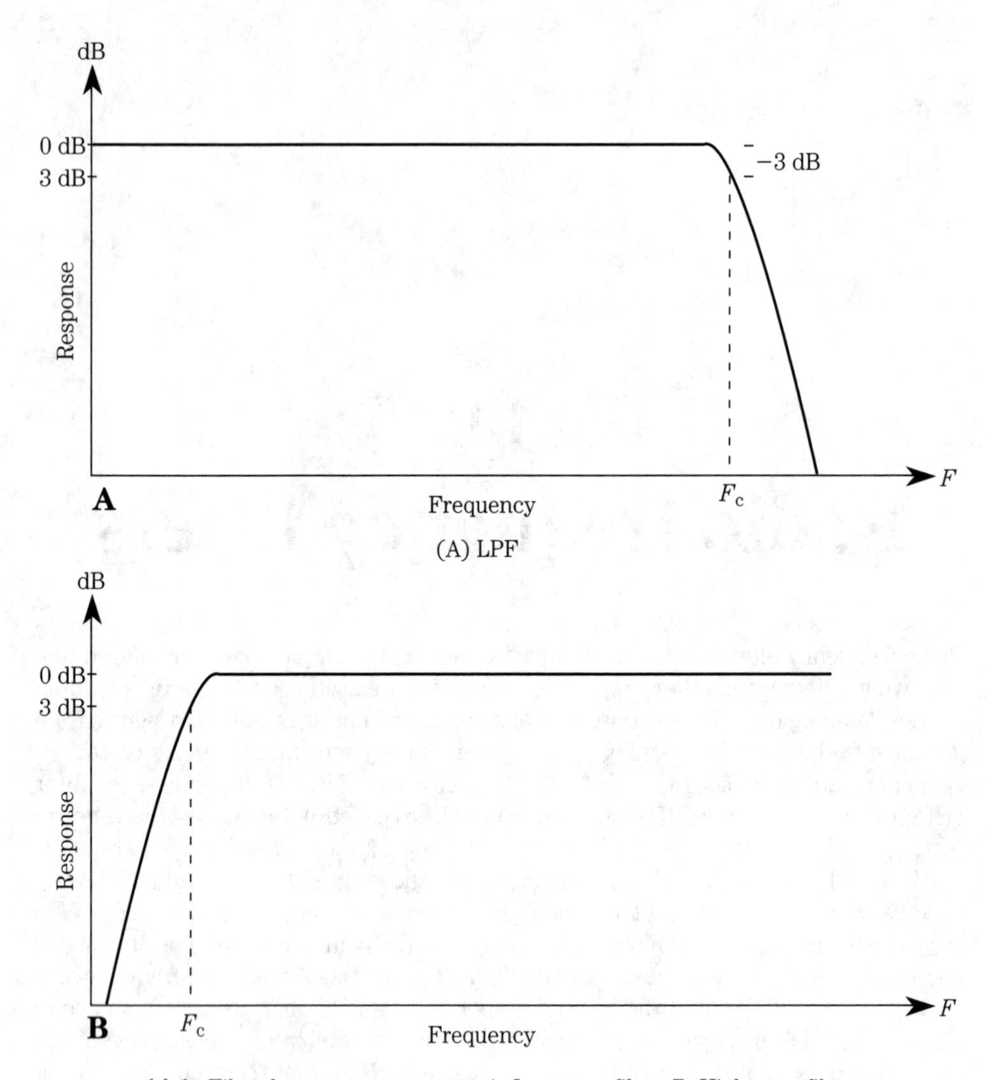

14-1 Filter frequency responses. A. Low-pass filter. B. High-pass filter.

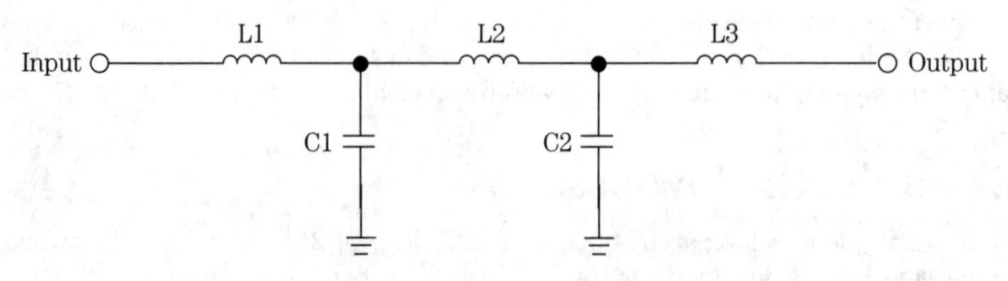

14-2 Low-pass filter circuit.

frequencies (e.g., 455 kHz). I needed it to facilitate a project that I am working on: a super AM DXers receiver (sorry, no details as yet). The normalized 1-MHz data values are shown in Table 14-1.

Table 14-1. Normalized 1 MHz capacitor/coil values.

Type of filter	C1	C2	C3	L1	L2	L3
LPF	4365	4365	---	9.13	15.7	9.13
HPF	2776	1662	2776	5.8	5.8	---

Assumes 0.1 dB ripple, 5-element filter

The *ripple factor* (0.1 dB) refers to the amount of variation in response (ripple) in the passband of the filter, and is expressed in decibels (dB). I selected the 0.1 dB figures because it makes a reasonable filter without making it difficult to find component values.

The table data are normalized to 1 MHz, so to find the values of inductance and capacitance needed for the actual filter, divide the values in the data table by the frequency in megahertz (MHz).

To find the values for my 3000 kHz (i.e., 3 MHz), 0.1 dB ripple LPF, then I divided the values by three: $C_1 = C_2 = 4364.7/3 = 1454.9$ pF; $L_1 = L_3 = 9.126/3 = 3.04$ µH; $L_2 = 5.24$ µH.

C_1: 1454.9 pF
C_2: 1454.9 pF
L_1: 3.04 µH
L_2: 5.24 µH
L_3: 3.04 µH

The coils are relatively easy to come by: wind them on Amidon Associates (2216 East Gladwick Street, Dominguez Hills, CA, 90220; Phones: (voice) 213-763-5770, (fax) 213-763-2250) toroid coil forms. The T-50-2 (RED) cores have an A_L value of 49, and operate from 2 to 30 MHz, while the T-50-15 (RED/WHT) have an A_L of 135 and operate over 0.1 to 2 MHz. In practice, I found that 3 MHz was not unreasonable for the –15 cores, so I opted to use them. Applying the formula below gave the number of turns:

$$N = \sqrt{\frac{L_{\mu H}}{A_L}} \qquad \textbf{[14-1]}$$

L_1, L_3: 3.04 µH, 15 turns, T-50-15 (RED/WHT)
L_2: 5.24 µH, 20 turns, T-50-15 (RED/WHT)

The capacitors are another matter: where in blazes do you get a 1454.9 pF capacitor? Well, one solution is to use a 0.0015 µF (1500 pF) and live with the slight frequency error. I did this and found that the filter had a cut-off frequency only slightly lower than 3000 kHz, and was acceptable. Otherwise, it is possible to select standard value capacitors that in some series or parallel combination total 1454.9 pF, or something close to it. For example, 75 pF, 560 pF, and 820 pF add to 1455 pF, and all of those values are easily available.

The capacitors used in the filter should be an NPO disk ceramic, silver-mica, or polyethylene. I bought several dozen of all types recently from Ocean State Electronics [P.O. Box 1458, 6 Industrial Drive, Westerly, RI, 02891; Phones: 1-800-866-6626 (orders), 401-596-3080 (inside RI), or 401-596-3590 (fax)]. Ask them for a copy of their catalog . . . you'll find a lot of ham building parts that you thought were "history" because other parts distributors no longer carry them.

Another approach is to use a combination of fixed-value and trimmer capacitors in the filter. This is a viable approach if you have a sweep generator and oscilloscope to align the filter, but can be a "bear" if you don't. A procedure for alignment of such filters is given in Hayward and DeMaw: *Solid-State Design for the Radio Amateur* (an ARRL publication).

Example: 3000 kHz low-pass filter

The 3000 kHz low-pass filter has an attenuation response such as shown in the *MathCAD 3.1* plot of Fig. 14-3. The attenuation (loss between input and output) is plotted along the vertical axis, while the frequency is plotted along the horizontal axis. The frequency marks are in terms of harmonics of the cut-off frequency (F, 2F, 3F, etc.). Notice that for frequencies quite far removed from the cut-off frequency, the attenuation is quite stiff. Little or no signal on those frequencies survives the filter . . . which is as it should be.

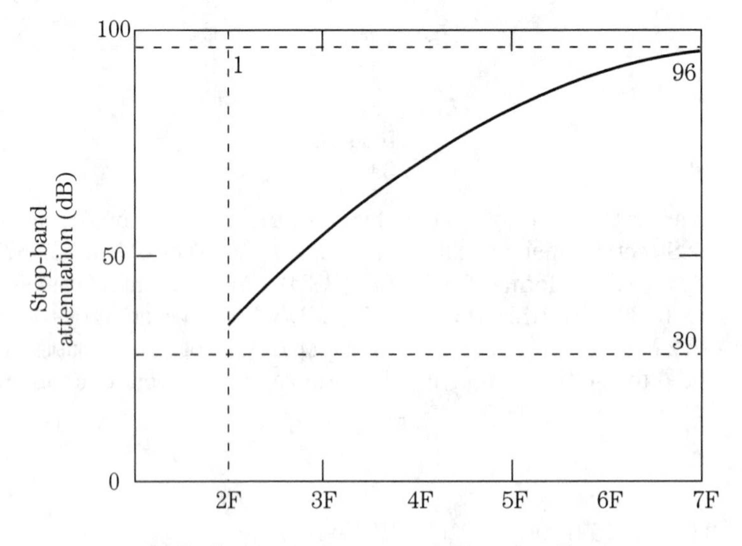

14-3 Frequency response of a low-pass filter.

High-pass filter

An example of a high-pass LC filter is shown in Fig. 14-4. Again, a total of five inductor and capacitor elements are used, but the order or role of each is reversed compared to the low-pass case. The normalized 1 MHz factors used in the calculations of components are taken from the HPF row in Table 14-1.

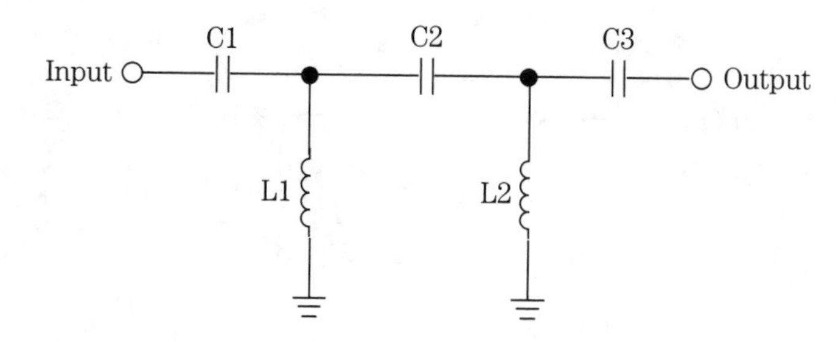

14-4 High-pass filter circuit.

The particular project was a high-pass filter with a lower end cut-off frequency of 2.5 MHz.This filter was to be used ahead of the front end of a direct conversion radio receiver project in order to keep out strong AM broadcast band signals. Those signals often come from very local (a few blocks away) AM transmitters that operate at power levels ranging from 500 watts to 50,000 watts, and can wreak havoc with HF receivers that are not very selective on the front end. Strong signals can break through the weak selectivity of the typical direct-conversion receiver, and either desensitize the front end or cause spurious signals to be generated (and received). Placing a shielded high-pass filter, which has a cut-off frequency between the upper end of the AM BCB (1700 kHz) and the lower end of the 75/80 meter band, will often cure these problems with no further "treatment" needed.

The simple circuit of Fig. 14-4 was selected. The normalized 1-MHz capacitor and inductor values from the table are divided by 2.5 (for 2500 kHz expressed in MHz) to obtain the values for our filter: $C_1 = C_3 = 2776/2.5 = 1110$ pF; $C_2 = 1662/2.5 = 665$ pF; and $L_1 = L_2 = 5.8/2.5 = 2.32$ μH. Capacitors C1 and C3 can best be made, I suspect, by paralleling a 1000-pF (0.001 μF) and 110-pF units (both are easily available in silver-mica form). Alternatively, a pair of 560-pF capacitors in parallel, or three 390-pF capacitors in parallel, can be substituted with only a small error. Capacitor C2 required 665 pF. This value can be achieved by selecting 680-pF units for the right value using a capacitance meter. Alternatively, a pair of 330-pF in parallel, or a 560-pF and 100-pF in parallel, with only a small error (indeed, the tolerances of the capacitors is larger). The coils can be made from T-50-2 toroid forms, wound with 22 turns of #26 enameled wire.

A collection of AM BCB RF filters and wavetraps

The problem of AM BCB interference to ham operators, shortwave listeners and the builders of many different RF electronic circuits is so difficult that I thought it might be nice to include a short section of circuits I've found successful over the past several decades.

The circuit in Fig. 14-5 is a simple five-element circuit such as the previous example, with a cut-off frequency of approximately 2 MHz. In this case, however, a little frequency response performance is sacrificed in favor of common "standard" value components. All of the capacitors are 1000 pF (or 0.001 μF) units. In order to

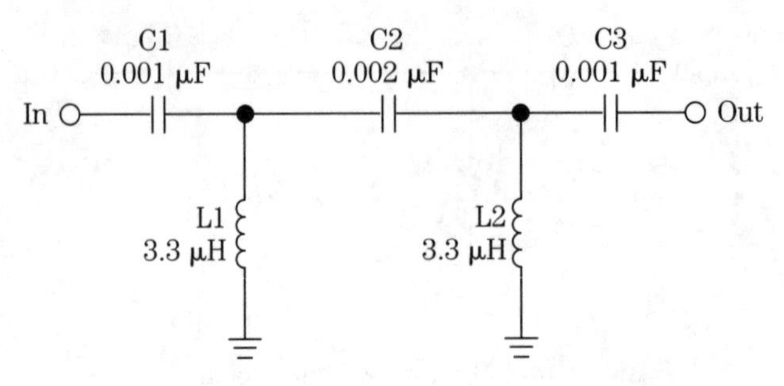

14-5 AM BCB HPF circuit (Type 1).

obtain 0.002 µF for C2, a pair of 0.001-µF units are connected in parallel. The inductors for the filter are 3.3 µH, and can be made by winding 26 turns of #26 enameled wire over a T-50-2 toroidal core. But the 3.3-µH value is also one of the standard values for both fixed and variable inductors made by *Toko*, and sold in the Digi-Key catalog. Thus, anyone can build this filter with rather ordinary components, and no large amount of effort. Of course, the filter works best when built inside of a small, shielded container.

A somewhat more complex high-pass filter is shown in Fig. 14-6. This filter is designed for a cut-off frequency of 1750 kHz, and has a steeper roll-off below the cut-off than the previous filters. Capacitors C1 and C2 are 0.001 µF units, while capacitors C3 and C4 are either three 0.001-µF units in parallel, or a single 0.003-µF unit. Both values are available in polyethylene capacitors. The inductors can be purchased as fixed or variable units, or made from T-50-2 toroidal cores. If the toroidal forms are used, L1 and L3 require 45 turns of #30 enameled wire, while L2 requires 22 turns.

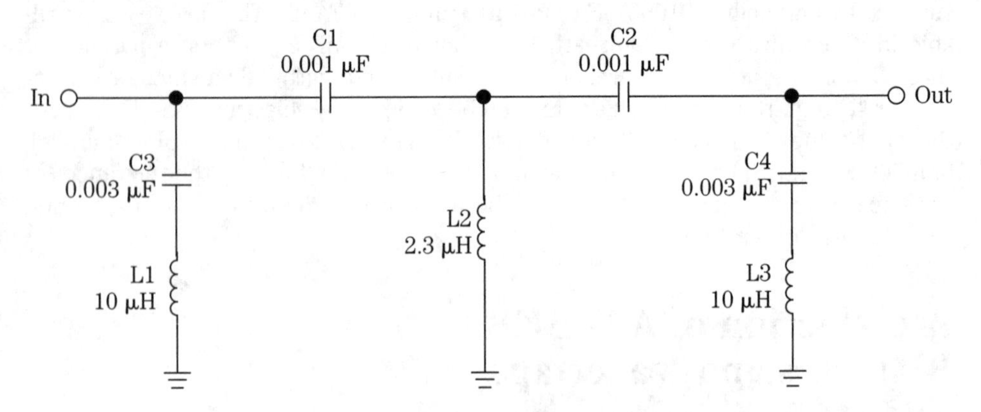

14-6 AM BCB HPF circuit (Type 2).

A still better filter is the 1800 kHz high-pass model shown in Fig. 14-7. In this filter circuit, there are ten LC elements used. As in the previous cases, standard value capacitors can be used in combination to form the unusual (nonstandard) capacitances needed for the filter. The inductors can be built on T-50-2 forms using the following winding data:

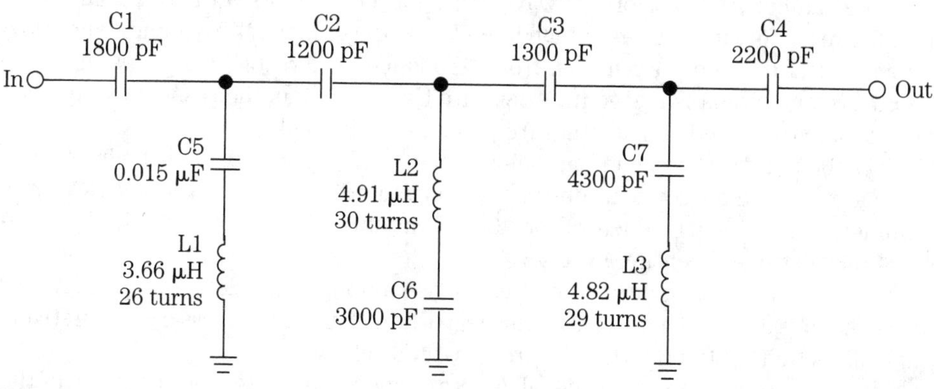

14-7 AM BCB HPF circuit (Type 3).

For really serious cases of AM BCB interference, you might require a special *wavetrap* circuit, such as Fig. 14-8. Cases where the wavetrap is needed are those where a single, high-powered AM station is located so close that it booms right through broadband filters such as the high-pass circuits shown earlier. To suppress these signals, you need a circuit that is tuned to the specific frequency of the offending AM broadcast station. Figure 14-8 shows such a circuit.

There are two basic forms of wavetrap: series and parallel. The series wavetrap circuit is a series-tuned LC resonant circuit that is placed across the signal line and ground. In Fig. 14-8, L1C1 is such a circuit. The series resonant LC tank circuit has a low impedance at the resonant frequency, and a high impedance at frequencies re-

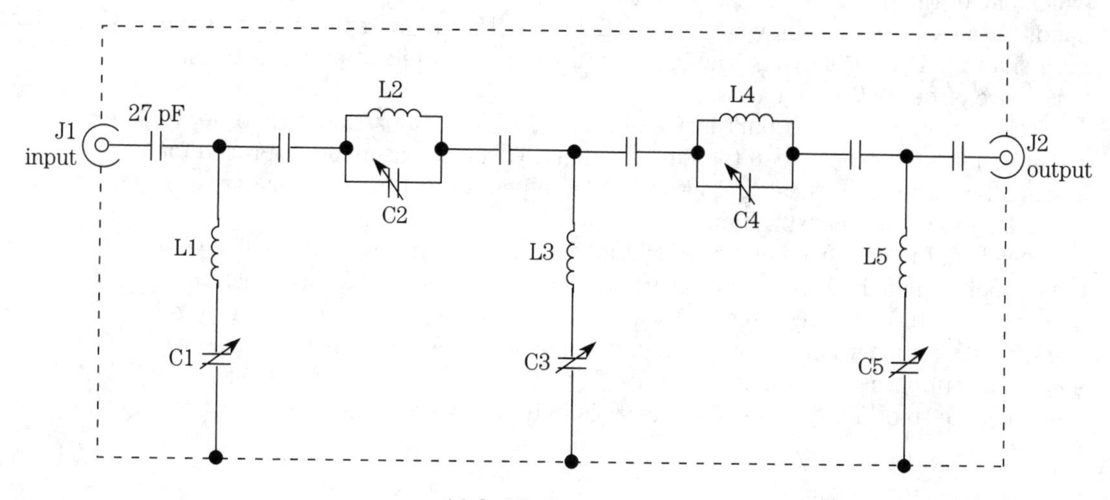

14-8 Wavetrap filter.

moved from resonance. Thus, when placed across the signal line in the manner of L1C1, the effect is to shunt a short circuit across the line at the resonant frequency, and leave it open at other frequencies.

The parallel-resonant form of wavetrap is placed in series with the signal line. A parallel-resonant circuit offers a high impedance at its resonant frequency, and a low impedance at frequencies removed from resonance. As a result, when a parallel-resonant LC tank circuit is placed in series with the signal line, the resonant frequency signals are attenuated and all other frequencies are passed.

Combining the series and parallel-resonant wavetraps into a single circuit provides a very large degree of attenuation of unwanted signals. For the AM BCB, it is common to use 220-μH inductors and 400-pF variable trimmer capacitors. This combination will tune the entire AM BCB.

A certain amount of interaction takes place amongst the elements when two or more sections are used in the same wavetrap filter. Thus, it is necessary to iteratively tune all sections until no further improvement is achieved.

Tuning can be done with a signal generator, which is set to the frequency of the offending station. Tune the trimmer capacitors, each one in its turn, until a drop in signal is seen at the output of the filter. Tune each section for a signal drop, then go back and do it again several times until no further improvement is seen. The filter can then be connected into the circuit with a receiver, and the results noted. There should be minimal attenuation of shortwave signals, but a large attenuation of the AM BCB signals.

2- to 30-MHz bandpass filter

Not all interference that afflicts shortwave receivers is from the AM BCB, although such stations are frequently the source of problems. In some cases, stations operating above the shortwave band also afflict shortwave receivers, especially nearby FM broadcast band (88-108 MHz) stations. In addition, there are many mixer circuits in which one of the frequencies that you wish to eliminate is above the HF shortwave bands. As a result, you might want to use a 2- to 30-MHz bandpass filter, such as the filter shown in Fig. 14-9, to pass the HF shortwave signals, while attenuating the signals above or below the HF bands.

The filter of Fig. 14-9 is built inside of three shielded compartments in order to minimize interactions between the filter sections. This is a reasonable approach for any multi-section filter. "Leakage" between elements is one of the principal deteriorating factors in filter construction.

Small multi-compartment boxes suitable for filters can be easily built using brass stock from hobby and model crafters stores. This stock comes in 1-inch to 3-inch wide by 10-inch long sizes and various sheet metal gauges. It can easily be worked with ordinary hand tools, although a pair of parallel jaws pliers from a jewelry hobby supplies store is a definite asset. Another asset is a low-cost sheet metal "bench brake" tool. These are sometimes sold in tool catalogs under the name "box formers."

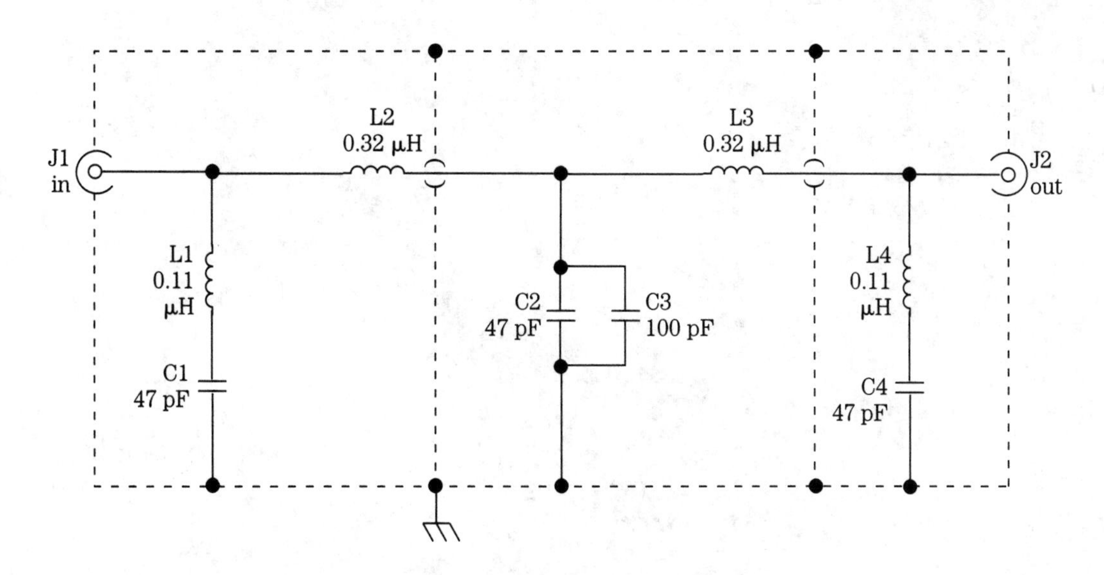

L1, L4: 5t., #26 enameled on T-50-2 (RED) toroid form

L2, L3: 8t., #26 enameled on T-50-2 (RED) toroid form

14-9 2- to 30-MHz bandpass filter with shielding.

15 Dealing with interference sources

AM-band, shortwave and VHF monitor receivers often produce output signals that are a terrible mess of weird squawks, squeals and other assorted objectionable noises. Shortwave receivers in particular show a lot of weird signals, but today a growing number of electronic products used throughout the community is making the cacophony louder and messier than ever before. Fortunately, many of the interference problems that you'll encounter on your receiver can be fixed, or at least reduced significantly in severity. Interference problems fit into two major categories: unwanted local radio station signals and unwanted radiation from other electronic devices.

Radio station interference

Radio signals from AM, FM, and TV stations are quite powerful, and can produce a number of problems for receiver owners in their immediate vicinity. Unfortunately, in many areas of the country, residences are located quite close to these types of stations (in a few areas, even shortwave stations are located near homes).

Typical problems from local transmitters include blanketing, desensitization, harmonic generation, and intermodulation. Let's look at each of these in turn.

Blanketing The blanketing problem occurs when a very strong local signal completely washes out radio reception all across the band. It generally affects the band that the signal is found in, but not other bands. The offending signal can be heard all across the band, or at numerous discrete points.

Desensitization The desensitization problem is a severely reduced receiver sensitivity due to the presence of a strong local signal. Desensitization can occur across a wide frequency spectrum—not only in the band where the offending signal is located. The strong offending signal might not be heard unless the receiver is tuned to or near its frequency.

Harmonic generation If the signal is strong enough to drive the RF amplifier of the receiver into nonlinearity, it might generate harmonics (integer multiples) of the strong signal. For example, if you live close to a 780 kHz AM broadcast band signal, and the receiver overloads from that signal, then you might be able to pick up the signal at $2f$ (1560 kHz), $3f$ (2340 kHz), $4f$ (3120 kHz) and so on throughout the shortwave bands. This problem is not the fault of the radio station (which must be essentially clean of harmonics because of FCC regulations), but rather is an inappropriate response on the part of your receiver.

Intermodulation This problem occurs when two signals of different frequencies (say, F_1 and F_2) mix together in a nonlinear element to produce a third frequency (F_3). The third signal is sometimes called a *phantom* signal. The frequencies involved might be (and probably are) the assigned fundamental frequency of a legitimate station, or its harmonics. The nonlinearity can be due to receiver overload, improper receiver design, or some other source (legend has it that rusted downspouts and corroded antenna connections can serve as a nonlinear pn junction). As they say in the new science of Chaos Theory, "... nonlinearity can arise throughout nature in subtle ways."

The possible new frequency (F_3) will be found from $F_3 = mF_1 \pm nF_2$, where m and n are integers, although not all possibilities are likely to occur in any given situation. Suppose one local ham is operating on 10,120 kHz (in the new 30-meter band), and another is operating on 21,390 kHz (in the 15-meter band). Both stations are operating normally, but at least one is close enough to overload your receiver and produce nonlinearity in the RF amplifier. When the second harmonic of 10,120 kHz (i.e., 20,240 kHz) combines with 21,390 kHz, the result is a third frequency at 21,390 kHz – 20,240 kHz = 1150 kHz ... right in the middle of the AM broadcast band.

With all of the signals that might exist in your locality, an extremely large number of possible "intermod" combinations can arise. In my hometown, there is a hill right in the middle of a densely populated residential neighborhood, on which there are two 50 kW FM broadcast stations, a 5 kW AM broadcast station, scores of VHF and UHF land mobile radio base station or repeater transmitters, and an assortment of paging systems, ham operators, a medical telemetry system and the microwave towers of an AT &T long-lines relay station—all within a city block or two. Only a few radio receivers work well unassisted in that neighborhood!

There are two approaches to overcoming these problems. First, either reject or somehow selectively attenuate the offending signal (or one of the signals in the case of an intermod problem). This is done using a wavetrap. Second, add a passive preselector to the front end of the radio receiver between the antenna line and the antenna input of the receiver.

Figure 15-1 shows three wavetrap circuits that can be used from the AM band up through VHF/UHF frequencies. The circuit in Fig. 15-1A is an *LC wavetrap* based on inductors and capacitors. There are two forms of LC resonant circuit shown in Fig. 15-1A. In series with the signal path is a *parallel-resonant* circuit (L2/C2). These circuits have a very high impedance at the resonant frequency, so will attenuate signals of that frequency trying to pass through the line between J1 and J2. At the same time, the impedance at all other frequencies is high, so those frequencies will pass easily from J1 to J2. Shunted across the line on both ends of the

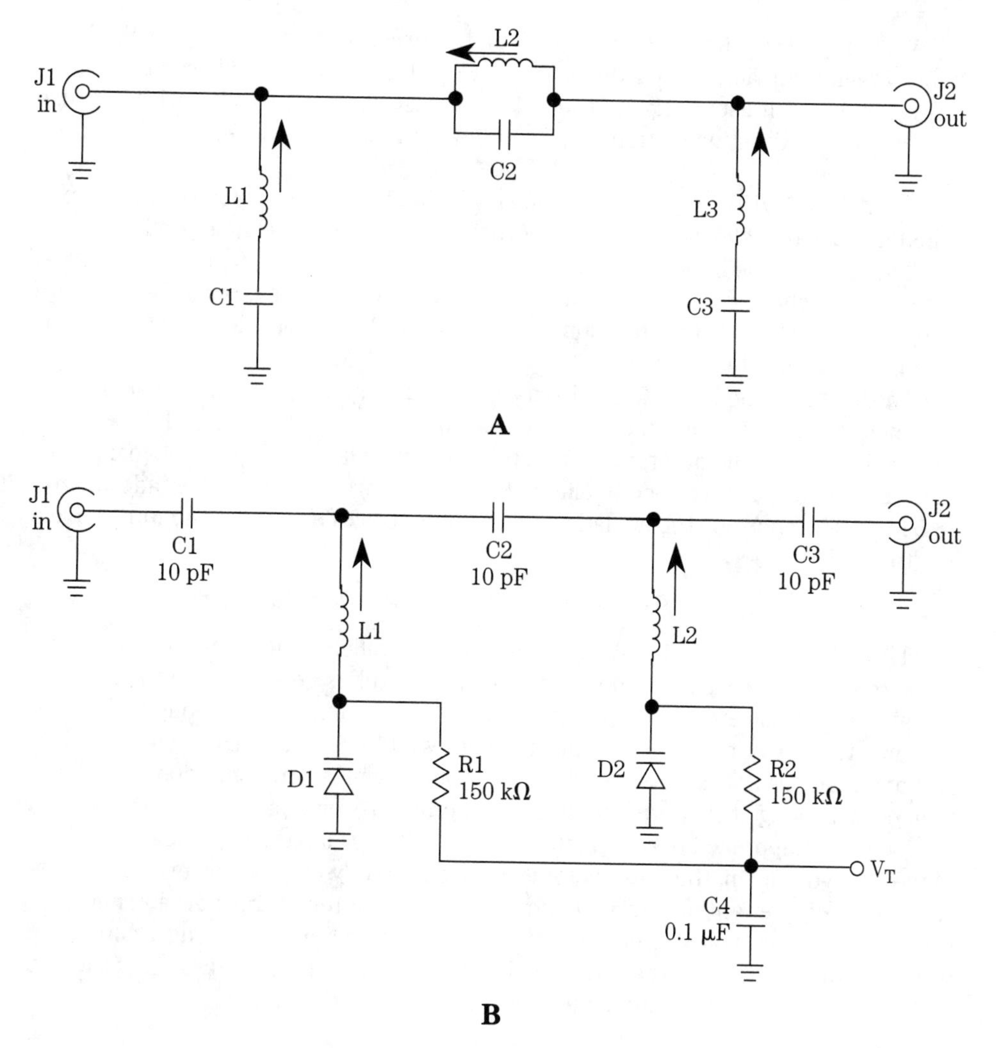

15-1 A. Series-parallel wavetrap. B. Voltage-tunable filter/trap.

signal path are a pair of series-resonant circuits. These LC circuits have a low impedance at the resonant frequency, and a high impedance at all other frequencies, so will shunt only the offending signals to ground.

The wavetrap of Fig. 15-1A can be built using either fixed inductors and variable capacitors, or vice versa. It can be built either for one, fixed frequency, or as a variable wavetrap that can be tuned from a front panel knob to attenuate the offending signal.

Figure 15-1B shows the circuit of a similar type of wavetrap, but is built with *voltage-variable capacitance diodes*, also called *varactors*. These diodes produce a capacitance across their terminals that is proportional to the applied reverse bias tuning voltage (V_T). In the circuit of Fig. 15-1B only series resonant circuits are used.

Wavetraps built from LC circuit elements are useful well into the VHF frequency region. In fact, many video and other electronics stores sell wavetraps such as Fig. 15-1A built especially for the FM broadcast band (88 to 108 MHz) because nearby FM broadcasting stations are frequent sources of interference to VHF television receivers.

Another VHF/UHF wavetrap is shown in Fig. 15-2A. This type of wavetrap is called a *half-wave shorted stub*. One of the properties of a transmission line is that it will reflect the load impedance every half-wavelength along the line. When the end of the stub is shorted, a "virtual" short-circuit will appear at every half-wavelength along the line. This length is frequency-sensitive, so the physical length of the line is a tuning factor for this wavetrap.

The length of the line is found from $L_{feet} = 492V/F_{MHz}$; where L is in feet and F is in megahertz. The V term is the *velocity factor* of the transmission line. For common coaxial lines, the value of V ranges from 0.66 (polyethylene) to 0.82 (polyfoam). For example, assume that we need to eliminate the signal from a local FM broadcaster on 88.5 MHz. The shorted stub is made from ordinary coax ($V = 0.66$), and must have a length of

$$[(492)\ (0.66)/88.5] = 3.669 \text{ feet (or 44 inches).}$$

The half-wave shorted stub is connected in parallel with the antenna input on the receiver. In the case shown in Fig. 15-2A, the stub is connected to the receiver and antenna transmission line through a coaxial "tee" connector. Although UHF connectors are shown, the actual connector you would use must match your receiver and antenna. Also, it is possible to use 300-ohm twinlead transmission line rather than coax, so long at that type of line is used on the receiver and antenna.

The other approach to solving the problem is to use a passive preselector ahead of the receiver. Again, the preselector is inserted directly in the transmission line between the receiver antenna input connector and the antenna transmission line. Figure 15-2B shows a typical circuit for this type of preselector. The tuning is controlled by C1A/L2 and C1B/L3, and are trimmed by C2 and C3 to permit tracking of the two LC circuits. The circuit should be built in a closed, shielded metal box.

Other interference

The world is a terrible place for sensitive radio receivers: a lot of pure crud (other than the programming material!) comes over the airwaves from a large variety of electrical and electronic devices. There are also some renegade transmitters out there. For example, in the HF shortwave spectrum, you will occasionally hear a "beka-beka-beka" pulsed signal that seems to hop around quite a bit. It will set down on your favorite listening frequency, and then go to another. Unfortunately, the nature of a pulsed signal is to spread out over several megahertz—wiping out large segments of the spectrum. That signal is an *over the horizon backscatter* (OTHB) radar in the USSR called by North American SWLs and hams the "Russian Woodpecker." That's one that we can do little about, unfortunately, so let's look at some interfering signals that we can affect.

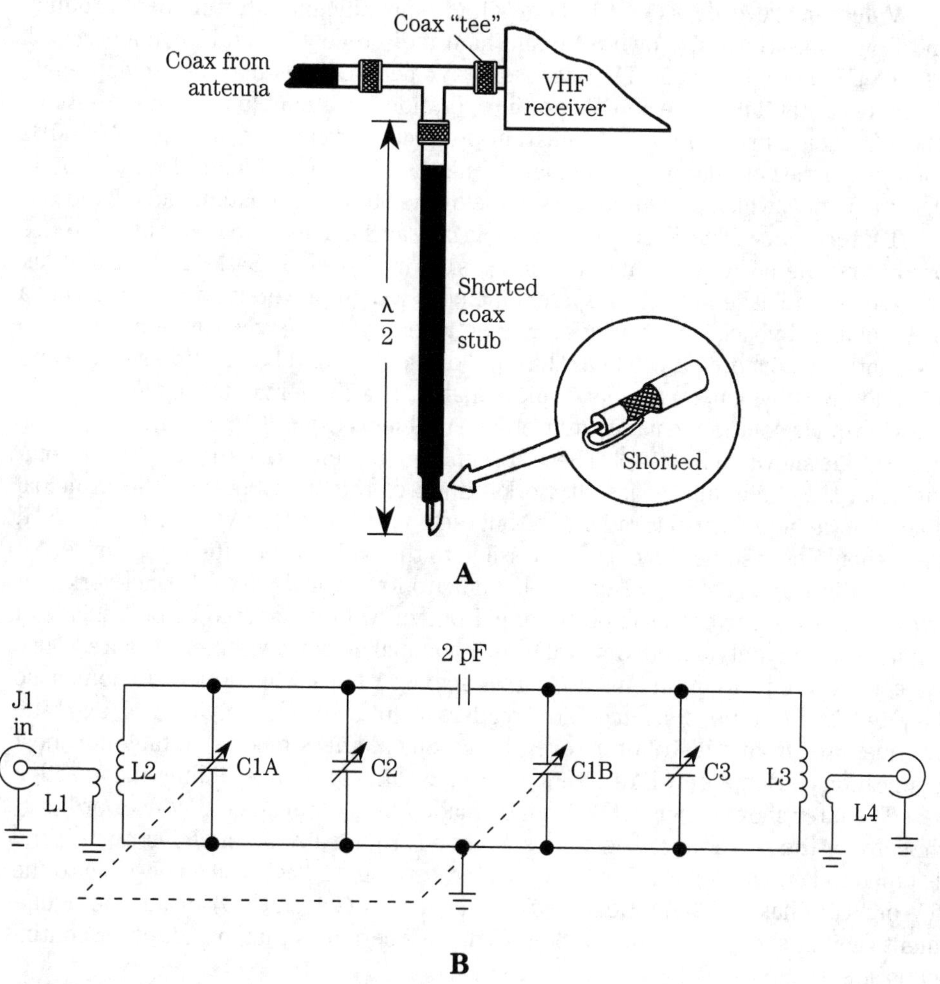

15-2 A. Halfwave shorted stub using coaxial cable. B. Preselector bandpass filter circuit.

Light dimmers A lot of homes are equipped with dimmers instead of switches to control the lighting. These devices are based on silicon-controlled rectifiers (SCR) that cut the ac sine wave off over part of its cycle, producing a harmonic-rich sharp waveform. These devices can produce a sound described as "frying eggs" well into the shortwave spectrum. Diagnosis is simple: turn off the light. If the noise stops when the light is turned off, then the dimmer is at fault. Although it is possible to install LC line noise filters at the dimmer, that approach is not usually feasible. It would be better to either:

- Remove the dimmer and replace it with an on-off switch
- Replace the dimmer with a special model that is designed to suppress radio noise
- Keep the darn light turned off when using the receiver

Videotape recorders (VCR) The VCR is a magnificent entertainment product, and I own one. But I also own SWL and ham radio receivers, and I can always tell when a popular movie is on TV. Because lots of people videotape the movie (legalities notwithstanding), their VCRs are in operation for a couple of hours at a time. The VCR contains a number of radiation-producing circuits, including a 3.58-MHz color subcarrier oscillator. On popular TV nights I can hear a load of trash around that frequency, which is right in the middle of the 80-meter amateur radio band.

TV receivers The TV set in your home contains at least two major interference producers: the 60-Hz vertical deflection system and the 15,734-kHz horizontal deflection system. The horizontal system includes a high-powered amplifier driving a high-voltage "flyback" transformer and a deflection yoke. As you tune up and down the shortwave band, you will hear "birdies" of the horizontal deflection signal every 15,734 kHz . . . which are the harmonic signals of the TV horizontal signal.

One quick solution to many interference problems from VCRs and TVs is the line noise filters shown in Fig. 15-3. These "EMI filters" are placed in the power line coming from the offending device. It works much of the time because the principal source of the noise signal is radiation from the power line of the VCR or TV. The EMI filter should be installed as close as possible to the body of the offending device.

The filter in Fig. 15-3A uses LC elements to form a low-pass filter network that is placed in series with the ac power line. Homebrew filters should be built inside of a quality, heavy-duty shielded metal box. If you make your own filter, be sure to use capacitors rated for continuous use across ac power lines. The inductors should also be rated for this type of service. Most readers should consider buying a ready-made LC line filter from a distributor (even Radio Shack offers models suitable for most applications). Homebrew EMI filters have a potential for danger if built incorrectly.

The filter shown in Fig. 15-3B can be made for any appliance that uses ordinary zipcord for the power line. The cord is wrapped around a ⅜ to ½-inch ferrite rod, and is taped to keep it in place. This filter should be mounted as close as possible to the TV or VCR chassis. The optional bypass capacitors (C1 and C2) inside the equipment cabinet should be 0.01 μF/1600 WVdc disk ceramics that are rated for continuous ac service.

Microwave ovens Today, most American homes are equipped with microwave ovens. These devices use a magnetron tube to produce several hundred watts of microwave power on a frequency of approximately 2450 MHz. The high voltage applied to the "maggie" is typically pulsating dc (i.e., it is rectified ac but not ripple filtered—as are true dc power supplies). This pulsating dc causes "hash" in radio receivers. Although better-quality microwave ovens are equipped with EMI filters, many are not. However, most manufacturers or servicers of microwave ovens can install EMI filters inside the oven. Alternatively, one of the EMI filters shown in Fig. 15-3 can be used.

Personal computer noise The proliferation of personal computers has greatly increased the amount of noise in the radio spectrum. The noise is caused by the internal circuits that operate using digital clock pulses. Older machines, which use internal clock frequencies of 1 MHz to 4.77 MHz (original IBM-PC), wipe out large portions of the AM and shortwave bands. If you doubt this, try using an AM radio near the computer! Later computers (XT-turbo and AT-class machines) used higher clock frequencies (e.g., 8, 10, 12, 16, 25, or 33 MHz), and these can wipe out the VHF

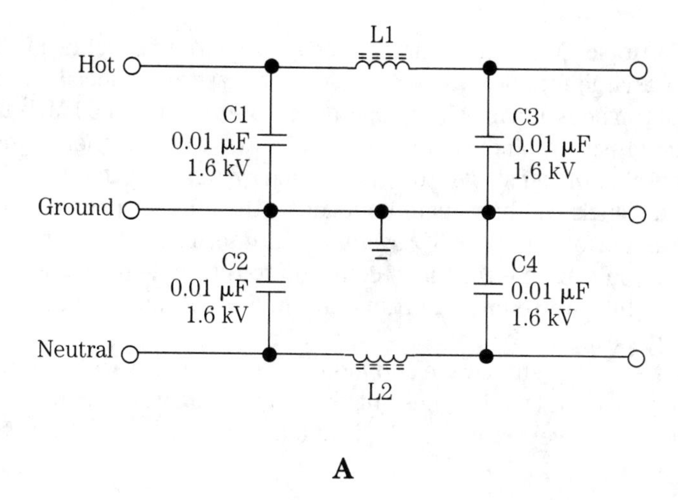

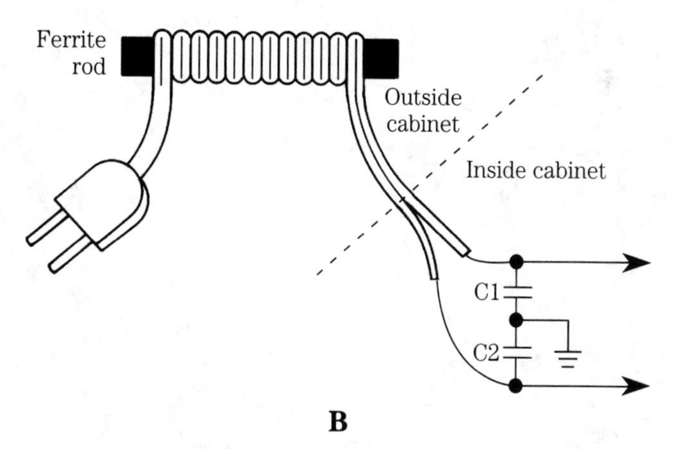

15-3 A. Power line LC EMF filter. B. Filter made by wrapping ac power line over ferrite rod.

bands as well . . . including the FM broadcast band. I've noticed that the XT-turbo machine that I use for word processing my technical articles and books wipes out the AM band in the regular mode, and the FM band in the turbo mode . . . but the turbo mode produces little interference on AM (a result of the higher clock frequency, I suspect).

Most of the noise produced by the computer is radiated from the power line or from the keyboard cable. In the latter case, make sure that a shielded keyboard cable is used. The power line noise is often susceptible to the same kinds of EMI filters discussed above.

Another source of noise is the printer, or more commonly the printer cable. If the cable between the computer and the printer is unshielded, replace it with a shielded type. Otherwise, you might want to consider a ferrite clamp-on filter bar such as the Amidon 2X-43 or equivalent.

Cable TV noise Many communities today are wired for cable TV. These systems transmit a large number of TV and FM broadcast and special service signals along a coaxial transmission line. They operate on frequencies of 54 MHz to 300 MHz in the 36-channel systems and 54 MHz to 440 MHz in the 55-channel systems. Whenever a large number of signals get together in one system, intermodulation is a possibility . . . and signals can be radiated outside of the official spectrum. This is how signals can leak out of the cable TV system, and interfere with your receiver. The only thing that you can do legally is to complain to the cable operator and insist he or she eliminate the interference. Fortunately, the FCC is your ally: by law, the operator must keep the signals home!

EMI interference to shortwave and monitor receivers is terribly annoying, and it sometimes get so bad that it wipes out reception altogether. Some of the "fixes" discussed above may well eliminate the problem, and make your radio hobby a lot more fun.

16
Radio transmission lines

Radio transmission lines are hard-wired conduits that are used for transporting RF signals between elements of a system. For example, in a radio transmitter, the transmission lines are typically used between the exciter output and final amplifier input, and between the transmitter output and the antenna. Although often erroneously characterized as a "length of shielded wire," transmission lines are actually complex networks containing the equivalent of all the three basic electrical components: resistance (R), capacitance (C), and inductance (L). Because of this, transmission lines must be analyzed in terms of an RLC network.

In this chapter, we will consider several types of radio transmission lines. Both step function and sine wave ac responses will be studied. Because the subject is both conceptual and analytical, we will use both analogy and mathematical approaches to examine the theory of transmission lines.

Figure 16-1 shows several basic types of transmission line. Perhaps the oldest and simplest form is the *parallel line* shown in Fig. 16-1A through 16-1D. In Fig. 16-1A we see an end view of the parallel conductor transmission line. The two conductors, of diameter d, are separated by an insulating dielectric (which might be air) by a spacing s. These designations will be used in calculations later.

Figure 16-1B shows a type of parallel line called *twinlead*. This is the old-fashioned television antenna transmission line. It consists of a pair of parallel conductors separated by a plastic dielectric. TV-type twinlead has a characteristic impedance (Z_0) of 300 ohms, while certain radio transmitting antenna twinlead has an impedance of 450 ohms. We will discuss Z_0 shortly, but for now accept that it is an attribute of the transmission line.

Another form of twinlead is *open line*, shown in Fig. 16-1C. In this case, the wire conductors are separated by an air dielectric, with support provided by stiff (usually ceramic) insulators. A tie wire (only one shown) is used to fasten each insulator end to the main conductor. Some users of open line prefer the form of insulator or sup-

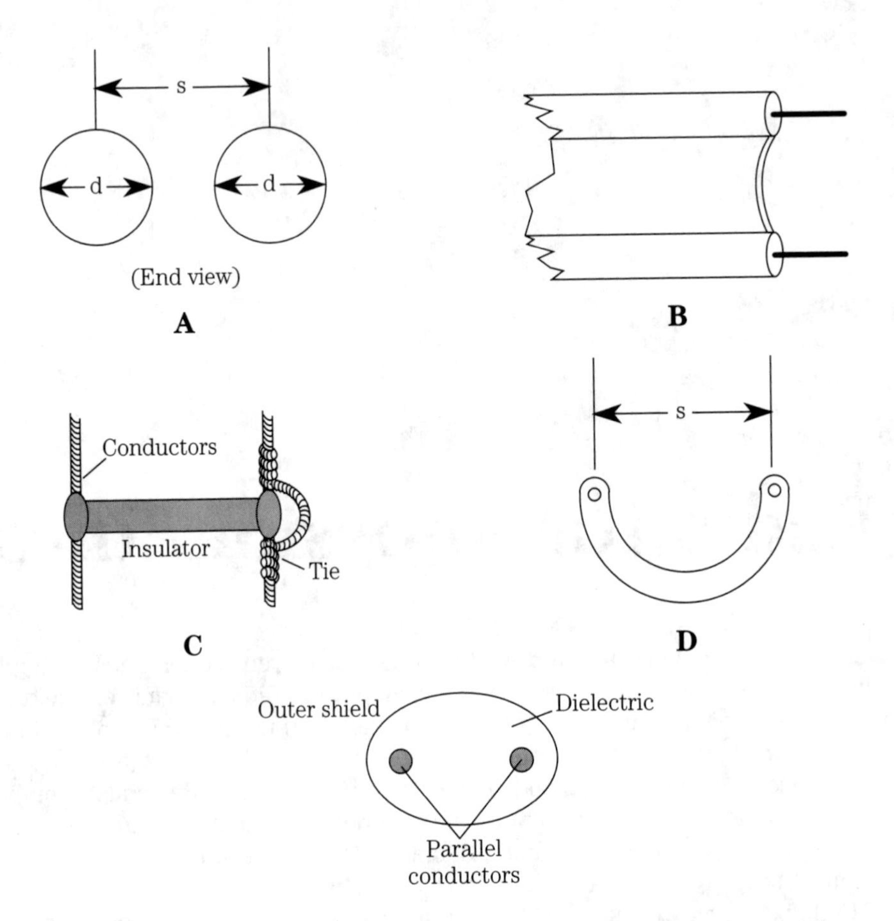

16-1 Transmission line. A. Parallel line end-view schematic. B. Twin-lead line. C. Ladder line construction detail. D. Long-path parallel line insulator. E. Shielded twin-lead.

porter shown In Fig. 16-1D. This form of insulator is made of either plastic or ceramic, and is in the form of a "U." The purpose of this shape is to reduce losses, especially in rainy weather, by increasing the leakage currents path length relative to spacing s.

Parallel transmission lines have been used at VLF, MW, and HF frequencies for many decades. Even antennas used into the low-VHF spectrum are sometimes found using parallel lines. The higher impedance of parallel lines (relative to coaxial cable) makes it show lower loss in high-power applications. For years, the VHF, UHF, and microwave application of parallel lines was limited to educational laboratories where they are well-suited to performing experiments (to about 2 GHz) with simple, low-cost instruments. Today, however, printed circuit and hybrid semiconductor packaging has given parallel lines a new lease on life, if not an overwhelming market presence.

Figure 16-1E shows a form of parallel line called *shielded twinlead*. This type of line uses the same form of construction as TV-type twinlead, but also has a braided shielding layer of metal surrounding it. This feature makes it less susceptible to noise and other problems.

The second form of transmission line, which finds considerable application at microwave frequencies, is *coaxial cable* (Fig. 16-2A through 16-2C). This form of line consists of two cylindrical conductors sharing the same axis (hence "coaxial") and separated by a dielectric (Fig. 16-2A). For low frequencies (in flexible cables), the dielectric may be polyethylene or polyethylene foam, but at higher frequencies Teflon and other materials are used. Dry air or dry nitrogen is also used in some applications.

Several forms of coaxial line are available. Flexible coaxial cable is perhaps the most common form. The outer conductor in such cable is made of either braid or foil (Fig. 16-2B). Again, television broadcast receiver antennas provide an example of such cable from common experience. Another form of flexible or semiflexible coaxial line is helical line (Fig. 16-2C), in which the outer conductor is spiral wound. Hardline (Fig. 16-2D) is coaxial cable that uses a thin-wall pipe as the outer conductor. Some hardline coax used at microwave frequencies uses a rigid outer conductor, and a solid dielectric.

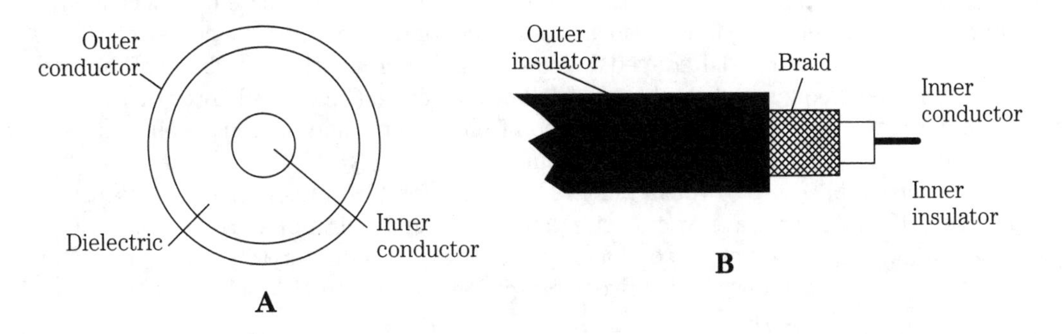

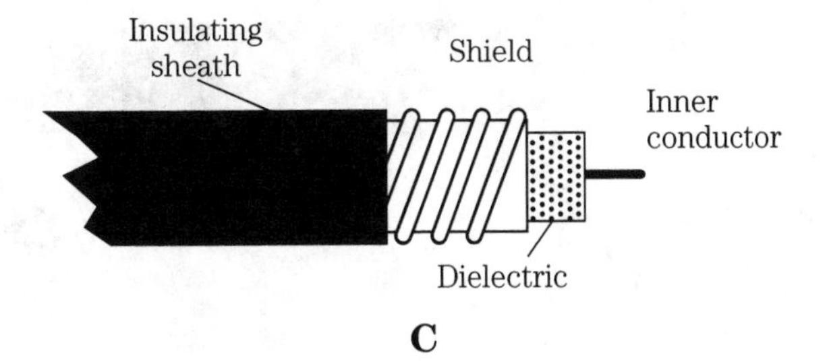

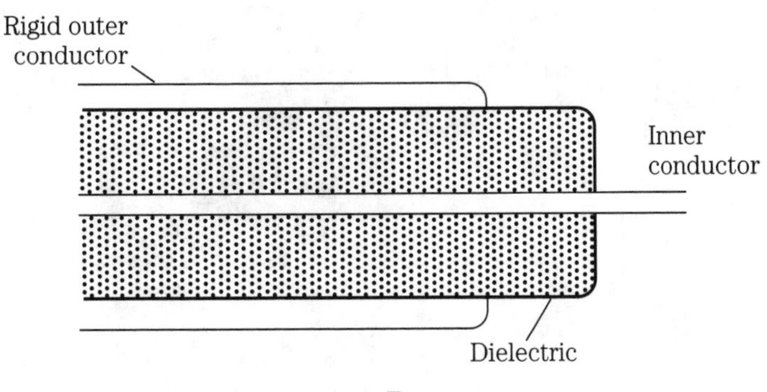

16-2 A. Schematic of coaxial cable. B. Standard coaxial cable (side view). C. Coaxial hardline. D. Rigid coaxial line.

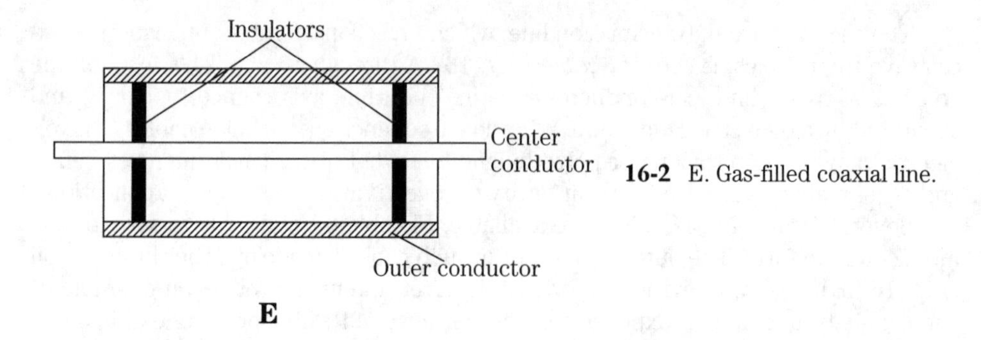

16-2 E. Gas-filled coaxial line.

Gas-filled line is a special case of hardline which is hollow (Fig. 16-2E), the center conductor being supported by a series of thin ceramic or Teflon insulators. The dielectric is either anhydrous (i.e., dry) nitrogen or some other inert gas.

Some flexible microwave coaxial cables use a solid "air-articulated" dielectric (Fig. 16-3A), in which the inner insulator is not continuous around the center conductor, but rather is ridged. Reduced dielectric losses increase the usefulness of the cable at higher frequencies. Double-shielded coaxial cable (Fig. 16-3B) provides an extra measure of protection against radiation from the line, and prevents EMI from outside sources from getting into the system.

Strip line, or *micro stripline*, (Fig. 16-3C) is a form of transmission line used at high UHF and microwave frequencies. The strip line consists of a critically sized conductor over a ground plane conductor, which is separated from it by a dielectric. Some strip lines are sandwiched between two ground planes, and are separated from each other by a dielectric.

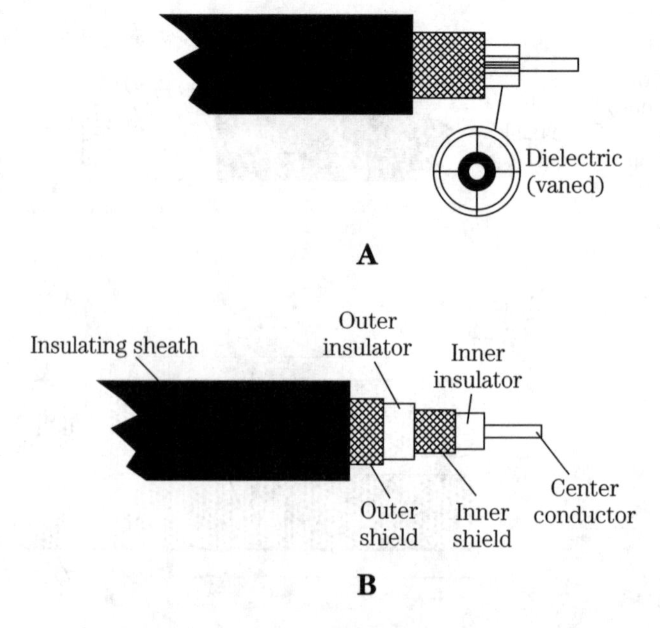

16-3 A. Articulated line. B. Double-shielded coaxial line.

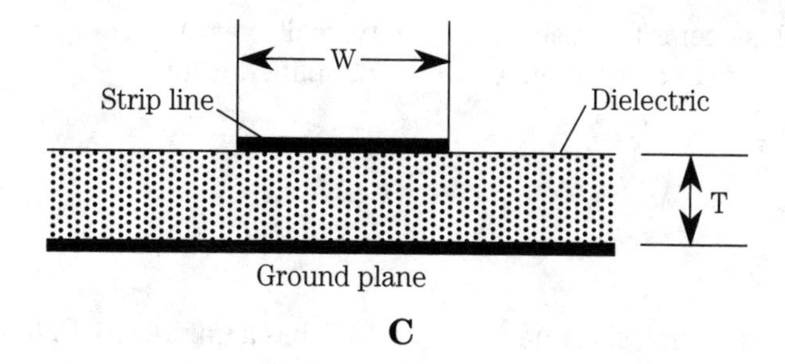

16-3 C. Strip-line.

Transmission line characteristic impedance (Z_o)

The transmission line is actually an RLC complex network (see Fig. 16-4), so has a *characteristic impedance*, Z_o, also sometimes called *surge impedance*. Network analysis will show that Z_o is a function of the per unit of length parameters resistance (R), conductance (G), inductance (L), and capacitance (C), and is found from:

$$Z_o = \sqrt{\frac{R + J\varpi L}{G + j\varpi C}} \quad [16\text{-}1]$$

where

Z_o is the characteristic impedance, in ohms (Ω)
R is the resistance per unit length, in ohms (Ω)
G is the conductance per unit length, in seimens (S)
L is the inductance per unit length, in henrys (H)
C is the capacitance per unit length, in farads (F)
ϖ is the angular frequency in radians-per-second ($2\pi F$)

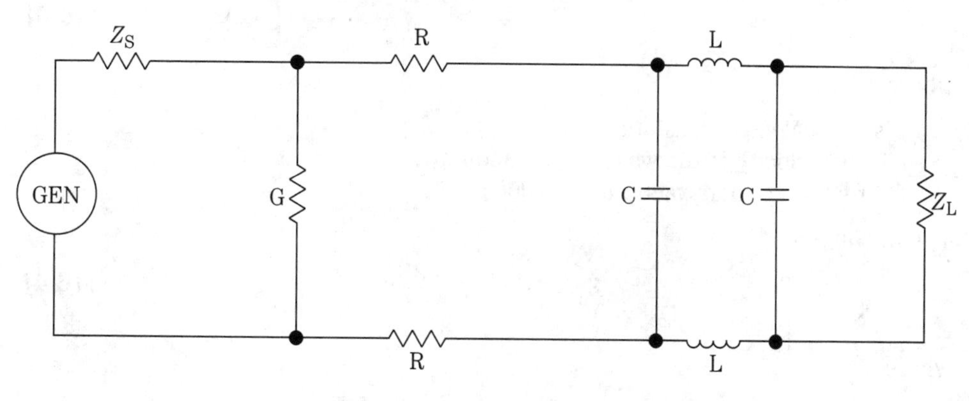

If $X \gg R$
Then: $Z_O = \sqrt{L/C}$

16-4 Equivalent circuit of transmission line.

In many RF systems the resistances are typically very low compared with the reactances, so Eq. 16-1 can be reduced to the simplified form:

$$Z_o = \sqrt{\frac{L}{C}} \qquad [16\text{-}2]$$

Example 16-1

A nearly lossless transmission line (R very small) has a unit length inductance of 3.75 nanohenrys, and a unit length capacitance of 1.5 pF. Find the characteristic impedance, Z_o.

Solution

$$Z_o = \sqrt{\frac{L}{C}}$$

$$Z_o = \sqrt{\frac{\left(3.75\ \text{nH} \times \dfrac{1\ \text{H}}{10^9\ \text{nH}}\right)}{\left(1.5\ \text{pF} \times \dfrac{1\ \text{F}}{10^{12}\ \text{pF}}\right)}}\ \text{ohms}$$

$$Z_o = \sqrt{\frac{3.75 \times 10^{-9}\ \text{H}}{1.5 \times 10^{-12}\ \text{F}}}\ \text{ohms} = 50\ \text{ohms}$$

The characteristic impedance for a specific type of line is a function of the conductor size, the conductor spacing, the conductor geometry (see again Fig. 16-1), and the dielectric constant of the insulating material used between the conductors.

The dielectric constant (ε) is equal to the reciprocal of the velocity (squared) of the wave when a specific medium is used

$$\varepsilon = \frac{1}{v^2} \qquad [16\text{-}3]$$

where

ε is the dielectric constant
v is the velocity of the wave in the medium
(Note: for a perfect vacuum $\varepsilon = 1.000$)

a) Parallel line

$$Z_o = \frac{276}{\sqrt{\varepsilon}} \log\left(\frac{2S}{d}\right) \qquad [16\text{-}4]$$

where

Z_o is the characteristic impedance, in ohms (Ω)
ε is the dielectric constant
S is the center-to-center spacing of the conductors

d is the diameter of the conductors
(S and d in same units)

b) Coaxial line

$$Z_o = \frac{138}{\sqrt{\varepsilon}} \text{ Log } \left(\frac{D}{d}\right) \quad \textbf{[16-5]}$$

where

D is the diameter of the outer conductor
d is the diameter of the inner conductor

c) Shielded parallel line

$$Z_o = \frac{276}{\sqrt{\varepsilon}} \text{ Log } \left(2A \times \frac{1 - B^2}{1 + B^2}\right) \quad \textbf{[16-6]}$$

where

$$A = s/d \text{ and } B = s/D$$

d) Stripline

$$Z_o = \frac{377}{\sqrt{\varepsilon_t}} \times \frac{T}{W} \quad \textbf{[16-7]}$$

where

ε_t is the relative dielectric constant of the printed wiring board (PWB)
T is the thickness of the printed wiring board
W is the width of the stripline conductor

The relative dielectric constant (ε_t) used above differs from the normal dielectric constant of the material used in the PWB. The relative and normal dielectric constants move closer together for larger values of the ratio W/T.

Example 16-2

A stripline transmission line is built on a 4-mm thick printed wiring board that has a relative dielectric constant of 5.5. Calculate the characteristic impedance if the width of the strip is 2 mm.

Solution

$$Z_o = \frac{377}{\sqrt{\varepsilon_t}} \times \frac{T}{W}$$

$$Z_o = \frac{377}{\sqrt{5.5}} \times \frac{4 \text{ mm}}{2 \text{ mm}} = 321 \text{ ohms}$$

In practical situations, we usually don't need to calculate the characteristic impedance of a stripline—but rather, we design the line to fit a specific system impedance (e.g., 50 ohms). We can make some selection choices of printed circuit material (hence dielectric constant) and thickness, but even these are usually limited in practice by the availability of standardized boards. Thus, stripline width is the variable parameter.

Equation 16-7 can be rearranged to the form:

$$W = \frac{377\,T}{Z_o\sqrt{\varepsilon}} \qquad \textbf{[16-8]}$$

The impedance of 50 ohms is accepted as standard for RF systems, except in the cable-TV industry (which is 75 ohms). The reason for the difference is that power handling ability and low-loss operation don't occur at the same characteristic impedance. For coaxial cables, for example, the maximum power handling ability occurs at about 30 ohms, while the lowest loss occurs at 77 ohms; 50 ohms is therefore a reasonable trade-off between the two points. In the cable-TV industry, however, the RF power levels are minuscule, but lines are long. The trade-off for TV is to use 75 ohms as the standard system impedance in order to take advantage of the reduced attenuation factor.

Velocity factor

In the discussion preceding this section we discovered that the velocity of the wave or signal in the transmission line is less than the free-space velocity, i.e., less than the speed of light. Further, we discovered in Eq. 16-3 that the velocity is related to the dielectric constant of the insulating material that separates the conductors in the transmission line. Velocity factor (v) is usually specified as a decimal fraction of c, the speed of light (3×10^8 m/s). For example, if the velocity factor of a transmission line is rated at 0.66, the velocity of the wave is 0.66c, or

$$(0.66)\ (3 \times 10^8 \text{ m/s}) = 1.98 \times 10^8 \text{ m/s}$$

Velocity factor becomes important when designing things like transmission line impedance transformers, phasing harnesses or any other device in which the length of the line is important. In most cases, the transmission line length is specified in terms of *electrical length*, which can be either an angular measurement (e.g., 180 degrees or π radians), or a relative measure keyed to wavelength (e.g., one-half wavelength, which is the same as 180 degrees). The physical length of the line is longer than the equivalent electrical length. For example, let's consider a 1-MHz half-wave transmission line.

A rule of thumb tells us that the length of a wave (in meters) in free-space is $300/F$, where frequency (F) is expressed in megahertz; therefore a half-wavelength line is $150/F$. At 1 MHz, the line must be 150/1 MHz, or 150 meters long. If the velocity factor is 0.80, then the *physical length* of the transmission line that will achieve the desired electrical length is $[(150 \text{ meters})\ (v)]\ /F = [(150 \text{ meters})\ (0.80)]\ /1$ MHz = 120 meters. The derivation of the "rule of thumb" is ". . . left as an exercise for the student" (hint: it comes from the relationship between wavelength, frequency and velocity of propagation for any form of wave).

Certain practical considerations regarding velocity factor result from the fact that the physical and electrical lengths are not equal. For example, in a certain type of phased array antenna design radiating elements are spaced half-wavelength apart, and must be fed at 180 degrees (half-wave) out of phase with each other. The simplest interconnect is to use a half-wave transmission line between the 0-degree element and the 180-degree element. According to the standard wisdom, the transmission line will create the 180-degree phase delay required for the correct operation of the antenna. Unfortunately, because of the velocity factor the physical length for a one-half electrical wavelength cable is shorter than the free-space half-wave distance between elements. In other words, the cable will be too short to reach between radiating elements by an amount proportional to the velocity factor!

Clearly, velocity factor is a topic that must be understood before transmission lines can be used in practical situations. Table 16-1 shows the velocity factors for several types of popular transmission line.

Table 16-1. Transmission line characteristics

Type of line	Z_o (ohms)	Vel. factor (v)
½ in. TV parallel line (air dielectric)	300	0.95
1 in. TV parallel line (air dielectric)	450	0.95
TV twin-lead	300	0.82
UHF TV twin-lead	300	0.80
Polyethylene coaxial cable	*	0.66
Polyethylene foam coaxial cable	*	0.79
Air-space polyethylene foam coaxial cable	*	0.86
Teflon	*	0.70

*Various impedances depending upon cable type.

Transmission line noise

Transmission lines are capable of generating noise and spurious voltages that the system interprets as valid signals. Several such sources exist. One source is the coupling between noise currents flowing in the outer conductor and the inner conductor. Such currents are induced by nearby electromagnetic interference and other sources (e.g., connection to a noisy ground plane). Although coaxial design reduces noise pick-up compared with a parallel line, the potential for EMI exists. Selection of a high-grade line, with a high degree of shielding, reduces the problem.

Another source of noise is the thermal noises in the resistances and conductances. This type of noise is proportional to resistance and temperature.

The mechanical movement of the cable also creates noise. One species of noise results from movement of the dielectric against the two conductors. This form of noise is caused by electrostatic discharges in much the same manner as the spark created by rubbing a piece of plastic against woolen cloth.

A second species of mechanically-generated noise is piezoelectricity in the dielectric. Although more common in cheap off-brand cables, you should be aware of it. Mechanical deformation of the dielectric causes electrical potentials to be generated.

Both species of mechanically generated noise can be reduced or eliminated by proper mounting that prevents motion of the cable. Although rarely a problem at lower frequencies, mechanical noise can be significant at microwave frequencies when signal levels are low.

Coaxial cable capacitance

A coaxial transmission line possesses a certain capacitance per unit of length. This capacitance is defined by:

$$C = \frac{24\,\varepsilon}{\text{Log}\,D/d}\;\frac{\text{pf}}{\text{meter}} \quad \textbf{[16-9]}$$

A long run of coaxial cable can build up a relatively large capacitance. For example, a common type of coax is rated at 65 pF/meter. A 150-meter roll thus has a capacitance of (65 pF/m) (150 m), or 9750 pF. When charged with a high voltage, as is done in performing cable breakdown voltage tests at the factory, the cable acts like a charged high voltage capacitor. Although rarely lethal to healthy humans, the stored voltage in new cable can deliver a nasty electrical shock and can irreparably damage electronic components.

Coaxial cable cut-off frequency (F_c)

The normal mode in which a coaxial cable propagates a signal is a *transverse electromagnetic* (TEM) *wave*, but other modes are possible—and usually undesirable. There is a maximum frequency above which TEM propagation becomes a problem, and higher modes dominate. Coaxial cable should *not* be used above a frequency of:

$$F = \frac{6.75}{(D + d)\sqrt{\varepsilon}}\ \text{GHz} \quad \textbf{[16-10]}$$

where

F is the TEM-mode cut-off frequency
D is the diameter of the outer conductor in inches
d is the diameter of the inner conductor in inches
ε is the dielectric constant

When maximum operating frequencies for cable are listed, the TEM-mode is cited. Beware of attenuation (loss), however, especially when making selections for VHF, UHF, and microwave frequencies. A particular cable may have a sufficiently high TEM-mode frequency, but still exhibit a high attenuation per-unit-length at frequencies that are of interest to you.

Transmission line responses

In order to understand the operation of transmission lines we need to consider two cases: step-function response and the steady-state ac response. The *step function* case involves a single event when a voltage at the input of the line snaps from zero (or a steady value) to a new (or non-zero) value, and remains there until all action dies out. This response tells us something of the behavior of pulses in the line, and in fact is used to describe the response to a single pulse stimulus. The steady-state ac response tells us something of the behavior of the line under stimulation by a sinusoidal RF signal.

Step-function response of a transmission line

Figure 16-5 shows a parallel transmission line with characteristic impedance (Z_o) connected to a load impedance (Z_L). The "generator" at the input of the line consists of a voltage source (V) in series with a "source impedance" (Z_S) and a switch (S1). Assume for the present that all impedances are pure resistances (i.e., $R + j_0$). Also, assume that $Z_s = Z_o$.

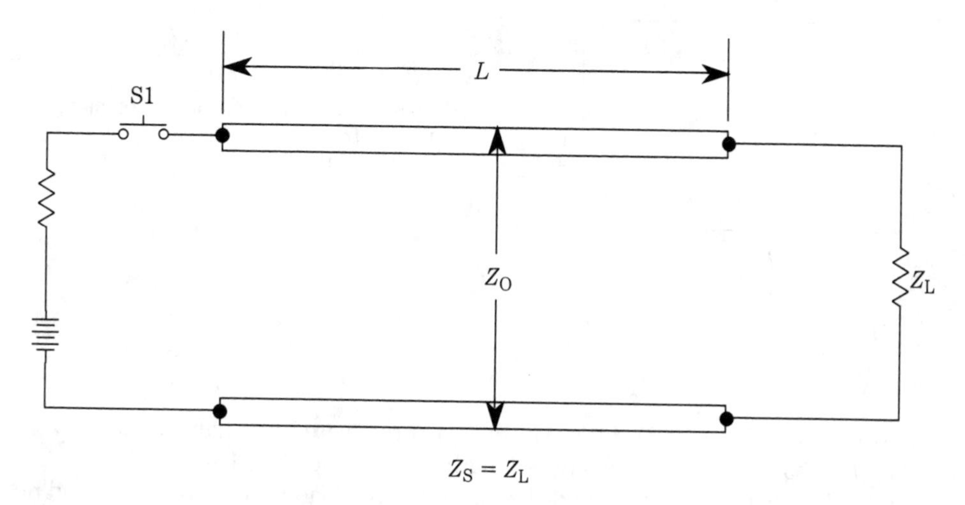

16-5 Schematic example of a transmission line.

When the switch is closed at time T_o (Fig. 16-6A), the voltage at the input of the line (V_{in}) jumps to $V/2$. When we discussed Fig. 16-4, you may have noticed that the LC circuit resembles a delay line circuit. As might be expected, therefore, the voltage wave front propagates along the line at a velocity (v) of:

$$v = \frac{1}{\sqrt{LC}} \qquad [16\text{-}11]$$

where

v is the velocity in meters per second (m/s)
L is the inductance in henrys (H)
C is the capacitance in farads (F)

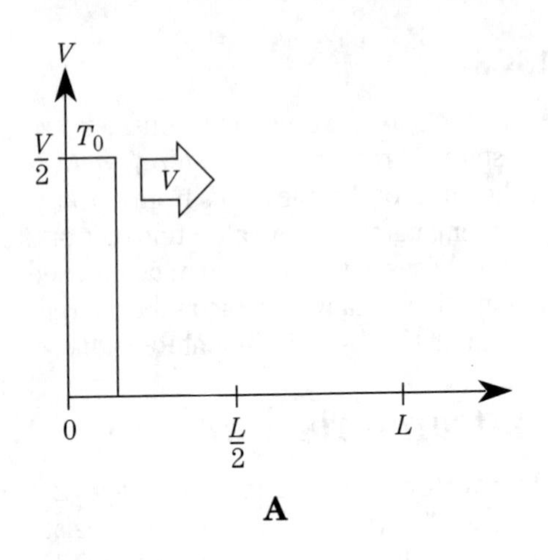

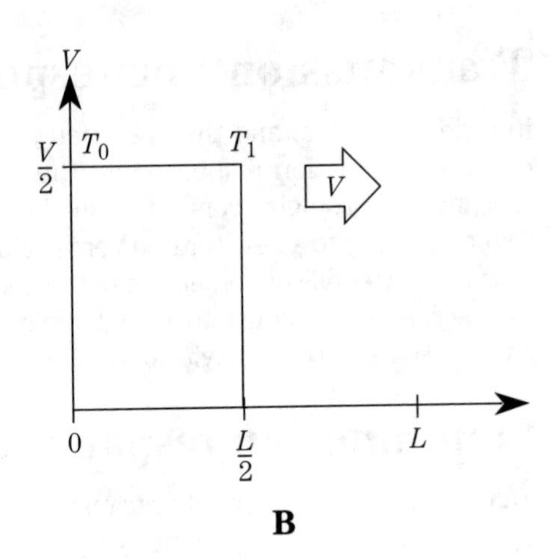

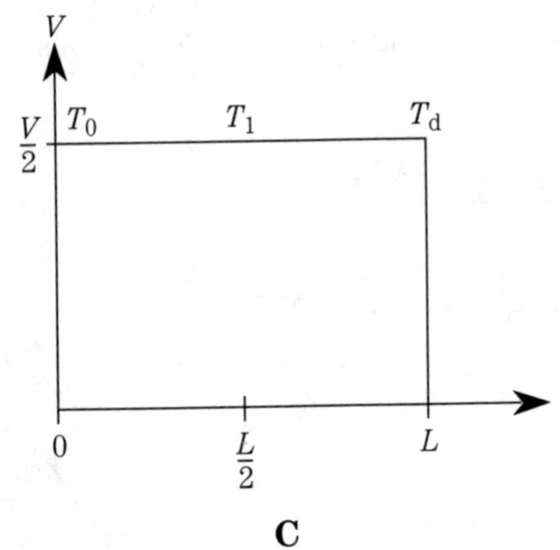

16-6 Step-function propagation along a transmission line at three points.

At time T_1 (Fig. 16-6B) the wavefront has propagated one-half the distance L, and by T_d it has propagated the entire length of the cable (Fig. 16-6C).

If the load is perfectly matched (i.e., $Z_L = Z_o$), the load absorbs the wave and no component is reflected. But in a mismatched system ($Z_L \neq Z_o$), a portion of the wave is reflected back down the line towards the generator.

Figure 16-7 shows the *rope analogy* for reflected pulses in a transmission line. A taut rope (Fig. 16-7A) is tied to a rigid wall that does not absorb any of the energy in the pulse propagated down the rope. When the free end of the rope is given a sudden vertical displacement (Fig. 16-7B) a wave is propagated down the rope at velocity v (Fig. 16-7C). When the pulse hits the wall (Fig. 16-7D), it is reflected (Fig. 16-7E) and propagates back down the rope towards the free end (Fig. 16-7F).

If a second pulse is propagated down the line before the first pulse dies out, then there will be two pulses on the line at the same time (Fig. 16-8A). When the two pulses interfere, the resultant will be the algebraic sum of the two. In the event that a pulse train is applied to the line, the interference pattern will set up *standing waves*, an example of which is shown in Fig. 16-8B.

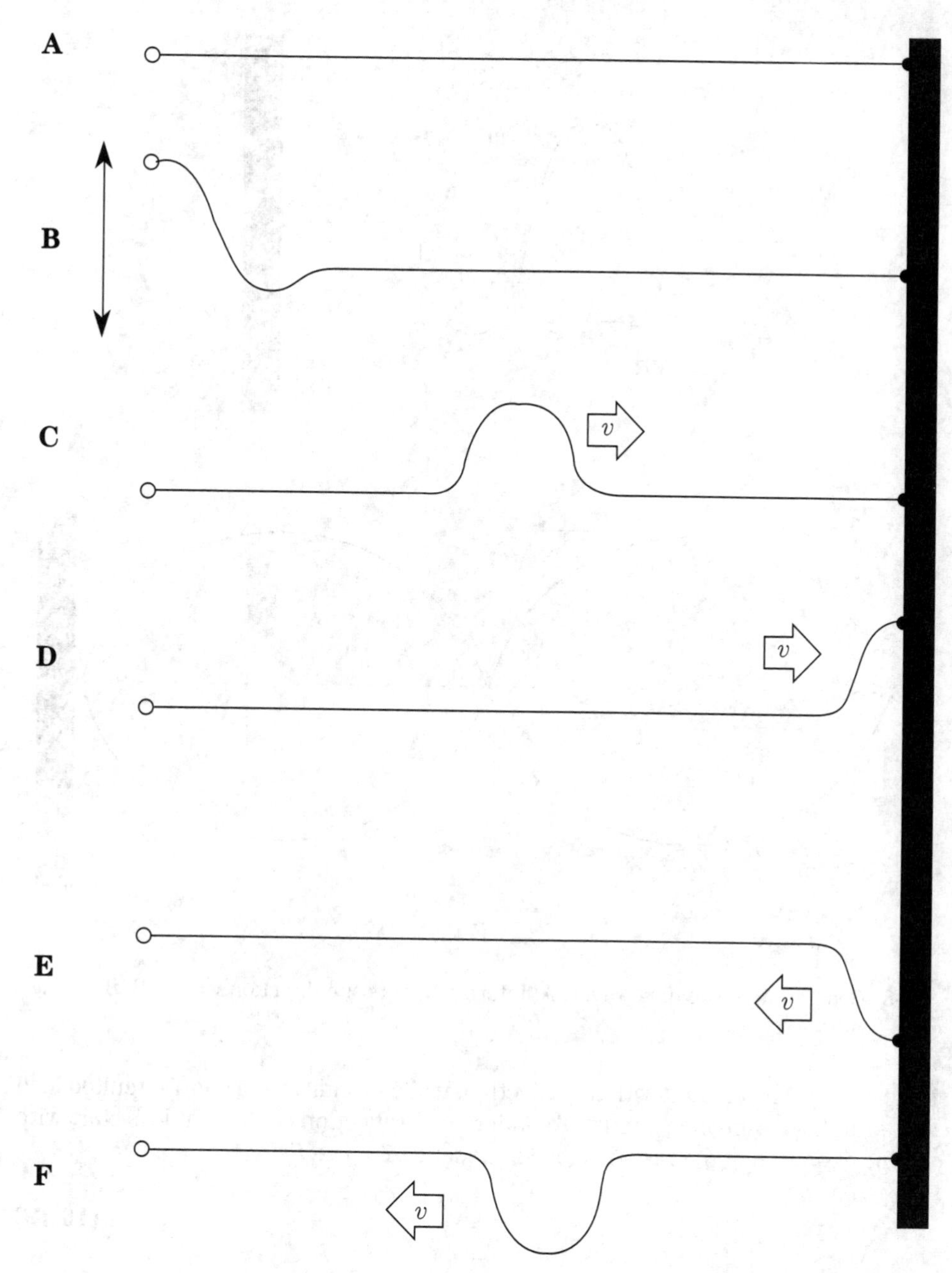

16-7 Rope analogy of a transmission line excited by ac.

Reflection coefficient (Γ)

The *reflection coefficient* (Γ) of a circuit containing a transmission line and load impedance is a measure of how well the system is matched. The value of the reflection coefficient varies from –1 to +1, depending upon the magnitude of reflection; Γ = 0

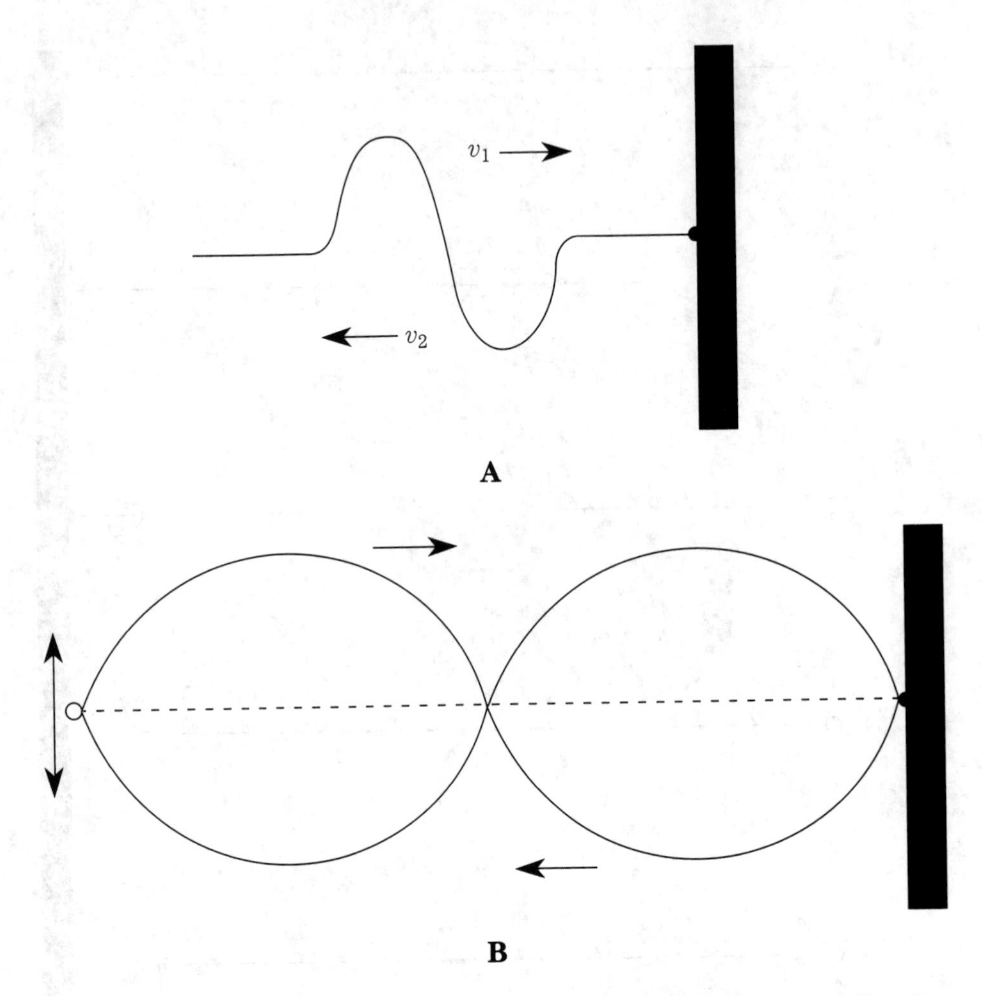

16-8 Generation of standing waves. A. Interfering opposite direction waves. B. Standing waves.

indicates a perfect match with no reflection, while –1 indicates a short circuited load and +1 indicates an open circuit. To understand reflection coefficient let's start with a basic definition of the resistive load impedance $Z = R + j_0$:

$$Z_L = \frac{V}{I} \tag{16-12}$$

where

Z_L is the load impedance $R + j_0$ in ohms (Ω)
V is the potential across the load in volts (V)
I is the current flowing in the load in amperes (A)

Because there are both reflected and incident waves, we find that V and I are actually the sums of incident and reflected voltages and currents, respectively. Therefore:

$$Z_L = \frac{V}{I} \qquad [16\text{-}13]$$

where

V_{inc} is the incident (i.e., forward) voltage

$$Z_L = \frac{V_{inc} + V_{ref}}{I_{inc} + I_{ref}} \qquad [16\text{-}14]$$

V_{ref} is the reflected voltage
I_{inc} is the incident current
I_{ref} is the reflected current

Because of Ohm's law, we may define the currents in terms of voltage, current and the characteristic impedance of the line:

$$I_{inc} = \frac{V_{inc}}{Z_o} \qquad [16\text{-}15]$$

and

$$I_{inc} = \frac{-V_{ref}}{Z_o} \qquad [16\text{-}16]$$

(The minus sign in Eq. 16-16 indicates that a direction reversal took place.)

The two expressions for current (Eqs. 16-15 and 16-16) may be substituted into Eq. 16-14 to yield:

$$Z_L = \frac{V_{inc} + V_{ref}}{\left(\frac{V_{inc}}{Z_o} + \frac{V_{ref}}{Z_o}\right)} \qquad [16\text{-}17]$$

The reflection coefficient (Γ) is defined as the ratio of reflected voltage to incident voltage (V_{ref}/V_{inc}), so by solving Eq. 16-17 for this ratio we find:

$$\Gamma = \frac{V_{ref}}{V_{inc}} = \frac{Z_L - Z_o}{Z_L + Z_o} \qquad [16\text{-}18]$$

Example 16-3

A 50-ohm transmission line is connected to a 30-ohm resistive load. Calculate the reflection coefficient, Γ.

Solution

$$\Gamma = \frac{Z_L - Z_o}{Z_L + Z_o}$$

$$\Gamma = \frac{50\text{ ohms} - 30\text{ ohms}}{50\text{ ohms} + 30\text{ ohms}} = 0.25$$

Example 16-4

In example 16-3 above, the incident voltage is 3 volts RMS. Calculate the reflected voltage.

Solution

if,

$$\Gamma = \frac{V_{ref}}{V_{inc}}$$

then,

$$V_{ref} = \Gamma V_{inc}$$

$$V_{ref} = (0.25)\ (3\ \text{volts}) = 0.75\ \text{volts}$$

The *phase* of the reflected signal is determined by the relationship of the load impedance and transmission line characteristic impedance. For resistive loads ($Z = R + j_0$): if the ratio Z_L/Z_o is 1.0, then there is no reflection; if Z_L/Z_o is less than 1.0 then the reflected signal is 180 degrees out-of-phase with the incident signal; if the ratio Z_L/Z_o is greater than 1.0 then the reflected signal is in-phase with the incident signal. Table 16-2 summarizes these relationships.

Table 16-2

Ratio	Angle of reflection
$Z_L/Z_o = 1$	No reflection
$Z_L/Z_o < 1$	180 degrees reflection
$Z_L/Z_o > 1$	0 degrees reflection

The ac response of the transmission line

When a CW RF signal is applied to a transmission line, the excitation is sinusoidal (Fig. 16-9), so it becomes useful for us to investigate the steady-state ac response of the line. The term *steady-state* implies a sine wave of constant amplitude, phase, and frequency. When ac is applied to the input of the line it propagates along the line at a given velocity. The ac signal amplitude and phase will decay exponentially in the manner shown by Eq. 16-19 below:

$$V_R = Ve^{(-\gamma l)} \qquad \textbf{[16-19]}$$

where

V_R is the voltage received at the far end of the line (V)
V is the applied signal voltage (V)
1 is the length of the line
γ is the propagation constant of the line

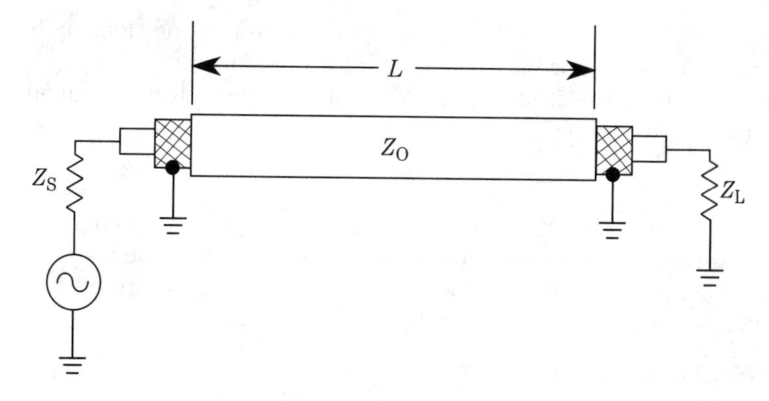

16-9 ac-excited transmission line.

The propagation constant (γ) is defined in various equivalent ways, each of which serves to illustrate its nature. For example, the propagation constant is proportional to the product of impedance and admittance characteristics of the line:

$$\gamma = \sqrt{ZY} \quad [16\text{-}20]$$

or, since $Z = R + j\omega L$ and $Y = G + j\omega C$, we may write

$$\gamma = \sqrt{(R + j\omega L)\,(G + j\omega C)} \quad [16\text{-}21]$$

We may also write an expression for the propagation constant in terms of the line attenuation constant (α) and phase constant (β):

$$\gamma = \alpha + j\,\beta \quad [16\text{-}22]$$

If we can assume that susceptance dominates conductance in the admittance term, and reactance dominates resistance in the impedance term (both usually true at microwave frequencies), then we may neglect the R and G terms altogether and write:

$$\gamma = j\omega\,\sqrt{LC} \quad [16\text{-}23]$$

We may also reduce the phase constant (β) to

$$\beta = \omega\,\sqrt{LC} \quad [16\text{-}24]$$

or

$$\beta = \omega\,Z_o C \quad \text{rad/m} \quad [16\text{-}25]$$

and, of course, the characteristic impedance remains

$$Z_o = \sqrt{LC} \quad [16\text{-}26]$$

Special cases

The impedance "looking-into" a transmission line (Z) is the impedance presented to the source by the combination of load impedance and transmission line characteristic impedance. Below are presented equations that define the "looking-in" impedance seen by a generator or source driving a transmission line.

The case where the load impedance and line characteristic impedance are matched is defined by:

$$Z_L = R_L + j0 \qquad \textbf{[16-27]}$$

In other words, the load impedance is resistive and equal to the characteristic impedance of the transmission line. In this case the line and load are matched, and the impedance looking-in will be simple $Z = Z_L = Z_o$. In other cases, however, we find different situations where Z_L is not equal to Z_o.

Z_L is not equal to Z_o in a random length lossy line:

$$Z = Z_o \left(\frac{Z_L + Z_o \operatorname{Tanh}(\gamma 1)}{Z_o + Z_L \operatorname{Tanh}(\gamma 1)} \right) \qquad \textbf{[16-28]}$$

where

Z is the impedance looking-in in ohms (Ω)
Z_L is the load impedance in ohms (Ω)
Z_o is the line characteristic impedance in ohms (Ω)
1 is the length of the line in meters (m)
γ is the propagation constant

Z_L not is equal to Z_o in lossless or very low loss random length line:

$$Z = Z_o \left(\frac{Z_L + jZ_o \operatorname{Tan}(\beta 1)}{Z_o + jZ_L \operatorname{Tan}(\beta L)} \right) \qquad \textbf{[16-29]}$$

Equations 16-28 and 16-29 serve for lines of any random length. For lines that are either integer multiples of half wavelength, or odd integer (i.e., 1, 3, 5, 7 . . . etc.) multiples of quarter wavelength, special solutions for these equations are found, and some of these solutions are very useful in practical situations. For example, consider . . .

Half wavelength lossy lines:

$$Z = Z_o \left(\frac{Z_L + Z_o \operatorname{Tanh}(\alpha 1)}{Z_o + Z_L \operatorname{Tanh}(\alpha 1)} \right) \qquad \textbf{[16-30]}$$

Above, we discover that the impedance looking into a lossless or very low-loss half wavelength transmission line is the load impedance:

$$Z = Z_L \qquad \textbf{[16-31]}$$

The fact that line input impedance equals load impedance is very useful in certain practical situations. For example, a resistive impedance is not changed by the line length. Therefore, when an impedance is inaccessible for measurement purposes, it can be measured through a transmission line that is an integer multiple of a half-wavelength.

Our next special case involves a quarter-wavelength transmission line, and those that are odd integer multiples of quarter wavelength (of course, even integer multiples of quarter wavelength obey the half wavelength criteria).

Quarter-wavelength lossy lines:

$$Z = Z_o \left(\frac{Z_L + Z_o \operatorname{Coth}(\alpha l)}{Z_o + Z_L \operatorname{Coth}(\alpha l)} \right) \tag{16-32}$$

and

Quarter-wavelength lossless or very low-loss lines:

$$Z = \frac{Z_o^2}{Z_L} \tag{16-33}$$

From Eq. 16-33 above we can discover an interesting property of the quarter-wavelength transmission line. First, divide each side of the equation by Z_o:

$$\frac{Z}{Z_o} = \frac{Z_o^2}{Z_L Z_o} \tag{16-34}$$

$$\frac{Z}{Z_o} = \frac{Z_o}{Z_L} \tag{16-35}$$

The ratio Z/Z_o shows an *inversion* of load impedance ratio Z_L/Z_o, or, stated another way:

$$\frac{Z}{Z_o} = \frac{1}{\frac{Z_L}{Z_o}} \tag{16-36}$$

Again, from Eq. 16-33 above we can deduce another truth about quarter-wavelength transmission lines:

if

$$Z = \frac{\sqrt{Z_o}}{Z_L} \tag{16-37}$$

then

$$Z_L = \sqrt{Z_o} \tag{16-38}$$

which means

$$Z_o = \sqrt{Z\,Z_L} \tag{16-39}$$

Equation 16-38 shows that a quarter-wavelength transmission line can be used as an *impedance-matching network*. Called a *Q-section*, the quarter-wavelength transmission line used for impedance matching requires a characteristic impedance Z_o if Z is the source impedance and Z_L is the load impedance.

Figure 16-10A shows a circuit in which an unmatched load is connected to a transmission line with characteristic impedance Z_o. The load impedance Z_L is of the form $Z = R \pm jX$, and in this case is equal to 50 – j20. A complex impedance load can be matched to its source by interposing the complex conjugate of the impedance. For example, in the case where Z = 50 – j20, the matching impedance network will require an impedance of 50 + j20 ohms. The two impedances combine to produce a result of 50 ohms. The situation of Fig. 16-10A shows a matching stub with a reactance equal in magnitude, but opposite sign, with respect to the reactive component of the load impedance. In this case, the stub has a reactance of +j20 ohms to cancel a reactance of –j20 ohms in the load.

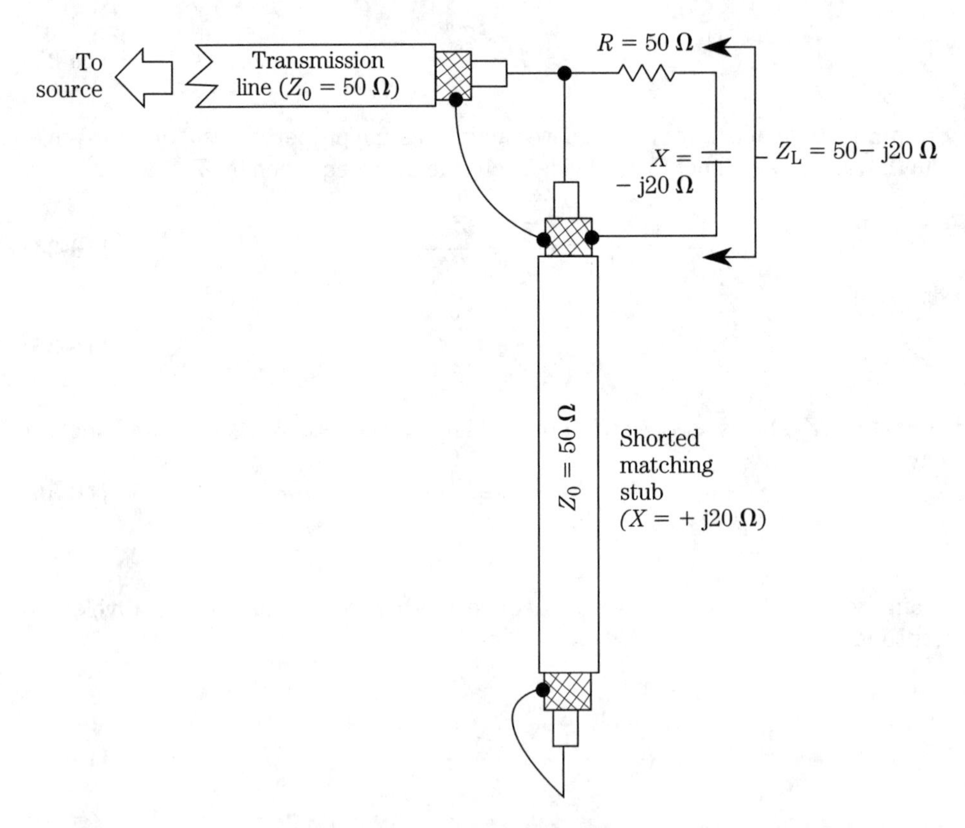

16-10 A. Stub impedance matching system.

A *quarter-wavelength shorted stub* is a special case of the stub concept that finds particular application in microwave circuits. Figure 16-10B shows a quarter-wave stub and its current distribution. The current is maximum across the short, but wave cancellation forces it to zero at the terminals. Because $Z = V/I$, when I goes to zero the impedance goes infinite. Thus, a quarter wave stub has an infinite impedance at its resonant frequency, and thus acts as an insulator. The concept may be hard to swallow, but the stub is a "metal insulator."

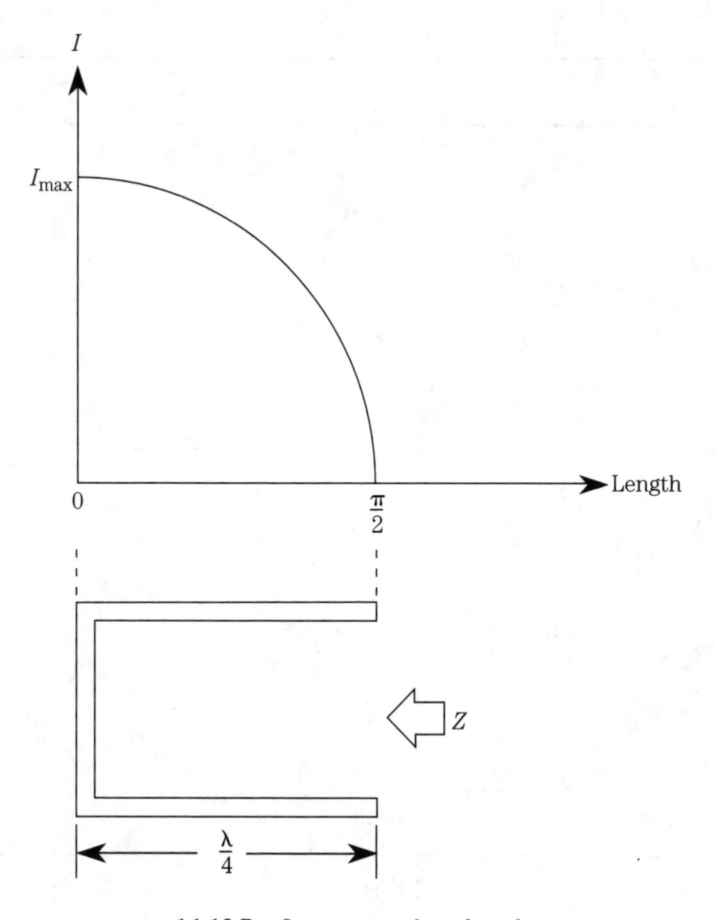

16-10 B. Quarter-wavelength stub.

Standing wave ratio

The reflection phenomenon was noted earlier when we discussed the step function and single-pulse response of a transmission line; the same phenomenon also applies when the transmission line is excited with an ac signal. When a transmission line is not matched to its load, some of the energy is absorbed by the load and some is reflected back down the line towards the source. The interference of incident (or "forward") and reflected (or "reverse") waves creates *standing waves* on the transmission line.

If the voltage or current is measured along the line, it will vary, depending on the load, according to Fig. 16-11. Figure 16-11A shows the voltage-vs-length curve for a matched line, i.e., where $Z_L = Z_o$. The line is said to be "flat" because the voltage (and current) is constant all along the line. But now consider Figs. 16-11B and 16-11C.

Figure 16-11B shows the voltage distribution over the length of the line when the load end of the line is *shorted*, i.e., $Z_L = 0$. Of course, at the load-end the voltage is zero, which results from zero impedance. The same impedance and voltage situation

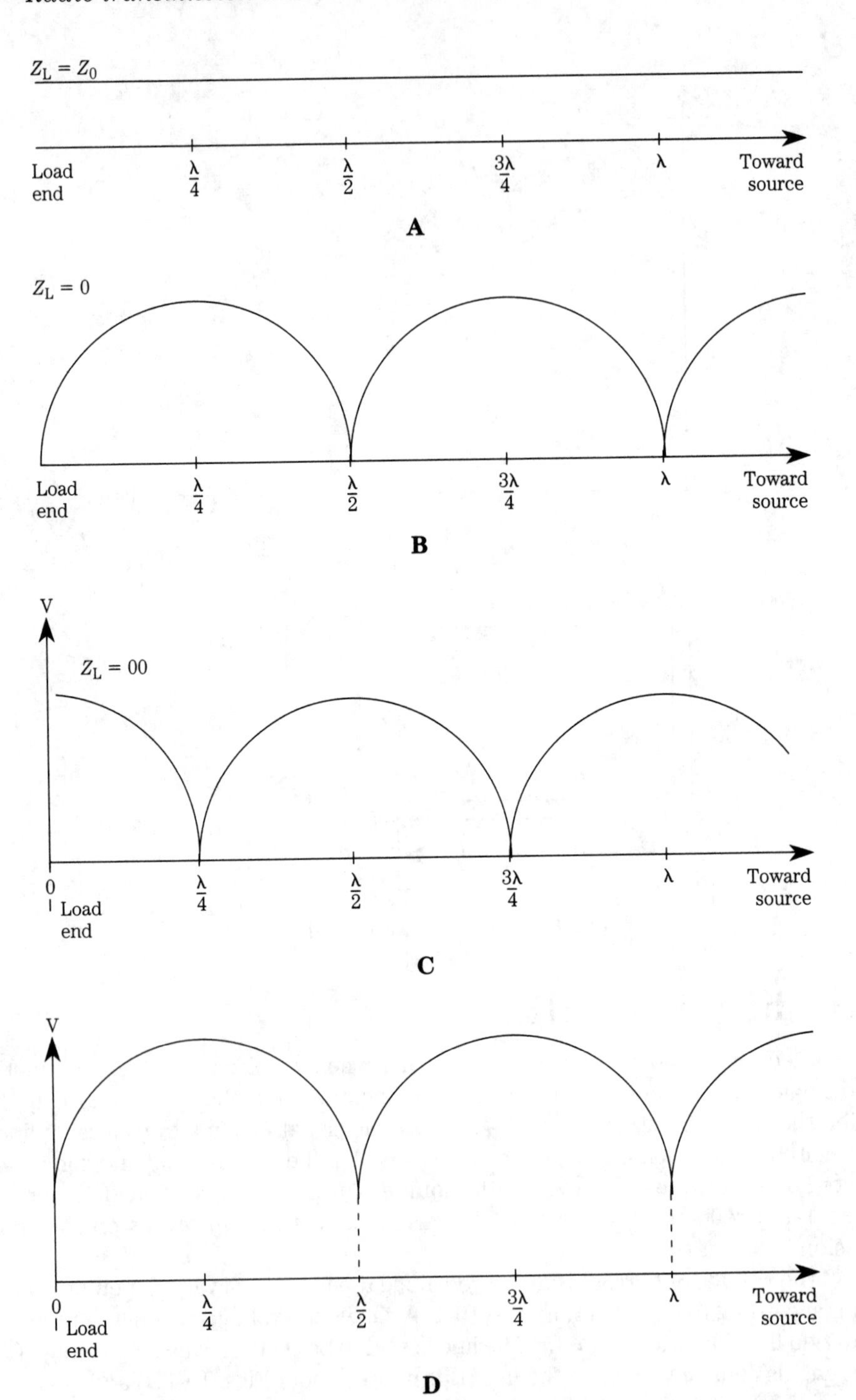

16-11 Voltage or current distribution along transmission line. A. Flat line (matched impedances). B. Shorted load. C. Open load. D. Load impedance not equal to line impedance.

is repeated every half wavelength down the line from the load end towards the generator. Voltage minima are called *nodes*, while voltage maxima are called *antinodes*.

The pattern in Fig. 16-11C results when the line is unterminated (open), i.e., $Z_L = \infty$. Note that the pattern is the same shape as Fig. 16-11B (shorted line), but phase-shifted 90 degrees. In both cases the reflection is 100 percent, but the phase of the reflected wave is opposite.

Figure 16-11D shows the situation in which Z_L is not equal to Z_o, but is neither zero nor infinite. In this case, the node represents some finite voltage, V_{min}, rather than zero. The *standing wave ratio* (SWR) reveals the relationship between load and line.

If the current along the line is measured, the pattern will resemble the patterns of Fig. 16-11. The SWR is then called *ISWR*, to indicate the fact that it came from a current measurement. Similarly, if the SWR is derived from voltage measurements it is called *VSWR*. Perhaps because voltage is easier to measure, VSWR is the term most commonly used in RF work. VSWR can be specified in any of several equivalent ways:

From incident voltage (V_i) and reflected voltage (V_r):

$$\text{VSWR} = \frac{V_i + V_r}{V_i - V_r} \qquad \textbf{[16-40]}$$

From transmission line voltage measurements (Fig. 16-11D):

$$\text{VSWR} = \frac{V_{max}}{V_{min}} \qquad \textbf{[16-41]}$$

From load and line characteristic impedance:

$Z_L > Z_o$:

$$\text{VSWR} = \frac{Z_L}{Z_o} \qquad \textbf{[16-42]}$$

$Z_L < Z_o$:

$$\text{VSWR} = \frac{Z_o}{Z_L} \qquad \textbf{[16-43]}$$

From incident (P_i) and reflected (P_r) power:

$$\text{VSWR} = \frac{1 + \sqrt{P_r/P_i}}{1 - \sqrt{P_r/P_i}} \qquad \textbf{[16-44]}$$

From reflection coefficient (Γ):

$$\text{VSWR} = \frac{1 + \Gamma}{1 - \Gamma} \qquad \textbf{[16-45]}$$

It is also possible to determine the reflection coefficient (Γ) from knowledge of VSWR:

$$\Gamma = \frac{VSWR - 1}{VSWR + 1} \qquad \textbf{[16-46]}$$

The relationship between reflection coefficient (Γ) and VSWR is shown in Fig. 16-12.

VSWR is usually expressed as a *ratio*. For example, when Z_L is 100 ohms and Z_o is 50 ohms, the VSWR is Z_L/Z_o = 100 ohms/50 ohms = 2, which is usually expressed as "VSWR = 2:1." VSWR can also be expressed in decibel form:

$$VSWR = 20 \text{ Log } (VSWR) \qquad \textbf{[16-47]}$$

The SWR is regarded as important in systems for several reasons. At the base of these reasons is the fact that the reflected wave represents energy lost to the load. For example, in an antenna system, less power is radiated if some of its input power is reflected back down the transmission line because the antenna feedpoint impedance does not match the transmission line characteristic impedance. In the next section, we will look at the problem of mismatch losses.

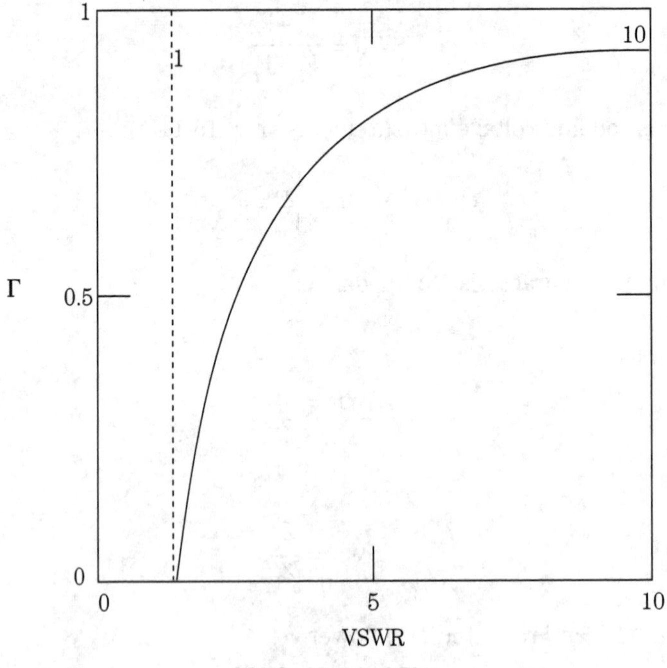

16-12 Mathcad Plot.

Mismatch (VSWR) losses

The power reflected from a mismatched load represents a loss, and will have implications that range from negligible to profound, depending on the situation. For example, the result might be a slight loss of signal strength at a distant point from an antenna, to the destruction of the output device in a transmitter. The latter problem so plagued early solid-state transmitters that designers opted to include shut-down circuitry to sense high VSWR and turn down output power proportionally.

In microwave measurements, VSWR on the transmission lines that interconnect devices under test, instruments and signal sources can cause erroneous readings—and invalid measurements.

Determination of VSWR losses must take into account two VSWR situations. In Fig. 16-9 we have a transmission line of impedance Z_o interconnecting a load impedance (Z_L) and a source with an output impedance Z_S. There is a potential for impedance mismatch at both ends of the line.

In the case where one end of the line is matched, the mismatch loss due to SWR at the mismatched end is:

$$\text{M.L.} = -10 \text{ Log}\left(1 - \left(\frac{\text{SWR} - 1}{\text{SWR} + 1}\right)^2\right) \qquad \textbf{[16-48]}$$

Which from the above equation you may recognize as:

$$\text{M.L.} = -10 \text{ Log}(1 - P^2) \qquad \textbf{[16-49]}$$

The equations above reflect the mismatch loss solution for low loss or "lossless" transmission lines. While a close approximation, there are situations where they are insufficient—namely when the line is lossy. While not very important at low frequencies, loss becomes higher at microwave frequencies. Interference between incident and reflected waves produces increased current at certain antinodes—which increases ohmic losses—and increased voltage at certain antinodes—which increases dielectric losses. It is the latter that increases with frequency. Equation 16-50 relates reflection coefficient and line losses to determine total loss on a given line.

$$\text{Loss} = 10 \text{ Log}\left(\frac{n^2 - \Gamma^2}{n - n\Gamma^2}\right) \qquad \textbf{[16-50]}$$

where

Loss is the total line loss in decibels
Γ is the reflection coefficient
n is the quantity $10^{(A/10)}$
A is the total attenuation presented by the line, in dB, when the line is properly matched ($Z_L = Z_o$)

17
Using pin diodes

Most modern radio transceivers (i.e., transmitter/receiver) use "relayless" switching to go back and forth between the RECEIVE and TRANSMIT states. In many cases, this switching is done with pin diodes. Similarly, IF filters or front-end bandpass filters are selected with a front panel switch that handles only direct current. How? Again, pin diodes. These interesting little components allow us to do switching at RF, IF, and audio frequencies without routing the signals themselves all over the cabinet. In this chapter we will see how these circuits work.

The P-I-N or "pin" diode is different from the pn-junction diode (see Fig. 17-1A): it has an insulating region between the P and N-type material (Fig. 17-1B). It is therefore a multi-region semiconductor device, despite having only two electrodes. The I-region is not really a true semiconductor insulator, but rather is a very lightly doped N-type region. It is called an "intrinsic" region because it has very few charge carriers to support the flow of an electrical current.

When a forward-bias potential is applied to the pin diode, charge carriers are injected into the I-region from both N and P regions. But the lightly doped design of the intrinsic region is such that the N- and P-type charge carriers don't immediately recombine (as in pn-junction diodes). There is always a delay period for recombination. Because of this delay phenomenon, there is always a small but finite number of carriers in the I-region that are uncombined. As a result, the resistivity of the I-region is very low.

One application that results from the delay of signals passing across the intrinsic region is that the pin diode can be used as an RF phase-shifter. In some microwave antennas, phase shifting is accomplished by the use of one or more pin diodes in series with the signal line. Although there are other forms of RF phase shifter (e.g., phase shifter) usable at those frequencies, the pin diode remains popular.

Figure 17-1C shows some of the package styles used for pin diodes at small-signal power levels. All but one of these shapes is familiar to most readers, although

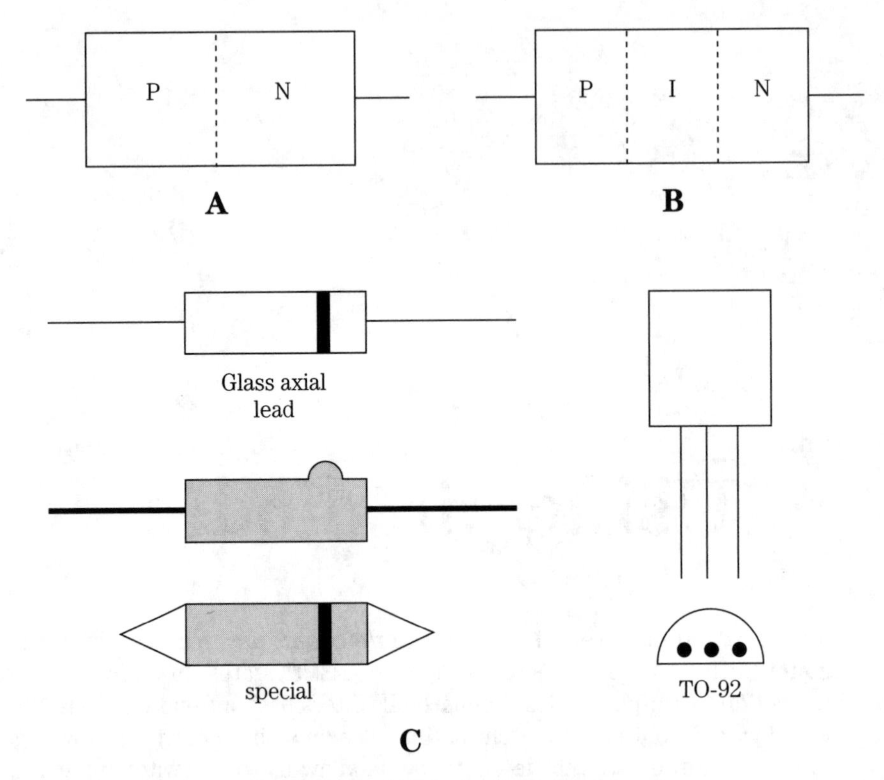

17-1 A. Pn diode. B. Pin diode. C. Pin diode packages.

the odd-shaped flat package is probably recognized only by people with some experience in UHF and up switching circuits. The NTE-553 and ECG-553 pin diode will dissipate 200 mW, and uses the standard cylindrical package style. The NTE-555 and ECG-555 device, on the other hand, uses the UHF flat package style and can dissipate 400 mW. I used these diodes for the experiments I performed to write this chapter because they are service shop replacement lines, and both ECG and NTE are widely distributed in local parts stores. An alternative that might be harder to come by is the MPN3404, which uses the TO-92 plastic package style shown in Fig. 17-1C.

Radio frequency ac signals can pass through the pin device, and under some circumstances, see it as merely a parallel-plate capacitor. We can use pin diodes as electronic switches for RF signals, an RF-delay line or phase-shifter, or as an amplitude modulator.

Pin diode switch circuits

Pin diodes can be used as switches in either the series or parallel modes. Figure 17-2 shows two similar switch circuits. In the circuit of Fig. 17-2A the diode (D1) is placed in series with the signal line. When the diode is turned on, the signal path has a low resistance, and when the diode is turned off, it has a very high resistance (thus provid-

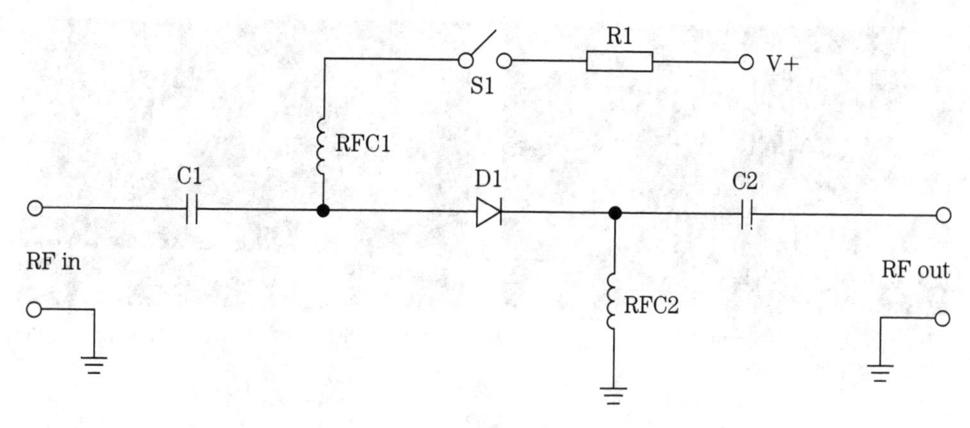

17-2 Series pin diode switching circuits. A. Resistor loaded.

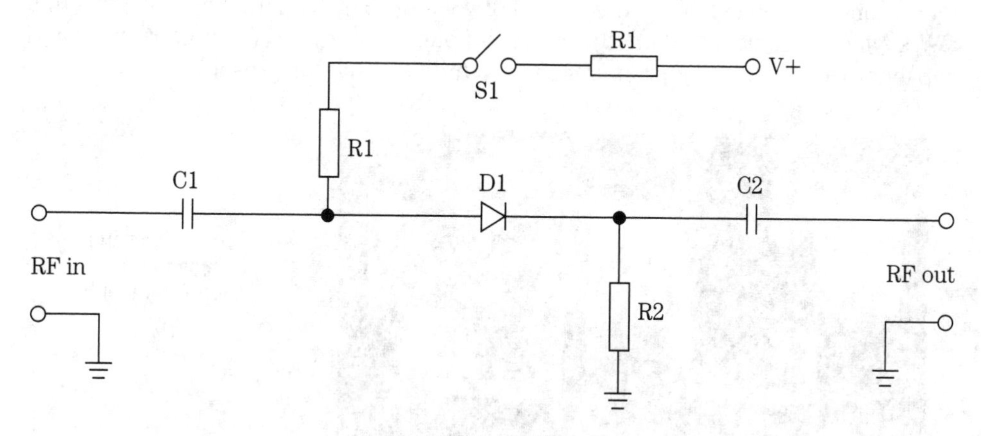

17-2 B. RF choke loaded.

ing the switching action). When switch S1 is open, the diode is unbiased, so the circuit is open by virtue of the very high series resistance. But when S1 is closed, the diode is forward biased and the signal path is now a low resistance. The ratio of off/on resistances provides a measure of the isolation provided by the circuit. A pair of radio frequency chokes (RFC1 and RFC2) are used to provide a high impedance to RF signals, while offering a low dc resistance.

Figure 17-2B is similar to Fig. 17-2A except that the RF chokes are deleted, and a resistor is added. Figure 17-3 shows a test I performed on the circuit of Fig. 17-2B using a 455 kHz IF signal (the 'scope was set to show only a few cycles of the 455 kHz). The oscilloscope trace in Fig. 17-3A shows the "on" position where +12 Vdc was connected through switch S1 to the pin diode's current-limiting resistor. This signal is 1200 mV peak-to-peak. The trace in Fig. 17-3B shows the same signal when the switch was "off" (i.e., +12 Vdc disconnected), but with the oscilloscope set to the same level. It appears to be a straight line. Increasing the sensitivity of the 'scope showed a level of 12 mV getting through. This means that this simplest circuit provides a 100:1 on/off ratio, which is 40 dB of isolation. (Note: for this experiment the ECG-555 and NTE-555 hot carrier pin diodes were used.)

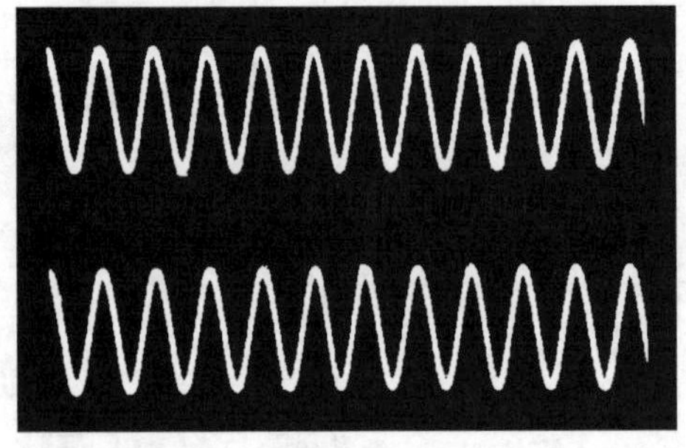

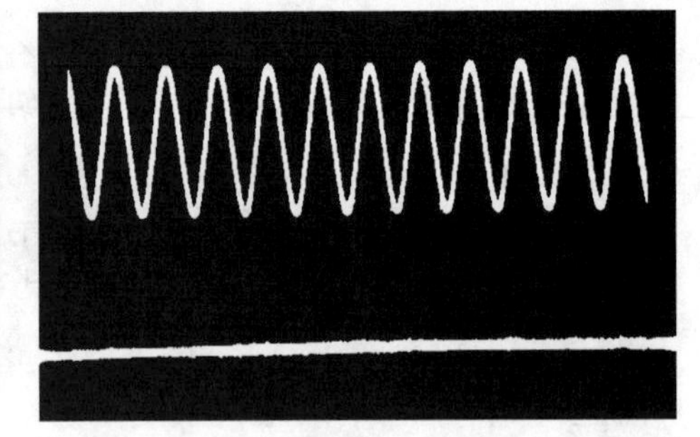

17-3 Pin diode switching action. A. Switch ON. B. Switch OFF (both cases: upper trace is input, lower trace is output).

Figure 17-4 shows the same switch when a square wave is used to drive the pin diode control voltage line, rather than +12 volts dc. This situation is analogous to a CW keying waveform. The photo represents one on/off cycle. With the resistor and capacitor values shown, a pronounced switching transient is present.

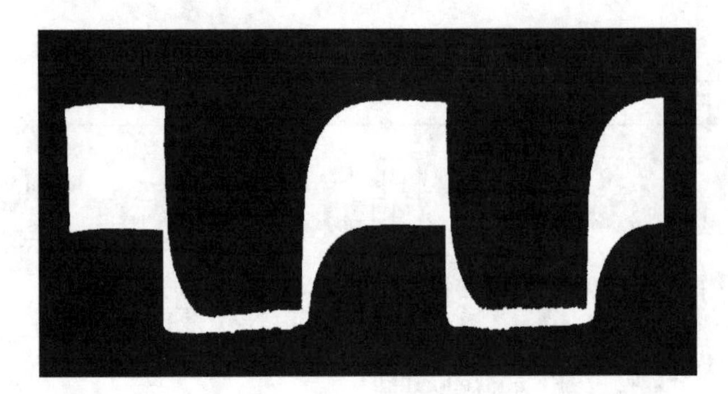

17-4 Pin diode switching using square wave chopper waveform.

Figure 17-5 shows the circuit for a shunt pin diode switch. In this case, the diode is placed across the signal line, rather than in series with it. When the diode is turned off, the resistance across the signal path is high, so operation of the circuit is unimpeded. But when the diode is turned on (S1 closed), a near-short-circuit is placed across the line. This type of circuit is turned off when the diode is forward biased. This action is in contrast to the series switch in which a forward-biased diode is used to turn the circuit on.

A combination series-shunt circuit is shown in Fig. 17-6. In this circuit, D1 and D2 are placed in series with the signal line, while D3 is in parallel with the line. D1 and D2 will turn on with a positive potential applied, while D3 turns on when a negative potential is applied. When switch S1 is in the ON position, a positive potential is applied to the junction of the three diodes. As a result, D1 and D2 are forward-biased, and thus take on a low resistance. At the same time, D3 is hard reverse-biased, so it has a very high resistance. Signal is passed from input to output essentially unimpeded (most pin diodes have a very low series resistance).

But when S1 is in the off position, the opposite situation obtains. In this case, the applied potential is negative so D1/D2 are reverse-biased (and take on a high series re-

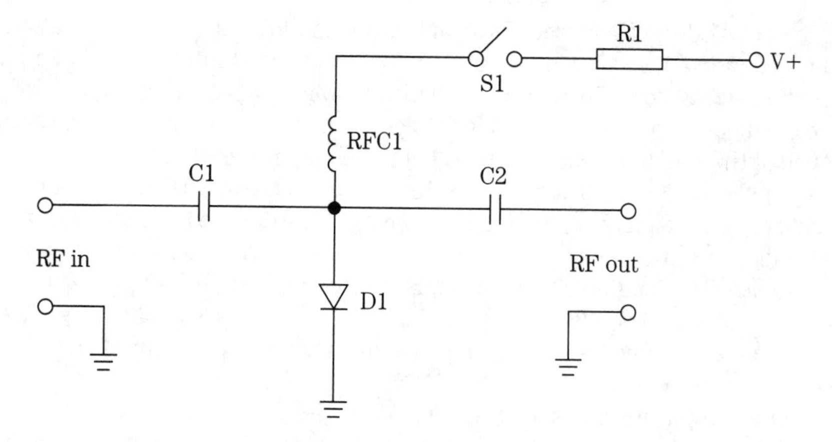

17-5 Shunt pin diode switch.

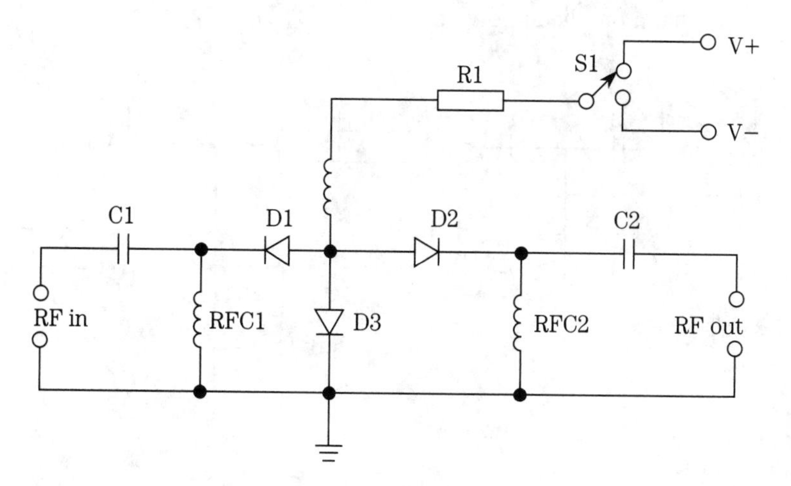

17-6 Series-shunt pin diode switch.

sistance), while D3 is forward-biased (and takes on a low series resistance). This circuit action creates a tremendous attenuation of the signal between input and output.

Pin diode applications

Pin diodes can be used as either a variable resistor or as an electronic switch for RF signals. In the latter case, the diode is basically a two-valued resistor, with one value being very high and the other being very low. These characteristics open several possible applications.

When used as a switch, pin diodes can be used to switch devices such as attenuators, filters and amplifiers in and out of the circuit. It has become standard practice in modern radio equipment to switch dc voltages to bias pin diodes rather than to directly switch RF/IF signals. In some cases, the pin diode can be used to simply short out the transmission path to bypass the device.

The pin diode will also work as an amplitude modulator. In this application, a pin diode is connected across a transmission line, or inserted into one end of a piece of microwave waveguide. The audio modulating voltage is applied through an RF choke to the pin diode. When a CW signal is applied to the transmission line, the varying resistance of the pin diode causes the signal to be amplitude modulated.

A popular police radar-jamming technique amongst another generation of "bandit techees" was to place a pin diode in a cavity, λ/4 from the rear, that was resonant at the radar gun frequency and fed at its open end with a horn antenna, and then modulate it with an audio sine wave equal to the doppler shift expected of either a 125-MPH or 0-MPH speeds. Don't try it, by the way, the cop's got better guns even if they do cause an—ahem—"personal problem" when left on the seat between their legs for any length of time.

Another application is shown in Fig. 17-7. Here, we have a pair of pin diodes used as a transmit-receive (TR) switch in a radio transmitter; models from low-HF to microwave use this technique. Where you see a so-called "relayless TR switch," it is almost certain that a pin diode network such as Fig. 17-7 is in use.

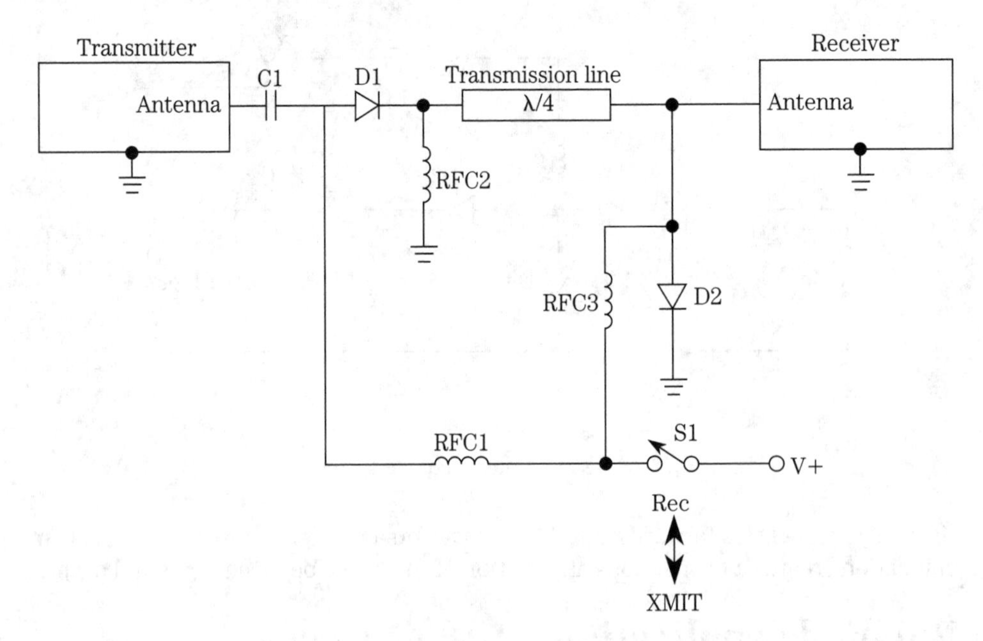

17-7 Pin diode T/R switching.

When switch S1 is open, diodes D1 and D2 are unbiased, and present a high impedance to the signal. Diode D1 is in series with the transmitter signal, and blocks it from reaching the antenna; diode D2, on the other hand, is across the receiver input and does not attenuate the receiver input signal at all. But when switch S1 is closed, the opposite situation occurs: both D1 and D2 are now forward biased. Diode D1 is now a low resistance in series with the transmitter output signal, so the transmitter is effectively connected to the antenna. Diode D2 is also a low resistance, is across the receiver input and causes it to short out. The isolation network can be either a quarter-wave transmission line, microstrip line designed into the printed circuit board, or an LC pi-section filter.

Transmitters up to several kilowatts have been designed using this form of switching, and almost all current VHF/UHF portable "handi-talkies" use pin diode switching. Higher-powered circuits require larger diodes than were discussed in this chapter, but they are easily available from industrial distributors.

Figure 17-8 shows how multiple IF bandpass filters are selected, with only dc being routed around the cabinet between circuitry and front panel. A set of input and output pin diode switches are connected as shown, and fed with a switch that selects either –12 Vdc or +12 Vdc alternately. When the switch is in the position shown, the +12 is to the filter 1 switches, so filter 1 is activated. When the switch is in the opposite position, the alternate filter is turned on. This same arrangement can be used in the front end of the receiver, or the local oscillator, to select LC components for different bands.

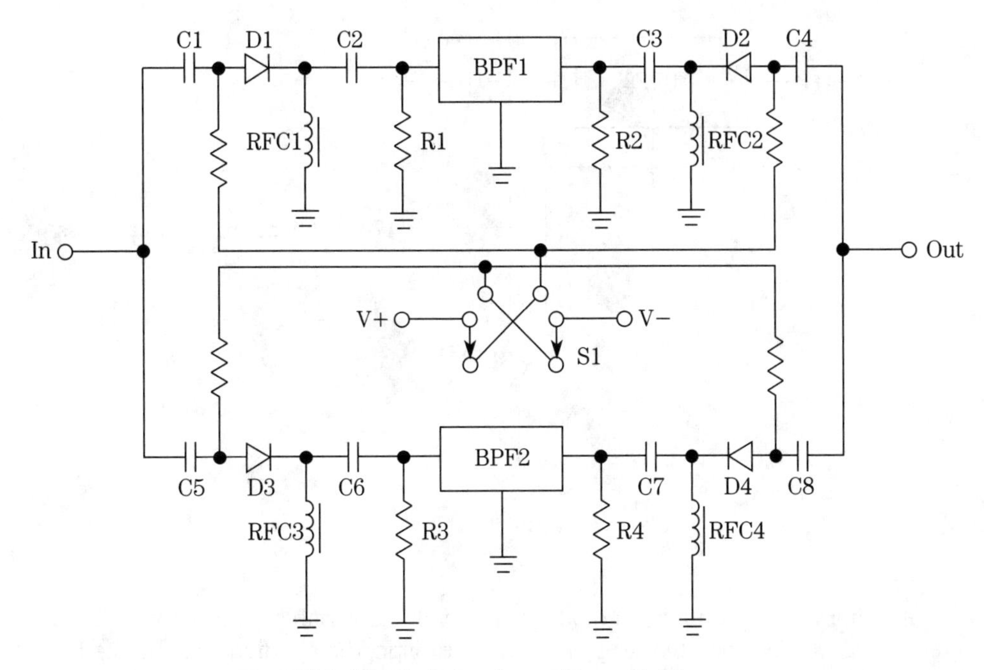

17-8 Pin diode bandpass filter switching.

Another filter selection method is shown in Fig. 17-9. This circuit is a partial representation of the front-end circuitry for the Heathkit SW-7800 general-coverage shortwave receiver. The entire circuit differs from this variant in that a total of six filter sections are used rather than just two, as shown in Fig. 17-9. In each bandpass filter (BPF1 and BPF2), a network of inductor and capacitor elements are used to set the center and both edge frequencies of the band to be covered.

The circuit of Fig. 17-9 is shown using a switch (S1) to apply or remove the +12 Vdc bias potential on the diodes, but in the actual receiver this potential is digitally controlled. The digital logic elements sense which of thirty bands are being used, and selects the input RF filter accordingly.

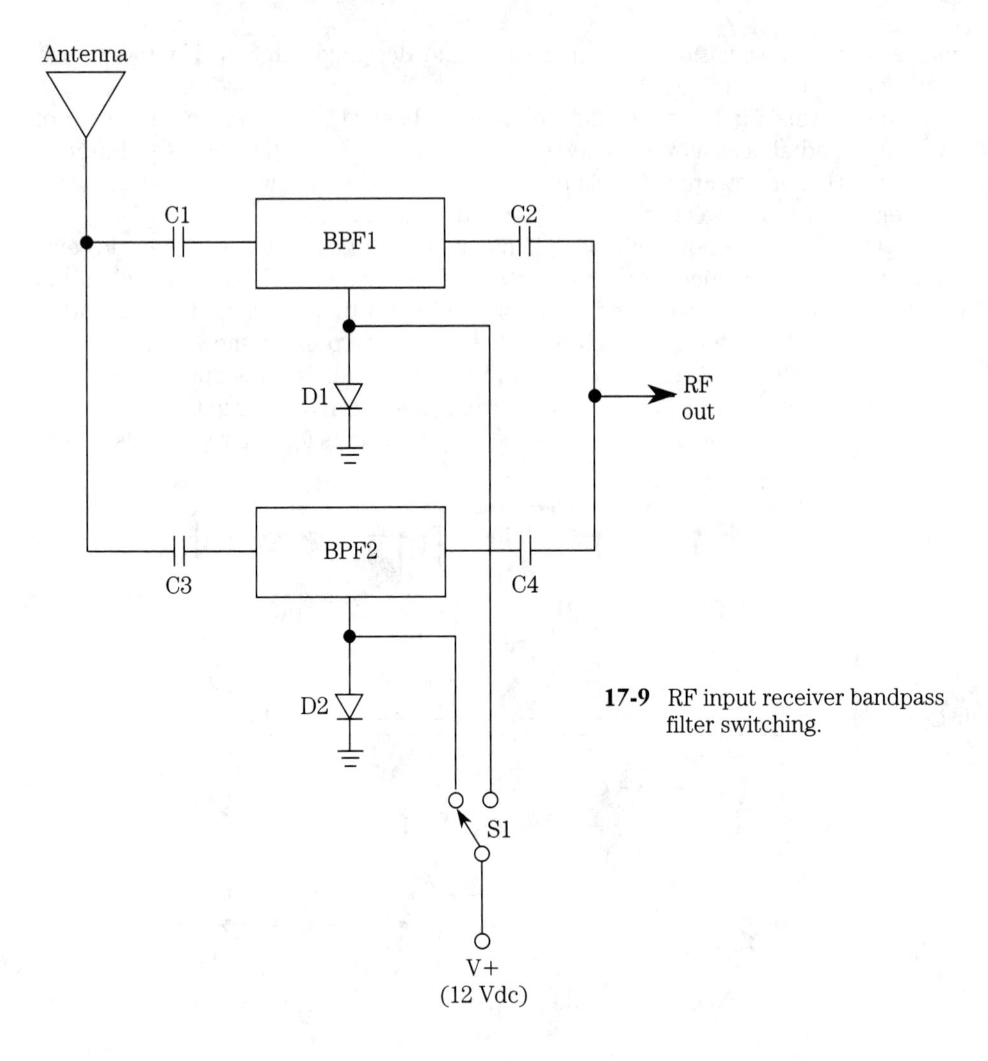

17-9 RF input receiver bandpass filter switching.

Another application for pin diodes is as a voltage-variable attenuator in RF circuits. Because of its variable resistance characteristic, the pin diode can be used in a variety of attenuator circuits. One of the simplest is the shunt attenuator of Fig. 17-10. The front end of this circuit is a bank of selectable bandpass filters per Fig. 17-9, so we will not dwell on that topic now. But at the output of the filter bank there is to ground a series combination of a capacitor (C1) and a pin diode (D1) shunted from the signal line.

The pin diode acts like an electronically variable resistor. The resistance across the diode's terminals is a function of the applied bias voltage. This voltage, hence the degree of attenuation of the RF signal, is proportional to the setting of potentiometer R1. The series resistor (R2) is used to limit the current when the diode is forward biased. This step is necessary because the diode becomes a very low resistance when a certain rather low potential is exceeded. This circuit is used as an RF gain control on some modern receivers.

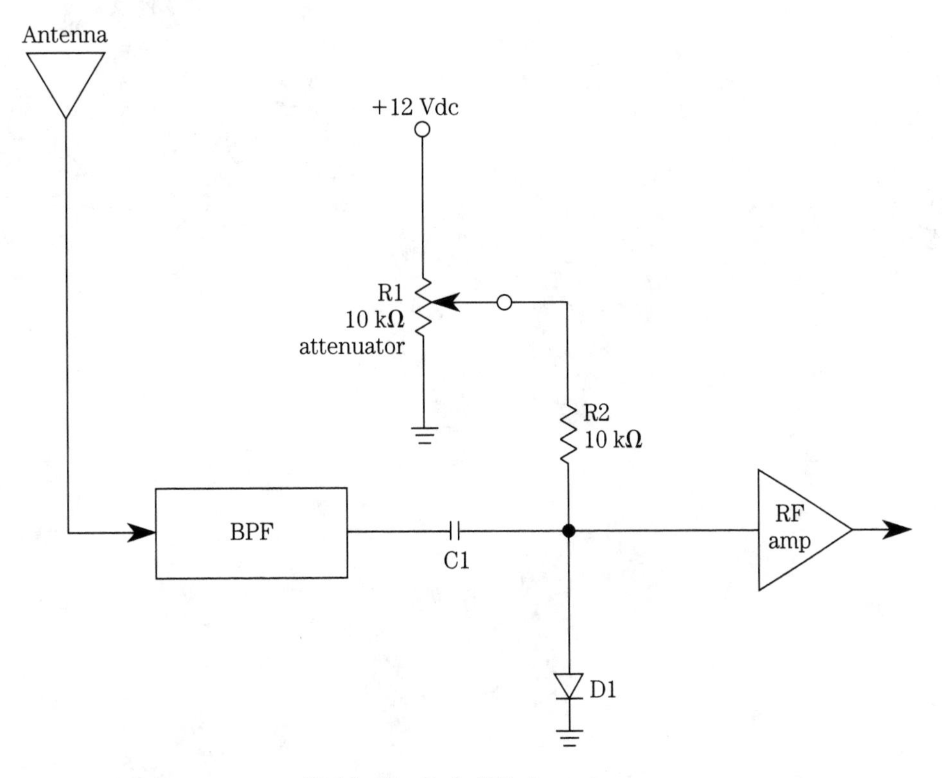

17-10 Pin diode RF attenuator.

Conclusion

Pin diodes are one of those components that are little known, but are nonetheless very useful. If you build any RF projects that could involve traditional switching, you will probably find that the pin diode makes a dandy substitute that avoids the problems associated with routing signal around switching circuits. Try it, you'll like it.

18 Microwave integrated circuits

Very wideband amplifiers have a bandpass (frequency response) of several hundred megahertz, or more, typically ranging from sub-VLF to the low end of the microwave spectrum. An example might be a range of 100 kHz to 1000 MHz (i.e., 1 GHz), although somewhat narrower ranges are more common. These circuits have a variety of practical uses: receiver preamplifiers, signal generator output amplifiers, buffer amplifiers in RF instrument circuits, cable television line amplifiers, and many other applications in communications and instrumentation. As valuable as they are, they were not found in many electronic hobbyist situations until very recently. Very wideband amplifiers are rarer than narrow band amplifier circuits because they were difficult to design and build until the advent of monolithic microwave integrated circuit (MMIC) devices.

Several factors contribute to the difficulty of designing and building very wideband amplifiers. For example, there are too many stray capacitances and inductances in a typical circuit layout, and they form resonances and filters that distort the frequency response characteristic. Circuit resistances also combine with the capacitances to effectively form low-pass filters that roll-off the frequency response at higher frequencies, sometimes drastically. If the RC phase shift of the circuit resistances and capacitances is 180 degrees at a frequency where the amplifier gain is ≥1 (and in very wideband circuits that is likely), and the amplifier is an inverting type (producing an inherent 180-degree phase shift), then the total end-to-end phase shift is 360 degrees—which is the criterion for self-oscillation.

If you ever tried to build a very wideband amplifier, it was probably a very frustrating experience. Until now. Because of new, low-cost devices called *silicon monolithic microwave integrated circuits* (MMICs), reportedly developed in large part for the benefit of the cable television industry, it is possible to design and build amplifiers that cover the spectrum from near-dc to about 2000 MHz, and use seven or fewer components. These devices offer gains of 13 to 30 dB of gain (see

Table 18-1), and produce output power levels up to 40 mW (+16 dBm). Noise figures range from 3.5 to 7 dB. Although several manufacturers offer products, those of Mini-Circuits (P.O. Box 350166, Brooklyn, NY, 11235-0003) are the most easily obtained by electronic hobbyists and amateur radio operators. In this chapter, we will examine the low-cost MAR-x series of MMIC amplifiers.

Table 18-1

Type number	Color dot	Gain (500 MHz) dB	Max. freq.
MAR-1	Brown	17.5	1000 MHz
MAR-2	Red	12.8	2000 MHz
MAR-3	Orange	12.8	2000 MHz
MAR-4	Yellow	8.2	1000 MHz
MAR-6	White	19	2000 MHz
MAR-7	Violet	13.1	2000 MHz
MAR-8	Blue	28	1000 MHz

Figure 18-1 shows the circuit symbol for the MAR-x devices. Note that it is a very simple device. The only connections are RF input, RF output and two ground connections. The use of dual grounds distributes the grounding, reducing the overall inductance and thereby improving the ground connection. Direct current (dc) power is applied to the output terminal through an external network. But more of that shortly.

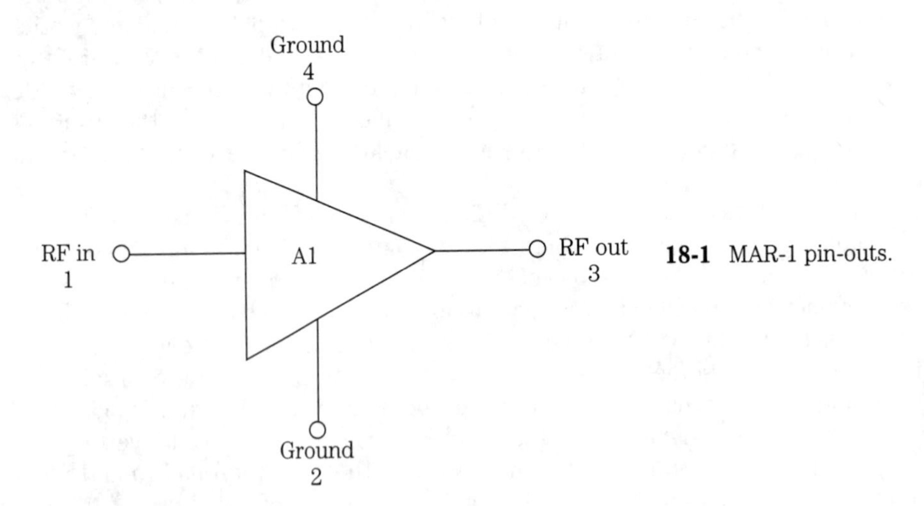

18-1 MAR-1 pin-outs.

The package for the MAR-x device is shown in Fig. 18-2. Although it is an IC, the device looks very much like a small UHF/microwave transistor package. The body is made of plastic, and the leads are wide metal strips (rather than wire) in order to reduce the stray inductance that narrower wire leads would exhibit. These devices are so small that handling them can be difficult; I found that hand forceps (tweezers) were necessary to position the device on a prototype printed circuit board. A magni-

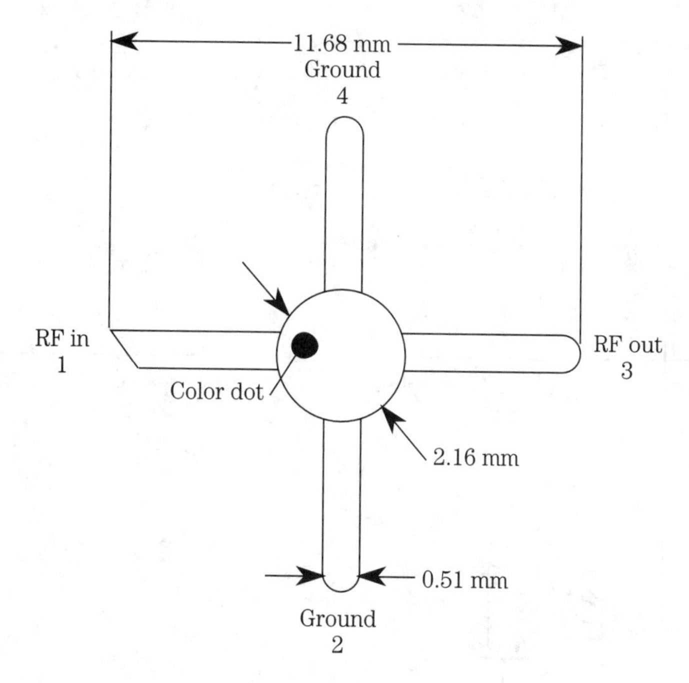

18-2 MAR-1 package.

fying glass or jeweler's eye loupe is not out of order for those with poor close-in eyesight. A color dot, and a beveled tip on one lead, are the keys that identify pin 1 (the RF input connection). When viewed from above, pin numbering (1,2,3, and 4) proceeds counterclockwise from the keyed pin.

Internal circuitry

The MAR-x series of devices inherently matches the 50-ohm input and output impedances without external impedance transformation circuitry, making it an excellent choice for general RF applications. Figure 18-3 shows the internal circuitry for the MAR-x devices. These devices are silicon bipolar monolithic ICs in a two transistor Darlington amplifier configuration. Because of the Darlington connection, the MAR-x devices act like transistors with very high gain. Because the transistors are biased internal to the MAR-x package, the overall gains are typically 13 to 33 dB, depending on the device selected and operating frequency. No external bias or emitter bias resistors are needed, although a collector load resistor to V+ is used.

The good match to 50 ohms for both input and output impedances (R) is due to the circuit configuration, and is approximately:

$$R = \sqrt{R_f R_E} \qquad \textbf{[18-1]}$$

If R_f is about 500 ohms, and R_E is about 5 ohms, then the square root of their product is the desired 50 ohms.

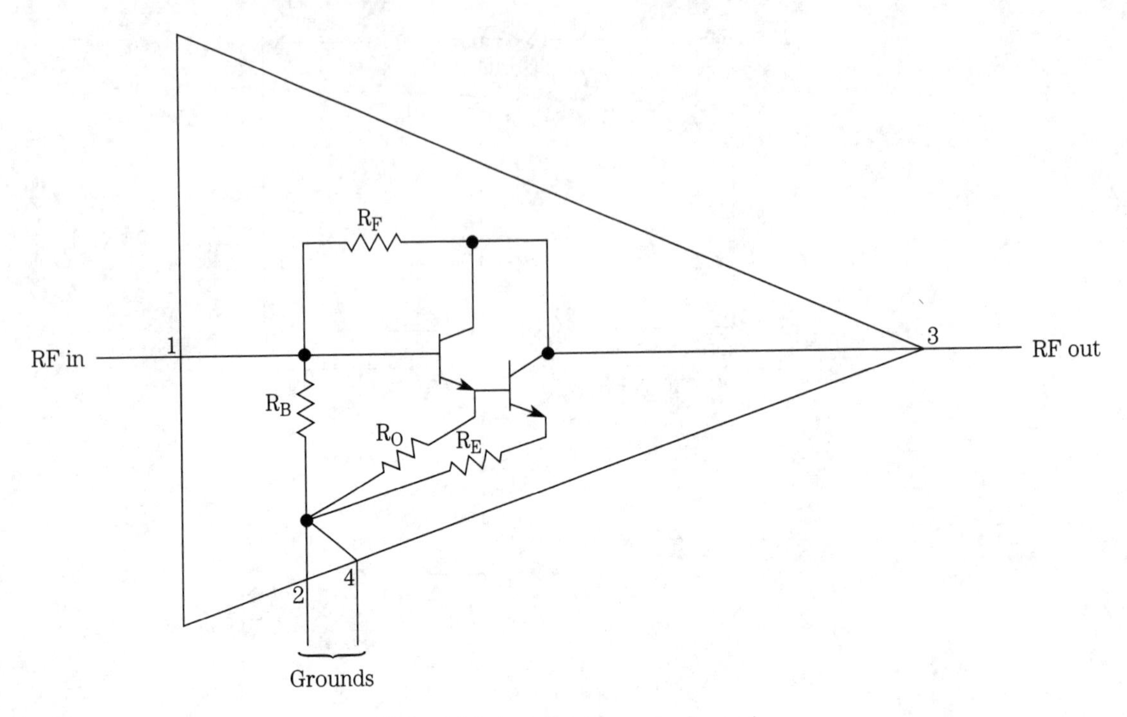

18-3 MAR-1 partial internal schematic.

Basic circuit

The basic circuit for a wideband amplifier project based on the MAR-x device is shown in Fig. 18-4. The RF IN and RF OUT terminals are protected by dc-blocking capacitors C1 and C2. For VLF and MW applications, use 0.01 μF disk ceramic capacitors, and for HF through the lower VHF (≤ 100 MHz) use 0.001 μF disk ceramic capacitors. But, if the project must work well into the high VHF through low microwave region (>100 MHz to 1000 MHz or so), then opt for 0.001 μF "chip" capacitors. If there is no requirement for lower frequencies, then chip capacitors in the 33 to 100 pF range can be used.

The capacitors used for C1 and C2 should be chip capacitors in all but the lower frequency (<100 MHz) circuits. Chip capacitors can be a bit bothersome to use, but their use pays ever greater dividends as operating frequency increases.

Capacitor C3 is used for two purposes. It will prevent signals from A1 from being coupled to the dc power supply, and from there to other circuits. It will also prevent higher frequency signals and noise spikes from outside sources from affecting the amplifier circuit. In some cases, a 0.001 μF chip capacitor is used at C3, but for the most part a 0.01 μF disk ceramic will suffice.

The other capacitor at the dc power supply is a 1 μF tantalum electrolytic. It serves to decouple low frequency signals, and smooth out short duration fluctuations in the dc supply voltage. Higher values than 1 μF may be required if the amplifier is used in particularly noisy environments.

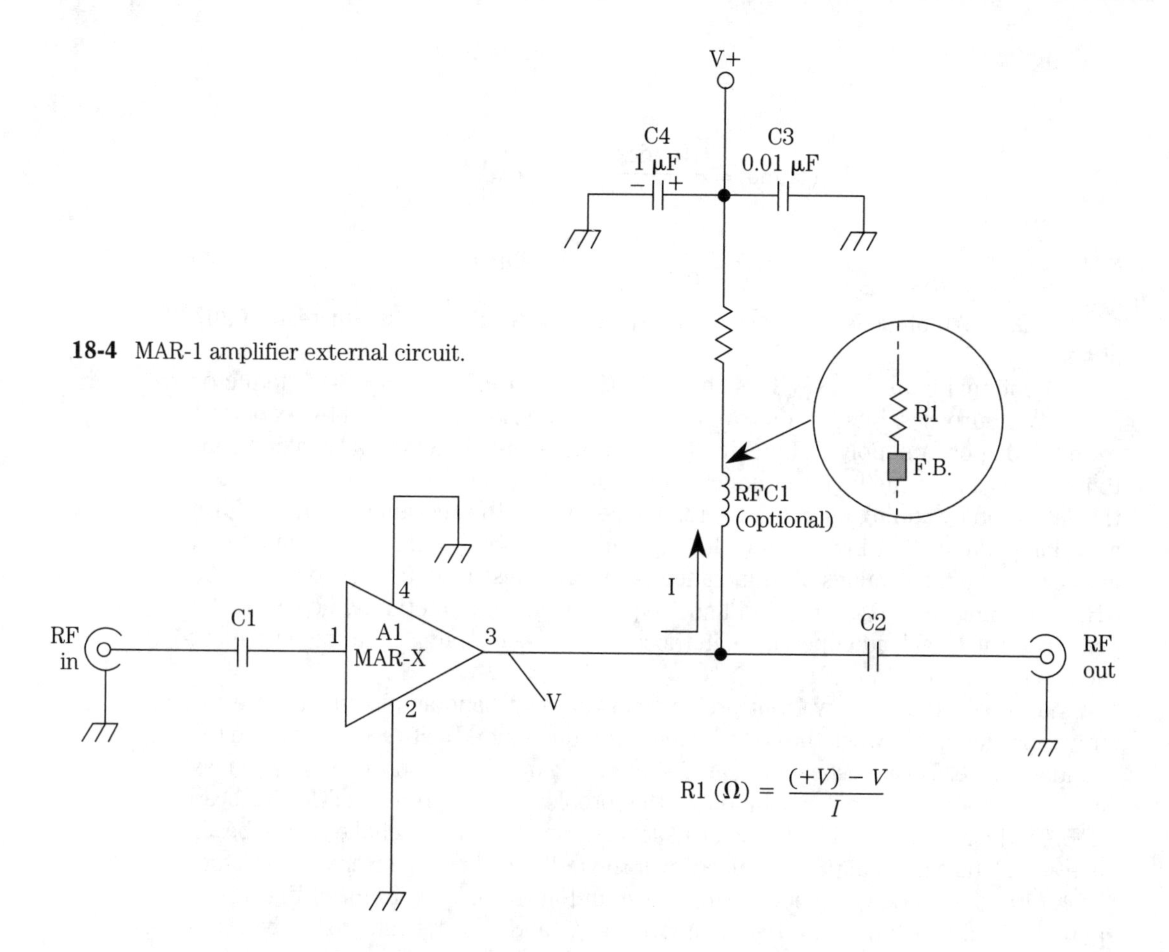

18-4 MAR-1 amplifier external circuit.

Direct current is fed to the amplifier through a current limiting resistor (R1), via the RF OUT terminal on the MAR-x (lead 3). The maximum allowable dc potential is +7.5 Vdc for MAR-8, +5 Vdc for MAR-1 through MAR-4, +4 Vdc for MAR-7, and 3.5 Vdc for MAR-6. If a minimum voltage V+ power supply is used, e.g., +5 Vdc for MAR-1, then make R1 a 47-ohm to 100-ohm resistor. Use only ¼-watt or ½-watt noninductive resistors, such as the carbon composition or metal film types. The use of higher V+ potentials (e.g., +9 to +12 Vdc) is necessary, then use a higher value noninductive resistor for R1. To determine the value of R1, decide on a current level (I), and do an Ohm's law calculation:

$$R1 \text{ (ohms)} = \frac{(V+) - V}{I} \qquad \textbf{[18-2]}$$

In most cases, a good operating current level for the popular MAR-1 is about 15 mA (or 0.015 A).

Example 18-1

Calculate a value for R_1 in a MAR-1 circuit when a +9 Vdc transistor radio battery is used for the V+ dc power supply. Assume I = 15 mA.

$$R_1 \text{ (ohms)} = \frac{(V+) - V}{I}$$

$$R_1 \text{ (ohms)} = \frac{(9 \text{ Vdc}) - (5 \text{ Vdc})}{0.015 \text{ A}}$$

$$R_1 \text{ (ohms)} = \frac{4 \text{ volts}}{0.015 \text{ A}} = 267 \text{ ohms}$$

Because 270 ohms is a nearby standard value, it would be used instead of 267 ohms.

An optional inductor, RFC1, is shown in the circuit of Fig. 18-4. This inductor serves two purposes. First, it improves the decoupling isolation of the MAR-x output from the dc power supply by blocking RF signals. Second, it acts as a "peaking coil" to improve gain on the high frequency end of the frequency response curve. It does this latter job by adding its inductive reactance (X_L) to the resistance of R1 to form a load impedance that increases with frequency because $X_L = 2\pi FL$. Depending on application, suitable values of inductance range from less than 0.5 μH to about 100 μH, depending on the application and frequency range. Sometimes, however, the coil forms the total load impedance. In those cases, a decoupling capacitor is used at the junction of RFC1 and R1.

Inductor coils are not without problems in very wideband amplifiers because the stray capacitances between the coil windings form unintended self-resonances with the coil inductance. These resonances can distort the frequency response curve and may cause oscillations. A popular solution to this problem is to use a small ferrite bead ("F.B." in the inset to Fig. 18-4). The bead acts as a small-value RF choke. These beads have a small hole in them that fits nicely over the radial lead of a quarter watt resistor.

Alternative dc power schemes are shown in Fig. 18-5. The circuit of Fig. 18-5A splits the load resistance into two components, R_1 and R_2. The value of R_1 will represent most of the required resistance, with R_2 typically being 33 to 100 ohms. This circuit, like the basic circuit, works well to V+ voltages of 7 to 9 Vdc, but is not recommended for V+ > 9 Vdc.

Power feed schemes that work well at V+ voltages greater than 9 Vdc are shown in Figs. 18-5B and 18-5C. Both use voltage regulation to stabilize the supply voltage to the MAR-x device. In Fig. 18-5B, a 68-ohm Vdc zener diode holds the voltage applied to R1 constant, and within an acceptable range, despite fluctuations in the source V+ potential.

Other MAR-x circuits

The simple circuit of Fig. 18-4 will work well in most cases, especially where the input and output impedance are reasonably stable. But if the input source or output load impedances vary, the amplifier may suffer a degradation of performance, or show some instability. One solution to the problem is to use resistive attenuator pads in the input and output signal lines. Attenuators in an amplifier circuit? Yes, that's right. A 1-dB or 2-dB attenuator in the input and output signal lines will pseudostabilize the impedances seen by the amplifier, but only marginally affects the

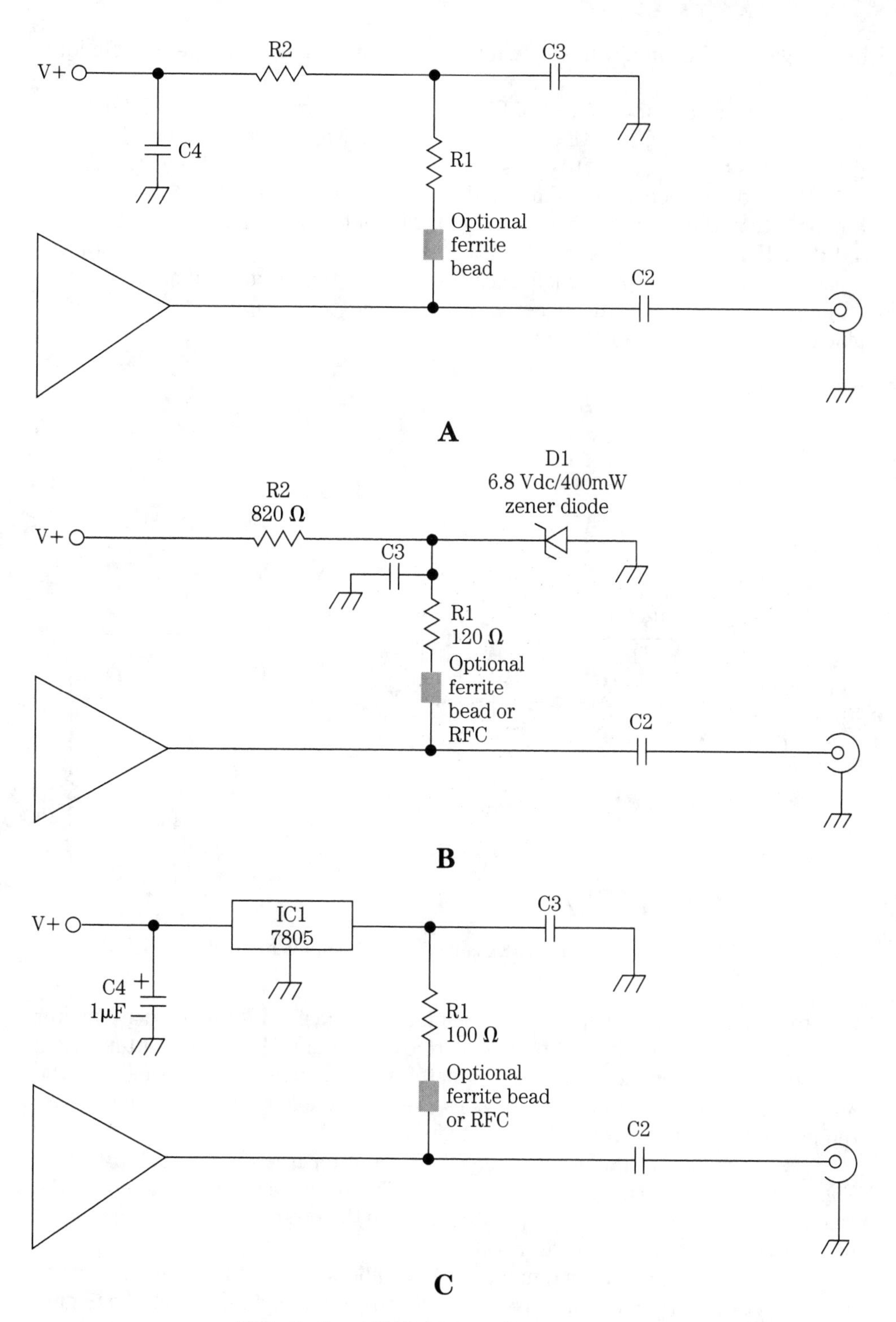

18-5 Optional MAR-1 output/dc power circuits.

overall gain of the circuit. In vacuum tube days, we called this type of technique "swamping."

Figure 18-6 shows the circuit of Fig. 18-4 revised to reflect the use of simple resistive attenuator pads in the input and output lines. With resistor values of 6.2 ohms for the series element, and 910 ohms for the two shunt lines, the attenuation factor is 1 dB. A 2-dB version uses 12 ohms and 470 ohms, respectively. If 1 dB attenuators are used, then the overall gain is the natural gain of the MAR-x device less 2 dB (or 4 dB if 2 dB attenuator pads are used). The resistors used for these attenuator pads must be noninductive types, such as carbon composition or metal film units. If the amplifier is to be used at the higher end of its range, then chip resistors are preferable to ordinary axial lead resistors.

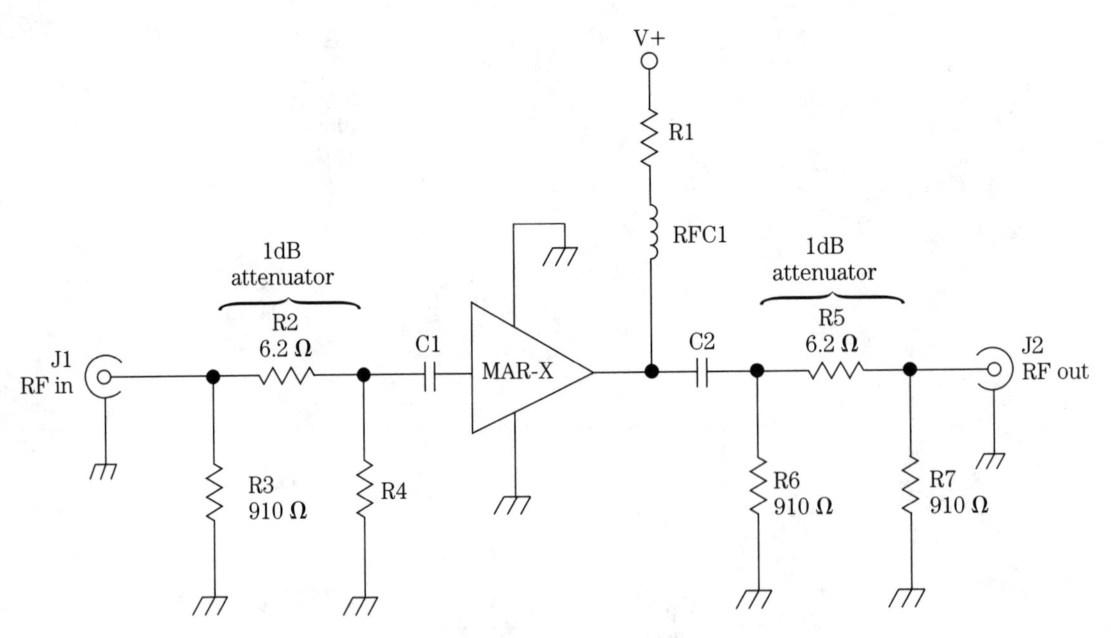

18-6 Expanded MAR-1 amplifier circuit.

An alternative approach is to use manufactured shielded RF 50-ohm attenuator pads. Another of Mini-Circuits' products are the AT-1 and MAT-1 1-dB attenuators; they are suitable for the purpose, and match the frequency range of most of the MAR-x products. These low-cost devices are similar except for size, and are intended for mounting on printed circuit boards.

Keep in mind that the use of attenuators is not for free (TANSTAFL principle: There Ain't No Such Thing As a Free Lunch). The resistive attenuators reduce the gain (as mentioned before), but they also increase the noise factor by an amount set by the loss factor of the attenuator pad.

For VHF, UHF, and low-end microwave amplifiers, it might be preferable to use a printed circuit strip-line transmission line for the input and output circuits. Figure 18-7A shows such a circuit with input (SL1) and output (SL2) strip lines, while Fig.

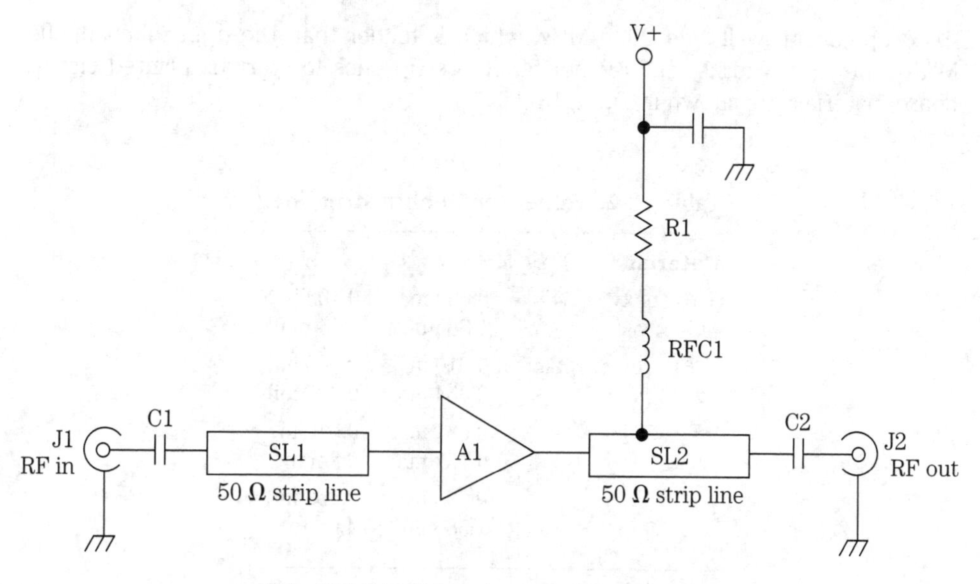

18-7 VHF/UHF MAR-1 preamplifier. A. Circuit.

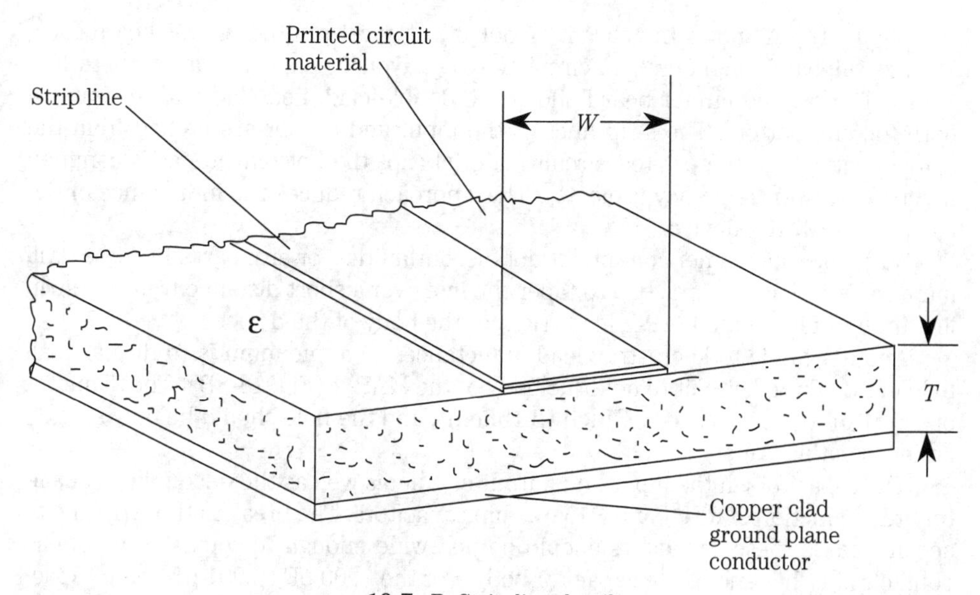

18-7 B. Strip-line detail.

18-7B shows detail of how these lines are made. The characteristic impedance (Z_o) of the line is a function of the relative dielectric constant of the printed circuit material (ε), the thickness of the material (*T* in Fig. 18-7B), and the width (*W* in Fig. 18-7B) of the strip line conductor. Common Epoxy G-10 printed circuit boards (ε ≈ 4.8) are usable to 1000 MHz and work well to about 300 MHz. Above 300 MHz, the losses increase significantly. PTFE-woven glass fiber printed circuit boards (ε ≈

2.55) operate to well over 2000 MHz, which is higher than the upper limit of the MAR-x devices. Widths required for 50-ohm strip lines for various printed circuit board materials are shown in Table 18-2.

Table 18-2. Values for 50-ohm strip line.

Material	ε	T	W
G-10 epoxy fiberglass	4.8	0.062 in 1.58 mm	0.108 in 2.74 mm
PTFE woven glass fiber	2.55	0.010 in 0.254 mm	0.025 in 0.635 mm
		0.031 in 0.787 mm	0.079 in 0.2 mm
		0.062 in 1.58 mm	0.158 in 4 mm

Figure 18-8A shows the circuit layout of a typical printed circuit board for a MAR-x wideband amplifier. The circuit for the layout is shown as an inset to Fig. 18-8A. The printed circuit board should be double-clad, i.e., clad with copper on both top and bottom. The strip lines at the input and output are etched from the component side of the printed circuit material (not the bottom side as is common practice in lower frequency projects). This approach reduces the inductance of the leads to the MAR-x device.

Strip lines should not contain abrupt discontinuities, or else parasitic losses will increase. It is common practice to taper the line over a short distance from the strip line to the width of the MAR-x leads right at the body of the device.

Another tactic to keep stray lead inductances to a minimum is to drill a small hole in the printed circuit to hold the body of the MAR-x (Fig. 18-8B). The diameter of the MAR-x package is 0.085 inch (2.15 mm), and the hole should be only slightly larger than this value.

The capacitors in the input and output circuit, as well as the decoupling capacitor at the junction of RFC1 and R1, are chip capacitors. The break in the strip line to accommodate these capacitors should be just wide enough to separate the ohmic contacts at either end of the capacitor body. For the 1000 pF (0.001 μF) chip capacitors that I used in making a model in preparation for this chapter, the insulated center section between contacts on the capacitors averaged 0.09 inch (2.3 mm) as measured on a vernier caliper set.

It is essential to keep ground returns as short as possible, especially when the amplifier will be operated in the higher end of its range. If you opt to use the ground plane cladding for the dc and signal return, plated through holes are required between the two sides of the board. These plated through holes must be placed directly below the ground leads of the MAR-x package.

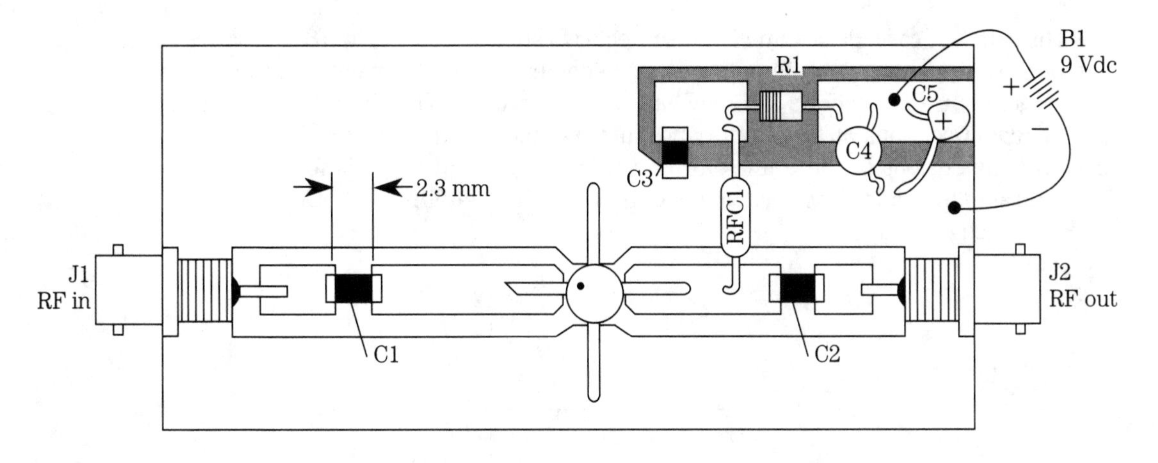

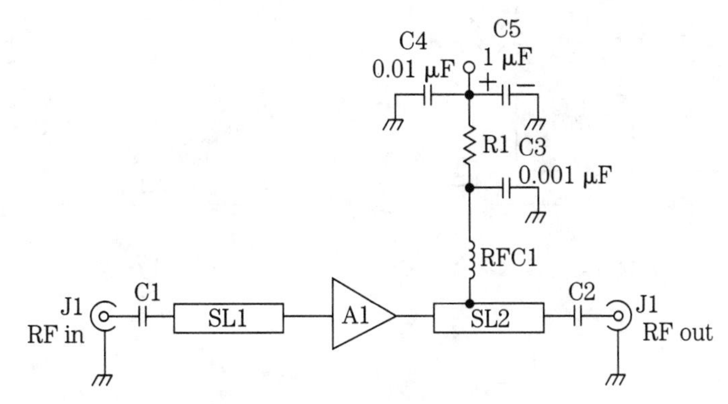

18-8 A. Printed circuit board layout.

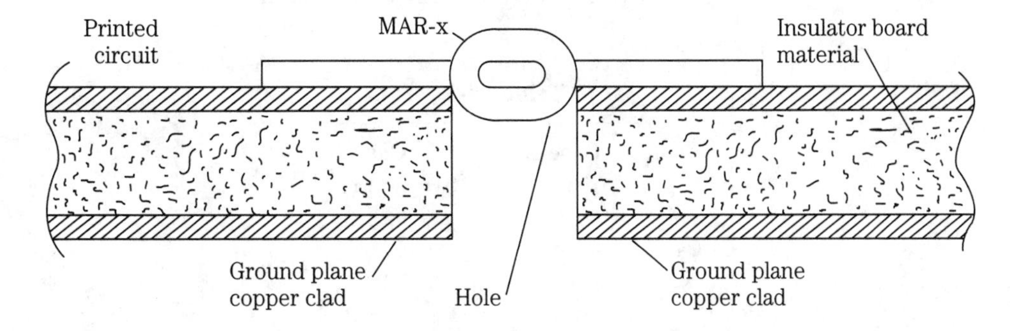

18-8 B. MAR-1 mounting.

Multiple device circuits

The MAR-x devices can be connected in cascade, parallel or push-pull. The cascade connection increases the overall gain of the amplifier, while the parallel and push-pull configurations increase the output power available.

The simplest cascade scheme is to connect two stages such as Fig. 18-4 in series so that the output capacitor of the first stage becomes the input capacitor of the second stage. Figure 18-9 shows a somewhat better approach. This circuit uses stripline matching sections at the inputs and outputs, and between stages. Table 18-3 gives the dimensions of these lines for two different cases: Case-A is for a 100- to 500-MHz amplifier, and Case-B is for a 500- to 2000-MHz amplifier. In both cases, the MAR-8 device is used.

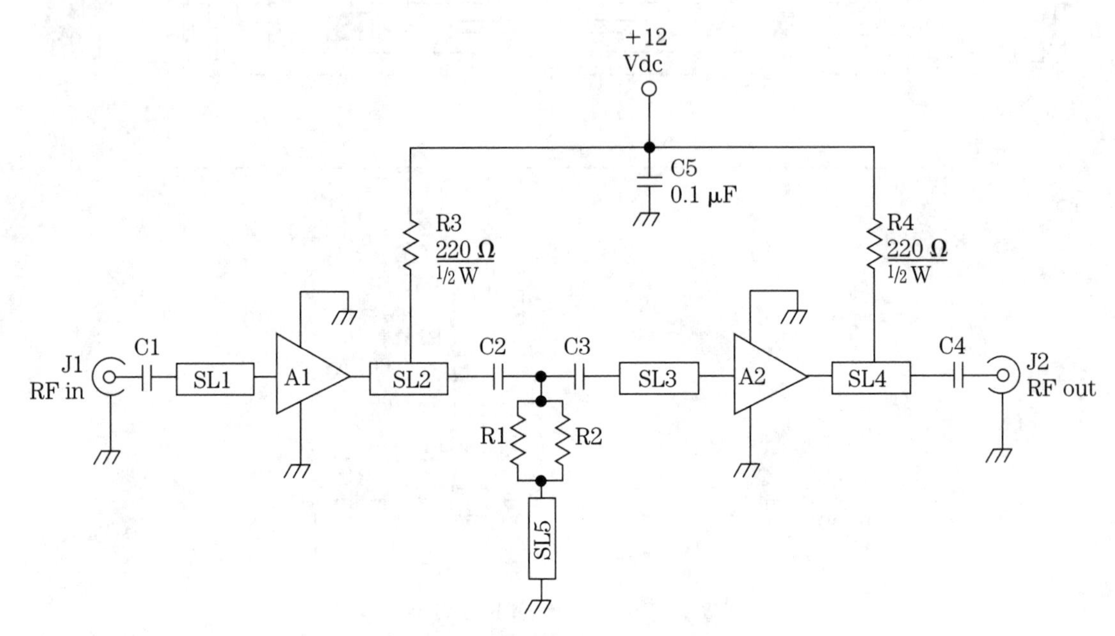

18-9 Cascade MAR-1 preamplifier.

Table 18-3

Component	Case-A	Case-B
R1	124 Ω	69.1 Ω
R2	69.8 Ω	69.1 Ω
C1,C4	470 pF	68 pF
C2	1.5 pF	2 pF
C3	7.5 pF	2 pF

Capacitors are chip type. Resistors are 1% chip type.

	W × L	
SL1	0.1 × 0.1 in 2.54 × 2.54 mm	0.04 × 0.1 in 1.02 × 2.54 mm
SL2	0.1 × 0.05 in 2.54 × 1.27 mm	0.04 × 0.1 in 1.02 × 2.54 mm

	W × L	
SL3	0.1 × 0.2 in 2.54 × 5.08 mm	0.04 × 0.1 in 1.02 × 2.54 mm
SL4	0.1 × 0.1 in 2.54 × 2.54 mm	0.04 × 0.1 in 1.02 × 2.54 mm
SL5	0.05 × 0.2 in 1.27 × 5.08 mm	0.05 × 0.2 in 1.27 × 5.08 mm

The parallel case is shown in Fig. 18-10. The MAR-x devices can be connected directly in parallel to increase the output power capacity of the amplifier. In the case of Fig. 18-10, there are four MAR-x devices connected in parallel. Other combinations are also possible. I built a two-up version for a signal generator output stage. The output power in Fig. 18-10 will be four times the power available from a single device.

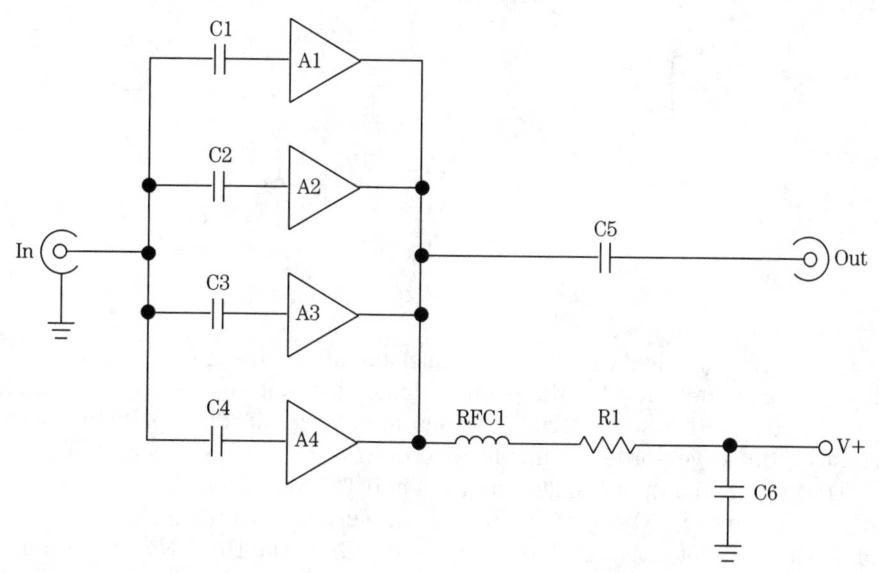

18-10 Parallel connection of MAR-x devices.

Unfortunately, in the parallel amplifier, the input and output impedances are no longer 50 ohms, but rather, it is 50/N where N is the number of devices connected in parallel. In the case shown in Fig. 18-10, there are four devices to the input and output impedances are 50/4 or 12.5 ohms. An impedance-matching device must be used to transform the lowered impedances to the 50-ohm standard for RF systems. Because most impedance transformation devices do not have the same wide bandwidth as the MAR-x devices, there is an obvious degradation of the bandwidth of the overall circuit.

The push-pull configuration is shown in Fig. 18-11. In this circuit there are two banks of two MAR-x devices each. The two banks are connected in push-pull, so this

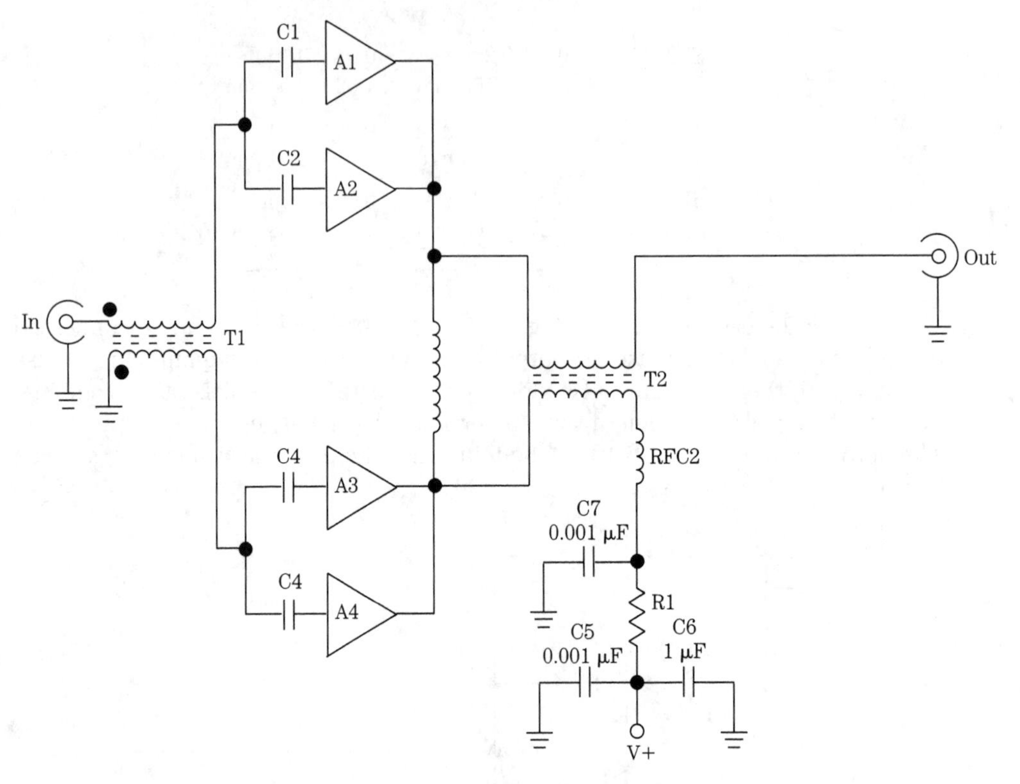

18-11 Parallel push-pull connection.

circuit is correctly called a push-pull parallel amplifier. This circuit retains gain and the increase in power level of the parallel connection, but improves the second harmonic distortion that some parallel configurations exhibit. Push-pull amplifiers inherently reduce even-order harmonic distortion.

The input and output transformers (T1 and T2) for the circuit of Fig. 18-11 are BALUN (BALanced UNbalanced) types, and are used to provide a 180-degree phase shift of the signals for the two halves of the amplifier. The BALUN transformers are typically wound on ferrite toroidal coil forms with #26 AWG or finer wire. Because the BALUN transformers are limited in frequency response, this circuit is typically used in medium wave and shortwave applications. A common specification for these transformers is to wind 6 to 7 bifilar turns on a toroidal form, the turns made of #28 enameled wire wrapped together to form a twisted pair of about five twists to the inch (≈ 2 twists per cm). Suitable cores (and a catalog) are available from Amidon Associates (P.O. Box 956, Torrance, CA 90508, USA).

Project 18-1: Universal, fixed-gain preamplifier

There is often a need for a universal preamplifier that has the standard 50-ohm input and output impedances. This preamplifier (Fig. 18-12) can be used ahead of radio receivers, as part of an RF instrumentation system, or wherever 15 dB of gain is

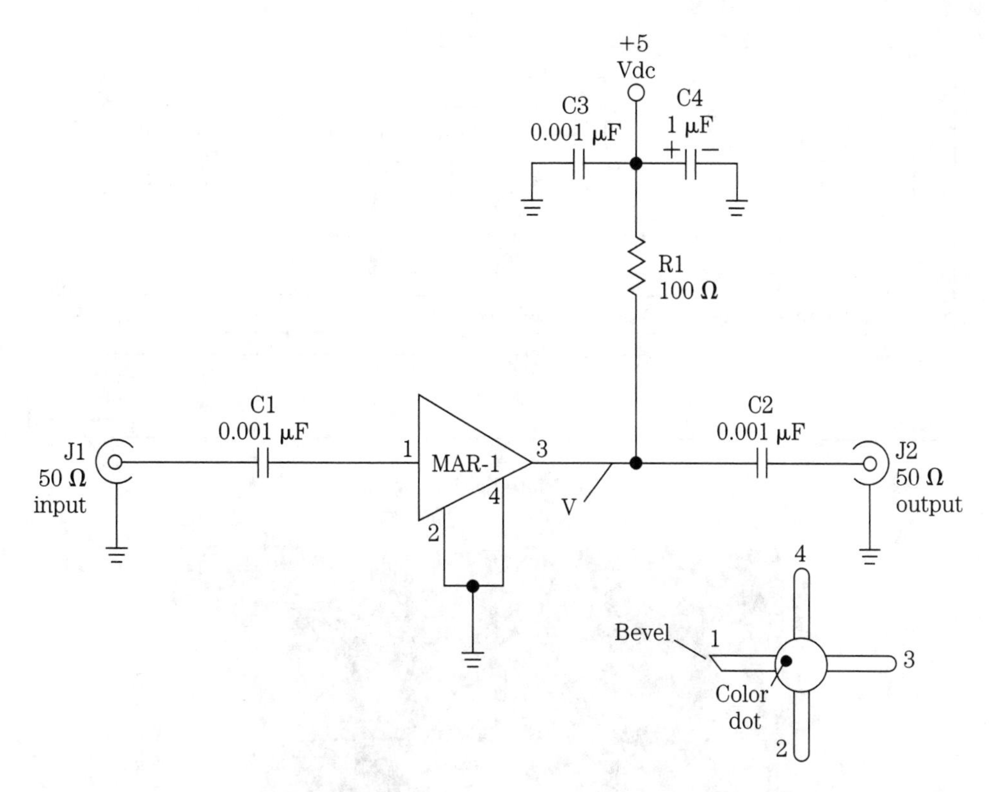

18-12 MAR-1 circuit (package symbol inset).

needed. It is based on the Mini-Circuits MAR-1 device. It will produce gain from the VLF region to a frequency of several hundred megahertz. With correct selection of capacitors, the gain will be available to frequencies in the 1000 MHz region . . . the upper limit of the MAR-1 device.

Input and output coupling is provided by a pair of 0.001 μF capacitors. For lower frequency devices (up to 150 MHz), these capacitors can be ordinary disk ceramic types. However, for all but the lowest frequency use, make sure that the leads of the disk ceramic capacitors are straightened, and that the excess ceramic material is cleaned from them.

Power is applied to the MAR-1 device from a +5 volt dc power supply through a 100-ohm resistor. This resistor can be a ¼-watt type, but must be of either carbon composition or metal film construction in order to reduce inductance; no wirewound resistors are allowed.

A sample printed circuit board template is shown in Fig. 18-13. This design preserves the strip-line needed for proper operation at higher frequencies. The dimensions assume that G-10 epoxy fiberglass printed circuit board is used. Other boards, with different dielectric constant, can also be used but the dimensions will change. A very crude prototype, which was designed for operation at less than 100 MHz, is shown in Fig. 18-14. This device was built on ordinary perforated Vectorboard overlaid with adhesive backed copper foil. The foil can be purchased in some electronic sup-

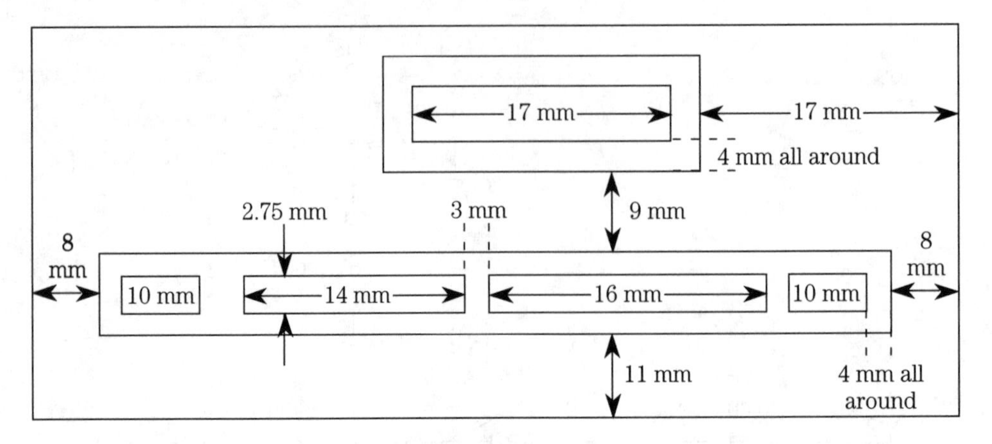

18-13 Printed circuit layout.

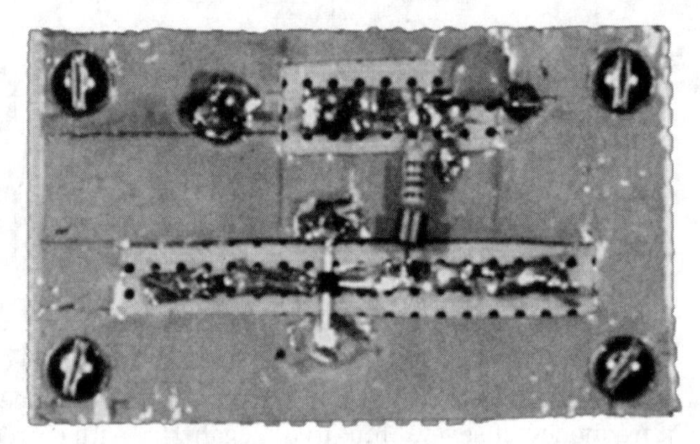

18-14 Actual layout using perfboard and press-on adhesive copper foil.

plies stores, but it is rare today. For crude projects like this one, it is possible to use non-adhesive backed foil, and glue it to the perfboard. Hobby shops, especially those that cater to doll house builders, sell this type of foil (you will want 40- or 44-gauge).

Project 18-2: Mast-mounted wideband preamplifier

Preamplifiers like that in Project 18-1 are intended for mounting close to the radio receiver, or other circuit, which they serve. This project describes a similar preamplifier, also based on the MAR-1 device, that will work at remote locations, and is powered from its own coaxial cable feedline.

Figure 18-15 shows the circuit of the remote portion. It consists of a standard MAR-1 device, but with optional –1 dB attenuators in series with input and output lines. The attenuators are used to enhance stability, but not all builders will want to use them. If you are willing to risk oscillation, and need the extra 2 dB of gain stolen by the attenuators, delete them.

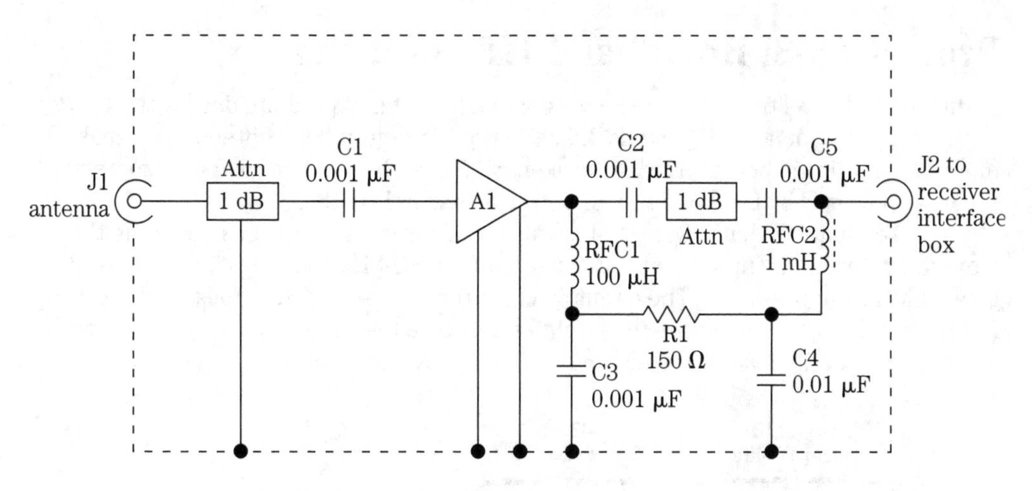

18-15 Powering preamplifier for remote mounting.

Note that the power supply end of resistor R1 is connected not to the power supply, but rather to the output coaxial connector (J2). A 1-mH RF choke (RFC2) isolates the dc circuit path from the RF flowing in the coaxial cable. Similarly, a dc-blocking capacitor (C5) isolates the MAR-1 and the attenuator from the dc voltage on the coaxial line.

The entire remote circuit must be built inside of a shielded box, and furthermore, that box should be weatherproof. If no weatherproof box is available, use an ordinary aluminum hobby box and coat the seams and joints with silicone seal or some other goop.

At the receiver end of the coaxial line, a dc/RF combiner box is needed; a design is shown in Fig. 18-16. The primary dc power in this case is a 9-Vdc battery. Again, an RF choke isolates the dc power supply from the RF in the circuit, and a dc blocking capacitor (C9) is used to keep the dc voltage from the receiver.

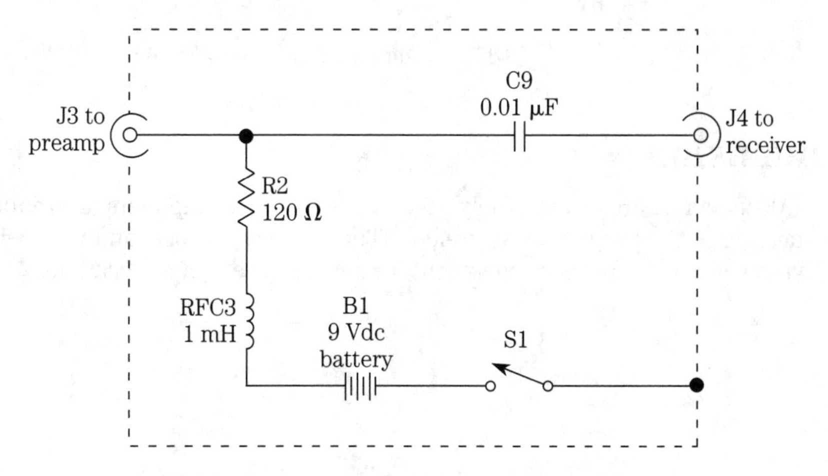

18-16 dc power supply for remote preamplifier.

Project 18-3: Broadband HF amplifier

Figure 18-17 shows the MAR-1 device used in a broadband preamplifier for the 3 to 30 MHz high-frequency (HF) bands. Like the previous circuit, it is intended for mast mounting, but if the power circuit is broken at "X" can be used other than remotely. It can be powered by the same sort of power box as in Fig. 18-16.

The key feature that differentiates this circuit from the previous circuit is the bandpass filter in the input circuit. It consists of a 1600-kHz high-pass filter followed by a 32-MHz low-pass filter. The circuit keeps strong, out-of-band signals from interfering with the operation of the preamplifier. Because the MAR-1 is a very wideband device, it will easily respond to AM and FM broadcast band signals.

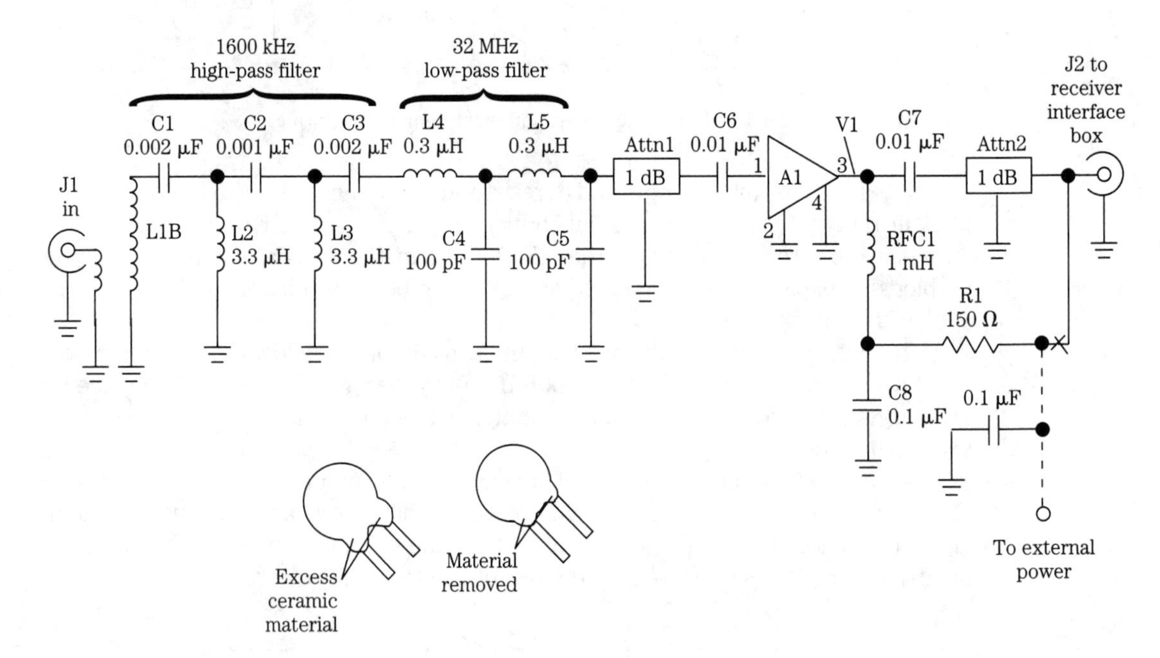

18-17 Circuit for a remote preamplifier.

Conclusion

The MAR-x devices are an extremely easy way to build RF amplifiers for frequencies from near-dc to the low microwave region. They are easy to use, and well-behaved. Hobbyists will find them very convenient for a wide variety of applications.

19 Electronic instrument projects

One of the basic assumptions of this book, indeed the entire *Mastering* . . . series, is that readers will like to build projects in order to learn, or put into practice, the principles discussed in the text. In this chapter, I will present a few RF construction projects. You can buy some of them in kit form from one source or another, while others must be put together from scratch. These projects are relatively sophisticated: an RF sweep generator, time domain reflectometer, spectrum analyzer, and so forth. They were selected to take your amateur radio or hobbyist RF endeavors a step beyond the ordinary to a mastery level.

Build an RF sweep generator

Building or experimenting with RF circuits can be done with a simple continuous wave (CW) signal generator, but it becomes apparent real quickly that an RF sweep generator would be a very handy thing to have on the bench. A sweep generator repetitively tunes through its set range of frequencies, thereby allowing you to examine the frequency response of the circuit under test on an oscilloscope. Circuits that can be better tested on a sweep generator than a CW signal generator include:

- High-pass and low-pass filters
- Bandpass filters
- LC tuning circuits
- Antenna tuners
- Crystal, ceramic or mechanical IF filters
- RF amplifiers
- Video amplifiers
- IF amplifiers

Sweep generators are used by service technicians to test and align radio and TV equipment. They are also used by engineers and technicians in laboratory settings. The service-grade sweep-signal generators are lower in cost than the laboratory models, but they often don't cover frequencies that amateurs and hobbyists need. The lab models are in the multi-kilobuck price range, so are out of the game as far as hams are concerned. A good alternative is to build a sweep generator from a kit.

A *73* magazine advertisement by Boyd Electronics (1998 Southgate Way, Grants, Pass, OR; Phone 503-476-9583) caught my eye. They are offering their Model RSG-30 RF sweep generator at a very attractive price. I contacted Jerry Boyd, and he shipped a kit-form RSG-30 (he also offers two assembled versions, but in order to fairly test the unit for most readers I wanted to assemble one myself. Besides, building stuff is the fun in technical writing).

RSG-30 features

The Boyd RSG-30 offers sweep from 2 to 30 MHz, and has the standard 50-ohm load needed for RF circuit testing. The output level is approximately 100 mV RMS, and sweep width is variable from 5 kHz to 30 MHz. A negative 12-volt, 20-ms trigger pulse is provided so the oscilloscope can be triggered in step with the sweep (which makes the presentation coherent). Three modes are offered in the Boyd RSG-30: CW, Video, and Symmetrical. The CW mode outputs a continuous, non-swept signal (Fig. 19-1A) on a single frequency set by the front panel FREQUENCY control. It can be used in the same manner as any RF signal generator. The VIDEO mode is a swept mode in which the RF frequency is swept from 2 to 30 MHz, while the SYMMETRICAL mode is a swept mode in which the width of the sweep (min to max frequency) is set by a SWEEP WIDTH control on the front panel (5 kHz to 30 MHz). Figure 19-1B shows the waveform on my oscilloscope when the SYMMETRICAL mode was selected and the SWEEP WIDTH was set close to maximum (so it is similar to the VIDEO mode).

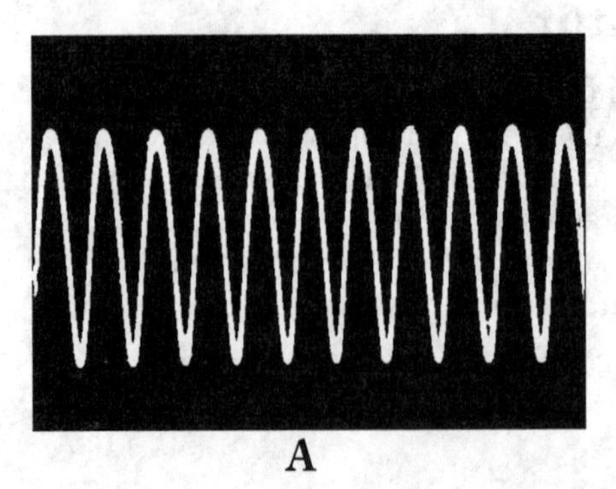

A

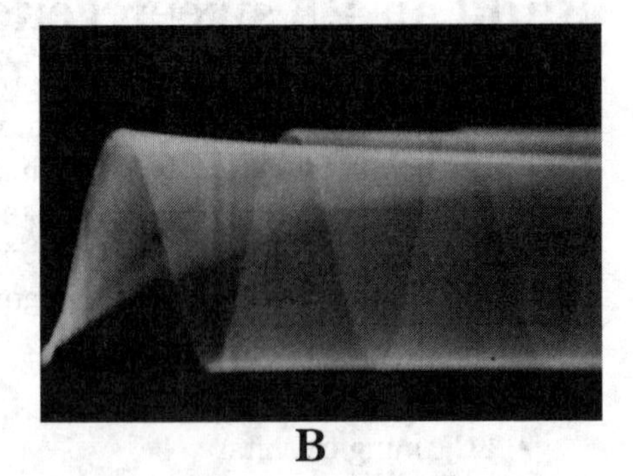

B

19-1 A. CW output signal. B. Symmetrical output signal.

RSG-30 internal circuitry

Figure 19-2 shows the block diagram of the Boyd RSG-30 circuitry. The RF signal is formed by mixing together two other signals: the output of a 50-MHz crystal oscillator, and a 52- to 80-MHz voltage tuned oscillator (VCO). A VCO is a circuit in which

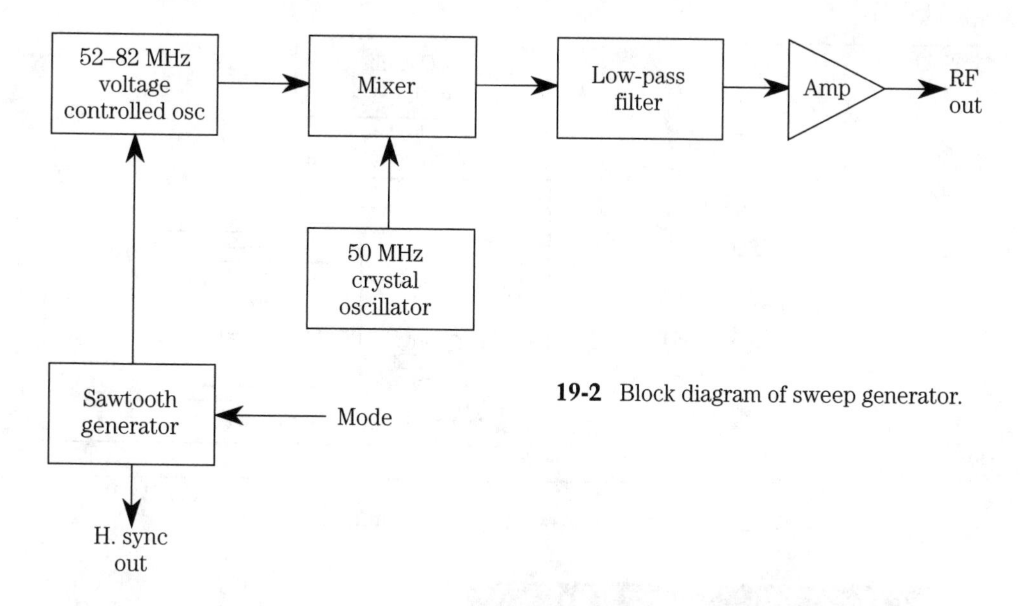

19-2 Block diagram of sweep generator.

the RF frequency is set by a tuning voltage applied to a voltage variable capacitance diode (or "varicap"). The output of the mixer is the 2- to 30-MHz difference between the crystal oscillator (XO) and VCO signals, and this signal is filtered in a low-pass filter to remove the remaining components of the XO and VCO. The 2- to 30-MHz filtered signal is amplified by a Mini-Circuits MAR-1 MMIC amplifier. These amplifier ICs have a natural 50-ohm output impedance.

The sweep and tuning voltage section contains a clock driven sawtooth generator, as well as dc offset circuitry for setting the center frequency. The sawtooth waveform allows the VCO to sweep linearly from a low to high frequency, and then snap back to the low-frequency end of the swept spectrum. Both the sweep width and center frequency controls are part of the sweep circuit.

Using the RSG-30 RF sweep signal generator

There are two basic ways to use the Boyd RSG-30 sweep signal generator. If you have an oscilloscope that has a 30-MHz (or greater bandwidth), or if you can be satisfied limiting the RSG-30 to a lower bandwidth (say, 10 MHz), then the RSG-30 can be hooked up as in Fig. 19-3. The SYNC connector on the rear panel of the RSG-30 is connected to the EXTERNAL TRIGGER input on the oscilloscope. The RF output of the RSG-30 is fed to the input of the circuit under test (CUT). The output of the CUT is fed to the high frequency vertical input of the oscilloscope.

Figure 19-4 shows the waveform to expect when the RF is viewed directly. This particular trace was taken when the CUT was a 10.7 MHz IF transformer (the type used in FM broadcast receivers). A narrower filter would produce a similar trace, but with less width. In some cases, users prefer to lower the trace to the oscilloscope baseline so that only the top portion of the symmetrical waveform shows.

Many amateurs have oscilloscopes these days, but most are low-frequency oscilloscopes that can be found at hamfests and surplused from the government, industry or local electronic repair shops. These oscilloscopes are not suitable for direct

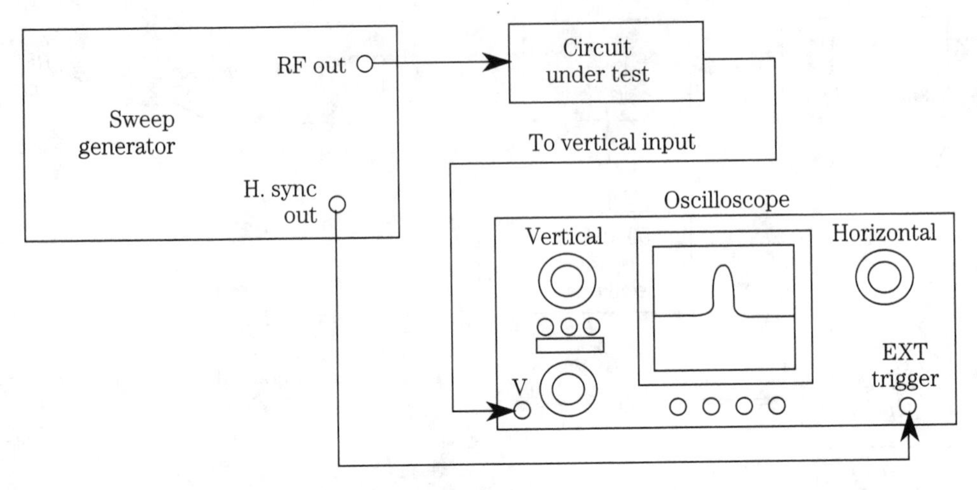

19-3 Connecting the sweep generator into the circuit.

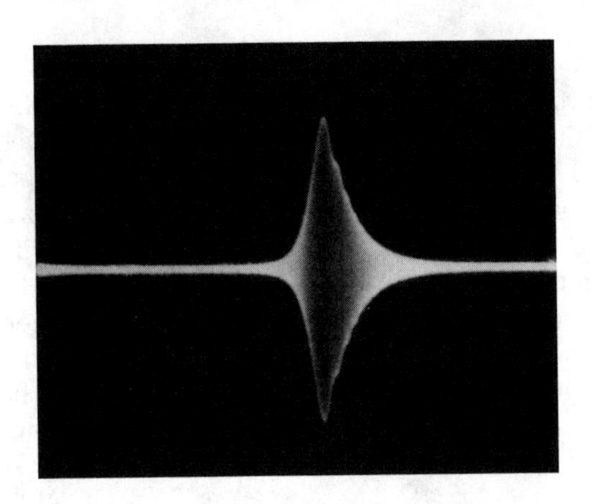

19-4 Waveform when 10.7 MHz transformer measured.

use with the RSG-30. If the hook-up of Fig. 19-3 is broken at "X," however, we can insert a detector circuit. Some people use a demodulator probe for the oscilloscope, while others use the circuit of Fig. 19-5. This circuit should be built inside a small shielded metal box. It consists of a diode detector (D1) and a filter capacitor (C2). The other capacitor is used for dc-blocking to prevent any dc from the CUT from messing up the diode.

The diode used in Fig. 19-5 is a germanium detector diode such as 1N34 or 1N60. These diodes are usually available at Radio Shack or at electronics parts places that sell Jim-Paks. Another source of the diodes is the service shop replacement lines of semiconductor such as NTE or ECG products. In those lines, the ECG-109 and NTE-109 are suitable.

Figure 19-6 shows the waveform to expect when using the diode detector. This waveform was taken from a 40-meter RF tuned circuit consisting of a disk ceramic capacitor and a 4.9 μH toroidal inductor. The diode detector and filter removes the residual RF, and presents just the instantaneous dc output of the detector.

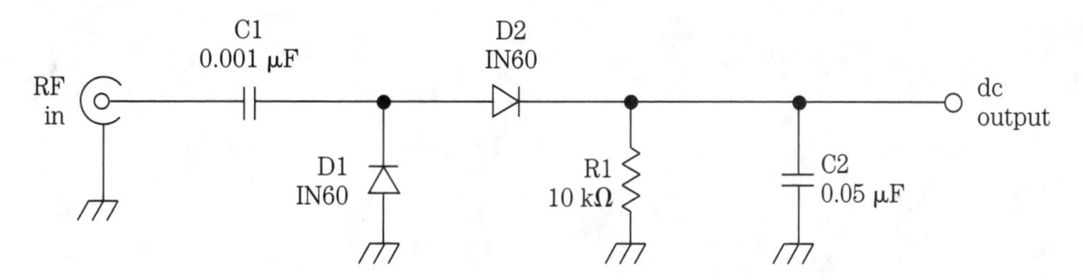

19-5 Demodulator circuit for use with sweep generator.

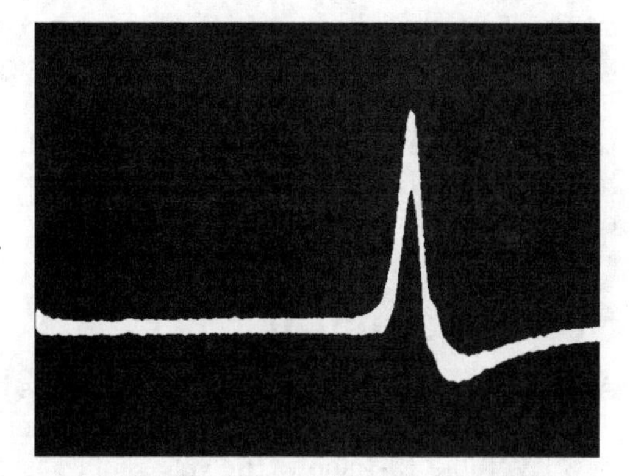

19-6 Diode detector output waveform.

Boyd Electronics offers the RSG-30 in several ways. First, you can buy a kit consisting of the printed circuit board and all parts needed for the PCB board (RSG-30K). They will also assemble and test this version for a higher price (order RSG-30A). For these options you will have to supply certain mechanical parts and the cabinet. Boyd Electronics gives you the Radio Shack part numbers. Boyd supplies a pair of adhesive templates that stick to the front and rear panels, and give the unit a real professional look. If you want the RSG-30 with a cabinet, then order RSG-30C.

Three things suggest themselves as improvements for any sweep generator project, and they apply to the Boyd unit as well: external step attenuator, a frequency translator for lower frequencies and a marker generator.

The *step attenuator* is needed because the Boyd sweep generator outputs a rather large signal level . . . too large for easy testing of receivers and amplifiers in most cases. While the signal level will work well with some tuned circuits and filters, it is inappropriate for nearly any application that has amplification associated with it. A step attenuator (Fig. 19-7) provides switch selectable levels of attenuation that can be in or out of the circuit as needed. In addition, the step attenuator will provide a swamping effect between the signal generator and the circuit under test in case the impedance of one or the other is not 50 ohms, or varies somewhat.

A *frequency translator* is needed because the Boyd sweep generator doesn't cover frequencies below 2 MHz. This limitation does not affect all hams because the IF frequencies in our HF rigs tend to be 8.83, 9.0, or 10.7 MHz . . . well within the range of the Boyd RF sweeper. But for those who need to sweep circuits below 2 MHz, including the once-standard 455 kHz IF frequency (used on Collins mechani-

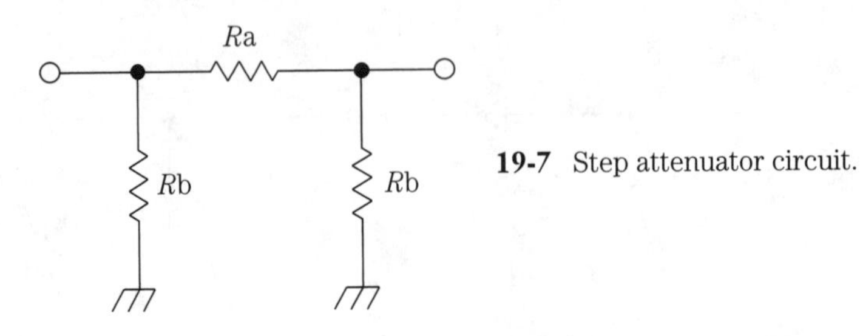

19-7 Step attenuator circuit.

cal filters even today), we need to be able to translate the Boyd sweeper's output to a lower frequency. We need a double-balanced mixer (DBM) and a crystal oscillator.

The *marker generator* is a standard crystal oscillator that allows known frequencies to be injected into the circuit for the purpose of calibrating certain spots on the band. For example, if you use a 9.0-MHz IF in your receiver, you might want to have a 9.000-MHz crystal oscillator to mark the spot on the oscilloscope presentation of the sweeper signal.

Step attenuator

A *step attenuator* such as Fig. 19-7 consists of several stages of pi-pad resistor networks, each of which can be switched into or out of the circuit with a DPDT switch or relay. Table 19-1 shows the values of resistors needed in the pi-attenuator for various popular levels of attenuation. Alternatively, if you want the attenuator to be a little more precise, then use Mini-Circuits AT-series fixed attenuators. These devices are designed to fit onto printed circuit boards and perfboards on the standard 0.100 inch center holes. The type number, AT-x, is formed by replacing the "x" with the level of attenuation desired; e.g., AT-1 is 1 dB, AT-6 is 6 dB, AT-10 is 10 dB, AT-12 is 12 dB, and AT-20 is 20 dB.

Table 19-1

Attenuation (dB)	Resistances Ra	(Ohms) Rb
1	6.2	910
2	12	470
3	18	300
5	33	200
10	75	100
20	270	68

In order to obtain higher orders of attenuation, you need only to series-connect several lower-order stages. For example, to obtain 40-dB attenuation, cascade two 20-dB attenuators, or a 20-dB and two 10-dB attenuators.

In some cases, you might want to use a barrel attenuator. These attenuators are in-line, fixed attenuators that have a male coaxial connector on one end and a female coaxial connector on the other. They can be placed anywhere in the transmission line from the signal source to the circuit under test, although in most cases, the preferred location is right at the signal generator output. The attenuator male connector is attached to the RF output connector of the signal generator, while the coaxial transmission line to the load is connected to the female connector on the attenuator. These devices are also available from Mini-Circuits, but at somewhat higher cost than the printed circuit variety.

One thing that you must do when building a multi-stage step attenuator is to use real good shielding between successive stages. Any signal leakage around the circuit detracts from the attenuation value selected. *The ARRL Handbook for Radio Amateurs* in most years has an attenuator project. In one version of that circuit, pieces of copper-clad printed circuit blank material are used to fashion the walls and sides of the step attenuator compartments . . . and only one stage is inside each compartment. You can also use brass stock from hobby shops to fashion shielding. Such stock can be worked with ordinary tin snips, scissors (if you don't care about dulling them) and hand tools.

Frequency translator

A *frequency translator* to make the RF sweeper work below 2 MHz is relatively easy to build. The Mini-Circuits (P.O. Box 350166, Brooklyn, NY, 11235-0003) passive double-balanced mixers such as the SRA-1, SBL-1 and SBL-1-1 are easily obtainable, and well-behaved (i.e., they do what they are advertised to do). We've discussed these devices in this column previously.

You can also use the Signetics NE-602 double-balanced mixer IC device for the translater. The NE-602 contains a Gilbert transconductance cell DBM and a local oscillator stage, and has been covered previously in these pages.

Four features are needed to make the translator work in this context: a mixer device, a crystal oscillator, a high-pass filter terminated in a 50-ohm dummy load and a low-pass filter that carries the output signal. The reason for the filters is that the output of the DBM will be the sum and difference of the RF sweeper and crystal oscillator signals ($F_1 + F_2$, and $F_1 - F_2$). The sum frequency is not needed, so it is passed through a high-pass filter to be absorbed in a 50-ohm dummy load (actually, a 51-ohm resistor will do). The difference frequency is passed through the low-pass filter to a 50-ohm output terminal. It is probably smart to use a matched amplifier at the output of the low-pass filter, because the mixer and the low-pass filter have insertion losses associated with them, and an amplifier will make up for that loss.

Figure 19-8 shows a circuit that can be used for this purpose. the RF sweeper input signal is fed to pin 1 of the DBM, which is its RF input. This signal must be kept below +1 dBm or the mixer might suffer harm. A series –3 dB attenuator is used to reduce the signal level. Even if the signal level is below the +1 dBm level, some people like to use the attenuator anyway because it provides a "swamping" effect against impedance variations. In those cases, a 1-dB attenuator can be used. Keep in mind, that for situations where the impedances are constant and the signal level within range (below +1 dBm), the attenuator is optional.

The local oscillator circuit is a standard crystal oscillator circuit with an output amplifier to boost the signal level. Ordinary npn silicon transistors can be used (2N2222, 2N4401, 2N3904, etc.). The mixer likes to see local oscillator signals in the +7 dBm range for proper conversion, which means, at 50 ohms, 5 mW power level,

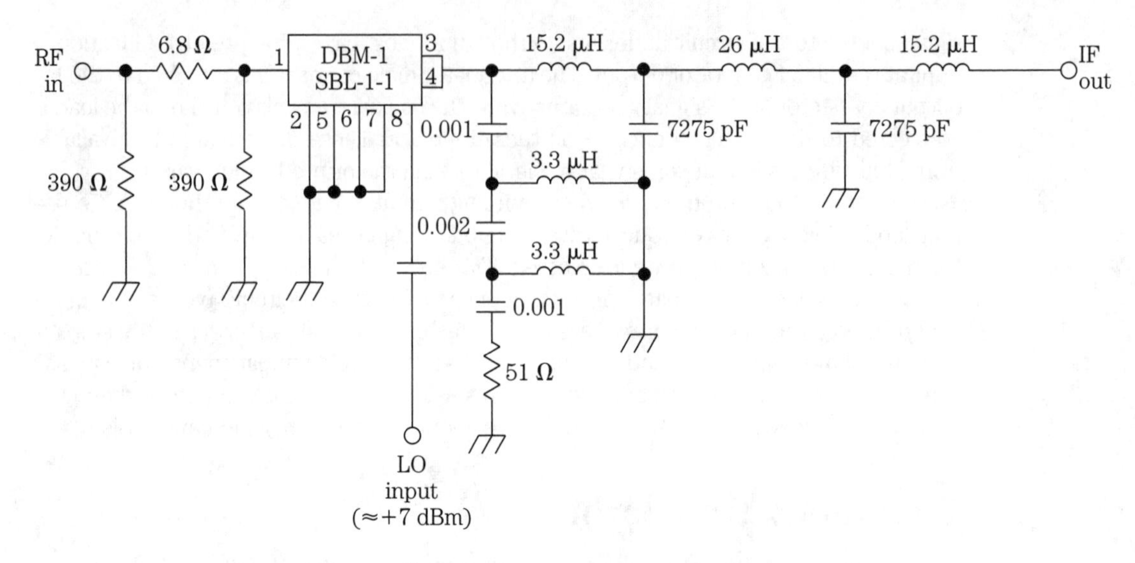

19-8 Frequency translator circuit.

or a peak-to-peak voltage of 700 mV. The crystal chosen can be anything in the 2- to 10-MHz region, so long as you can adjust the sweep generator to be within the difference frequency of the lowest sweep generator output frequency. I chose a 6-MHz crystal because it is one of the standard "microprocessor clock" crystals that is available at low cost from local parts sources. Crystal suppliers can make any exact frequency you need, or you can use one of the computer clock standard frequencies, or a 3.579-MHz color TV "color burst" crystal . . . all at low cost.

The output filters can be easily made from toroid coil forms, or, if you prefer, use standard coils obtained from parts suppliers. If you opt to use the toroidal cores, then use T-50-2 (RED) cores. These devices have an A_L value of 49, so the following turns counts will suffice:

L1, L2	3.14 µH	25 turns
L3, L5	4.9 µH	32 turns
L4	8.5 µH	42 turns

The capacitors in the filter should be either silver mica or NPO ceramic devices, with the latter being preferred over the former. Because the values of the capacitors are not standard in all cases, combinations of two or three capacitors can be used to achieve the desired value.

The output RF amplifier, a MAR-1 device by Mini-Circuits, is connected in the standard configuration for that part. Note that there are input, output and two ground terminals, but no V+ terminal. In this style of amplifier, the dc power is fed to the MAR-1 through the output terminal.

Marker generator

A *marker generator* is a crystal oscillator on a standard frequency that is used to identify points on the swept curve seen on the oscilloscope. Any reasonably accurate signal generator can be used as long as the method of combining the signals is provided. Two approaches are taken: either the resistor "star" combiner (Fig. 19-9A) or the hybrid transformer (Fig. 19-9B). The star resistor combiner can be made for any practical number of inputs. Each resistor must be $1/n$ the system impedance, where n is the number of ports. For this circuit, the system impedance is 50 ohms (standard for RF circuits), and there are three ports, so each resistor is 16.66 ohms. As a practical matter, selected 18-ohm carbon composition resistors will suffice. Use an ohmmeter to find the 18-ohm, 5% resistors that are closest to 16.7 ohms. The hybrid combiner is based on a ferrite transformer. Use 12 turns on an FT-23-72 ferrite toroid. Alternatively, write to Mini-Circuits for their catalog of RF parts, and select a commercially made hybrid combiner.

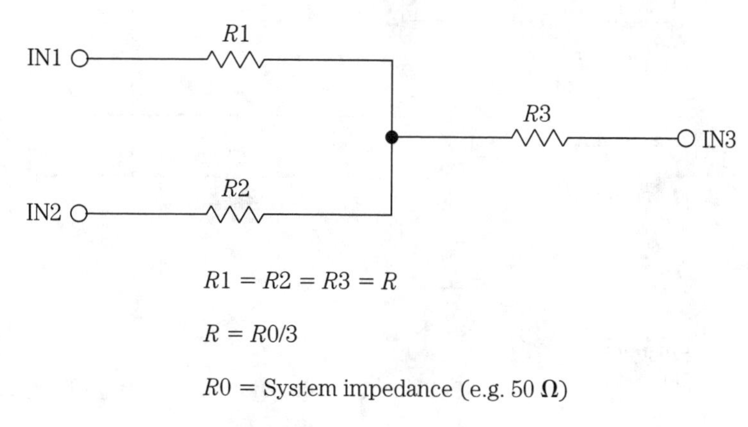

19-9 A. Star combiner.

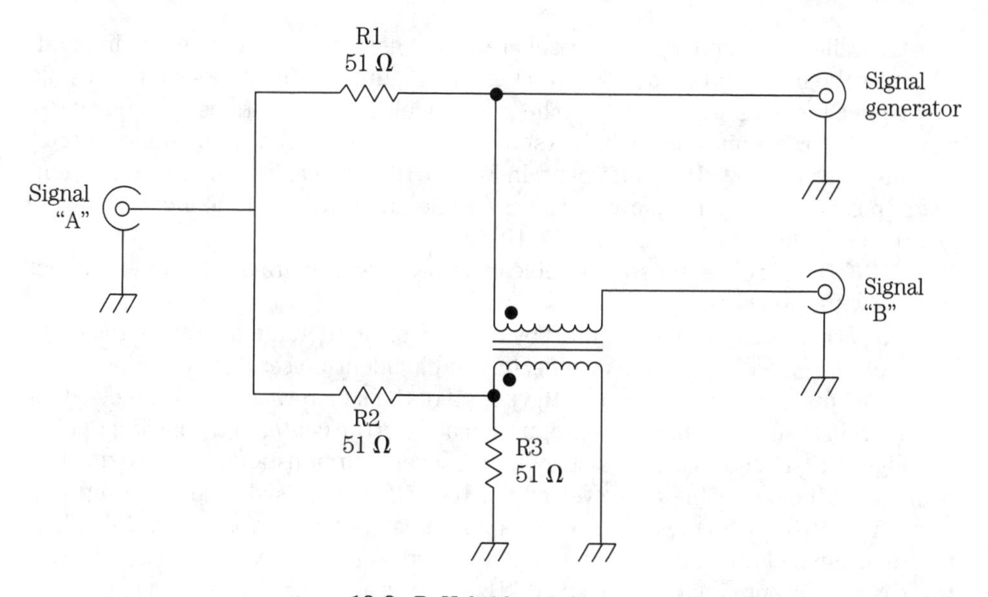

19-9 B. Hybrid transformer.

Build a time domain reflectometer (TDR)

The step function (or pulse) response of the transmission line leads to a powerful means of analyzing the line and its load on an oscilloscope. Figure 19-10A shows (in schematic form) the test set-up for *time domain reflectometry* (TDR) measurements.

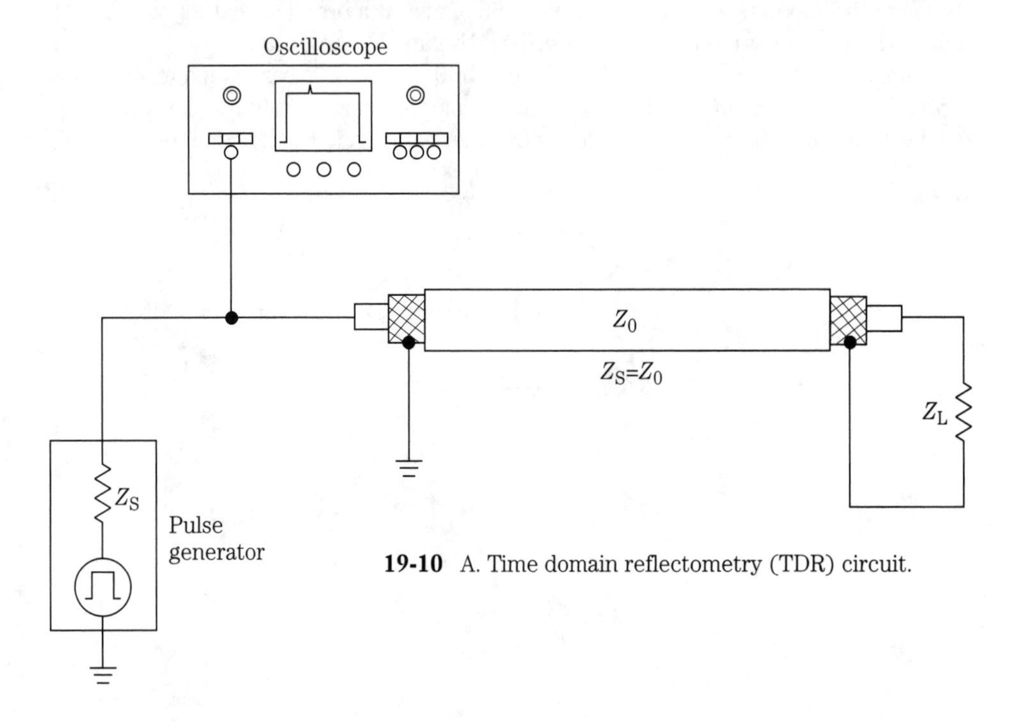

19-10 A. Time domain reflectometry (TDR) circuit.

An oscilloscope and a pulse (or square wave) generator are connected in parallel across the input end of the transmission line. Figure 19-10B shows a pulse test jig that I built for testing lines at HF. The small shielded box contains a TTL squarewave oscillator circuit. Although a crystal oscillator can be used, an RC-timed circuit running close to 1000 kHz is sufficient. In Fig. 19-10B you can see the test pulse generator box is connected in parallel with the cable under test and the input of the oscilloscope. A closer look is seen in Fig. 19-10C.

A BNC "Tee" connector and a double male BNC adapter are used to interconnect the box with the 'scope.

If a periodic waveform is supplied by the generator, then the display on the oscilloscope will represent the sum of reflected and incident pulses. The duration of the pulse (i.e., pulse width), or one-half the period of the squarewave, is adjusted so that the returning reflected pulse arrives approximately in the center of the incident pulse.

Figure 19-11 shows a TDR display under several circumstances. Approximately 30 meters of coaxial cable with a velocity factor of 0.66 was used in a test set-up similar to Fig. 19-10. The pulse width was approximately 0.9 microseconds (µS). The horizontal sweep time on the 'scope was adjusted to show only one pulse, which in this case represented one-half of a 550-kHz squarewave (Fig. 19-11B).

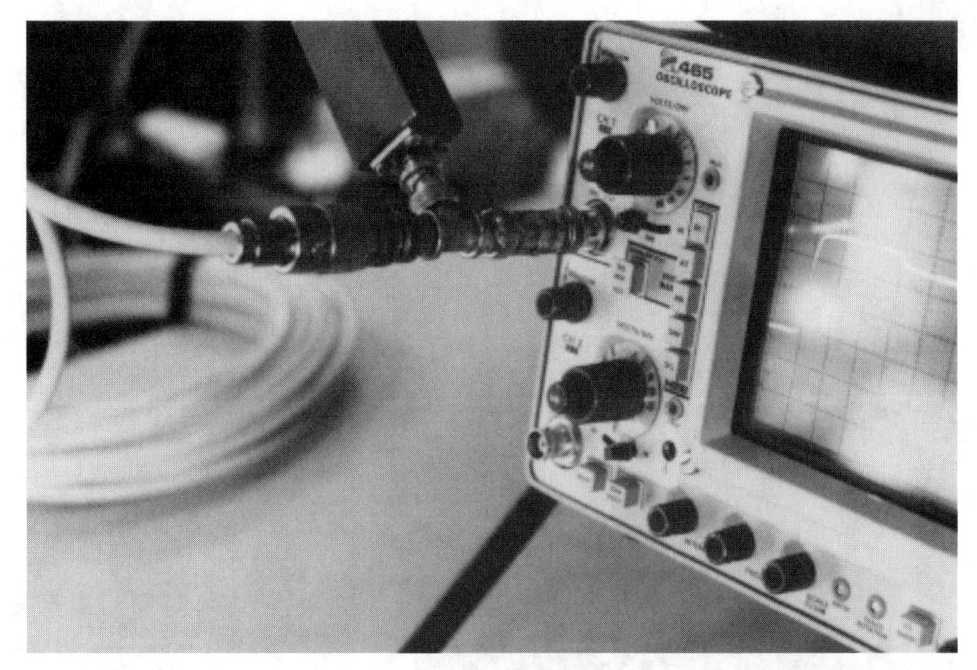

19-10 B. Practical TDR set-up.

19-10 C. Close-up detail of TDR set-up.

The displayed trace in Fig. 19-11B shows the pattern when the load is matched to the line, or in other words, $Z_L = Z_o$. A slight discontinuity exists on the high side of the pulse, and this represents a small reflected wave. Even though the load and line were supposedly matched, the connectors at the end of the line presented a slight impedance discontinuity that shows up on the 'scope as a reflected wave. In general, any discontinuity in the line, any damage to the line, any too-sharp bend or other anomaly causes a slight impedance variation, hence a reflection.

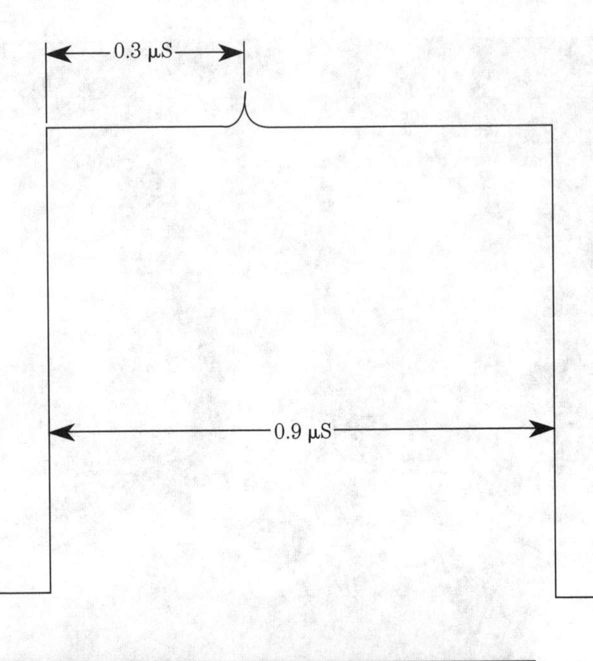

19-11 A. Idealized TDR pulse. Small "pip" on top is reflected signal interfering with forward pulse.

19-11 B. TDR pulse with no significant reflection.

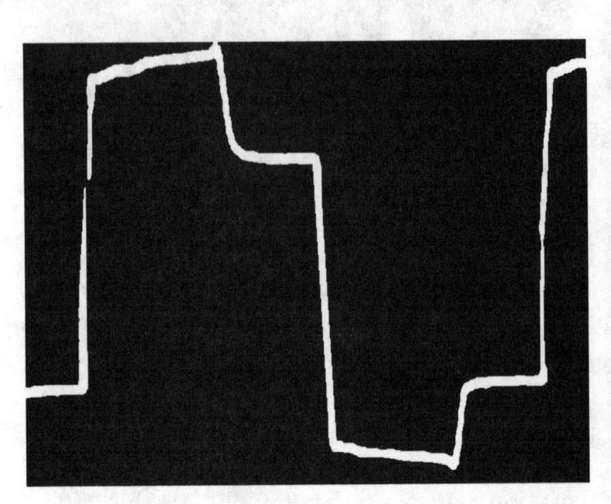

19-11 C. Various TDR reflections.

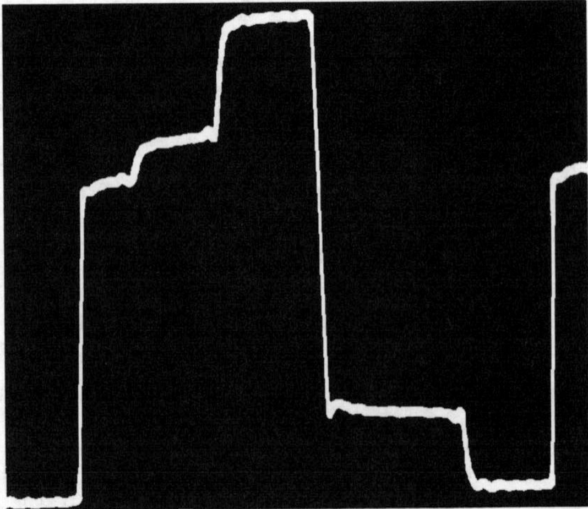

19-11 D. Various TDR reflections.

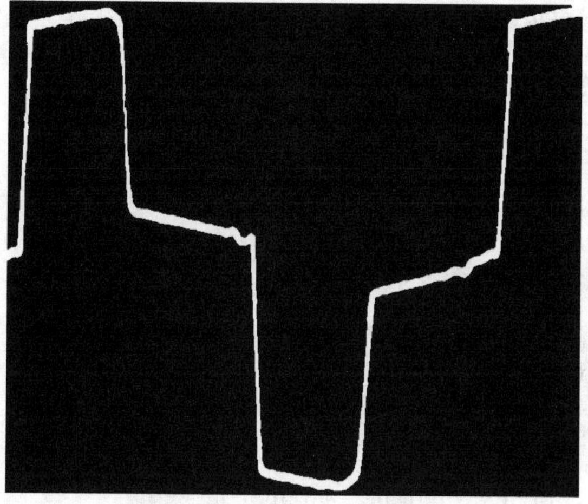

19-11 E. Various TDR reflections.

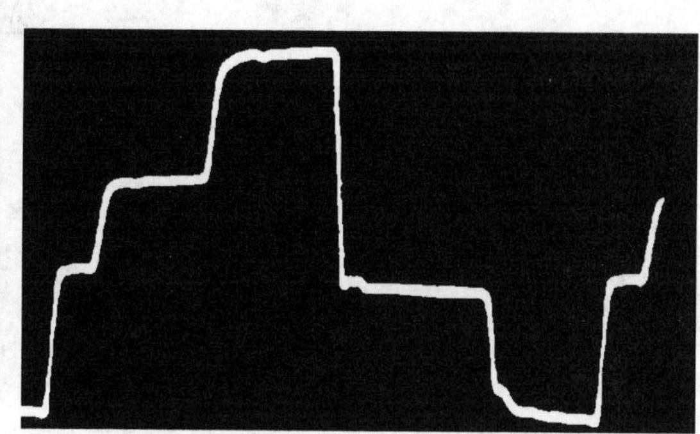

19-11 F. Various TDR reflections.

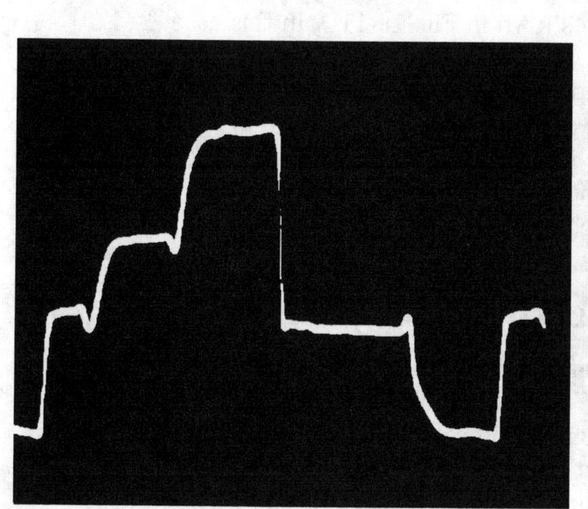

19-11 G. Various TDR reflections.

19-11 H. Various TDR reflections.

Notice that the anomaly occurs approximately one-third of the 0.9 μS duration (or 0.3 μS) after the onset of the pulse. This fact tells us that the reflected wave arrives back at the source 0.3 μS after the incident wave leaves. Because this time period represents a round-trip, we can conclude that the wave required 0.3 μS/2, or 0.15 μS to propagate the length of the line. Knowing that the velocity factor is 0.66 for that type of line we can calculate its approximate length:

$$Length = cvT \qquad [19\text{-}1]$$

$$Length = \frac{3 \times 10^8 \text{ meters}}{\text{sec}} \times (0.66) \times (1.5 \times 10^{-7} \text{ sec}) = 29.7 \text{ meters}$$

. . . which agrees "within experimental accuracy" with the 30 meters actual length prepared for the experiment ahead of time. Thus, the TDR set-up (or a TDR instrument) can be used to measure the length of a transmission line. A general equation is:

$$L_{\text{meters}} = \frac{cvT_d}{2} \qquad [19\text{-}2]$$

where

L is the length in meters (m)
c is the velocity of light (3×10^8 m/s)
v is the velocity factor of the transmission line
T_d is the round-trip time between the onset of the pulse and the first reflection

Figures 19-11C through 19-11H show the behavior of the transmission line to the step-function when the load impedance is mismatched to the transmission line (Z_L not equal to Z_o). In Fig. 19-11C we see what happens when the load impedance is less than the line impedance (in this case $0.5Z_o$). The reflected wave is inverted, and sums with the incident wave along the top of the pulse. The reflection coefficient can be determined by examining the relative amplitudes of the two waves.

The opposite situation, in which Z_L is $2Z_o$, is shown in Fig. 19-11D. In this case the reflected wave is in-phase with the incident wave, so adds to the incident wave as shown. The cases for short circuited load and open circuited load are shown in Figs. 19-11E and 19-11F, respectively. The cases of reactive loads are shown in Figs. 19-11G and 19-11H. The waveform shown in Fig. 19-11G resulted from a capacitance in series with a 50-ohm (matched) resistance; the waveform in Fig. 19-11H resulted from a 50-ohm resistance in series with an inductance.

Building the poor man's spectrum analyzer

In September 1986 an exciting article on spectrum analyzers appeared in *Ham Radio* magazine ("Low-Cost Spectrum Analyzer With Kilobuck Features"). Having been in both communications servicing and engineering school, I have used spectrum analyzers on several occasions. But the cost prohibited all but the rich amateur from owning his or her own. In fact, most professionals couldn't afford them: when I priced a plug-in spectrum analyzer to fit our existing biomedical electronics laboratory oscilloscope mainframe, it costed out at over $12,000! Then came W4UCH and his article on the WA2PZO/Science Workshop "Poor Man's Spectrum Analyzer." As little as the kits cost, I decided to buy'n'build my own spectrum analyzer.

The WA2PZO concept is based on the fact that modern TV tuners, especially the "cable ready" variety, are varactor-tuned. The familiar switched inductor tuner is replaced by a voltage-tuned varactor oscillator. Two types are available. One type, which was used in the W4UCH article, has separate low-VHF, high-VHF and UHF bands. A switch is used to select which band is tuned. The second type is a wide-range "cable ready" tuner that tunes from low-VHF through UHF television bands in one 0–35 volt (some are 0–30 volt) range. Obviously, if you can modulate the tuning voltage with a sawtooth waveform (see "Practically Speaking" column, *Ham Radio*, Jan. 1987), you have a swept tuner. Demodulate its amplified IF output and display it on a 'scope, and you have a spectrum analyzer. Sheer genius. I bought both forms of tuner from Science Workshop; Fig. 19-12 shows the cable-ready wide-range model.

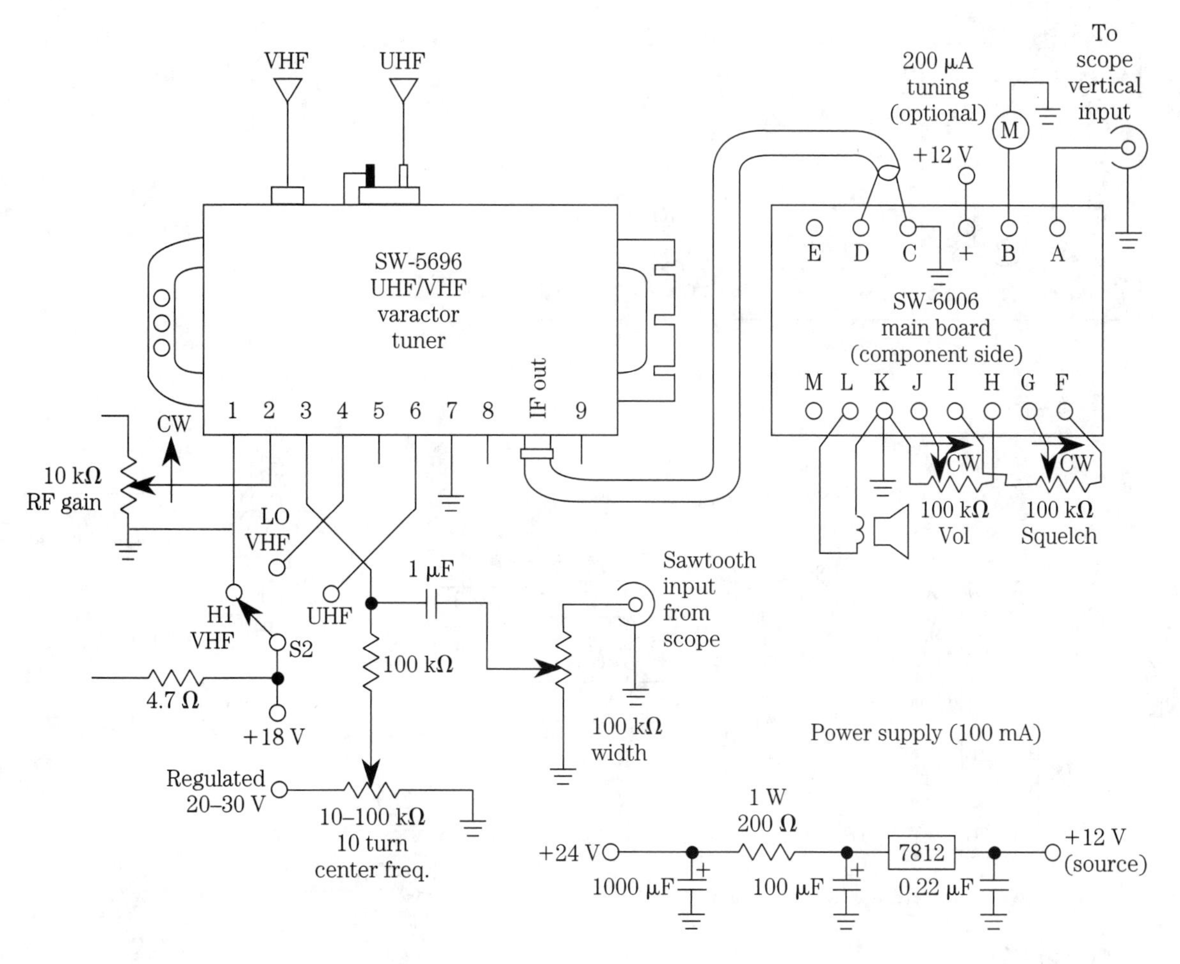

19-12 Science Workshop "Poor Man's Spectrum Analyzer."

The IF board used in the W4UCH article, and sold by Science Workshop, is shown in Fig. 19-13. The term "IF" used here actually means a fixed frequency, single-conversion superheterodyne FM receiver tuned to 45.75 MHz (the TV tuner's IF output frequency), and down-converted to the standard 10.7 MHz used for FM receiver IF amplifiers. Because the IF strip is actually a single-conversion receiver, the overall spectrum analyzer is a dual-conversion superhet. And in fact, it can be used as a VHF receiver if the sweep is turned off.

The literature that came with the IF board suggested that it be well-shielded, and that feedthrough capacitors be used on all leads except the IF input. The shielded enclosure is a standard chassis box with foldover flanges. Beware of many

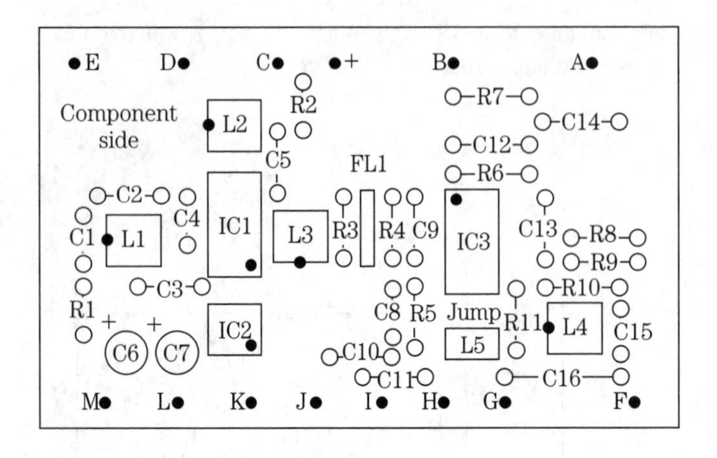

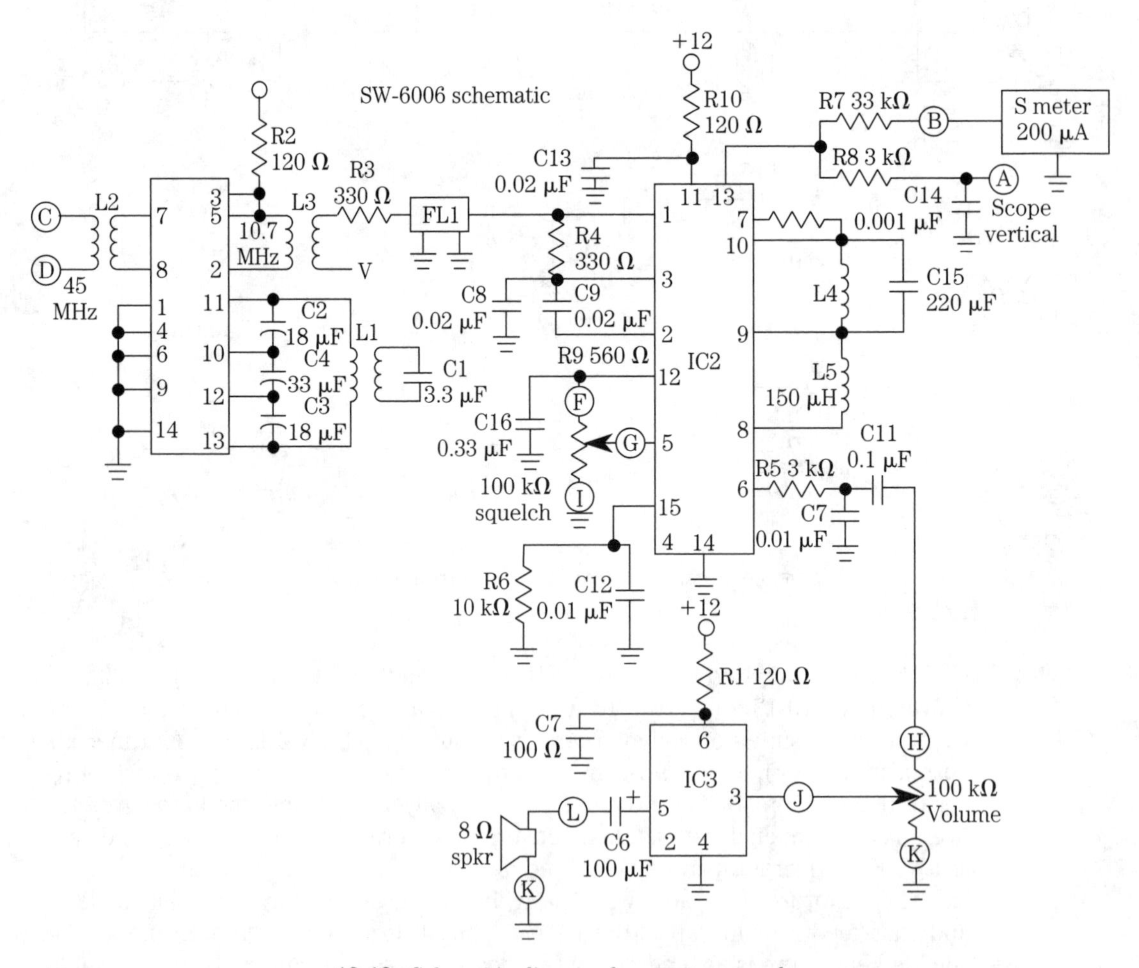

19-13 Schematic diagram for spectrum analyzer.

"shielded boxes" now on the market. The flanged type shown is minimally acceptable for shielded projects. The type of box that does not have overlapping flanges is not acceptable at all. Some LMB boxes use little dimples on each edge for support, and will not adequately shield most RF projects. While they are fine for audio and dc projects, at RF, they leave a great deal to be desired.

Being an "older guy" in radio, I still called the feedthrough capacitors by the name "feedthrough capacitors," so had a dickens of a time finding them locally; it seems that they are now called "EMI filters." Luckily, a local number for Newark Electronics was found in the Yellow Pages, so I was able to buy them direct from the source called for in the article.

In retrospect, "next time" I might try using a single connector for all leads other than the IF, and 0.002 μF disk ceramic capacitors on each lead at the connector. A good chassis mounted connector costs about $5 (or less), and disk capacitors cost about $0.80 in high quality units and are dirt-cheap in bargain packs. The EMI filters called for in the article are about $4 each, and about 12 are needed ($48!).

Adding sweep

A significant problem with the W4UCH article for many readers is the lack of a sawtooth circuit. W4UCH used the sawtooth output of his ancient Heath OL-1 oscilloscope to sweep the tuner. That approach is great if you have an ancient oscilloscope. But modern oscilloscopes rarely have the sawtooth available on the front or rear panels. Also, many do not have a horizontal input. Look at your own oscilloscope's front panel. Some two-channel oscilloscopes have an "X-Y" mode on the vertical selector. If that is the case, one of the vertical channels can be re-configured as a horizontal channel at the flick of a switch.

If you do not have a horizontal input, or X-Y capability, then you can still build the "Poor Man's Spectrum Analyzer" if you have either an EXTERNAL TRIGGER input (most 'scopes do) or a TRIGGER GATE output. The former allows an external signal, such as the falling edge of an external sawtooth, to trigger the sweep. The latter outputs a narrow pulse every time the oscilloscope triggers. By allowing the 'scope to self-trigger, you get a string of pulses that can be used to trigger certain types of sawtooth generator.

Science Workshop makes a sweep board (Fig. 19-14) available that can be used for generating and controlling an external sawtooth. Although I am not totally happy with the design, it suffices at this point. There are two problems with the design, as I see it. First, the sawtooth is not very linear, and its fall time is too long. Second, the sawtooth clips at various settings of the center frequency and sweep rate controls.

Proof of the pudding is in the display

Aside from the fractured wisdom above, we all know that no collection of parts'n'things is of any use whatever unless it works. Figure 19-14 shows an oscilloscope photo of my brand-new Poor Man's Spectrum Analyzer. The center frequency was adjusted to the low end of the FM broadcast band. The large center spike is the signal from my elderly Measurements Model 80 signal generator set to about 85 MHz. The small spike to its right is WAMU-FM (88.5 MHz), my favorite bluegrass public radio station; the other spikes are other signals in the FM band. The large signal barely visible off the left is, I suspect, Channel 5 TV in Washington, DC.

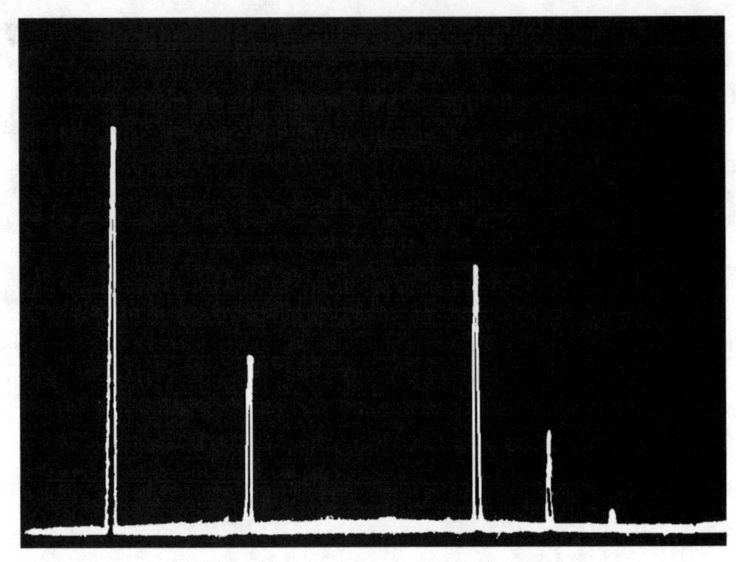

19-14 Typical display on an oscilloscope.

Users who don't have a horizontal input must use the sawtooth to trigger the sweep through the EXTERNAL TRIGGER input. I recommend using the negative trailing edge of the sawtooth waveform for this purpose (set TRIGGER SLOPE—or equivalent switch—to the negative position). Also, be sure to make the sweep time across the entire horizontal aspect of the 'scope graticule equal to the period of the sawtooth leading edge. Otherwise, the 'scope and sawtooth won't sweep in sync and the display will "double clutch."

Improvements

The spectrum analyzer project has spurred a few ideas for changes or improvements (time permitting). First, I plan to redesign the sawtooth generator (possibly generating the sawtooth digitally). Second, I plan to add an amplifier/attenuator based on Mini-Circuits fixed attenuators and a Signetics NE-5205 amplifier. The range will be –60 to +19 dB. Third, there may be a converter for HF, and tuners to band-limit the spectrum analyzer at will to certain VHF Amateur bands. This modification will punch out certain local signals that tend to drive receivers into intermod problems at my QTH. Fourth, WA2PZO is working on a tracking oscillator circuit, and in fact has a tentative approach to its design. A tracking oscillator produces an output at the spectrum analyzer's center frequency. Besides its obvious use as a signal source, it is also useful to drive a frequency counter. Presently, tuning indications is by seat-of-the-pants calibration of the voltage control. I plan to buy the WA2PZO tracking oscillator kit if, and when, it becomes available. Varactor tuners are inherently nonlinear in their voltage-vs-frequency characteristic. Although I haven't run an analysis to fit the curve to data, the V-vs-F curve looks parabolic. One reason for making the sawtooth by digital means is that there is a digital means for linearizing the tuning. I learned this trick in a biomedical electronics laboratory, and it is normally used to linearize transducers, thermocouples and the like.

A digital frequency counter module for home constructors

Every now and then a product comes out that excites me. One of these products is a digital frequency counter module that you can use to make your own counter (good up to 2 meters) . . . in your own configuration . . . in the manner that you prefer. The Optoelectronics, Inc. (5821 N.E. 14 Avenue, Fort Lauderdale, FL, 33334; Phone 305-771-2050, or toll-free on 1-800-327-5912, or fax on 305-771-2052) Model UTC-150 counter-timer module (Fig. 19-15) is a high-performance digital counter designed in "panel meter" format. That is, it is treated as a component that can be installed on a panel in a cabinet. The module only measures 3.55 inches wide, 1.765 inches high and 0.85 inches thick (0.44 inches if connector not included). What makes it possible to pack a lot of performance into such a small package is the special applications specific integrated circuit (ASIC) designed by Optoelectronics for the UTC-150.

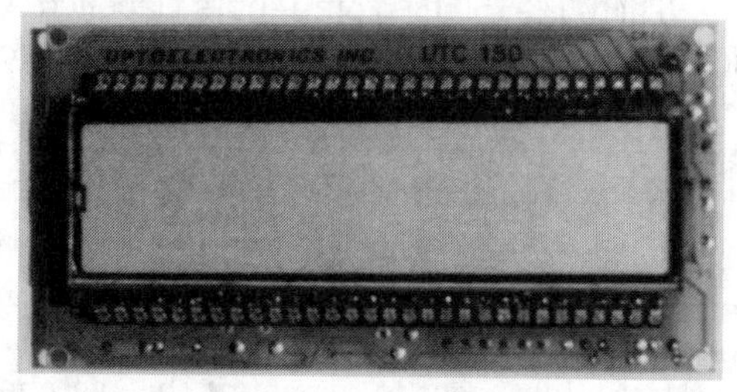

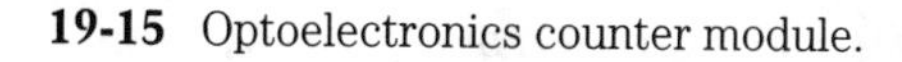

19-15 Optoelectronics counter module.

The Optoelectronics UTC-150 will perform the following functions: frequency measurement, period measurement ($T = 1/F$), time interval measurement, ratio, and average. Only the most costly counters usually have all of these features.

Applications for the UTC-150 include a free-standing or hand-held (battery powered) instrument, an output meter for a signal generator, an add-on frequency measurement accessory for an oscilloscope (must have a VERT OUT connection), an output frequency indicator for an RF signal generator (either newly built or retrofitted into an older instrument), or a digital frequency dial for an older ham transceiver that has an analog dial. Homebrewers have long wanted a module like the UTC-150. Only a minimal amount of external circuitry is needed to engage all of the functions. While other circuits can be added to extend the usefulness of the UTC-150, it will operate as specified with the circuit shown. A +5-Vdc power supply @50 mA is needed. Only 250 mW of power are needed, so the unit can be battery powered.

The UTC-150 will cover the guaranteed frequency range of 0.1 Hz to 150 MHz. Mine clocked on the bench to greater than 210 MHz on an unmodulated signal from my elderly Measurements Model 80 signal generator, but that performance is not guaranteed by the manufacturer. You can get frequency coverage to more than 1 GHz (1000 MHz) if you buy an out-board pre-scaler for the counter. These devices are very high-speed frequency dividers. A 10:1 prescaler will make a 1.3 GHz signal down-convert to 130 MHz. Alternatively, you can also homebrew a heterodyne down-converter. Various editions of the *ARRL Radio Amateur's Handbook* have circuits that will work for this purpose.

The display on the UTC-150 is a 10 digit, 120 segment, liquid crystal (LCD) assembly. The large display area makes it possible to display VHF signals with as fine as 0.1 Hz resolution (10 digits displayed). It will also resolve 1 Hz at 150 MHz (9 digits) when the time-base is one second. In addition to the digits, the display has GATE, FUNCTION, and INPUT annunciators. It will tell you whether the counter is measuring frequency, period, ratio, interval, or average, and will indicate the units being used to place the decimal point.

The time base of the UTC-150 is an on-board 1-PPM, 10-MHz crystal oscillator. It provides gate times of 0.01, 0.1, 1.0 and 10 seconds. In the average mode, the UTC-150 will average your selection of 1, 10, 100, or 1000 cycles.

Another little application of the UTC-150 display is an analog bar graph on the lower-right quadrant. This horizontal bar graph is driven by an eight-bit analog-to-digital converter (A/D), and because the bar graph input is available to the user, it can be used as a signal level display (the bar graph is a 16-segment LCD bar, 3 dB per step, over a range of 0 to 2.5 Vdc).

The UTC-150 has four inputs (labelled A, B, C, and D), all of which are TTL logical levels. An RF sensitivity of 600 mV is provided.

Optoelectronics also offers a variation on the theme. Another product is the UTC-150 Engineering Evaluation Kit. This kit includes a printed circuit board (Fig. 19-16A for the PCB, Fig. 19-16B with the UTC-150 mounted) and dc power supply to run the thing. Although it is intended for operation on the engineer's bench, it is also useful for some readers as a project. Note, however, that the printed circuit board is not easily mounted in a box so that the switches are in the right spot. Some readers might want to dismount the switches and install them elsewhere, retaining the power supply and input connector portions of the printed board. Otherwise, just buy the UTC-150 straight—not the evaluation kit.

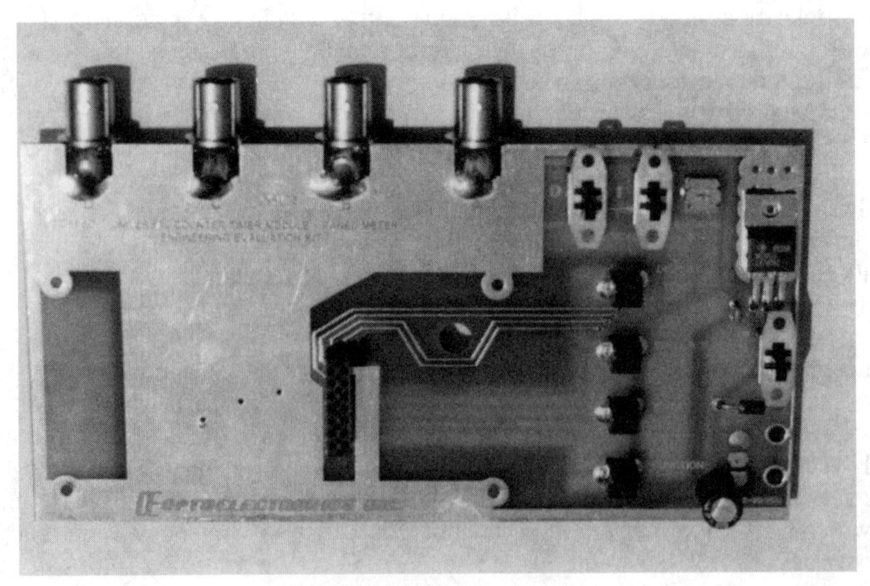

19-16 A. Evaluation board.

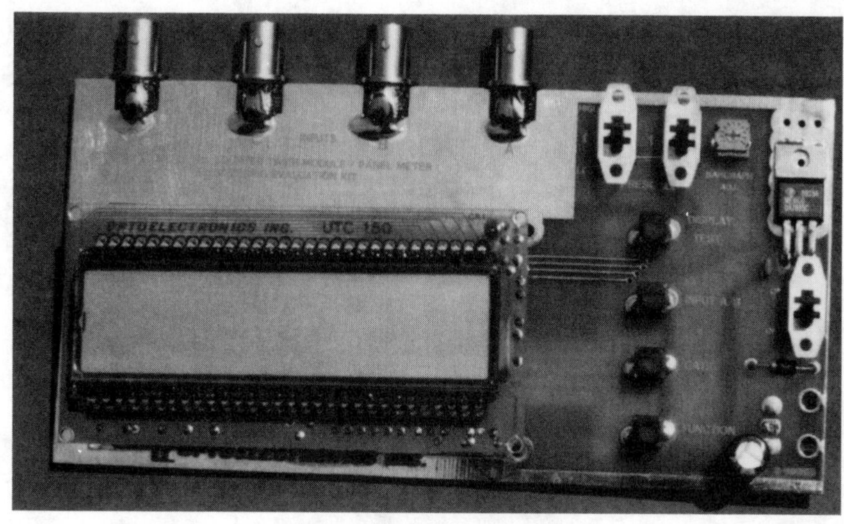

19-16 B. UTC-150 mounted.

A

Vectors

A *vector* (Fig. A-1A) is a graphical device that is used to define the *magnitude* and *direction* (both are needed) of a quantity or physical phenomena. The *length* of the arrow defines the magnitude of the quantity, while the direction in which it is pointing defines the direction of action of the quantity being represented.

Vectors can be used in combination with each other. For example, in Fig. A-1B we see a pair of displacement vectors that define a starting position (P_1) and a final position (P_2) for a person who travelled from point P_1 12 miles north and then 8 miles east to arrive at point P_2. The *displacement* is this system is the hypotenuse of the right triangle formed by the "north" vector and the "east" vector. This concept was once illustrated pungently by a university bumper sticker's directions to get to a rival school: *"North 'til you smell it, east 'til you step in it."*

Figure A-1C shows a calculations trick with vectors that is used a lot in engineering, science and especially electronics. We can *translate* a vector parallel to its original direction, and still treat it as valid. The "east" vector (E) has been translated parallel to its original position so that its tail is at the same point as the tail of the "north" vector (N). This allows us to use the *Pythagorean theorem* to define the vector. The magnitude of the displacement vector to P_2 is given by:

$$P_2 = \sqrt{N^2 + E^2} \qquad \textbf{[A-1]}$$

But recall that the magnitude only describes part of the vector's attributes. The other part is the direction of the vector. In the case of Fig. A-1C the *direction* can be defined as the angle between the "east" vector and the displacement vector. This angle (θ) is given by:

$$\theta = \arccos\left(\frac{E_1}{P}\right) \qquad \textbf{[A-2]}$$

In generic vector notation there is no "natural" or "standard" frame of reference, so the vector can be drawn in any direction so long as the users understand what it means. In the system above, we have adopted—by convention—a method that is basically the same as the old-fashioned *Cartesian coordinate system* X-Y graph. In the example of Fig. A-1B the X axis is the "east" vector, while the Y axis is the "north" vector.

In electronics, the vectors are used to describe voltages and currents in ac circuits are standardized (Fig. A-2) on this same kind of cartesian system in which the inductive reactance (X_L), i.e., the opposition to ac exhibited by inductors, is graphed in the north direction, the capacitive reactance (X_c) is graphed in the south direction and the resistance (R) is graphed in the east direction. Negative resistance (west direction) is sometimes seen in electronics. It is a phenomenon in which the current *decreases* when the voltage increases. RF examples of negative resistance include tunnel diodes and Gunn diodes.

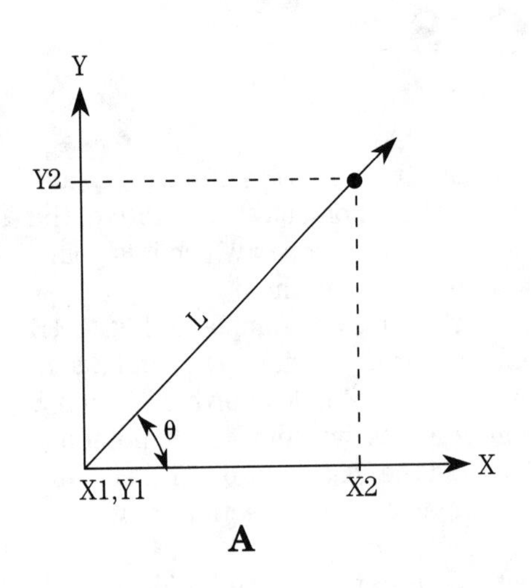

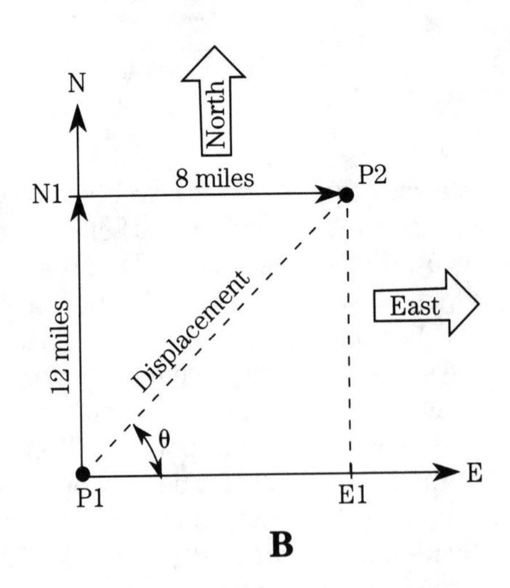

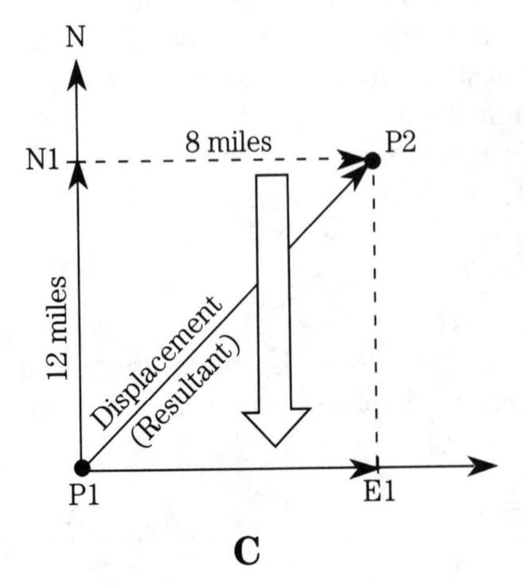

A-1 A. Vector has both direction (θ) and magnitude (L). B. Magnitude represents direction of displacement, regardless of the route taken. C. Equivalent vector.

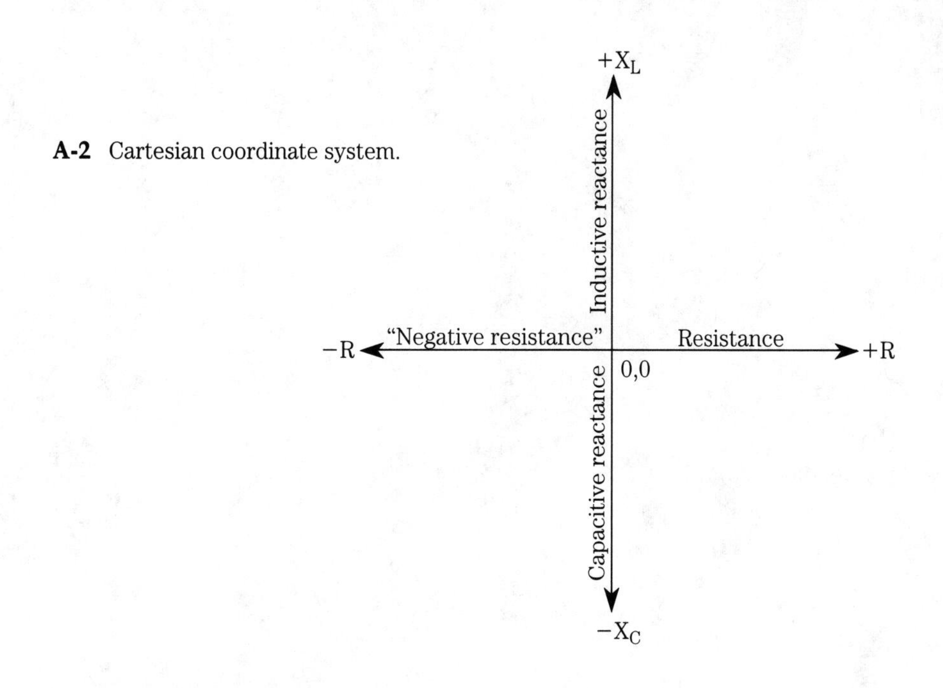

A-2 Cartesian coordinate system.

B

BASIC program to calculate RF tank circuit component values

This listing was taken by converting the BASIC to ASCII immediately after a successful run of the program. Both executable (".EXE") and listable files are available from the author on MS-DOS diskettes. Contact him for price at

P.O. Box 1099
Falls Church, VA 22041

```
100 'PROGRAM TO CALCULATE LC TANK CIRCUIT VALUES
110 'A color monitor is needed for this program. If no color monitor is
120 'is used, then delete Line Nos. 140 and 150.
130   CLS:KEY OFF
140   SCREEN 9
150   GOSUB 1030:'Get graphic of circuit diagram
160 'Start of program
170     CLS:KEY OFF:SCREEN SCR:PI = 22/7
180 IF SCR > 0 THEN COLOR 1,7
190  LOCATE 6,15:PRINT "Calculate LC tank circuit values from knowledge"
200  LOCATE  7,15:PRINT "of desired frequency range and variable capacitor"
210  LOCATE  8,15:PRINT "values (in picofards)."
220  LOCATE 10,15:PRINT "Enter MINIMUM frequency (F1)in kilohertz (kHz):";
230      INPUT F1$:' Obtain value of lowest frequency
240        F1 = VAL(F1$):' Test input of F1 to see if zero
250          IF F1 = 0 THEN BEEP
260          IF F1 = 0 THEN 220 ELSE 270
270  LOCATE 12,15:PRINT "Enter MAXIMUM frequency (F2) in kilohertz (kHz):";
280      INPUT F2$:' Obtain value of highest frequency
290        F2 = VAL(F2$):' Test input of F2 to see if zero
300          IF F2 = 0 THEN BEEP
310          IF F2 = 0 THEN 270
320            IF F2 = F1 THEN GOSUB 1440:'Test for incorrect freq. input
330            IF F1 > F2 THEN GOSUB 1440
340            IF F2 = F1 THEN 160 ELSE 350
350            IF F1 > F2 THEN 160 ELSE 360
360  LOCATE 14,15:PRINT "Select Capacitor Values..."
```

```
370  LOCATE 16,15:PRINT "Enter MINIMUM value of capacitor in picofarads (pF):";
380      INPUT CMIN$
390        CMIN = VAL(CMIN$)
400          IF CMIN = 0 THEN BEEP
410          IF CMIN = 0 THEN 360 ELSE 420
420  LOCATE 18,15:PRINT "Enter MAXIMUM value of capacitor in picofarads (pF):";
430      INPUT CMAX$
440        CMAX = VAL(CMAX$)
450          IF CMAX = 0 THEN BEEP
460          IF CMAX = 0 THEN 420 ELSE 470
470            IF CMIN = CMAX THEN GOSUB 1480:'Test for incorrect cap. input
480            IF CMIN > CMAX THEN GOSUB 1480:'and display error message
490            IF CMIN = CMAX THEN 360 ELSE 500
500            IF CMIN > CMAX THEN 360 ELSE 510
510  CLS:'Clear screen and print all entered values
520  LOCATE 6,15:PRINT "Minimum frequency (F1):";F1;" kHz"
530  LOCATE 7,15:PRINT "Maximum frequency (F2):";F2;" kHz"
540   LOCATE 9,15:PRINT "Minimum capacitance in C1: ";CMIN;" pF"
550   LOCATE 10,15:PRINT "Maximum capacitance in C1: ";CMAX;" pF"
560      FRATIO = F2/F1:'Calculate frequency ratio
570      CRATIO = FRATIO^2:'Calculate capacitance ratio
580      CDELTA = CMAX-CMIN
590   LOCATE 11,15:PRINT "Frequency ratio: ";
600     PRINT USING "##.###";FRATIO;:PRINT ":1"
610   LOCATE 12,15:PRINT "Required capacitance ratio: ";
620     PRINT USING "##.###";CRATIO;:PRINT ":1"
630   LOCATE 13,15:PRINT "Capacitance differential: ";
640     PRINT USING "####.##";CDELTA;:PRINT " pF"
650   XMIN = CDELTA/(CRATIO - 1):'Calculate min. total cap. needed
660   XMAX = CDELTA + XMIN:' Calculate max. total cap. needed
670   TRIMCAP = XMIN - CMIN
680     IF TRIMCAP < 0 THEN GOSUB 1580 ELSE 690
690     IF TRIMCAP < 0 THEN 160 ELSE 700
700   LOCATE 15,15:PRINT "Trimmer capacitance needed: ";
710   PRINT USING "####.##";TRIMCAP;:PRINT " pF"
720     LUH = 1/(4*PI^2*F1^2*XMAX*10^-12)
730   LOCATE 16,15:PRINT "Inductance needed: ";
740   PRINT USING "####.##";LUH;:PRINT " uH"
750   LOCATE 18,15:PRINT "Allocate a few pF for strays and then make up"
760   LOCATE 19,15:PRINT "remaining capacitance with either fixed or variable"
770   LOCATE 20,15:PRINT "trimmer capacitors."
780   LOCATE 22,15:GOSUB 1630:'Get press any key subroutine
790   CLS:'Clear screen for new message
800   LOCATE 10,15:PRINT "The trimmer capacitance will include all capacitances"
810   LOCATE 11,15:PRINT "seen by the inductor, and that could include a large"
820   LOCATE 12,15:PRINT "capacitance in other parts of the circuit. In certain"
830   LOCATE 13,15:PRINT "oscillator circuits, especially Colpitts and Clapp"
840   LOCATE 14,15:PRINT "oscillators, the other capacitances are large."
850   LOCATE 15,15:PRINT "Check a good oscillator design handbook for further"
860   LOCATE 16,15:PRINT "information."
870   LOCATE 18,15:GOSUB 1630
880   CLS:LOCATE 12,15:PRINT "(D)o Another Problem?"
890   LOCATE 13,15:PRINT "(E)nd Program?"
900   CHOICE$ = INPUT$(1)
910   IF CHOICE$ = "D" THEN CHOICE = 1
920   IF CHOICE$ = "d" THEN CHOICE = 1
930   IF CHOICE$ = "E" THEN CHOICE = 2
940   IF CHOICE$ = "e" THEN CHOICE = 2
950      IF CHOICE < 1 THEN 880 ELSE 960
960      IF CHOICE > 2 THEN 880 ELSE 970
970      IF INT(CHOICE) = CHOICE THEN 980 ELSE 880
```

```
980   ON CHOICE GOTO 160,990
990 CLS:LOCATE 12,20:PRINT "Program Ended...Goodbye"
1000  TIMELOOP=TIMER:WHILE TIMER < TIMELOOP + 2:WEND
1010 CLS
1020 END
1030 'Subroutine to draw circuit picture on screen
1040 CLS:LINE (450,100)-(100,100):'Draw connecting rails
1050 LINE (450,200)-(100,200)
1060 LINE (175,180)-(165,120),,B:'Draw inductor L1
1070 LINE (170,200)-(170,180)
1080 LINE (170,120)-(170,100)
1090 LINE (175,180-ZZ)-(165,180-ZZ)
1100 ZZ = ZZ + 5
1110 IF ZZ =  60 GOTO 1130
1120 GOTO 1090
1130 LINE (250,200)-(250,153):'Draw C1 (main tuning capacitor)
1140 LINE (250,147)-(250,100)
1150 LINE (265,153)-(235,153)
1160 LINE (265,147)-(235,147)
1170 LINE (265,165)-(235,135)
1180 LINE (350,200)-(350,153):'Draw C2 (trimmer capacitor)
1190 LINE (350,147)-(350,100)
1200 LINE (365,153)-(335,153)
1210 LINE (365,147)-(335,147)
1220 LINE (365,165)-(335,135)
1230 LINE (450,200)-(450,153):'Draw stray capacitances
1240 LINE (450,147)-(450,100)
1250 LINE (465,153)-(435,153)
1260 LINE (465,147)-(435,147)
1270 LINE (102,102)-(98,98),1,B
1280 LINE (102,202)-(98,198),1,B
1290 LINE (252,202)-(248,198),1,BF
1300 LINE (252,102)-(248,98),1,BF
1310 LINE (352,202)-(348,198),1,BF
1320 LINE (352,102)-(348,98),1,BF
1330 LINE (172,202)-(168,198),1,BF
1340 LINE (172,102)-(168,98),1,BF
1350 LOCATE 11,19:PRINT "L1":'Add alphabetic labels
1360 LOCATE 11,27:PRINT "C1"
1370 LOCATE 11,40:PRINT "C2"
1380 LOCATE 11,49:PRINT "Stray"
1390 LOCATE 16,15:PRINT "C1 is main tuning capacitor"
1400 LOCATE 17,15:PRINT "C2 is trimmer + fixed capacitors"
1410 LOCATE 18,15:PRINT "Stray is wiring capacitance plus other"
1420 LOCATE 19,15:PRINT "capacitors in the circuit."
1430 LOCATE 21,15:GOSUB 1630:CLS:RETURN:'End of Subroutine
1440 ' Error Message Subroutine (Incorrect Frequency Input)
1450   BEEP:LOCATE 17,15:PRINT "ERROR!  F2 must be greater than F1"
1460   LOCATE 19,15:GOSUB 1630:' Go get press any key subroutine
1470 RETURN:'End of Subroutine
1480 'Error Message Subroutine (Incorrect Capacitance Input)
1490  BEEP:LOCATE 20,15:PRINT "ERROR! Cmax must be greater than Cmin"
1500   LOCATE 22,15:GOSUB 1630:' Go get press any key subroutine
1510   GOSUB 1530:'Clear screen and reprint F1 and F2
1520  RETURN:' End of Subroutine
1530 ' Subroutine to clear screen and reprint values of F1 and F2
1540  CLS
1550  LOCATE 6,15:PRINT "Minimum frequency (F1):";F1;" kHz";
1560  LOCATE 7,15:PRINT "Maximum frequency (F2):";F2;" kHz"
1570 RETURN:'End of Subroutine
1580 'Subroutine for Too Small Trimmer Capacitance
```

```
1590  CLS
1600  LOCATE 12,15:PRINT "ERROR!!! Not a viable combination...try again"
1610  LOCATE 14,15:GOSUB 1630:'Go get press any key subroutine
1620  RETURN
1630 ' Press Any Key... Subroutine
1640 PRINT "Press Any Key To Continue..."
1650 A$ = INKEY$:IF A$ = "" THEN 1650 ELSE 1660
1660 RETURN:'End of Subroutine
```

Index

E

F

G

H

N

O

P

Q

W

X

About the Author

An experienced electrical engineer, technician, and ham radio operator, Joseph J. Carr holds a Certified Electronics Technician (CET) certificate in both consumer electronics and communications. He is a columnist for *Popular Electronics*, *Popular Communications*, and *73 Amateur Radio Today*, and author of *Practical Antenna Handbook—2nd edition*.